Industrial Heating

Principles, Techniques, Materials, Applications, and Design

MECHANICAL ENGINEERING

A Series of Textbooks and Reference Books

Founding Editor

L. L. Faulkner

Columbus Division, Battelle Memorial Institute
and Department of Mechanical Engineering
The Ohio State University
Columbus, Ohio

1. *Spring Designer's Handbook*, Harold Carlson
2. *Computer-Aided Graphics and Design*, Daniel L. Ryan
3. *Lubrication Fundamentals*, J. George Wills
4. *Solar Engineering for Domestic Buildings*, William A. Himmelman
5. *Applied Engineering Mechanics: Statics and Dynamics*, G. Boothroyd and C. Poli
6. *Centrifugal Pump Clinic*, Igor J. Karassik
7. *Computer-Aided Kinetics for Machine Design*, Daniel L. Ryan
8. *Plastics Products Design Handbook, Part A: Materials and Components; Part B: Processes and Design for Processes*, edited by Edward Miller
9. *Turbomachinery: Basic Theory and Applications*, Earl Logan, Jr.
10. *Vibrations of Shells and Plates*, Werner Soedel
11. *Flat and Corrugated Diaphragm Design Handbook*, Mario Di Giovanni
12. *Practical Stress Analysis in Engineering Design*, Alexander Blake
13. *An Introduction to the Design and Behavior of Bolted Joints*, John H. Bickford
14. *Optimal Engineering Design: Principles and Applications*, James N. Siddall
15. *Spring Manufacturing Handbook*, Harold Carlson
16. *Industrial Noise Control: Fundamentals and Applications*, edited by Lewis H. Bell
17. *Gears and Their Vibration: A Basic Approach to Understanding Gear Noise*, J. Derek Smith
18. *Chains for Power Transmission and Material Handling: Design and Applications Handbook*, American Chain Association
19. *Corrosion and Corrosion Protection Handbook*, edited by Philip A. Schweitzer
20. *Gear Drive Systems: Design and Application*, Peter Lynwander
21. *Controlling In-Plant Airborne Contaminants: Systems Design and Calculations*, John D. Constance
22. *CAD/CAM Systems Planning and Implementation*, Charles S. Knox
23. *Probabilistic Engineering Design: Principles and Applications*, James N. Siddall
24. *Traction Drives: Selection and Application*, Frederick W. Heilich III and Eugene E. Shube
25. *Finite Element Methods: An Introduction*, Ronald L. Huston and Chris E. Passerello

26. *Mechanical Fastening of Plastics: An Engineering Handbook*, Brayton Lincoln, Kenneth J. Gomes, and James F. Braden
27. *Lubrication in Practice: Second Edition*, edited by W. S. Robertson
28. *Principles of Automated Drafting*, Daniel L. Ryan
29. *Practical Seal Design*, edited by Leonard J. Martini
30. *Engineering Documentation for CAD/CAM Applications*, Charles S. Knox
31. *Design Dimensioning with Computer Graphics Applications*, Jerome C. Lange
32. *Mechanism Analysis: Simplified Graphical and Analytical Techniques*, Lyndon O. Barton
33. *CAD/CAM Systems: Justification, Implementation, Productivity Measurement*, Edward J. Preston, George W. Crawford, and Mark E. Coticchia
34. *Steam Plant Calculations Manual*, V. Ganapathy
35. *Design Assurance for Engineers and Managers*, John A. Burgess
36. *Heat Transfer Fluids and Systems for Process and Energy Applications*, Jasbir Singh
37. *Potential Flows: Computer Graphic Solutions*, Robert H. Kirchhoff
38. *Computer-Aided Graphics and Design: Second Edition*, Daniel L. Ryan
39. *Electronically Controlled Proportional Valves: Selection and Application*, Michael J. Tonyan, edited by Tobi Goldoftas
40. *Pressure Gauge Handbook*, AMETEK, U.S. Gauge Division, edited by Philip W. Harland
41. *Fabric Filtration for Combustion Sources: Fundamentals and Basic Technology*, R. P. Donovan
42. *Design of Mechanical Joints*, Alexander Blake
43. *CAD/CAM Dictionary*, Edward J. Preston, George W. Crawford, and Mark E. Coticchia
44. *Machinery Adhesives for Locking, Retaining, and Sealing*, Girard S. Haviland
45. *Couplings and Joints: Design, Selection, and Application*, Jon R. Mancuso
46. *Shaft Alignment Handbook*, John Piotrowski
47. *BASIC Programs for Steam Plant Engineers: Boilers, Combustion, Fluid Flow, and Heat Transfer*, V. Ganapathy
48. *Solving Mechanical Design Problems with Computer Graphics*, Jerome C. Lange
49. *Plastics Gearing: Selection and Application*, Clifford E. Adams
50. *Clutches and Brakes: Design and Selection*, William C. Orthwein
51. *Transducers in Mechanical and Electronic Design*, Harry L. Trietley
52. *Metallurgical Applications of Shock-Wave and High-Strain-Rate Phenomena*, edited by Lawrence E. Murr, Karl P. Staudhammer, and Marc A. Meyers
53. *Magnesium Products Design*, Robert S. Busk
54. *How to Integrate CAD/CAM Systems: Management and Technology*, William D. Engelke
55. *Cam Design and Manufacture: Second Edition; with cam design software for the IBM PC and compatibles*, disk included, Preben W. Jensen
56. *Solid-State AC Motor Controls: Selection and Application*, Sylvester Campbell
57. *Fundamentals of Robotics*, David D. Ardayfio
58. *Belt Selection and Application for Engineers*, edited by Wallace D. Erickson
59. *Developing Three-Dimensional CAD Software with the IBM PC*, C. Stan Wei
60. *Organizing Data for CIM Applications*, Charles S. Knox, with contributions by Thomas C. Boos, Ross S. Culverhouse, and Paul F. Muchnicki
61. *Computer-Aided Simulation in Railway Dynamics*, by Rao V. Dukkipati and Joseph R. Amyot

62. *Fiber-Reinforced Composites: Materials, Manufacturing, and Design*, P. K. Mallick
63. *Photoelectric Sensors and Controls: Selection and Application*, Scott M. Juds
64. *Finite Element Analysis with Personal Computers*, Edward R. Champion, Jr. and J. Michael Ensminger
65. *Ultrasonics: Fundamentals, Technology, Applications: Second Edition, Revised and Expanded*, Dale Ensminger
66. *Applied Finite Element Modeling: Practical Problem Solving for Engineers*, Jeffrey M. Steele
67. *Measurement and Instrumentation in Engineering: Principles and Basic Laboratory Experiments*, Francis S. Tse and Ivan E. Morse
68. *Centrifugal Pump Clinic: Second Edition, Revised and Expanded*, Igor J. Karassik
69. *Practical Stress Analysis in Engineering Design: Second Edition, Revised and Expanded*, Alexander Blake
70. *An Introduction to the Design and Behavior of Bolted Joints: Second Edition, Revised and Expanded*, John H. Bickford
71. *High Vacuum Technology: A Practical Guide*, Marsbed H. Hablanian
72. *Pressure Sensors: Selection and Application*, Duane Tandeske
73. *Zinc Handbook: Properties, Processing, and Use in Design*, Frank Porter
74. *Thermal Fatigue of Metals*, Andrzej Weronski and Tadeusz Hejwowski
75. *Classical and Modern Mechanisms for Engineers and Inventors*, Preben W. Jensen
76. *Handbook of Electronic Package Design*, edited by Michael Pecht
77. *Shock-Wave and High-Strain-Rate Phenomena in Materials*, edited by Marc A. Meyers, Lawrence E. Murr, and Karl P. Staudhammer
78. *Industrial Refrigeration: Principles, Design and Applications*, P. C. Koelet
79. *Applied Combustion*, Eugene L. Keating
80. *Engine Oils and Automotive Lubrication*, edited by Wilfried J. Bartz
81. *Mechanism Analysis: Simplified and Graphical Techniques, Second Edition, Revised and Expanded*, Lyndon O. Barton
82. *Fundamental Fluid Mechanics for the Practicing Engineer*, James W. Murdock
83. *Fiber-Reinforced Composites: Materials, Manufacturing, and Design, Second Edition, Revised and Expanded*, P. K. Mallick
84. *Numerical Methods for Engineering Applications*, Edward R. Champion, Jr.
85. *Turbomachinery: Basic Theory and Applications, Second Edition, Revised and Expanded*, Earl Logan, Jr.
86. *Vibrations of Shells and Plates: Second Edition, Revised and Expanded*, Werner Soedel
87. *Steam Plant Calculations Manual: Second Edition, Revised and Expanded*, V. Ganapathy
88. *Industrial Noise Control: Fundamentals and Applications, Second Edition, Revised and Expanded*, Lewis H. Bell and Douglas H. Bell
89. *Finite Elements: Their Design and Performance*, Richard H. MacNeal
90. *Mechanical Properties of Polymers and Composites: Second Edition, Revised and Expanded*, Lawrence E. Nielsen and Robert F. Landel
91. *Mechanical Wear Prediction and Prevention*, Raymond G. Bayer
92. *Mechanical Power Transmission Components*, edited by David W. South and Jon R. Mancuso
93. *Handbook of Turbomachinery*, edited by Earl Logan, Jr.

94. *Engineering Documentation Control Practices and Procedures*, Ray E. Monahan
95. *Refractory Linings Thermomechanical Design and Applications*, Charles A. Schacht
96. *Geometric Dimensioning and Tolerancing: Applications and Techniques for Use in Design, Manufacturing, and Inspection*, James D. Meadows
97. *An Introduction to the Design and Behavior of Bolted Joints: Third Edition, Revised and Expanded*, John H. Bickford
98. *Shaft Alignment Handbook: Second Edition, Revised and Expanded*, John Piotrowski
99. *Computer-Aided Design of Polymer-Matrix Composite Structures*, edited by Suong Van Hoa
100. *Friction Science and Technology*, Peter J. Blau
101. *Introduction to Plastics and Composites: Mechanical Properties and Engineering Applications*, Edward Miller
102. *Practical Fracture Mechanics in Design*, Alexander Blake
103. *Pump Characteristics and Applications*, Michael W. Volk
104. *Optical Principles and Technology for Engineers*, James E. Stewart
105. *Optimizing the Shape of Mechanical Elements and Structures*, A. A. Seireg and Jorge Rodriguez
106. *Kinematics and Dynamics of Machinery*, Vladimír Stejskal and Michael Valásek
107. *Shaft Seals for Dynamic Applications*, Les Horve
108. *Reliability-Based Mechanical Design*, edited by Thomas A. Cruse
109. *Mechanical Fastening, Joining, and Assembly*, James A. Speck
110. *Turbomachinery Fluid Dynamics and Heat Transfer*, edited by Chunill Hah
111. *High-Vacuum Technology: A Practical Guide, Second Edition, Revised and Expanded*, Marsbed H. Hablanian
112. *Geometric Dimensioning and Tolerancing: Workbook and Answerbook*, James D. Meadows
113. *Handbook of Materials Selection for Engineering Applications*, edited by G. T. Murray
114. *Handbook of Thermoplastic Piping System Design*, Thomas Sixsmith and Reinhard Hanselka
115. *Practical Guide to Finite Elements: A Solid Mechanics* Approach, Steven M. Lepi
116. *Applied Computational Fluid Dynamics*, edited by Vijay K. Garg
117. *Fluid Sealing Technology*, Heinz K. Muller and Bernard S. Nau
118. *Friction and Lubrication in Mechanical Design*, A. A. Seireg
119. *Influence Functions and Matrices*, Yuri A. Melnikov
120. *Mechanical Analysis of Electronic Packaging Systems*, Stephen A. McKeown
121. *Couplings and Joints: Design, Selection, and Application, Second Edition, Revised and Expanded*, Jon R. Mancuso
122. *Thermodynamics: Processes and Applications*, Earl Logan, Jr.
123. *Gear Noise and Vibration*, J. Derek Smith
124. *Practical Fluid Mechanics for Engineering Applications*, John J. Bloomer
125. *Handbook of Hydraulic Fluid Technology*, edited by George E. Totten
126. *Heat Exchanger Design Handbook*, T. Kuppan
127. *Designing for Product Sound Quality*, Richard H. Lyon
128. *Probability Applications in Mechanical Design*, Franklin E. Fisher and Joy R. Fisher

129. *Nickel Alloys*, edited by Ulrich Heubner
130. *Rotating Machinery Vibration: Problem Analysis and Troubleshooting*, Maurice L. Adams, Jr.
131. *Formulas for Dynamic Analysis*, Ronald L. Huston and C. Q. Liu
132. *Handbook of Machinery Dynamics*, Lynn L. Faulkner and Earl Logan, Jr.
133. *Rapid Prototyping Technology: Selection and Application*, Kenneth G. Cooper
134. *Reciprocating Machinery Dynamics: Design and Analysis*, Abdulla S. Rangwala
135. *Maintenance Excellence: Optimizing Equipment Life-Cycle Decisions*, edited by John D. Campbell and Andrew K. S. Jardine
136. *Practical Guide to Industrial Boiler Systems*, Ralph L. Vandagriff
137. *Lubrication Fundamentals: Second Edition, Revised and Expanded*, D. M. Pirro and A. A. Wessol
138. *Mechanical Life Cycle Handbook: Good Environmental Design and Manufacturing*, edited by Mahendra S. Hundal
139. *Micromachining of Engineering Materials*, edited by Joseph McGeough
140. *Control Strategies for Dynamic Systems: Design and Implementation*, John H. Lumkes, Jr.
141. *Practical Guide to Pressure Vessel Manufacturing*, Sunil Pullarcot
142. *Nondestructive Evaluation: Theory, Techniques, and Applications*, edited by Peter J. Shull
143. *Diesel Engine Engineering: Thermodynamics, Dynamics, Design, and Control*, Andrei Makartchouk
144. *Handbook of Machine Tool Analysis*, Ioan D. Marinescu, Constantin Ispas, and Dan Boboc
145. *Implementing Concurrent Engineering in Small Companies*, Susan Carlson Skalak
146. *Practical Guide to the Packaging of Electronics: Thermal and Mechanical Design and Analysis*, Ali Jamnia
147. *Bearing Design in Machinery: Engineering Tribology and Lubrication*, Avraham Harnoy
148. *Mechanical Reliability Improvement: Probability and Statistics for Experimental Testing*, R. E. Little
149. *Industrial Boilers and Heat Recovery Steam Generators: Design, Applications, and Calculations*, V. Ganapathy
150. *The CAD Guidebook: A Basic Manual for Understanding and Improving Computer-Aided Design*, Stephen J. Schoonmaker
151. *Industrial Noise Control and Acoustics*, Randall F. Barron
152. *Mechanical Properties of Engineered Materials*, Wolé Soboyejo
153. *Reliability Verification, Testing, and Analysis in Engineering Design*, Gary S. Wasserman
154. *Fundamental Mechanics of Fluids: Third Edition*, I. G. Currie
155. *Intermediate Heat Transfer*, Kau-Fui Vincent Wong
156. *HVAC Water Chillers and Cooling Towers: Fundamentals, Application, and Operation*, Herbert W. Stanford III
157. *Gear Noise and Vibration: Second Edition, Revised and Expanded*, J. Derek Smith
158. *Handbook of Turbomachinery: Second Edition*, Revised and Expanded, edited by Earl Logan, Jr. and Ramendra Roy
159. *Piping and Pipeline Engineering: Design, Construction, Maintenance, Integrity, and Repair*, George A. Antaki

160. *Turbomachinery: Design and Theory*, Rama S. R. Gorla and Aijaz Ahmed Khan
161. *Target Costing: Market-Driven Product Design*, M. Bradford Clifton, Henry M. B. Bird, Robert E. Albano, and Wesley P. Townsend
162. *Fluidized Bed Combustion*, Simeon N. Oka
163. *Theory of Dimensioning: An Introduction to Parameterizing Geometric Models*, Vijay Srinivasan
164. *Handbook of Mechanical Alloy Design*, edited by George E. Totten, Lin Xie, and Kiyoshi Funatani
165. *Structural Analysis of Polymeric Composite Materials*, Mark E. Tuttle
166. *Modeling and Simulation for Material Selection and Mechanical Design*, edited by George E. Totten, Lin Xie, and Kiyoshi Funatani
167. *Handbook of Pneumatic Conveying Engineering*, David Mills, Mark G. Jones, and Vijay K. Agarwal
168. *Clutches and Brakes: Design and Selection, Second Edition*, William C. Orthwein
169. *Fundamentals of Fluid Film Lubrication: Second Edition*, Bernard J. Hamrock, Steven R. Schmid, and Bo O. Jacobson
170. *Handbook of Lead-Free Solder Technology for Microelectronic Assemblies*, edited by Karl J. Puttlitz and Kathleen A. Stalter
171. *Vehicle Stability*, Dean Karnopp
172. *Mechanical Wear Fundamentals and Testing: Second Edition, Revised and Expanded*, Raymond G. Bayer
173. *Liquid Pipeline Hydraulics*, E. Shashi Menon
174. *Solid Fuels Combustion and Gasification*, Marcio L. de Souza-Santos
175. *Mechanical Tolerance Stackup and Analysis*, Bryan R. Fischer
176. *Engineering Design for Wear*, Raymond G. Bayer
177. *Vibrations of Shells and Plates: Third Edition, Revised and Expanded*, Werner Soedel
178. *Refractories Handbook*, edited by Charles A. Schacht
179. *Practical Engineering Failure Analysis*, Hani M. Tawancy, Anwar Ul-Hamid, and Nureddin M. Abbas
180. *Mechanical Alloying and Milling*, C. Suryanarayana
181. *Mechanical Vibration: Analysis, Uncertainties, and Control, Second Edition*, Revised and Expanded, Haym Benaroya
182. *Design of Automatic Machinery*, Stephen J. Derby
183. *Practical Fracture Mechanics in Design: Second Edition, Revised and Expanded*, Arun Shukla
184. *Practical Guide to Designed Experiments*, Paul D. Funkenbusch
185. *Gigacycle Fatigue in Mechanical Practive*, Claude Bathias and Paul C. Paris
186. *Selection of Engineering Materials and Adhesives*, Lawrence W. Fisher
187. *Boundary Methods: Elements, Contours, and Nodes*, Subrata Mukherjee and Yu Xie Mukherjee
188. *Rotordynamics*, Agnieszka (Agnes) Muszn´yska
189. *Pump Characteristics and Applications: Second Edition*, Michael W. Volk
190. *Reliability Engineering: Probability Models and Maintenance Methods*, Joel A. Nachlas
191. *Industrial Heating: Principles, Techniques, Materials, Applications, and Design*, Yeshvant V. Deshmukh

Industrial Heating

Principles, Techniques, Materials, Applications, and Design

Yeshvant V. Deshmukh

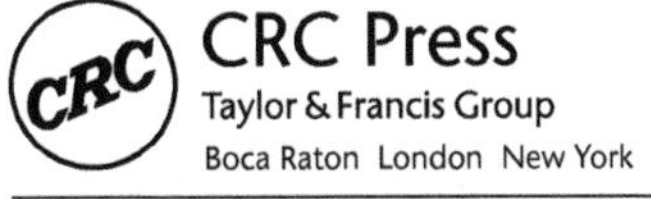

CRC Press is an imprint of the
Taylor & Francis Group, an **informa** business
A TAYLOR & FRANCIS BOOK

CRC Press
Taylor & Francis Group
6000 Broken Sound Parkway NW, Suite 300
Boca Raton, FL 33487-2742

First issued in paperback 2019

© 2005 by Taylor & Francis Group, LLC
CRC Press is an imprint of Taylor & Francis Group, an Informa business

No claim to original U.S. Government works

ISBN-13: 978-0-8493-3405-4 (hbk)
ISBN-13: 978-0-367-39284-0 (pbk)
Library of Congress Card Number 2004062059

This book contains information obtained from authentic and highly regarded sources. Reasonable efforts have been made to publish reliable data and information, but the author and publisher cannot assume responsibility for the validity of all materials or the consequences of their use. The authors and publishers have attempted to trace the copyright holders of all material reproduced in this publication and apologize to copyright holders if permission to publish in this form has not been obtained. If any copyright material has not been acknowledged please write and let us know so we may rectify in any future reprint.

Except as permitted under U.S. Copyright Law, no part of this book may be reprinted, reproduced, transmitted, or utilized in any form by any electronic, mechanical, or other means, now known or hereafter invented, including photocopying, microfilming, and recording, or in any information storage or retrieval system, without written permission from the publishers.

For permission to photocopy or use material electronically from this work, please access www.copyright.com (http://www.copyright.com/) or contact the Copyright Clearance Center, Inc. (CCC), 222 Rosewood Drive, Danvers, MA 01923, 978-750-8400. CCC is a not-for-profit organization that provides licenses and registration for a variety of users. For organizations that have been granted a photocopy license by the CCC, a separate system of payment has been arranged.

Trademark Notice: Product or corporate names may be trademarks or registered trademarks, and are used only for identification and explanation without intent to infringe.

Library of Congress Cataloging-in-Publication Data

Deshmukh, Yeshvant V.
Industrial heating : principles, techniques, materials, applications, and design / Yeshvant V. Deshmukh.
p. cm.
Includes bibliographical references and index.
ISBN 0-8493-3405-5 (alk. paper)
1. Heating. 2. Furnaces. I. Title.

TH7121.D37 2004
621.402'2--dc22 2004062059

Visit the Taylor & Francis Web site at
http://www.taylorandfrancis.com

and the CRC Press Web site at
http://www.crcpress.com

Dedicated to

The memory of my Guru,
(Late) Prof. G.K. (Nana) Ogle

Formerly
The Principal and Head of Metallurgy Department
College of Engineering—Pune, India

Contents

Preface

Heating is an integral part of many processes. It is used in diverse processes such as heat treating, shaping, casting, moulding, and joining fabrication, in one form or another. Food processing, drying, waste disposal, desalining, and many other operations depend on heating. The materials processed are metals, alloys, semiconductors, polymers, textiles, farm products minerals, city and industrial garbage, and so on. Each material and process requires heating methods suitable to its properties and the desired end products.

I came in contact with heating processes and the design field in the early 1960s by chance. Since then the association has graduated to consultation practice involving a wide range of heating problems. Except a few books published in the 1950s, on furnace design for mainly metallurgical industries, I found that there is no book on general design techniques for heating. In the last half-century, there have been substantial developments in heating techniques and related materials. Laser, electron beam, and microwave heating; fiber and fiber-based refractories; and powerful high vacuum pumps are some of the notable advances, that were practically nonexistent in the 1950s. The present book is an attempt to provide design information on the traditional and modern heating processes and auxiliary techniques.

The book is mainly aimed at designers engaged in the design and manufacture of furnaces, laboratory apparatus, and material processing equipment. It will also help the end users or buyers of such equipment to formalize their requirements and arrive at specifications. Research workers can design and specify their experimental setups involving heating. As the coverage of heating processes and allied techniques included in this book is very wide, almost all industries will find it as a useful design guide and source book. It will be advantageous if the reader has some basic knowledge of physics, chemistry, and mathematics upto the under graduate level in science or engineering.

The science of "heat transfer" is at the core of heating processes. The analysis of related heat transfer and estimation of the effective heat transfer coefficient is the first step toward a successful design. The book opens with a review of selected topics in steady state and transient heat transfer. A qualitative description of some topics in fluid mechanics and aerodynamics is also included because of their influence on heat transfer. Mathematical treatment is kept at the minimum possible and more attention is given to concepts underlying the phenomenon. It is my experience that due to years of separation from the academic field, the industrial community, in general, is out of touch with the basic concepts; hence the need for these reviews.

This is followed by fuels, their combustion and combustion devices. Solid fuels are excluded as they are not used in small or medium heating applications.

Garbage and waste incineration is treated in detail. I feel that the growing problem of waste disposal in urban and industrial areas all over the world is bound to make incineration on large scale a necessity in future. All heating-process designers must have at least a preliminary knowledge of incineration combustion and the problems with practical incinerators.

For the same reasons biogas generation and its combustion is included with other traditional fuels.

Electricity as a "fuel" is used in industries on a large scale and in many forms. All these forms are discussed in detail without going deep into electrical engineering. Electric arc is not covered as it is used in large-scale industries.

Auxilliary techniques related to heating, such as vacuum technology, pyrometry, protective atmosphere, and heat exchangers are discussed in sufficient detail. Refractory, ceramic, and metallic materials used in the construction of furnaces are dealt with a view to bring about their useful properties and limitations.

A large number of solved problems are included at each stage. They should help the designer in understanding the underlying principles. The appendices are meant mainly for *clearing* some basic concepts which could not be included in the text. This need arises from still continuing usage of diverse units. This book has used SI units throughout.

The presentation of extensive data, about material properties, in tabular form has been avoided. It is felt that selected data given in various tables and figures should be sufficient for the preliminary design. Abundant data in more precise numerical forms are available in almost all handbooks listed in the references. It is also suggested that for problems involving graphical methods used in transient heating, humidity etc., enlarged accurate graphs available elsewhere should be referred for better precision. What are included in text are reasonable outlines limited by resolution in reproduction.

All the design problems involved in the preliminary estimation of heating can be satisfactorily solved by a hand held scientific calculator. If a number of reiterations are called for, they can be worked out on a computer. Many softwares are available for specialized areas in heat transfer but they are (almost) all dedicated to particular situations. The designer is expected to have access to computational facilities.

An interdisciplinary book of this type is never complete. It is quite likely that some aspects of heating are either left out or not sufficiently covered. I will be thankful to receive constructive suggestions about any errors, omissions, and improvements to make the book more useful.

A second volume is in the planning stage. It will open with a discussion on the estimation of the heat transfer coefficient in practical situations followed by some typical construction features. The major part will consist of a number of fully worked-out designs. Any suggestions for inclusions in this volume will be highly appreciated. Your comments may be directly communicated to me at the address given below.

The preface will remain incomplete if I forget to mention my wife Sumitra. She has been my counselor for all these years. The book was a challenging job due to its complexity and my age. Her constant support, encouragement, and help has made this task a pleasure and fulfillment.

Mangesh Limaye has converted my sketches to computer drawings. He has also typed the manuscript. He has done both jobs with great skill and patience.

I also take this opportunity to express my thanks to the staff of Marcel Dekker for the design and organization; and Sam (Samik Roy Chowdhury) and his staff at ITC (Ashish Bhatnagar, Bhavinder Singh, and Subir Saha) for editorial services and production of the book.

Dr. Yeshvant V. Deshmukh

About the Author

Dr. Yeshvant V. Deshmukh received graduateships in metallurgical and mechanical engineering from the Pune University. Subsequently, he received his doctorate in mechanical and production engineering.

He has over forty years of combined experience in teaching, research, and consultancy in engineering. He has taught undergraduate and graduate classes in metallurgy and mechanical engineering at the Government Polytechnic, Pune and B.V. College of Engineering, Pune where he was a professor and the chairperson of the mechanical engineering department.

His consultancy was mainly in design of furnaces and heating, heat treatment, and other processes.

He is an associate member of the Institution of Engineers (India) and a Fellow of the Institute of the Mechanical Engineers (India). He has written about 10 technical books and contributed extensively to metallurgical and mechanical fields.

Abstract

Heating is an important operation in almost all industrial and domestic processes. A large variety of heating techniques is available at the designer. Some examples are fuel burning, electrical heating, radiative heating, and so on.

There is no book, presently available, which discusses all these diverse heating processes, their principles, choice, design materials, and limitations. This book attempts at providing information about complimentary topics such as vacuum technique, temperature measurement and control, fuels, and protective atmospheres. This makes the book self-contained in all respects.

Heating is an offshoot of the science of "heat transfer" and "fluid mechanics." This serves as a refresher course for the reader and develop an understanding about heating processes.

Waste incineration and biogas generation are specially included topics. They are discussed in detail and will lead to incinerator design. In coming years, these two topics will assume high importance in environment protection.

A large number of solved problems at each stage will help develop confidence in the designer in application of theory-to-practical situations.

Instead of giving large number of tables and data, the book encourages the use of standard handbooks and development of the designer's personal database. However, sufficient data on properties of materials are presented in both graphical and numerical form.

The book should be useful to manufacturers, designers, sales personnel, and users of heating.

Acknowledgments

I am thankful to the following for their help in preparation of this book.

M.S. Gajendragadkar, Dr. P.K. Roy, S.K. Paknikar,
R.S. Marathe of the British Library, PUNE, India
Technical Book Services, PUNE, India

M/s Kanthal AB, Sweden,
Ircon Inc, U.S.A.

Hauck Manufacturing Co. U.S.A.
Ecoflam, Italy

Chapter 1

Introduction

CONTENTS

1.1 IN THE BEGINNING

Fire is perhaps the first natural element that mankind discovered and mastered. Forest fires or volcanoes exhibited this power. It was then used for heating in winter and for cooking. Man was so awed by fire that earlier civilizations deified fire and sun.

In the millions of years that followed, we learned a lot about the production and use of heat. In the last few hundred years the sciences of thermodynamics, heat transmission, heat absorption, and generation were formulated and became the basis of modern industrial growth. Almost all industrial and domestic processes depend on the generation and use of heat. Electricity generation, production, processing and shaping of metals, manufacture of chemicals, and processing and cooking of food all depend on heat. Heating in winter and cooling in summer has made life comfortable and habitable all over the globe.

In this book we will explore some methods of heat generation, and principles of heat transfer and heat absorption. The choice of heat-generation methods is necessarily limited to small and medium heating processes. Large scale or heavy melting, and extraction and refinement of metals and minerals are excluded.

It is our aim to develop sufficient understanding of the underlying principles of heat generation and transmission by various methods that will lead to the design and estimation of the process to suit the intended application.

Before we launch into the details of processes and mechanisms of heat, it is necessary to bear in mind a few underlying thermodynamic principles common to all processes under investigation.

1. Heat will always flow or be transmitted from a higher to a lower temperature.
2. The "state" of heat in a body or in a given region of a body is completely given by the temperature of that point. We will use the Celsius (°C) and the Kelvin (K = °C + 273) scales throughout.
3. Heat generation, transmission, and absorption are basically "inefficient" processes as some heat is always irrecoverably lost. If we consider Q_1 as the heat-generated energy input to the generator and Q_2 as the energy absorbed by the object, the efficiency of the process will be Q_2/Q_1 which will, in any practical process, rarely exceed 30%.
4. It will not be possible to perform exact heat-flow calculations at all stages of even a simple heating process because of losses that will be incurred at every point.

Hence, heat calculations, in a sense, are "imprecise." The design philosophy will be to ensure that the required heating takes place in the designated time with minimum possible losses.

5. We will divide a given process into two primary classes, steady state and transient. In the steady state process the temperatures at all points are stationary, i.e., they do not change with time. In a transient process the temperature changes with time.

   A process will be transient when heating is going on. On reaching the required temperature, it may become steady. Usually we will be interested in only one aspect, either steady or transient.
6. The fraction of generated heat incident on the surface of a body is not totally absorbed. Some is reflected, some transmitted, and only some will be absorbed and will heat the body. Thus, only the last (absorbed) fraction is the "useful" heat. We will discuss these phenomena later in detail. Most of the objects that we are interested in will be opaque and there will be no transmission. However, the medium between the heat generator and the work (air, gases) may also absorb and transmit the heat passing through them.

1.2 HEATING SYSTEM CLASSIFICATION

Any heating system will have two main components, the heat generator G and the work or the object P that is to be heated. Heat will be transferred from the generator to the object by a heat transfer process T. Based on these three, a general classification system can be proposed as shown in Figure 1.1.

1. Figure 1.1(A) shows a system that has the generator and the work separated by a distance. Heat is transmitted through the medium in between, as shown by the marked arrows. This is obviously an inefficient system as a considerable portion of the generated heat will be wasted (as shown by the unmarked arrows).

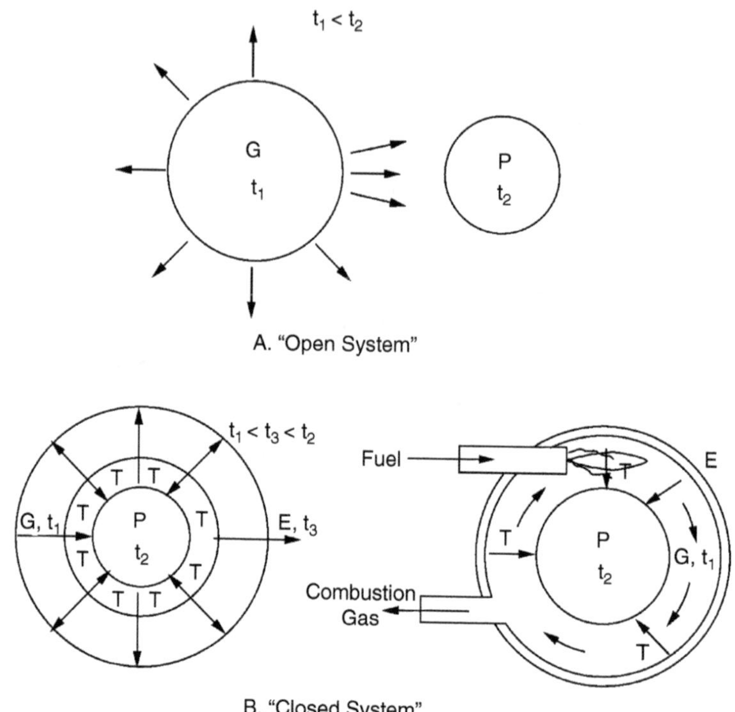

Figure 1.1 Systems in which the generator *G* and the work *P* are separate.

A pot or a piece of metal placed on an electric plate or gas burner belongs to this system. We call this an "open" system.

The above system can be converted into a "closed" system as shown. We now surround the generator and the work by providing an "insulated" enclosure E. Now all the generated heat will reach the work either directly or via reflection from the enclosure. It is presumed that the enclosure is 100% reflecting.

Most of the indirectly heated resistance furnaces (Figure 1.1(B)) belong to this class. Heat transfer from the generator to the enclosure to the work is an important design factor for this type. Fuel-fired furnaces also belong to this class. An important difference in this

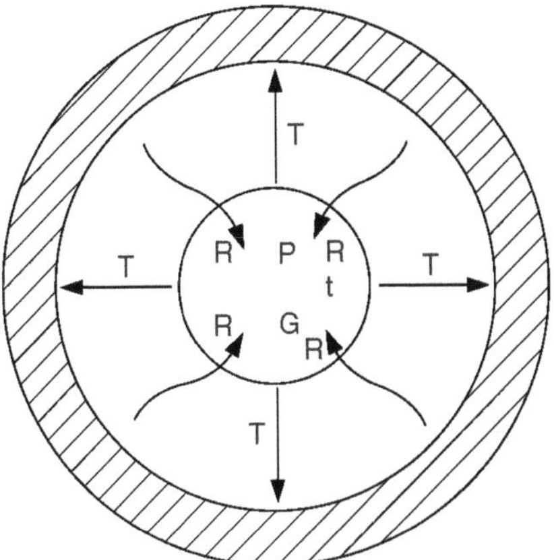

Figure 1.2 Heating system in which "nonthermal" radiation is used to create heat in the work. Heat generator and work are not separate.

type is the heat lost from combustion gases that leave the furnace enclosure. This heat loss is also quite considerable and reduces efficiency despite the presence of the enclosure.

2. Other types of heating systems do not have separate generator and work. By using certain techniques, heat is directly generated in the work as shown in the Figure 1.2. The heat is created by radiation *R* generated in a source *S*. An enclosure may or may not be required. Appreciable amount of heat may be lost from the work if it is heated much above the surrounding. These processes are usually very fast and such heat loss can be minimized. It may appear that these systems are highly efficient. However, the generation of radiation in the source *S* is very inefficient (~ 5–10%) but there are several other advantages which will be discussed later.

   Induction heating, direct resistance heating, and microwave heating belong to this class (Figure 1.3). In induction heating, high frequency (10^3–10^5 Hz) electromagnetic oscillation is created in the work by

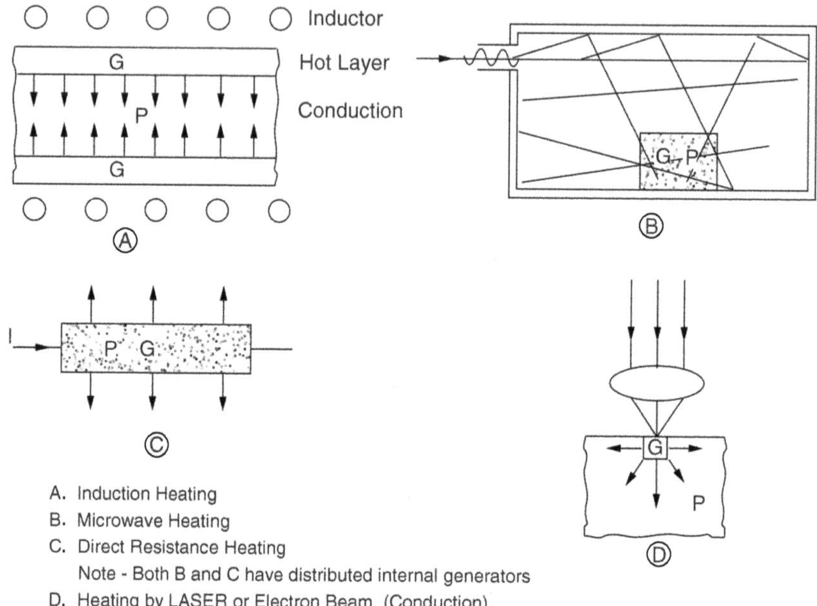

Figure 1.3 Special heating modes.

using an oscillator and an inductor. The work gets heated by the induced eddy current. In microwave heating the work is placed in an electromagnetic field of very high frequency (10^{11}–10^{14} Hz). The molecules of the work vibrate and create heat. In direct electric resistance or capacitive heating the work is made a component of an electric circuit. The power I^2R or dielectric losses in it due to current I produce heat. Later we will discuss all these processes in detail to highlight their individual features.

Heat transfer in all furnaces of this type is directly related to the surfaces of the generator and to the work taking part in the transfer, and the ratio F_P/F_G becomes a decisive factor in the design.

The heat transfer from the generator to the work takes place over paths through the medium. Thus, the properties of the medium decide the mode and efficiency of transfer. The medium may be air or a special gas or combustion gases, or vacuum and is called the "atmosphere." Thus a classification based on atmosphere arises, such as vacuum furnace, protective atmosphere furnace, salt bath furnace and the like.

3. Two modern heating processes that need to be specially mentioned are laser and electron beam heating. To some extent they belong to both systems mentioned above. A laser is a light beam having a single wavelength and excellent collimation. When focused on the work surface, it produces a spot with very high energy density (~10^5–10^8 W/cm^2) which melts or evaporates the material in the spot. Penetration below the surface is low but deeper penetration is possible (Figure 1.3(D)).

   An electron beam is a stream of high-energy electrons which can be focused on the work with similar power density. Heat is produced when the electrons lose their kinetic energy on impact. The penetration is deeper than that with laser beam heating.

   Both techniques are useful for precision heating/melting at a spot (~ 0.1–2.0 mm diameter) that can be easily controlled. They are useful heat sources for cutting, welding, and drilling operations requiring precision. Energywise, they are highly inefficient but offer many other advantages than any other process can. These techniques are now in commercial use and are discussed in detail.

1.3 CLASSIFICATION OF HEATING MODES

In the heating systems of the first type, i.e., systems in which the generator and the work are separated and enclosed in an enclosure, the heat transfer process shows multiple modes.

If the heat generator and the work are in good contact, the transfer is only by conduction (see Figure 1.4(A)). The conductivities of the two are the decisive factors. However, such direct contact is possible only in rare cases.

The space between the generator and the work is usually filled with air or some process gas (Figure 1.4(B)). Gases are always in natural circulation due to the buoyancy effect. In some

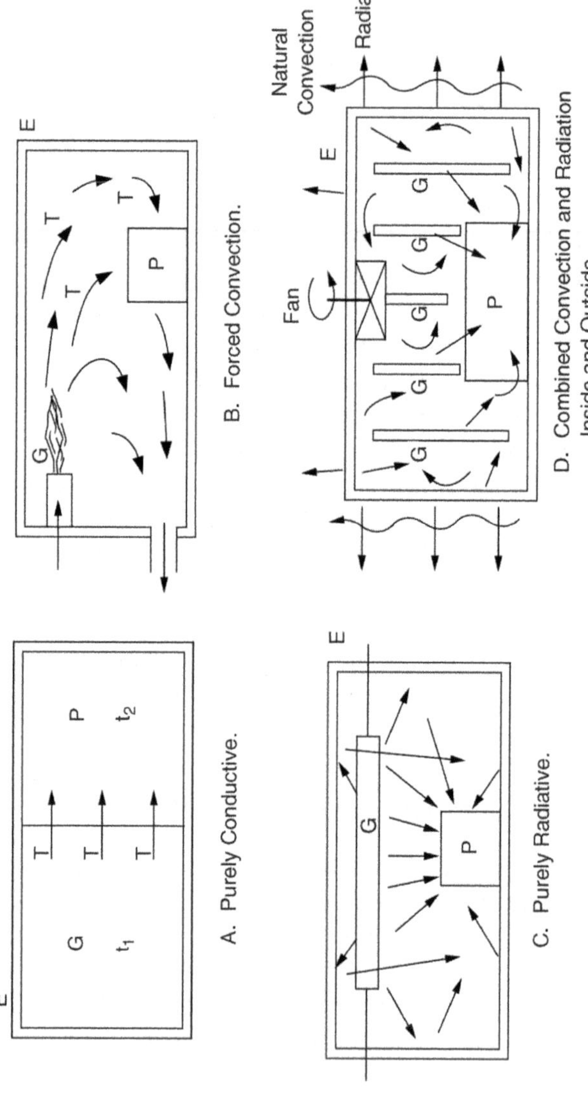

Figure 1.4 Heat transfer modes.

cases, the gases may be in forced circulation induced by a fan or pushed by a flame. The gases pick up heat from the generator and pass it to the work during circulation. This then becomes the main transfer mode and is called "convection" (natural or forced). In such cases, fluid dynamic properties such as viscosity, velocity, and flow channel geometry become the deciding factors in heat transfer.

Hot bodies radiate heat in the space around them. Thus a generator at a higher temperature than the work will radiate heat toward the work (Figure 1.4(C)). No physical contact or intervening gas is necessary. In the system under consideration, heat transfer will take place by radiation in addition to convection. If there is no gas in the intervening space (i.e., there is vacuum), radiation will still take place. Thus radiation is the third mode of heat transfer. It takes place alone or accompanies convection.

This discussion shows that ignoring pure conduction as a rare case, heat transfer between the generator and the work will take place by convection and/or radiation. One such furnace is shown in Figure 1.4(D). Note that in addition to internal radiation and convection (assisted by fan) the exterior is also transferring heat to the surrounding by both processes. We will quantify these modes in the next chapter. In case the transfer takes place simultaneously by both convection and radiation, the temperature of the hot body decides which mode will be dominant. Later we will show that radiation is dominant above about 600°C and convection below this temperature.

In high-temperature furnaces heat transfer is mainly by radiation. Low-temperature furnaces are convective furnaces and proper circulation of gases in the enclosure and around the work are their main design features.

In fuel-fired furnaces the flame is the main radiator and the inside enclosure surface is a secondary radiator. Both contribute toward radiant heat transfer to the work. The combustion gases evolving from the flame circulate in the enclosure and contribute to heat transfer by convection. The placement of the burners and the location of the gas exit port, with respect to the work and the enclosure, become important for obtaining maximum heat transfer to the work.

Vacuum furnaces are purely radiative. As the radiation from the generator also strikes the vessel it is necessary to use reflectors and cool the vessel wall with water circulation.

Heat transfer in laser and electron beam heating is mainly by conduction but has some peculiarities. Heat is created at the focal spot on the surface and is conducted to lower layers (Figure 1.3(D)). A considerable amount of incident energy is lost by reflection. The temperature at the spot is very high and some of the surface layer is melted, evaporated, and splashed. This makes the beam go deeper but the bulk transfer is by conduction. Because of high energy density the process is very fast and complete (through) heating requires a fraction of a minute. As intense heat is created on a very small spot, laser and electron beams are known as "concentrated heat sources."

In direct contact resistance heating, capacitance heating, and microwave heating (Figure 1.3(B) and Figure 1.3(C)), heat is created uniformly throughout the work on an atomic or molecular scale. Hence, transfer phenomena are not significant.

Induction heating heats a relatively thin surface layer (0.5–3.0 mm) (Figure 1.3(A)). The heat is then conducted to the inner layers. Due to concentration of heat generation on the surface and conduction to the interior, induction and laser heating offers a possibility of heating only to the desired depth by controlling the time of radiation.

Note that once the heat flux reaches the surface of the object it is carried inside by conduction only.

1.4 AUXILLIARY TECHNIQUES

While discussing convection in Section 1.3, we have seen the importance of proper circulation of gases in the furnace enclosure. The stream of gases may be smooth (laminar) or with eddies or recirculation (turbulent). These two types exhibit different flow patterns when they pass over walls, single or bulk work objects, and ducts. Consequently the heat transfer from such streams is also significantly affected. Gases in furnaces are generally at or near atmospheric pressure. When at extremely low pressure, such as in vacuum, they exibit an altogether different behavior.

Many parts of furnaces require water-cooling to keep their temperature low. This is usually achieved by circulating water through these parts. Here our object is to remove the heat. We have to consider heat transfer to the water flowing through the cooling channels. We again come across a type of flow and its effects on heat transfer. This will help us to decide the quantity of cooling water and the pump capacity required.

Considering the importance of gas and water circulation in heating processes we will review the underlying principles of related "fluid dynamics" in a separate chapter.

To protect metal from oxidation at high temperature and to bring about some changes in the composition of surface layers (e.g., carburizing, nitriding, etc.) special protective atmospheres or vacuum are used along with many heating processes. The generation and control of these atmospheres and vacuum are discussed in separate chapters.

The success of a heating process is determined solely by measuring and monitoring the temperature. A wide range of measuring techniques for high temperatures (pyrometry) are available. It is of paramount importance to choose a proper pyrometer for measurement and control of the given process. Pyrometry and temperature control are discussed separately.

Let us begin our discussions with heat transfer processes and proceed to auxiliary techniques. Solved examples at each stage will make the underlying design technique clear.

Chapter 2

Fluid Dynamics

CONTENTS

2.1 INTRODUCTION

In heating processes we come across gases and liquids. For example, when a fuel is combusted a large volume of gases are evolved as combustion products. These gases carry a considerable amount of heat and can be used for heating the work or for preheating of combustion air, thus increasing the efficiency of heat utilization.

Water is extensively used for cooling furnace parts such as doors and walls. The cooling is achieved by circulating water at high velocity through tubes. Correct flow rate and pumping power is required to be estimated to achieve optimal cooling effect.

In this chapter we will review the characteristics of fluids (gases and liquids) in motion, i.e., principles of fluid dynamics. Fluid mechanics is a separate branch of engineering sciences. Only that part which is essential and complementary to heat transfer is reviewed. For detailed discussions, reference should be made to specialized books [3, 4].

The gases that we come across in heating are almost always at or near atmospheric pressure but at higher temperatures. Under such conditions (of constant pressure) they behave like fluids, hence, the discussion that follows applies equally to gases and liquids. Extremely low pressure or rarefied gases occur in vacuum and are dealt with in Chapter 12.

2.2 SOURCES OF GASES IN FURNACES

Following are the main sources of gases in furnaces:

1. *Combustion products* — Burning fuels produce combustion gases which contain N_2, CO, CO_2, and H_2O. In excess-air-assisted combustion they contain some O_2. The amount of gas produced by burning (combusting) one unit quantity of fuel (kg, m^3) can be calculated as

discussed in Chapter 5. These are usually referred to as volume at standard temperature and prenure (STP) or normal temperature and prenure (NTP). Due to expansion at combustion (flame) temperature the volume of the gas increases enormously and must be considered in the calculation of the furnace volume, ducts stack, etc.

2. *Protective atmospheres* — These are prepared by the controlled combustion of gases like ammonia. Composition is similar to combustion gases (see Chapter 13 for detail). However, their volume in process is much less than the combustion products mentioned above. Both combustion and protective gases require safe disposal after exit; so a proper design of ducts, stacks, etc. is required.
3. *Inert or neutral gases* — These are used to provide nonreacting atmospheres and are costly. He, A, and N_2 are mainly used. Their volume is small and they do not require elaborate disposal equipment.
4. Many furnaces (mainly electrical) have air as trapped or confined atmosphere and do not pose disposal problems.

2.3 FLOW OF GASES

Gases expand on heating—The volume increases and density decreases with the increase in temperature. At constant pressure the volume of a gas is directly proportional, and its density inversely proportional, to the temperature.

$$V_t = V_o(1+\beta t) = V_O \frac{T}{273} \qquad (2.1)$$

$$\text{or} \quad Vt_2 = V_t \frac{T_2}{T_1}$$

Similarly

$$\rho_t = \rho_o/(1+\beta t) = \rho_o \frac{273}{T} \qquad (2.2)$$

where t is the temperature in °C, and T the temperature in K.

$$\beta = \text{coefficient of volume expansion} = \frac{1}{273}\ {}^{\circ}\mathrm{C}^{-1} \tag{2.3}$$

Thus, a hot gas is lighter than a cold gas. Under gravity a cold gas moves down and hot gas rises up, setting up a current or flow called "natural convection" or the "buoyancy effect." In an enclosure such as a furnace, air currents will be set up and assist the heat transfer process.

The same effect can be used to regulate the outflow of gases from a furnace and maintain the required constant pressure inside a gas-producing (fuel fired) furnace. This is known as the "stack effect" and will be discussed in detail later.

Another interesting property of gases is their kinematic viscosity υ that increases with increasing temperature. For almost all liquids, υ decreases with temperature. Due to the increase in viscosity the shear resistance of gases increases and hot gases move sluggishly. This effect becomes appreciable beyond about 500–600°C. This is one of the reasons why convection heat transfer is not effective above these temperatures and radiation dominates the process. Consequently, low-temperature ovens and appliances (T < 500°C) depend on the convective mode of transfer and require attention to the movement of gases. This effect is displayed by air and all other gases and their mixtures (fine gas).

Gas flow can also be set up by forced convection. In a fuel-burning furnace the flame (or jet) constantly pushes fuel, air, and combustion products forward thus setting a forward flow. This happens at all flame temperatures. At T > 500–600, the major contribution is still radiation but the available hot-gas stream can be used to obtain additional transfer by convection. We will discuss ways for using this source.

Some ovens and furnaces use a fan to set up gas flow inside. This is true forced convection. The flow thus set up has a relatively low velocity and is mainly used to obtain a homogeneous atmosphere throughout. Protective atmosphere furnaces, drying or baking ovens, and precipitation hardening (aging) processes use a fan. Besides achieving a uniform atmosphere, the fan does assist in heat transfer if positioned properly with respect to the work.

High-temperature furnaces will definitely benefit by using a fan but there are two problems. First, the increased viscosity will require more power. Second, materials that can handle (or circulate) gases at such temperatures are costly and have a limited life. Jet engines routinely handle high volumes of gas at very high temperatures but it is economically impossible to use such arrangements in furnaces.

Heat transfer with forced and natural convection is discussed in detail in Chapter 3. Here we are more concerned about the flow characteristics of these two modes.

2.4 IMPORTANCE OF FLUID FLOW IN HEATING

Consider a quiescent layer of fluid (e.g., a gas) on a metal surface. Let T_g be the temperature of the gas and T_m that of the metal (surface) with the condition $T_g > T_m$. Heat will be transferred from the gas to the metal by conduction from bulk gas to the interface and from there to the metal. Typical conductivities of the gas and the metal are 5×10^{-2} and 50.0 w/m°C, respectively. This shows that heat transfer by pure conduction will be very low.

Now consider the same system with the gas flowing over the metal surface. The flow will continuously bring hot gas to the interface and will pick up "cold" gas to the bulk. In these circumstances the heat transfer will be very fast and be in the order of 5–1000 w/m°C.

The above example brings out the importance of gas flow (or fluid flow) in heating. Note that the same effect will be observed if $T_g \ll T_m$, i.e., when the interface is cooled by the fluid.

Some important conclusions can be drawn from the above discussion:

1. Heat transfer in a system with fluid in motion will be dependent on the velocity of the fluid or more correctly, on the type of flow.

2. Heat transfer will take place only where there is contact between the fluid and the surface. If the fluid (for whatever reason) separates there will be practically no heat transfer.
3. At or very near the surface the heat transfer will still be by conduction.
4. If the flow is not uniform over the surface the heating will also not be uniform.

A detailed discussion of the relation between fluid flow and heat transfer (convective heat transfer will be presented in Chapter 3 and Chapter 4.)

In the present chapter we will only consider the behavior of the fluid as it passes over objects and through tubes or ducts.

2.5 CLASSIFICATION OF FLUID FLOW

When a fluid flows over a surface or inside a tube or duct, the flow exhibits many patterns that can be used for the classification of the flow. Heat transfer between the surface and the fluid depends on the type of flow. Similarly, the loss of pressure due to friction depends on the flow pattern.

Velocity V, kinematic viscosity υ, and a critical dimension d determine the flow pattern. The critical dimension d is the internal diameter of the tube or the width of the surface. Hence, the Reynolds number Re is a convenient dimensionless parameter for classification.

$$\mathrm{Re} = \frac{Vd}{\upsilon} \tag{2.4}$$

Based on the Reynolds number, the main types of flow patterns that we will come across are the laminar and turbulent flows (Figure 2.1).

Laminar flow is a smooth flow having parallel flow lines. (Figure 2.1(B)). There is no mixing and the lines are continuous. The fluid layer in contact with the surface has zero velocity. Subsequent layers have increasing velocities until at some distance (y, r) the velocity is equal to the original value V. The velocity

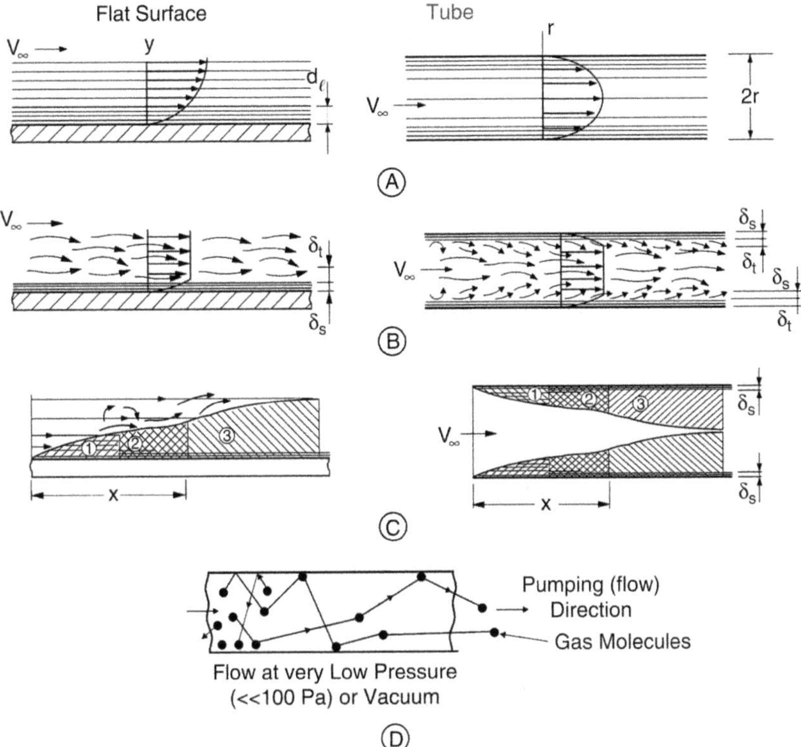

Figure 2.1 Classification of fluid flow.

profile in a well-established or well-developed laminar flow is parabolic. The fluid layer with low or zero velocity near the surface is known as the "boundary layer," having a thickness δ_p.

If the fluid is flowing through a tube the boundary layer is all over the perimeter. Laminar flow occurs at Re <2500 (approximately).

Turbulent flow exhibits discontinuous and zig-zag or wavy flow lines (Figure 2.1(A)). There is mixing with fluctuations in velocity and pressure, and the flow is unsteady. A thin fluid layer near the surface has a laminar pattern and is known as the boundary sublayer δ_s. Beyond the sublayer is the boundary layer δ_t in which the velocity goes on increasing until it acquires the value V_∞. If the fluid is flowing through a pipe, the boundary layers δ_t, δ_s exist all over the perimeter.

In a well-developed turbulent flow, the velocity pattern has a trapezoidal shape as shown in Figure 2.1(B).

Turbulent flow is well established at Re >10^5 (approximately). All other conditions (v, d) remaining constant, velocity in turbulent flow is much higher than that in laminar flow.

In between Re = 2500 and Re = 10^5 the flow pattern is mixed, i.e., it is partly laminar and partly turbulent and is called transitional flow (Figure 2.1(C)). The Reynolds number at which the flow changes from laminar to turbulent is called the critical Reynolds number Re_{cr}. There is a range representing Re_{cr} but for convenience it is usually taken as 2500–3000.

Whenever a fluid with a free flow velocity V_∞ meets the frontal edge of a surface or enters a tube, the velocity pattern changes from laminar → transitional → turbulent. This change takes place over a certain length x (Figure 2.1(C)) called the entry length.

Reaching x, the flow assumes a well-developed or fully-developed pattern. If the tube is shorter than x the flow is not fully developed. Thus, if a small (or short) object is being heated by a fluid stream, the flow pattern over it is difficult to predict.

Flow at very high velocities becomes transonic, supersonic, or hypersonic. The classification of these flows is done on the basis of Mach number M.

$$M = \frac{V}{V_S}$$

where V is the velocity of fluid and V_S is the velocity of sound in the fluid.

All these ultrasonic flow patterns occur at $M \geq 1$. We will not come across such high velocities in our subject matter and hence these are not discussed further.

At very low pressures (<<100 Pa) the stream lines lose their identity because the flow is molecular (Figure 2.1(D)). There is no boundary layer, viscosity, or velocity. We have to talk about mass or volume pumped in unit time. Flow under low pressure or vacuum is discussed in Chapter 12.

2.6 FLOW OVER OBJECTS

Consider a cylinder of diameter d placed across a stream of hot air at various temperatures. Kinematic viscosity υ (m²/sec) at different temperatures is available in the standard table (see Appendix F). We will assume velocity (V) of the air stream as 1.0 m/sec and the diameter $d = 0.1$ m. The product $V \times d = 1 \times 0.1 = 0.1$. Since $\mathrm{Re} = Vd/\upsilon$, substituting the above values we get

$$\mathrm{Re} = \frac{0.1}{\upsilon} \tag{2.5}$$

The results of calculation for Re at various temperatures are shown in Figure 2.2. It shows that the Reynolds number decreases with increasing temperature. This is expected as the kinematic viscosity of gases increases with increasing temperature with a rate change in the range of 600–800°C. For example, Re = 0.43 × 104 at 100°C and 0.044 at 1200°C but the range is between 10^3 and 10^4. Increasing the diameter changes the range to 10^3–10^5. This exercise shows that at gas

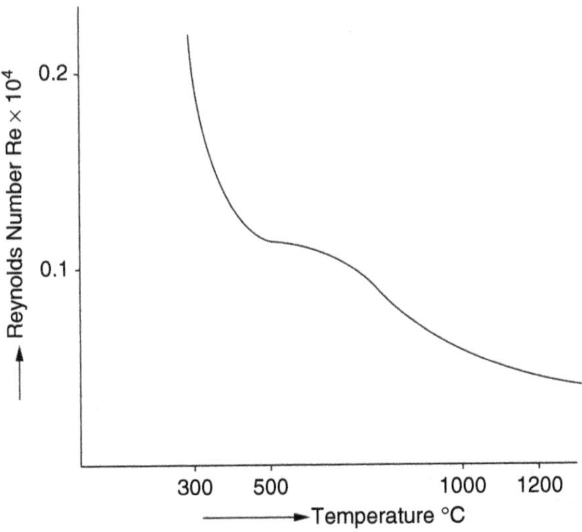

Figure 2.2 Variation of Reynolds number of air with temperature.

velocities normally encountered in furnaces the flow is generally of transient nature.

Experimental results for flow patterns across a cylinder are shown in Figure 2.3, in the range Re = 10^3–10^5. There are turbulent wakes and eddies on the backside. This increases the local heat transfer on the backside.

If the velocity is very low (~0.1 m/sec), the Reynolds number will be in the range 10^2–10^3, the stream will cover a larger surface (by its boundary layer) and there will be no turbulent wake. This gives a more uniform heat transfer but at a lower rate.

Very uniform heating will be obtained at a still lower Reynolds number (1–10) but the heating rate will be very low. These conditions may be suitable for backing or drying operations.

Turbulent flow requires Re > 10^5. This requires articles with small diameters (0.01 ~ 0.05 m) and high velocities (~100–1000 m/sec). The first condition (i.e., small diameter) is not always satisfied. The second condition can be sometimes satisfied by using a high-pressure convective burner. Generally more than one burner is used (Figure 2.4). They are arranged so that the gas stream strikes the work tangentially. This covers all the surface uniformly.

A vortex is a rotating gas stream that has a spiral structure, high gas velocities, and a uniform axial motion. A vortex is used in some furnaces.

Except the direct use of high-velocity flames, turbulent streams cannot be handled by using fans or blowers owing to materials problems.

2.7 FLOW SEPARATION

The necessary condition for convection heat transfer is a good contact between the fluid and the surface of the solid.

In both types of flow (laminar and turbulent) there exists a boundary layer through which the heat transfer takes place by conduction.

Under certain combinations of velocity and surface geometry, the boundary layer (and hence the flow) separates

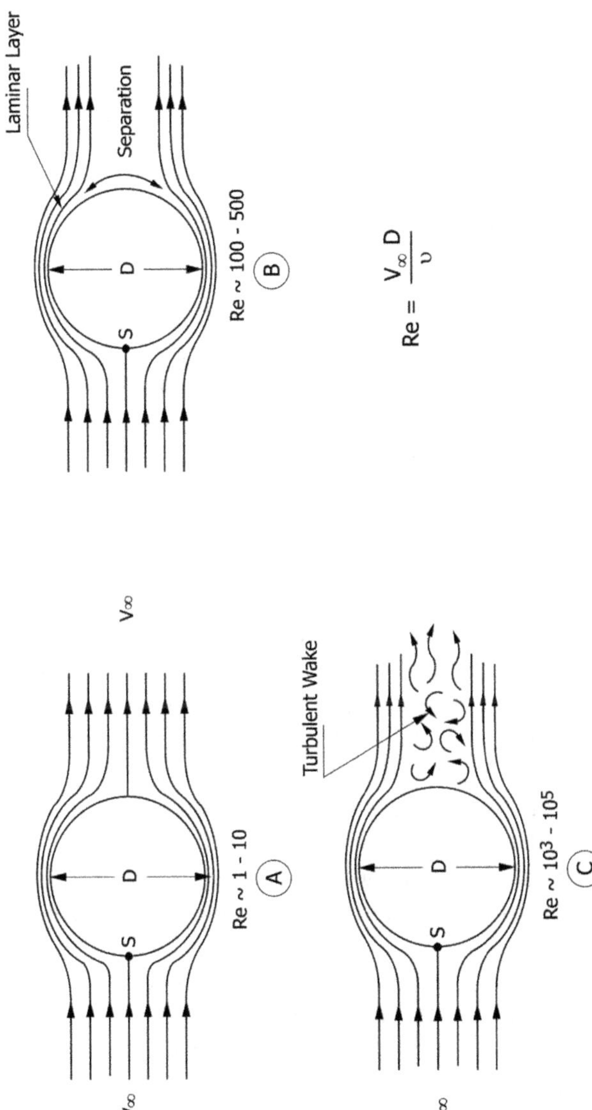

Figure 2.3 Flow around a cylinder at various Reynolds numbers.

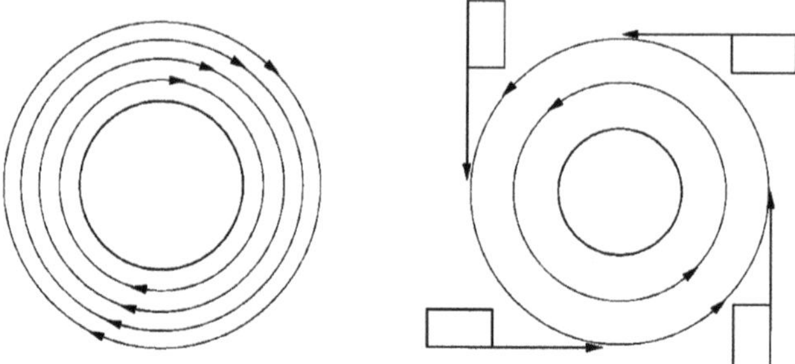

Figure 2.4 Connective heating at high velocities.

from the surface as shown in Figure 2.5(A). This impedes the heat transfer. Such conditions usually occur at high velocities and when there are obstacles in the flow path. The flow attempts to jump over the obstacle or take a shortcut over an edge or corner. Some examples are shown in Figure 2.5(B), Figure 2.5(C), and Figure 2.5(D). If the velocity is high, eddies and vortices are created below the separation. The separation and the vortices create a pressure drop and reduce the velocity heat.

Separation affects the flow around work, which can have any shape. Similarly the flow over walls, corners, parts, and ports is affected. Some typical patterns of flow occurring at expansions, contractions, and right-angled bends are shown in Figure 2.5(E), Figure 2.5(F), and Figure 2.5(H). The loss of head at such obstacles can be calculated, as discussed in the subsequent section.

Flow over or about cylindrical bodies is shown in Figure 2.3. The separation and the formation of turbulent wake is clearly seen. The flow over a rectangular body divides as shown forming a turbulent wake and separation at the edges.

Flow separation should be avoided as far as possible. For the work, the shape is not under control but the orientation relative to gas stream can be changed in the furnace enclosure. The corners can be rounded but they will soon get eroded by the gas stream. Rounding of corners is not possible in small furnaces constructed with refractory bricks.

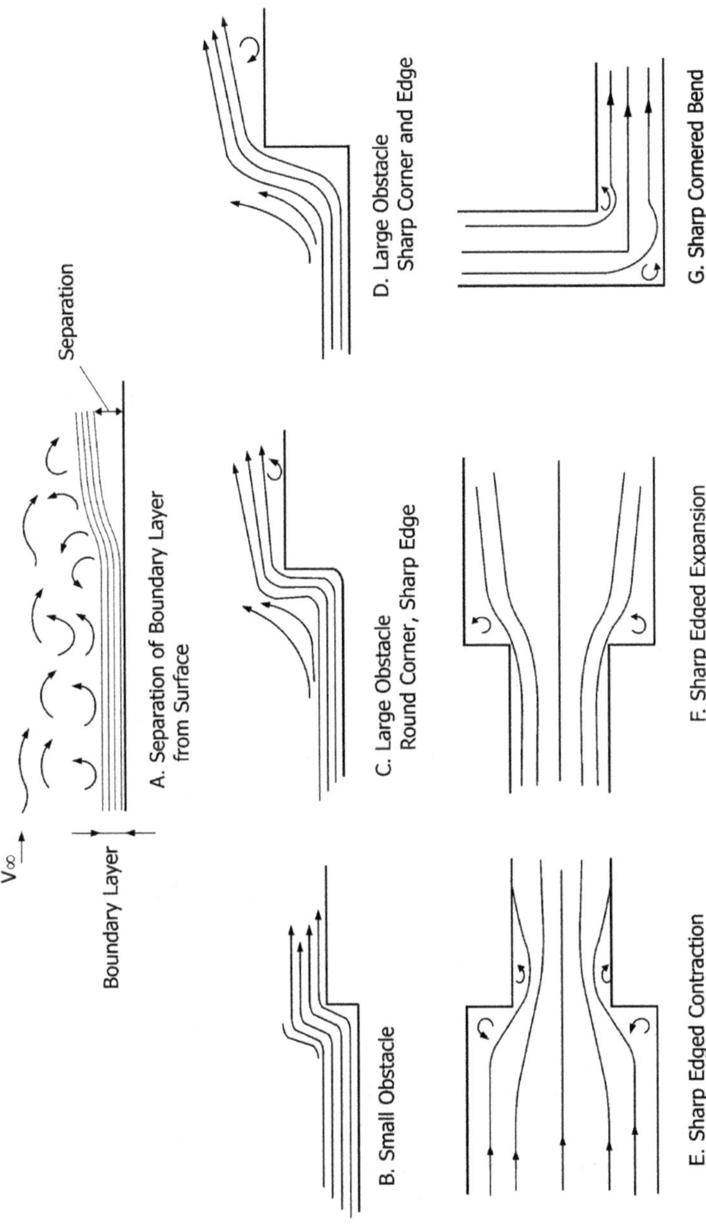

Figure 2.5 Separation of flow.

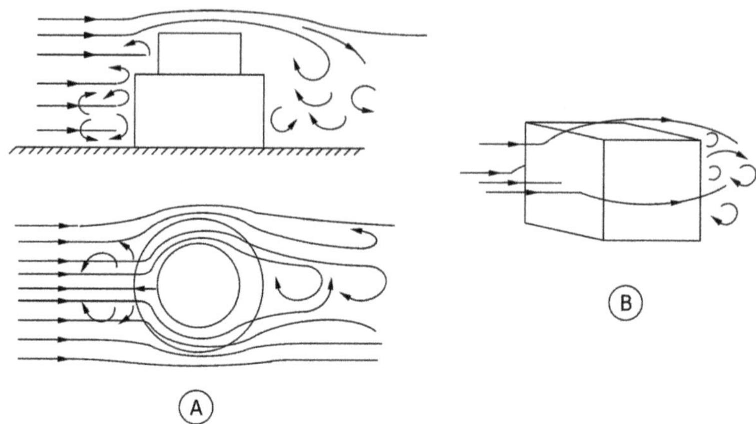

Figure 2.6 Fluid-convection currents around various bodies in laminar/transitional conditions.

As the possibility of flow separation is more at high velocities, the flow velocities used in convective furnaces are low and generally in a range of 1–20 m/sec.

Typical flow patterns around a stepped cylindrical body and a cubic body are shown in Figure 2.6(A) and Figure 2.6(B).

In many instances the work is kept in a basket. A typical basket shape is cubic. The method of stacking workpieces in the basket should be such that it promotes the flow through it. A random packing will impede the flow as shown in Figure 2.7(A) and Figure 2.7(B).

If there are a number of workpieces loaded in a furnace, the distance between two pieces should be wide and uniform. A narrow and random separation (Figure 2.7(C)) will prevent hot gases from forming turbulent wake. Wide and even spacing will lead to the formation of wakes or eddies and improve heat transfer in the whole load.

When a gas stream is stopped by a wall it is dispersed sideways. There may be separation of dispersed stream if the velocity is high. If a stream strikes a relatively short obstacle it will try to climb over with some separation (see Figure 2.8(A) and Figure 2.8(B)).

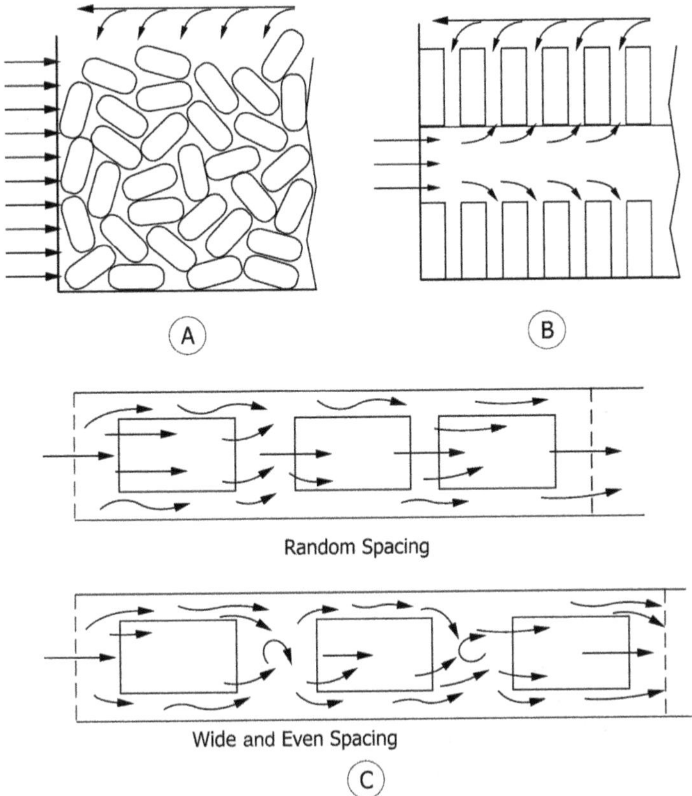

Figure 2.7 Air flow around charges in a furnace.

2.8 FORCED CIRCULATION IN ENCLOSURES

Fuel-fired furnaces have one or more flames inside the enclosure. These emit (produce) a high-speed stream of combustion gases which are at or near flame temperature (1000–2200°C). Before they exit through the port, they circulate around the interior and work. The main heat transfer to the work is by radiation from flames and walls. However, if the gases are made to circulate properly some of their heat is recovered before exit. The gas circulation is "forced" and will depend on

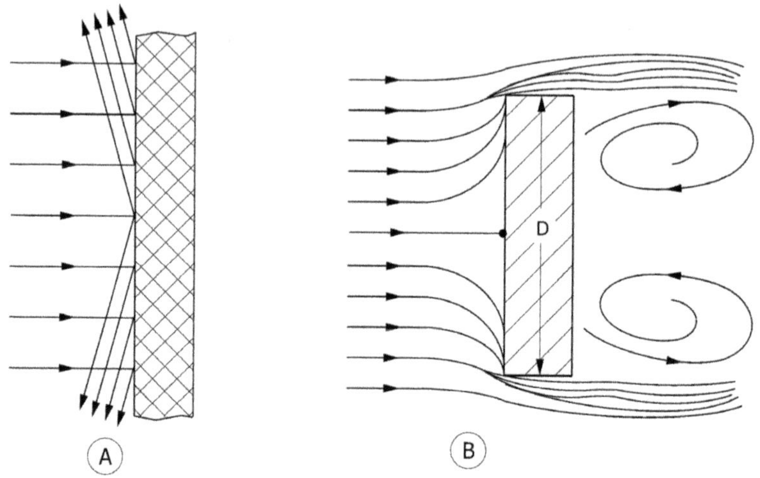

Figure 2.8 Fluid flow and obstacles.

the relative placement of flame work and exit port. Some examples of this are shown in Figure 2.9.

Circulation patterns create vortices and pockets of hot and cold gases; small vortices at corners are unavoidable and unusable. Work placed in a large vortex will show better heating.

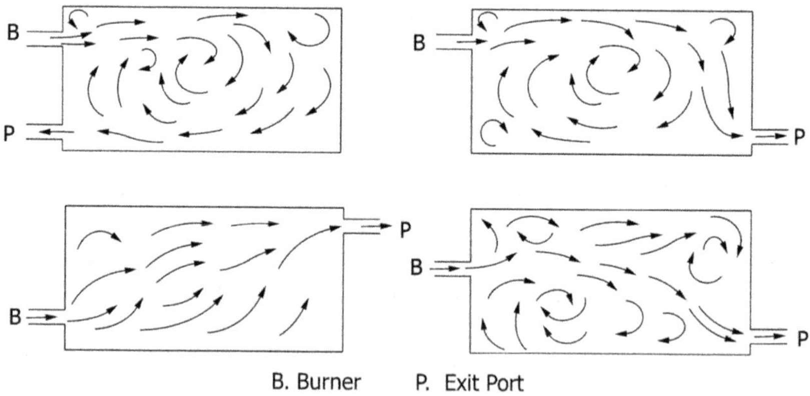

Figure 2.9 Circulation of combustion gases in fuel-fired furnaces.

2.9 USE OF FANS

Forced gas flow can be created in enclosures by using a fan. Due to possible damage by temperature and gas attack, the motor and drive components are mounted outside and only a part of the drive shaft and rotor (propeller) are inside. The fans used are centrifugal, propeller type, and rotated at a low speed (200–400 RPM). The parts in contact with gases and high temperatures (>350–400°C) are made of nickel alloys. For lower temperatures they can be made of stainless steel.

Several possible locations of the fan are shown in Figure 2.10. Higher or synchronous speeds impose several stresses on the blades and the shaft. There is no significant contribution of fans toward heat transfer at high speeds.

The main purpose of fans is to distribute the gas evenly so as to reduce concentration gradients. Hence, they are mainly used in protective or special atmosphere furnaces. Banking and drying ovens exit steam (moisture) and carbon dioxide that can be removed by a fan. Quenching oil can be stirred by using a fan (or more correctly, by a stirrer) properly located outside the quench area.

Very small (~100 mm diameter) fans are used in electronic equipment to remove heat generated in operations.

As the size of the enclosure is determined by other considerations, the maximum size of the fan and the gas that can be moved at a given speed is automatically decided.

2.10 NATURAL GAS CIRCULATION INSIDE FURNACES

Electrically heated resistance furnaces or other indirectly heated enclosures are filled with air that gets heated. Currents are set up inside due to natural or free convection. As stated before, natural convection arises because of the density difference between hot and cold gases. The former has less density, i.e., it is light-weight and rises up while the latter sinks down. In high-temperature furnaces the contribution

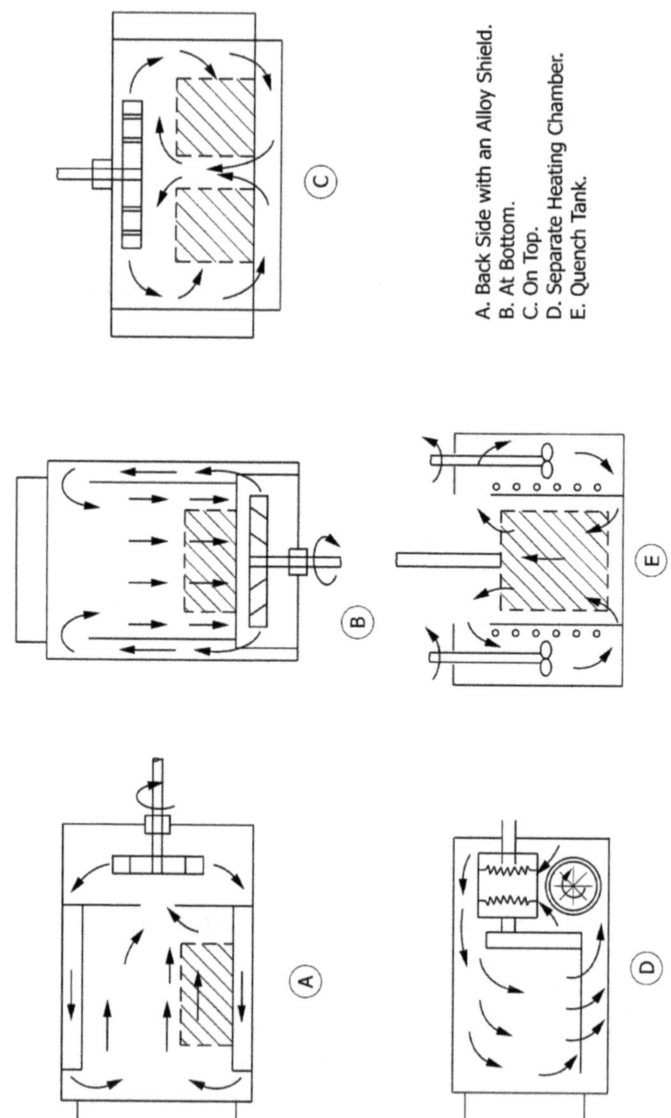

Figure 2.10 Use of fans in heating and cooling.

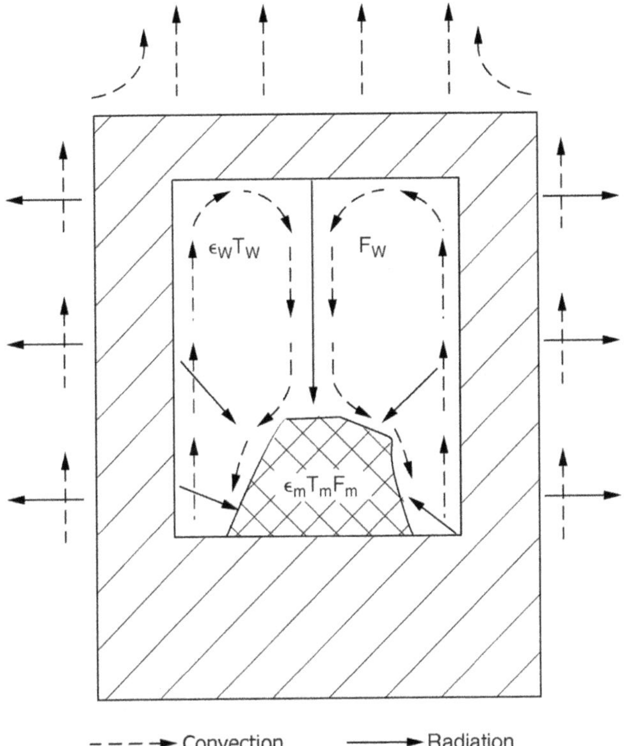

Figure 2.11 Radiation and convection heat transfer inside and outside of a furnace.

of convective currents toward heat transfer is small but significant.

The outside enclosure or wall temperature is usually 50–80°C while the environment may be at 20–35°C. This temperature difference gives rise to external natural convection. These currents make a significant contribution toward the heat loss to the surroundings.

Both external and internal convection currents are shown in Figure 2.11. We will discuss the calculation of heat transfer by natural convection in Chapter 3. Note that this heat transfer (due to gas flow) is in addition to radiation.

2.11 BERNOULLI'S THEOREM OF FLUID FLOW

Bernoulli's theorem is essentially the law of conservation of energy as applied to a fluid stream.

At constant density and constant temperature a fluid stream possesses three types of energy or three pressures at any chosen point:

1. Kinetic energy due to its velocity V (m/sec) is $\rho V^2/2$ (N/m^2)
2. Static or piezometric pressure P (N/m^2)
3. Potential energy Z due to the height h (m) of the stream above a chosen datum

$$Z = \rho g h \ \text{N/m}^2 \tag{2.6}$$

thus

$$Z + \frac{\rho V^2}{2} + P = \text{constant} \quad (\text{N/m}^2 \text{ or Pa}) \tag{2.7}$$

If the point is on datum level $Z = 0$ and if the fluid is stationary $V = 0$ the above equation can be written in other units. If pressure is measured in a column or "head" units (e.g., mm H_2O, etc.)

$$h + \frac{V^2}{2g} + \frac{P}{\rho g} = \text{constant} \quad (\text{m}) \tag{2.8}$$

In energy units J, the equation becomes

$$h_{mg} + \frac{mV^2}{2} + \frac{P_m}{\rho} = \text{constant} \quad (\text{J or N}\cdot\text{m}) \tag{2.9}$$

The above equations can be applied to both liquids and gases, provided that the density is constant. These equations do not consider the irreversible losses in the stream.

Consider two points, ① and ②, in a fluid stream as shown in Figure 2.12(A). The cross section and velocity at ① is A_1 and V, and that at ② is A_2, V_2, respectively.

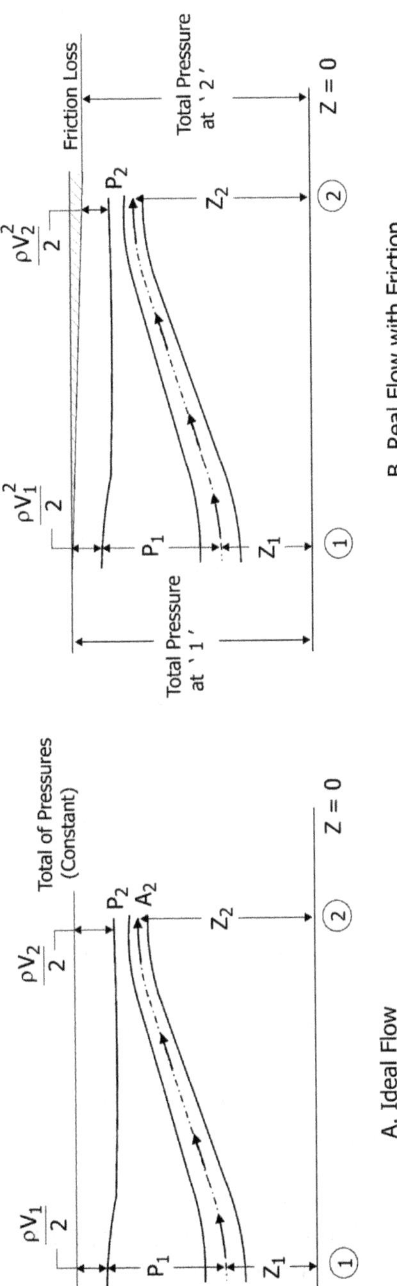

Figure 2.12 Application of Bernoulli's theorem.

Since $A_1 > A_2$ and the mass flow rate m is constant, $m = A_1V_1 = A_2V_2$ and $V_1 < V_2$. Applying Bernoulli's theorem at ① and ②

$$Z_1 + P_1 + \frac{\rho V_1^2}{2} = Z_2 + P_2 + \frac{\rho V_2^2}{2} = \text{constant} \quad (\text{Pa}) \tag{2.10}$$

Z_1 and Z_2 are the stream elevations and P_1 and P_2 the pressures at ① and ②.

For a horizontal pipe $Z_1 = Z_2$ hence

$$P_1 - P_2 = \frac{\rho}{2}\left(V_2^2 - V_1^2\right) \tag{2.11}$$

For a real situation some energy will be lost due to friction between the tube wall and the fluid and Equation (2.11) will not be applicable (Figure 2.12(B)). It will now have the form

$$Z_1 + P_1 + \frac{\rho V_1^2}{2} = Z_2 + P_2 + \frac{\rho V_2^2}{2} + P_f \tag{2.12}$$

where P_f is the pressure loss due to friction (Pa).

Pressure loss due to friction is called "frictional" or "major" loss. This loss depends on the characteristic of the flow and the geometry of the surface.

The frictional loss of pressure occurs in all ducting and pipes and is discussed in detail in Section 2.12. Practical ducts and pipe lines contain many components such as expansions, contractions, flow regulators, and junctions. These are called local features. All such local features cause a pressure drop (h_t) and are discussed in the subsequent sections. They are called "local" or "minor losses," but can be quite substantial losses.

There is one more form of energy in a fluid. It is called the internal energy "m" (Joules). It is proportional to the temperature of the fluid. On heating or cooling the internal energy changes by $Cv\ dT$, where Cv is the specific heat at constant volume (J/kg°C), and dT is the change in temperature. Liquids absorb heat at constant volume. Our applications generally involve constant volume. If the change in the internal energy is from u_1 to u_2, a full expression of Bernoulli's equation will be

$$u_1 + P_1 + \frac{\rho V_1^2}{2} + Z_1 = u_2 + P_2 + \frac{\rho V_2^2}{2} + Z_2 \tag{2.13}$$

2.12 FRICTIONAL LOSSES IN FLOW

While discussing the flow of a fluid over a surface, we have seen that a thin layer of fluid in contact with the surface is stationary. The bulk fluid over this layer slides over it and undergoes shear. Some of the energy of the fluid is irreversibly lost in this shearing or friction. The thin layer in which the velocity increases from zero to bulk value is the boundary layer (Figure 2.13(A)). The frictional loss by shear is a function of velocity (V m/sec) and kinematic viscosity (υ m²/sec) and can be correlated to the Reynolds number (Re).

For laminar flow (Re < 2300), the dimensionless frictional loss coefficient (f_ℓ) is given by

$$f_\ell = \frac{64}{\text{Re}} \tag{2.14}$$

For transitional and turbulent flow in the range 2300 < Re < 10^5 the loss coefficient is given by

$$f_t = \frac{0.3164}{\sqrt[4]{\text{Re}}} \tag{2.15}$$

The loss of head due to friction (h_f) is given by

$$h_f = f_{(\ell,t)} \frac{L}{d} \frac{V^2}{2g} \quad \text{m} \tag{2.16}$$

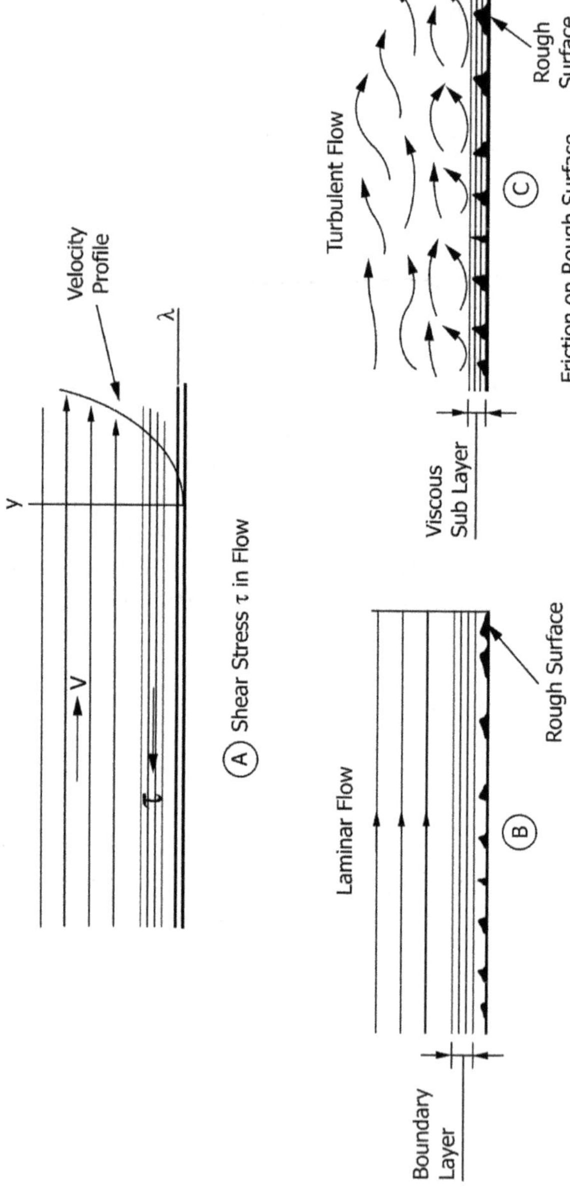

Figure 2.13 The nature of friction in fluid flow.

where L is the length and d the diameter of the pipe. The pressure loss due to friction is

$$p_f = f_{(\ell,t)} \frac{L}{d} \frac{\rho V^2}{2} \quad \text{Pa} \tag{2.17}$$

where ρ is the density of the fluid at bulk temperature. The above discussion applies only to smooth pipes in which the frictional loss is only due to shear.

A common feature of all surfaces over which the fluid flows is roughness. A rough surface with pronounced projections will offer higher resistance of friction to the flow. Surface roughness is thus the second cause of frictional head loss.

In laminar, and to some extent in transitional flow, there exists a slow moving boundary layer of appreciable thickness. Surface projections due to roughness (if sufficiently small) will be covered or drawn by the boundary layer. The main body of the flow will not be affected by the projections (Figure 2.13(B)).

The boundary layer in turbulent flow is very thin and does not cover the projections. The stream lines are obstructed, affecting the flow (Figure 2.13(C)). Hence, for turbulent flow the frictional loss depends on both viscosity and roughness and no single expression can be derived for the determination of the frictional coefficient f.

The surface roughness in absolute terms is the height of projection (mm). However, this height is not uniform all over and we use relative roughness k/d, where k is the mean height of the projection and d is the pipe diameter (both mm).

There are many empirically developed expressions for determining the frictional coefficient f for flow in rough pipes at Re > 10^3 and various relative roughnesses. A graphical plot for this purpose, developed first by Nikurdse and later supplemented by Moody and others, is shown in Figure 2.14.

More information can be obtained from handbooks Listed in the *Bibliography*.

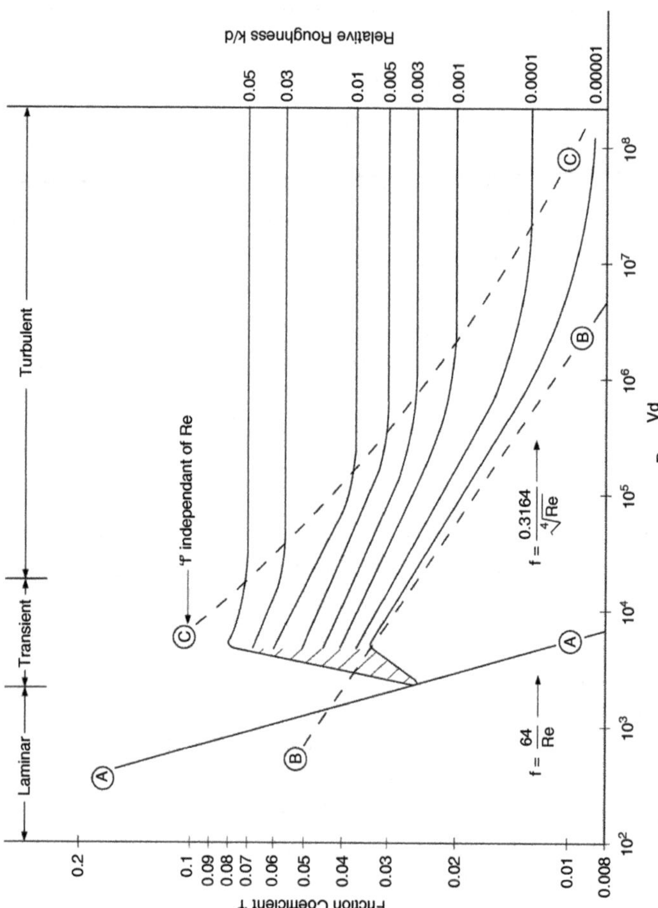

Figure 2.14 Friction factor for fluid flow inside pipes of various relative roughness and Reynolds number.

The figure shows laminar, transitional, and turbulent regions. The following features of the plot are noteworthy:

1. In the laminar region the friction is independent of surface roughness (*k*/*d*).
2. The critical Reynolds number (~ 2500) is independent of roughness.
3. The friction factor increases linearly with Reynolds number but is mostly independent of roughness.
4. For small (<0.015) values of roughness and Reynolds number ($<10^5$), the friction factor follows the line $f = 0.3164/\sqrt[4]{\text{Re}}$.
5. At higher Reynolds number and higher roughness the friction factor is constant, as indicated by the horizontal portion of curves.
6. In the region between line B and the curve C, the friction factor depends on both the Reynolds number and roughness.

For the flow of gases inside furnaces the Reynolds number is usually of the order of 10^3–10^5 and we can safely use the relation $f = 0.3164/\sqrt[4]{\text{Re}}$.

The flow of water in cooling channels is usually turbulent. To reduce frictional losses we use smooth tubes. The flow conditions will be between lines B and C.

Absolute roughness values (k) of many types of surfaces and materials are available in handbooks listed in the *Bibliography*. A short list a materials of interest is given in Table 2.1.

EXAMPLE 2.1

Combustion gases are flowing through a duct 0.8 × 0.6 m section and 5.0 m long at a volume rate 3.0 m^3/sec. Inlet and outlet temperatures of the gas are 650 and 860°C, respectively. The density of gas at 0°C is 1.3 kg/m^3. The average kinematic viscosity is 69.8×10^{-6} m^2/sec. The frictional coefficient for the duct is 0.05.

Determine the pressure loss in the duct.

TABLE 2.1 Absolute Surface Roughness k (mm) for Selected Materials

Material	k (mm)
Clean, new brass or copper pipes	0.0015
Seamless steel pipes	0.05
Used steel pipes	0.14–0.19
Rusty, dirty pipes	0.6–0.75
Ceramic pipes	0.2–0.4
Glazed brick	0.45–3.0
Brickwork with morter	0.8–6.0

Note: The friction factor offered by a surface will change (usually increase) during service due to abrasion and corrosion.

Solution

As the duct is rectangular, the hydraulic diameter (d_h) is given by

$$= \frac{2a \times b}{a+b} \text{ (where } a \text{ and } b \text{ are the sides)}$$

$$= \frac{2 \times 0.8 \times 0.6}{(0.8 + 0.6)} = 0.686 \text{ m}$$

$$\text{Velocity of gas } V = Q/A$$

$$= \frac{3}{(0.8 \times 0.6)} = 6.25 \text{ m/sec}$$

Reynolds number (Re)

$$= \frac{V \times d}{\upsilon}$$

$$= \frac{6.25 \times 0.686}{69.8 \times 10^{-6}} = 6.15 \times 10^4$$

As Re < 10^5, the flow is transitional.

Density at 0°C (ρ_0) is 1.3 kg/m^3

$$\rho_t = \frac{\rho_0}{1+\beta t} \text{ where } \beta = \frac{1}{273}$$

$$\rho_{800} = \frac{1.2}{1+\dfrac{800}{273}} = 0.329 \text{ kg/m}^3$$

$$\rho_{650} = \frac{1.3}{1+\dfrac{650}{273}} = 0.383 \text{ kg/m}^3$$

Average density ρ is (0.329 + 0.383)/2 = 0.356 kg/m^3
The loss of pressure due to friction is

$$p_f = f \times \frac{L}{d} \times \frac{\rho V^2}{2}$$

$$= 0.05 \times \frac{5}{0.686} \times \frac{0.356 \times 6.25^2}{2}$$

$$= 2.73 \text{ Pa}$$

EXAMPLE 2.2

A furnace panel in Figure 2.15 is cooled by circulating water through the pipes welded to the panel. The pipe diameter is 12.0 mm and water velocity is 3 m/sec. Average water temperature is 40°C at which the density is 992.2 kg/m^3 and viscosity is 0.659×10^{-6}.

Calculate the pressure drop in the pipe between the inlet and outlet headers.

Solution

At a speed 3 m/sec and pipe diameter 0.012 m, the volumetric flow rate is

$$Q = \frac{\pi \times 0.012^2}{4} \times 3$$

$$= 3.39 \times 10^{-4} \text{ m}^3\text{/sec}$$

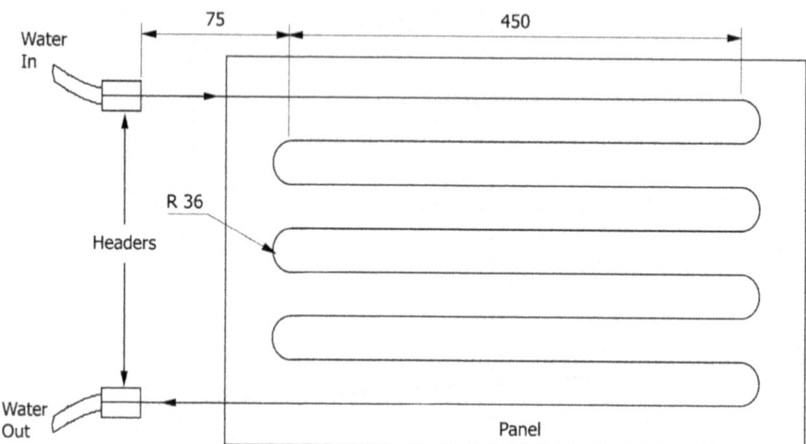

Figure 2.15 Cooling panel (pipe diameter 12.0 mm).

The mass flow rate is

$$m = \frac{Q}{\rho} = \frac{3.39 \times 10^{-4}}{992.2} = 3.42 \times 10^{-7} \text{ m}^3/\text{sec}$$

There are four head losses:

1. Loss in bends
2. Loss in straight lengths (including bends)
3. Loss at entry
4. Loss at exit

$$\text{Length of straight pipe} = 450 \times 8 + 75 \times 2 = 3750 \text{ mm } (3.75 \text{ m})$$

$$\text{Length of bends} = \pi \times 36 \times 7 = 792 \text{ mm } (0.792 \text{ m})$$

$$\frac{R}{d} \text{ for bends} = \frac{36}{12} = 3$$

The loss coefficient for smooth bends having $R/d > 1$ is 0.11. There are 14 right angle bends (7 × 2). The total loss in head due to bends is

$$h_{\text{bend}} = 0.11 \times 14 \times \frac{3^2}{2 \times 9.81} = 0.706$$

The Reynolds number is

$$\mathrm{Re} = \frac{Ud}{\upsilon} = \frac{3 \times 0.012}{0.659 \times 10^{-6}}$$

$$= 5.46 \times 10^4$$

The flow is nearly turbulent.
The friction factor (using the Blasius equation) is

$$\frac{\lambda}{t} = \frac{0.3164}{\sqrt[4]{\mathrm{Re}}} = \frac{0.3164}{\sqrt[4]{5.46 \times 10^4}}$$

$$= 0.0207$$

Loss of head in friction is

$$h_f = \lambda_t \frac{\ell}{d} \frac{V^2}{2g}$$

$$= 0.0207 \frac{3.75 + 0.792}{0.012} \times \frac{3^2}{2 \times 9.81} = 3.60$$

Loss of head at entry

$$h_{\text{entry}} = 0.5 \times \frac{V^2}{2g} = 0.5 \frac{3^2}{2 \times 9.81} = 0.229$$

Loss of head at exit

$$h_{\text{exit}} = \frac{V^2}{2g} = \frac{3^2}{2 \times 9.81} = 0.458$$

Total loss of head

$$h = \begin{bmatrix} \text{loss in} \\ \text{straight pipe} \end{bmatrix} + \begin{bmatrix} \text{loss in} \\ \text{bends} \end{bmatrix} + \begin{bmatrix} \text{entry} \\ \text{loss} \end{bmatrix} + \begin{bmatrix} \text{exit} \\ \text{loss} \end{bmatrix}$$

$$= 3.60 + 0.706 + 0.229 + 0.458$$

$$= 4.99 \text{ m } H_2O$$

Pressure drop

$$P = h\rho = 4.99 \times 492.2$$

$$= 4.95 \times 10^3 \text{ kg/m}^2 \simeq 5 \text{ kPa}$$

2.13 LOCAL LOSSES

In Section 2.12, we discussed the loss of pressure, or head, due to friction that occurs when a fluid flows over straight rough surfaces. In a practical situation we use several components in a pipeline that change the cross section and the direction of flow. These components are called "local disturbances." They cause further losses of pressure, which are called minor or local losses. Several components that cause local disturbances and minor losses are shown in Figure 2.16. The pressure loss from these components is mainly due to the turbulence set up by the change in cross section and direction.

Three types of local disturbance are most common for our purpose. They are sudden contraction, sudden expansion, and a hyphen (generally right-angled) bend which may be sharp or smooth. The pressure (local) loss (h_ℓ) at these features can be calculated by applying Bernoulli's theorem to the stream just before and after the local feature; the proofs are not discussed here. The loss is given in the form of a coefficient ζ to be applied to the dynamic, or, velocity head $V^2/2q$. Thus

$$p_\ell = \zeta \frac{V^2}{2g} \text{ (m)} \tag{2.18}$$

$$\text{or } h_\ell = \zeta \frac{V^2}{2g} \text{ (Pa)} \tag{2.19}$$

Determination of the friction factor for some commonly used local features follows.

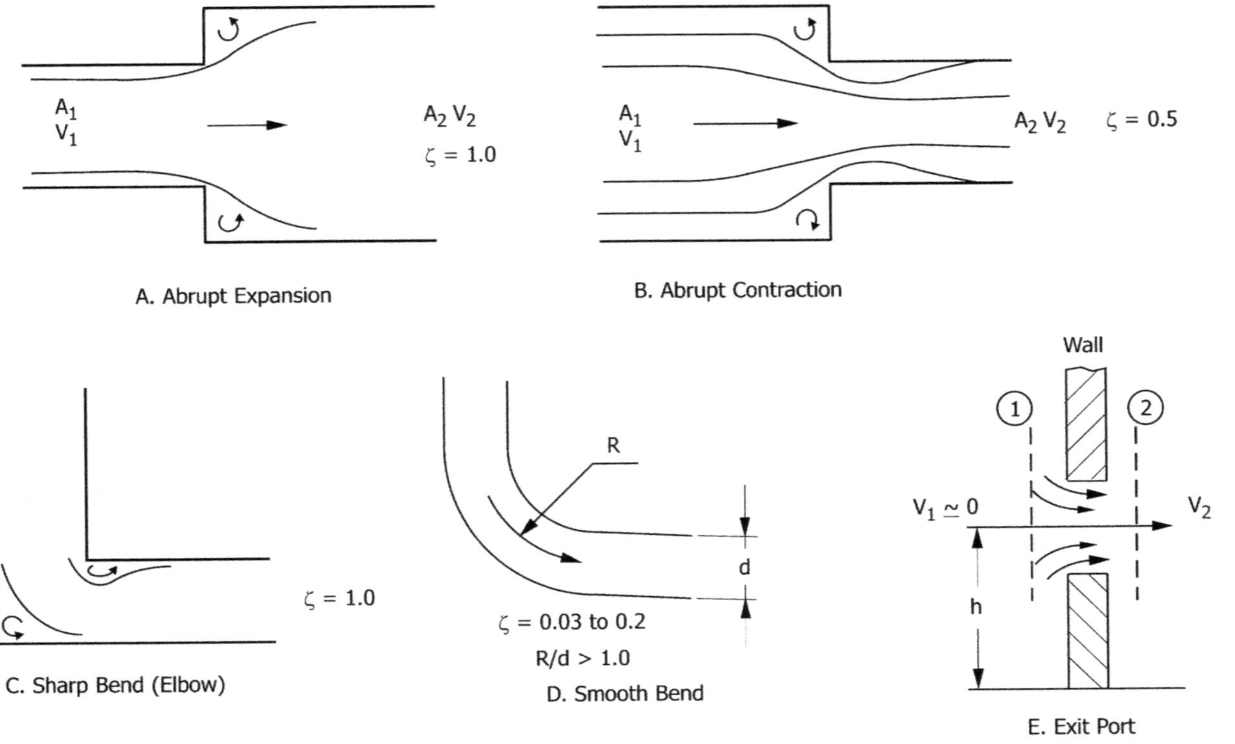

Figure 2.16 Loss factors (ζ) for some common local features.

2.13.1 Common Local Features

Sudden expansion (Figure 2.16(A))

$$h_\ell = \frac{V_1^2}{2g} \text{ if } A_2 >> A_1 \tag{2.20}$$

or $\zeta = 1$ and the dynamic head is completely lost as $V_2 = 0$
More generally

$$\left.\begin{aligned} h_\ell &= \left(1 - \frac{A_1}{A_2}\right)^2 \frac{V_1^2}{2g} \\ \text{i.e., } \zeta &= \left(1 - \frac{A_1}{A_2}\right)^2 \end{aligned}\right\} \tag{2.21}$$

Sudden contraction (Figure 2.16(B))

$$h_\ell = 0.5 \frac{V_2^2}{2^g} \text{ if } A_1 >> A_2 \tag{2.22}$$

or $\zeta = 0.5$

More generally

$$\left.\begin{aligned} h_\ell &= 0.5\left(1 - \frac{A_2}{A_1}\right)^2 \frac{V_2^2}{2g} \\ \text{i.e., } \zeta &= 0.5\left(1 - \frac{A_2}{A_1}\right) \end{aligned}\right\} \tag{2.23}$$

Sharp Bend (Figure 2.16(C))
The coefficient ζ depends on angle of bend α.
For $\alpha = 90°$ $\zeta = 1$ for $\alpha = 120°$ $\zeta = 1.15$

Sharp bends cause severe head loss and are not used for liquids. For gas flow through refractory channels, they are unavoidable.

Smooth bends (Figure 2.16(D))

Here the head loss depends on R/d ratio where R is the bend radius and d the inner pipe diameter.

For a 90° bend and $1 < R/d < 15$,
So that,

h_ℓ varies from 0.3 to 0.2

that is,

ζ varies from 0.2 to 0.3

The above formulae apply for round passages. For rectangular passages and other bend angles, reference should be made to handbooks listed in the *Bibliography.*

There are many other local features or components used in gas ducting and water pipelines. The coefficient ζ for these are experimentally determined. These can be found out from handbooks listed in the *Bibliography.* For special features such as recuperates (heat exchangers) and gas-cleaning devices, the coefficient can be obtained from the manufacturers.

Flow regulators (Figure 2.17)

Two types of regulating devices are used to control the fluid flow. For gases, a gate valve is used. The resistance offered by this depends on the opening available. When fully open, $\zeta = 0.1$; at 50% opening, $\zeta = 4.0$; and for 20% or less opening, $\zeta \geq 40$.

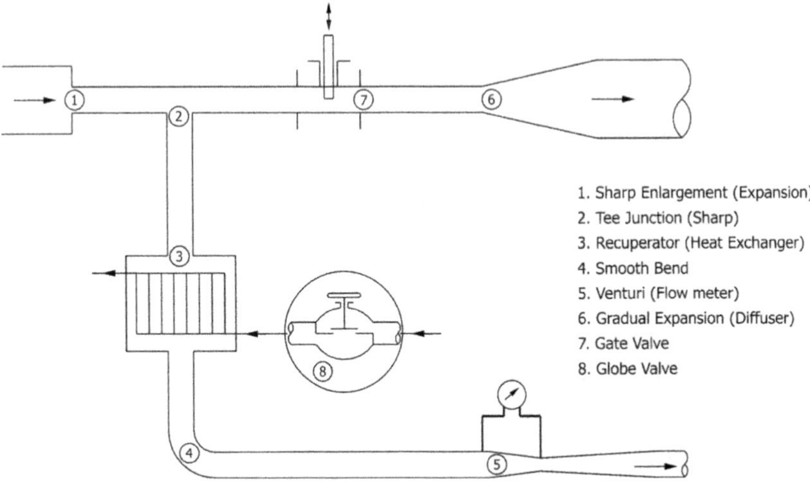

Figure 2.17 Some typical local features.

For liquids (water), globe valves are used. The resistance factor ζ varies between 0.2 to 30 for 100 to 20% opening. Globe valves of generally standard size and applicable value of coefficient can be easily obtained from the manufacturer. The coefficient also depends on the Reynolds number.

2.13.2 Gas Flow through Ports

The gases generated in furnaces exit to discharge systems through one or more openings or ports. The pressure inside the furnace is positive (higher than atmospheric) but low so that the exit velocity is nearly zero (Figure 2.16(E)). If we apply Bernoulli's equation to planes just before (1) and just in front of the port (2),

$$V_2 = \sqrt{\frac{2(P_2 - P_1)}{\rho}} \quad V_1 = 0 \tag{2.24}$$

or more correctly

$$V_2 = \phi\sqrt{\frac{2(P_2 - P_1)}{\rho}} \tag{2.25}$$

where ϕ is discharge coefficient. For thin wall $\phi = 0.98\text{–}1.0$. The volume flow rate through the port will be

$$= V_2 \times A = \phi A\sqrt{\frac{2(P_1 - P_2)}{\rho}} \quad \text{m}^3\text{/sec} \tag{2.26}$$

The mass flow rate is

$$= \rho V_2 A = \phi A\sqrt{2(P_1 - P_2)\rho} \quad \text{kg/sec} \tag{2.27}$$

where A is the area of opening.

These experiments will help us to decide the port area for a desired discharge.

If the port is located above bottom level the applicable pressure difference is

$$P_1 - P_2 = \rho g h$$

where h (m) is the height of the port from the bottom.

2.13.3 Pump Power

When a fluid circulates through a typical circuit the total loss of pressure from entry to exit Δp will consist of the sum entry loss, loss through all local features such as valves, gates, bends, loss due to friction, and loss at the exit.

$$\Delta p = \sum \text{Loss at all stages}$$

The quantity of fluid required to be circulated (G m^3/sec) is calculated from process variables such as heat transfer.

Usually we use a pump for the circulation.

The power required (N) for the pump (W) is given by

$$N = \frac{V\Delta p}{\eta} = \frac{G\Delta p}{\rho\eta} \tag{2.28}$$

where

V = Volume rate of circulation (m^3/sec)
G = Mass rate of flow (kg/sec)
Δ_p = Total pressure loss through circuit (N/m^2, Pa)
ρ = Density of fluid (kg/m^3)
η = Efficiency of pump and motor

For air and other gases we use a fan or blower and a similar equation applies.

EXAMPLE 2.3

A furnace burns 5.0 l of light oil per hour. The following data are available.

Oil density	850 kg/m^3 (25°C)
Combustion products	11.50 m^3/kg (25°C)
Density of combustion products	1.28 kg/m^3, Air 1.22 kg/m^3
Excess air	10%
Theoretical air	10.7 m^3/kg
Furnace temperature	1000°C

Assume a pressure difference of 1.5 Pa on the two sides of the port and the velocity coefficient = 0.8.

If the port is located on wall at 0.5 m from the bottom or located at bottom level, determine the size of the exit port in each case.

Solution

Weight of oil consumed/h

$$= \frac{850}{1000} \times 3 = 4.25 \text{ kg/h}$$

Combustion gases generated

$$= 4.25 \times 11.5 = 48.9 \text{ m}^3\text{/h (25°C)}$$

Excess air (at 10%)

$$= 0.1 \times 10.7 \times 4.25 = 4.55 \text{ m}^3\text{/h (25°C)}$$

Total gas volume

$$V_0 = 48.9 + 4.55 = 53.45 \text{ m}^3$$

At 1000°C the volume is

$$V_{1000} = V_0(1 + \beta t)$$

$$= 53.45\left(1 + \frac{1000}{273}\right) \simeq 250 \text{ m}^3\text{/h}$$

The volume flow rate$= \dfrac{250}{3600} = 0.07 \text{ m}^3\text{/sec}$

The density of combustion products and air is nearly same

$$\rho_0 = 1.28 \text{ m}^3\text{/kg}$$

$$\rho_{1000} = \rho_0 \frac{1}{1+\beta t} = 1.28 \times \frac{273}{1000+273} = 0.275 \text{ kg/m}^3$$

$$\rho_0 - \rho_{1000} = 1.28 - 0.275 = 1.005$$

$$V_2 = \phi\sqrt{\frac{2(P_2 - P_1)}{\rho}} P_1 - P_2 = 1.5 \text{ Pa}$$

$$= 0.8\sqrt{\frac{2 \times 1.5}{0.275}}$$

$$= 2.64 \text{ m/sec}$$

$$\text{Volume flow rate } = \text{Velocity} \times \text{Area}$$

$$0.07 = 2.64 \times A$$

$$A = 0.0265 \text{ m}^2$$

If the duct is circular with diameter d (m)

$$d = \sqrt{\frac{A \times 4}{\pi}} = \sqrt{\frac{0.0265 \times 4}{\pi}}$$

$$= 0.184 \text{ m } (18.4 \text{ cm})$$

If the duct is rectangular, the size is about 20 × 13 cm.

If the discharge port is located 0.5 m above the furnace bottom (at atmospheric pressure) the pressure difference $P_1 - P_2 = g \times h \times (\rho_0 - \rho_{1000})$, i.e., $9.8 \times 0.5 \times 1.005 = 4.53\ \rho_0$

Substituting this in the above expression we get

$$\text{duct area} = 0.0153 \text{ m}^2 \quad \text{or approximately } 15 \text{ cm} \times 10 \text{ cm}$$

2.14 STACK EFFECT

Combustion gases leave the furnace through the exit port. They then pass through the flues or ducts and other local features before being let out into the atmosphere through the stack or chimney.

Consider a simple system as shown in Figure 2.18(A). At the port ①, the gas pressure p_1 is nearly atmospheric (p_a) and temperature t_1 equal to that of the furnace (t_f).

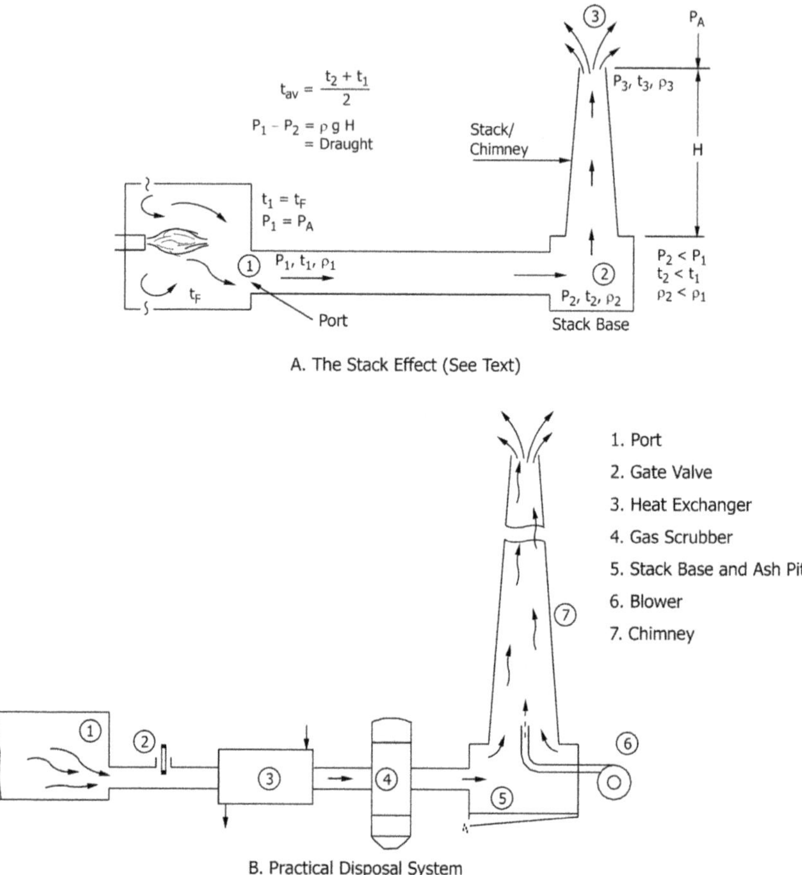

Figure 2.18 Combustion gas disposal.

During passage from port ①, through the duct to point ② at the base of the stack, the pressure and temperature will continuously decrease because of friction and heat transfer through the duct walls. Let the pressure and temperature at point ② be p_2 and t_2, respectively.

The difference of pressure between the points ① and ② is called the "excess pressure" and has a negative value (i.e., less than p_a).

$$\Delta p = p_2 - p_1$$

To enable the gas to be released into the atmosphere the pressure must be brought to or slightly more than atmosphere.

This is achieved by passing the gas through a tall (usually tapering) chimney of height H. The chimney contains a column of hot gas of height H. The pressure exerted by the column is $\rho_{av}gH$ where ρ_{av} is the average density of stack gases. This way of balancing excess pressure by using a stack is called "stack effect."

If ρ, ρ_2, and ρ_a are the densities of gas at the port, stack, and atmosphere.

$$\Delta p = p_1 - p_2 = (\rho_a - \rho_2)gH \tag{2.29}$$

From Equation (2.30), the minimum required stack height can be calculated. There are a number of factors not considered here, such as changing cross sections, and changing temperature and pressure through the stack. More elaborate treatment can be found in the books listed in the *Bibliography.*

The chimney height as calculated from the above equation can be increased by 15–20% to obtain more practical results.

The term Δp (excess pressure) will be left to be calculated by detailed calculations of friction and local losses through the duct work. Considering the relation between densities $(\rho_t = \rho_0 \frac{1}{1+\beta t})$ and $g = 9.81$ m/sec, the practical form of the above equation is

$$\Delta p = 354\, g\, H\left(\frac{1}{T_1} - \frac{1}{T_2}\right) \tag{2.30}$$

It is generally believed that a taller chimney will give better draft and reduce the environmental hazards of releasing gases at a lower level. This is not always true.

2.15 PRACTICAL FLUE SYSTEM

The flue disposal system shown in Figure 2.18(A) is highly simplified. A practical system (Figure 2.18(B)) may have many components such as bends. Regulators will produce local loss and the actual pressure difference or excess pressure will be much more, requiring a taller chimney.

There are practical limitations in erecting or building very tall chimneys. To obtain the necessary draught or pressure balance a mechanical blower, ejector, or booster is used (Figure 2.18(B)) at the base of the stack. These are required for large and complicated plants, such as power generating stations, and are not discussed here.

Electrical furnaces do not produce combustion gases, hence a gas disposal system is not required. A simple vent tube is sufficient to take care of minor gas envission.

Controlled atmosphere furnaces have to dispose of the depleted atmosphere gas as it is combustible and poisonous. It is drawn out through a vent and burned off by a pilot flame. For details refer to Chapter 13.

EXAMPLE 2.4

A circular steel pipe of 0.15 m diameter carries hot air at 0.07 m³/sec. Pipe length is 2.5 m. One end of the pipe is at 1000 °C and the other at 600 °C.

Take friction factor = 0.02

Determine the required height of the chimney at exit.

Solution

Mean air temperature

$$= \frac{1000 + 600}{2} = 800\,°\text{C}$$

Velocity of air

$$U = \frac{\text{Mean volume flow rate}}{\text{Area}}$$

$$= \frac{0.07}{\frac{\pi \times 0.15^2}{4}} = 4 \text{ m/sec}$$

Density of air (ρ) at 800 °C = 0.329 kg/m^3
Kinematic viscosity of air at 800 °C γ = 134.8 × 10^{-6} m^2/sec
Average Reynolds number

$$= \frac{U \times d}{\upsilon}$$

$$= \frac{4 \times 0.15}{134.8 \times 10^{-6}} = 4.45 \times 10^3$$

The air flow is transitional
Friction factor $f = 0.02$
Friction loss

$$P_f = f \times \frac{L}{d} \times \frac{\rho V^2}{2}$$

$$= 0.02 \times \frac{2.5}{0.15} \times 0.329 \frac{4^2}{2}$$

$$= 0.877 \text{ Pa}$$

Pressure loss at entry $p_e = \zeta \dfrac{\rho V^2}{2}$ and $\zeta = 0.5$

$$= \frac{0.5 \times 0.329 \times 4^2}{2}$$

$$= 1.32 \text{ Pa}$$

Pressure loss at exit $= \dfrac{\rho V^2}{2}$ and $\zeta = 1.0$

$$= \frac{0.329 \times 4^2}{2}$$

$$= 2.63 \text{ Pa}$$

Total loss of pressure

$$= \text{Entry loss} + \text{Friction loss} + \text{Exit loss}$$

$$\Delta p = 1.32 + 0.877 + 2.63$$

$$= 4.827 \text{ Pa} = \text{Excess pressure at exit (–ve)}$$

This excess pressure is to be balanced by the chimney

$$\rho_1 = 1.3 \text{ (atmospheric)}, \quad \rho_2 = 0.404 \text{ (at } 600°\text{C)}$$

$$\Delta p = (\rho_1 - \rho_2) g H$$

$$4.827 = (1.3 - 0.404) \times 9.81 \times H$$

$$H = 0.549 \text{ m}$$

at 20% extra

$$H = 0.66 \text{ m or } \sim 1 \text{ m}$$

Chapter 3

Steady State Heat Transfer

CONTENTS

3.1 INTRODUCTION

Industrial heating is based on the science of "heat transfer." Heat ($Q \cdot q$) is produced or is available from a heat source such as a combustion flame, a hot surface, and a hot filament or wire. From these sources it is transferred to a cold surface or medium such as an object or work, a cold wall, or the surrounding cold environment. Studying the rate ($dq/d\tau$) of heat transfer and obtaining the quantity of heat transferred ($\int q \, d\tau$) are the principal objectives of studying heat transfer. The flow of heat is always from a higher temperature to a lower one, i.e., under a temperature gradient. The thermal state of a body or surface is indicated by its temperature (T, t). Heat will always flow from a higher to lower temperature, i.e., a positive temperature gradient.

Ideally the available heat should be totally transferred to the desired object or site. However, like all practical processes, some heat is always taken up by surroundings, i.e., it is lost. This gives rise to the concept of heating efficiency. The "design" for heating thus involves attempts to assure maximum and proper heat transfer to the desired site or object and reduces the loss to a minimum, under given conditions.

There are three basic modes of heat transfer viz. conduction, convection, and radiation.

Conduction is without any movement or flow (no momentum transfer). Convection involves flow of one medium over another steady (solid) medium (simultaneous heat and moment transfer). Radiation is a contactless process where a hot object heats a cold surface by electromagnetic radiation. These three basic processes of heat transfer are reviewed in the subsequent sections.

There are two aspects of heat transfer with conduction, convection, and radiation. Steady state transfer is when the temperatures of all the involved bodies are constant with time., i.e.,

$$\frac{dT}{d\tau} = 0 \qquad (3.1)$$

Transient or time-dependent heating involves a change of temperature with time (τ). Whenever an object is being heated (e.g., an object in furnace) the transient conditions are important. Some of the transient conditions important for design are reviewed next, the aim being to establish the relationship between time and temperature at a particular place (T, τ, X).

Later we will see that the three processes (conduction, convection, and radiation) of heat transfer rarely operate individually. They take place simultaneously and in design we have to consider their contribution to the whole process. So, we accept some and neglect some.

Understanding the process of heat transfer and the evolution strategy of a design for a given problem is the essence of this book. What follows in this chapter is a review of the principles and applications of the science of "heat transfer" to the processes in the scope of this book.

It is presumed that the reader has undergone some undergraduate course on heat (in physics) or heat transfer (in engineering). Only bare and essential principles are reviewed as a refresher.

For details and derivations, reference to specialized text is recommended.

3.2 STEADY STATE CONDUCTION

Single Layer Wall

Consider a solid (e.g., a wall) of thickness δ (m) as shown in Figure 3.1 (A). The left-hand face of the wall is heated by a constant heat flow q (w). The hot face at 1 is at temperature t_1 (°C) and the other (cold) face is at a lesser temperature t_2. The heating has reached a steady state so that both t_1 and t_2 are constant with respect to time. The heat conducted through the wall is proportional to the temperature gradient $(t_1 - t_2)/\delta$. If A is the area of walls (m^2), the basic equation for

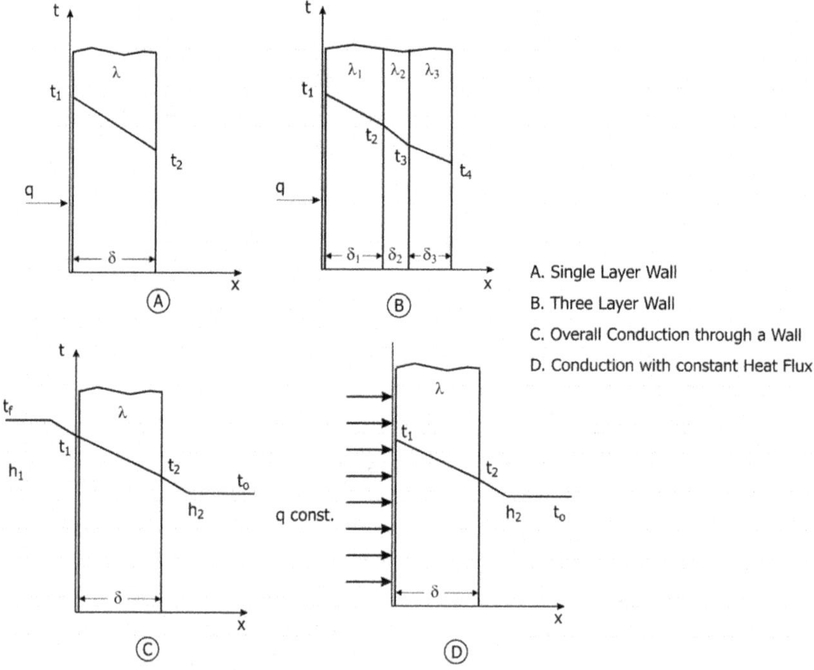

Figure 3.1 Steady state conduction.

conduction is

$$q = \lambda \frac{(t_1 - t_2)}{\delta} A \text{ w} \tag{3.2}$$

or

$$\frac{q}{A} = \lambda \frac{(t_1 - t_2)}{\delta} \text{ w/m}^2 \tag{3.3}$$

where λ is the constant of proportionality and is called the "thermal conductivity." It has dimensions w/m°C.

Thermal conductivity thus represents the amount of heat conducted through a unit thickness of material under a unit temperature difference and is a property of the material. It is mostly applicable to solids. Liquids and gases do have conductivity but the heat transfer through them is dominated by convection, as discussed in the subsequent sections.

The order of thermal conductivities of various materials is shown in Figure 3.2. For exact values applicable to a particular material, refer to data given in various tables in this book or special handbooks.

Thermal conductivity depends on temperature, as shown in Figure 3.3, for some materials. In general it decreases for pure metals and increases for alloys. In most of the calculations involved in design we neglect the dependence and assume a constant conductivity λ. For numerical calculations we may use an average value.

Note that in conduction there is no macro movement of atoms or molecules of the solid.

Equation (3.2) can be rearranged and compared with Ohm's law.

$$I = \frac{V}{R}, \qquad q = \frac{(t_1 - t_2)}{\delta/\lambda} \tag{3.4}$$

This shows that the heat flows by conduction under a thermal potential difference $t_1 - t_2$ and a "thermal resistance" δ/λ.

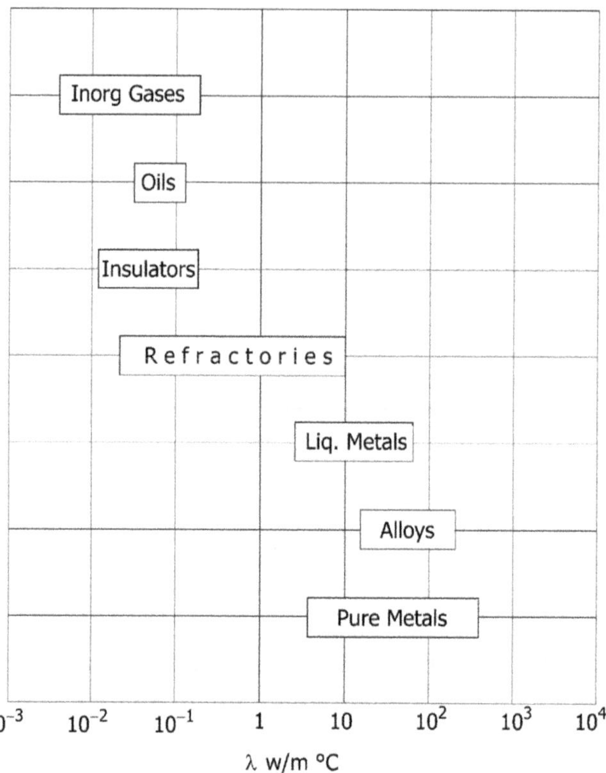

Figure 3.2 Comparative thermal conductivities (λ) of some materials.

The temperature dependence of thermal conductivity is often given as

$$\lambda_t = \lambda_1(1 + bt) \tag{3.5}$$

where b is a constant.

Equation (3.2) and Equation (3.3) can be extended to multilayer walls having different conductivity for each layer.

Figure 3.1(B) shows a three-layer flat wall with conductivities λ_1, λ_2, and λ_3. It is assumed that all the layers are in perfect contact with t_2 and t_3 as interlayer temperatures. At steady state all temperatures are constant and the heat passing

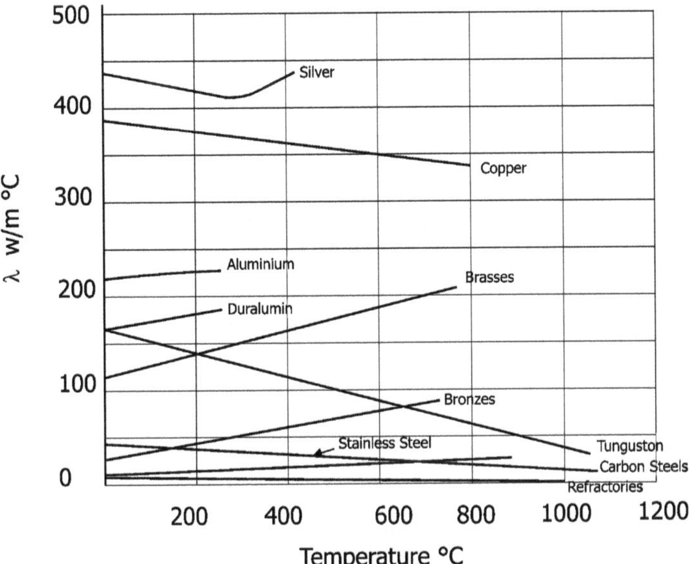

Figure 3.3 Variation of thermal conductivity with temperature for some metals and alloys.

through the wall and each layer is equal. Hence

$$q = \lambda_1 \frac{(t_1 - t_2)}{\delta_1} = \lambda_2 \frac{(t_2 - t_3)}{\delta_2} = \lambda_3 \frac{(t_3 - t_4)}{\delta_3} \text{ w} \tag{3.6}$$

$$q = \frac{(t_1 - t_4)}{\frac{\delta_1}{\lambda_1} + \frac{\delta_2}{\lambda_2} + \frac{\delta_3}{\lambda_3}} \text{ w} \tag{3.7}$$

The term in the denominator can be written as $\Sigma_1^3 \frac{\delta}{\lambda}$ and be considered as the equivalent thermal resistance of the wall.

If the heat flow q is known, Equation (3.6) can be used to calculate interlayer temperatures, e.g.,

$$\left.\begin{aligned} t_2 &= t_1 - \frac{q\delta_1}{\lambda_1} \\ t_3 &= t_2 - \frac{q\delta_2}{\lambda_2} \text{ etc.} \end{aligned}\right\} \tag{3.8}$$

These equations will be useful in deciding the thickness of various layers so that the temperatures across them are within safe limits.

Referring again to Figure 3.1(C), the temperature (t_1) of the hot wall and that (t_2) of the cold face will be determined respectively by the furnace temperature (t_f) or the heating mode and by the environment or outside temperature (t_o).

The wall may be heated by a high-temperature source such as a hot filament or coil by radiation. In many instances, e.g., when hot gases circulate along the wall, heating is by convection. The heat transfer is then governed by Newton's law, i.e.,

$$q = h(t_f - t_1) \text{ w/m}^2 \tag{3.9}$$

where h is the "heat transfer coefficient" and has the dimensions w/m^2°C. Similar heat transfer will take place on the cold side. Thus if h_1 and h_2 are the heat transfer coefficients applicable to the inner and outer surfaces of the wall Figure 3.1(C).

$$q = h_1(t_f - t_1) = h_2(t_2 - t_0) \tag{3.10}$$

combining this with the heat transfer through the wall (e.g., Equation (3.10),

$$q = \frac{(t_f - t_o)}{\frac{1}{h_1} + \frac{\delta}{\lambda} + \frac{1}{h_2}} \tag{3.11}$$

or

$$q = k(t_f - t_o) \tag{3.12}$$

where

$$k = \frac{1}{\frac{1}{h_1} + \frac{\delta}{\lambda} + \frac{1}{h_2}} \tag{3.13}$$

Comparing Equation (3.12) with Equation (3.13), k can be identified as the "overall heat transfer coefficient."

Similar expression can be developed for a multilayer wall

$$q = k\,(t_f - t_o)$$

where now

$$k = \frac{1}{\frac{1}{h_1} + \sum_1^n \frac{\lambda}{\delta} + \frac{1}{h_2}} \tag{3.14}$$

It is important to understand that the heat transfer coefficient (h or k) is not a material constant like the thermal conductivity λ. It will depend on a very large number of properties of the medium such as its density ρ and viscosity μ. We will discuss the nature of the heat transfer coefficient in subsequent sections on convection.

The heat transfer to the wall may take place by a constant heat flow, i.e., q will be constant (w/m^2) on the hot side. The heat transfer on the cold side is still governed by Newton's law and a heat transfer coefficient h_2. As q is now known (see Figure 3.1(D))

$$\left.\begin{aligned} q &= \frac{\lambda(t_1 - t_2)}{\delta} \\ &= h_2(t_2 - t_0) \end{aligned}\right\} \tag{3.15}$$

From which

$$\left.\begin{aligned} t_2 &= \frac{q}{h_2} + t_0 \frac{1}{h_2} + \frac{\delta}{\lambda} \\ t_1 &= t_0 + q \end{aligned}\right\} \tag{3.16}$$

Equation (3.15) and Equation (3.16) can be used to calculate the wall temperatures t_1 and t_2.

A similar expression can be developed for a multilayered wall to calculate the surface and interlayer temperatures.

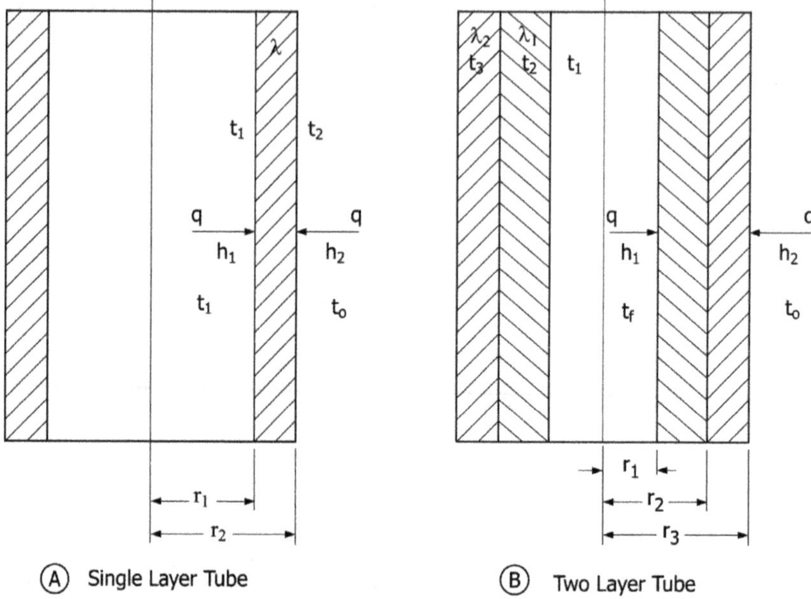

Figure 3.4 Steady state conduction through tubes.

Alternatively, it can be used to determine the thickness δ of layers if the temperatures are known.

We will come across circular surfaces such as tubes or enclosures in equipment design. The conduction equations previously developed can be applied to these shapes.

Consider a single layer circular tube shown in Figure 3.4(A). The heat flowing radially out is given by

$$Q = \frac{2\pi\lambda\ell}{\ln\frac{r_2}{r_1}}(t_1 - t_2) \text{ w} \tag{3.17}$$

or

$$q = \frac{Q}{\ell} = \frac{2\pi\lambda}{\ln\frac{r_2}{r_1}}(t_1 - t_2) \text{ w/m} \tag{3.18}$$

For a two-layered tube having conductivities λ_1 and λ_2 (Figure 3.4(A))

$$q = \frac{(t_1 - t_3)}{\frac{1}{2\pi\lambda}\ln\frac{r_2}{r_1} + \frac{1}{2\pi\lambda}\ln\frac{r_3}{r_2}} \text{ w/m} \tag{3.19}$$

If h_1 and h_2 are the heat transfer coefficients and t_f and t_o the temperatures inside and outside the tube, then for a single layer tube

$$q = \frac{(t_f - t_o)}{\frac{1}{h_1} + \frac{1}{2\pi\lambda}\ln\frac{r_2}{r_1} + \frac{1}{h_2}} \tag{3.20}$$

where the denominator is the overall heat transfer coefficient.

EXAMPLE 3.1

The hot face and cold face temperatures of a furnace wall are 1250 and 80°C, respectively. The heat transfer coefficient on the cold face is 18 w/cm^2°C. Atmospheric temperature is 30°C.

The following materials are being considered for wall construction:

1. Fire clay brick ($\lambda = 0.85$ w/m°C)
2. Light weight fire clay ($\lambda = 0.50$ w/m°C)
3. Ceramic fiber board ($\lambda = 0.18$ w/m°C)
4. Ceramic fiber blanket ($\lambda = 0.08$ w/m°C)

Now calculate:

The heat loss per unit area from wall to atmosphere.

The wall thickness required for each material under consideration.

If the furnace temperature is 1350°C, what is the transfer coefficient on the hot face?

Solution

Assuming steady state conditions, the heat loss from unit area of wall to atmosphere is

$$\begin{aligned} q &= \text{heat transfer coefficient } (h) \times \text{temperature difference} \\ &= 18(80 - 30) \\ &= 900 \text{ w/m}^2 \end{aligned}$$

The conduction equation is

$$q = \frac{\lambda}{\delta}(t_1 - t_2)$$

$$\delta = \frac{\lambda(t_1 - t_2)}{q}$$

$$= \frac{\lambda(1250 - 80)}{900} = 1.30\lambda$$

Substituting the values of λ for each material

$$\begin{aligned} &\delta \text{ (fire clay)} &&= 1.30 \times 0.85 = 1.100 \text{ m} \\ &\delta \text{ (light wt.)} &&= 1.30 \times 0.50 = 0.650 \text{ m} \\ &\delta \text{ (fiber board)} &&= 1.30 \times 0.18 = 0.234 \text{ m} \\ &\delta \text{ (fiber blanket)} &&= 1.30 \times 0.08 = 0.104 \text{ m} \end{aligned}$$

Let the hot face transfer coefficient be h_1

$$q = h_1(t_f - t_1)$$

$$h_1 = \frac{q}{(t_f - t_1)} = \frac{900}{1350 - 1250} = 9 \text{ w/m}^2$$

As a check $$q = \frac{(t_f - t_0)}{\frac{1}{h_1} + \frac{\delta}{\lambda} + \frac{1}{h_2}}$$

$$= \frac{1350 - 30}{\frac{1}{9} + \frac{1.1}{0.85} + \frac{1}{18}} = 907 \text{ w/m}^2$$

EXAMPLE 3.2

The wall in the previous example is to be constructed with two layers to reduce the heat loss.

The hot face is made of ceramic fiber (λ = 0.18 w/m°C). The outer layer is to be made from rock wool (λ = 0.33 w/m°C).

The fiber board layer is 0.12 m thick. Determine the thickness of rock wool layer. The maximum temperature for rock wool is 700°C. Check if this condition is satisfied.

What will be the effect of a 0.005 m steel sheet used to cover the wall from outside.

Solution

All temperature and conditions remaining the same as the previous example

$$q = 900 \ \ \text{w/m}^2$$

For a two-layer wall

$$q = \frac{\lambda_1(t_1 - t_2)}{\delta_1} = \frac{\lambda_2(t_2 - t_3)}{\delta_2} \ \ \text{w/m}^2$$

δ_1 and δ_2 are thicknesses of the fiber board and rock wool, respectively, t_2 is the interlayer temperature.

The overall transfer equation is

$$q = \frac{t_f - t_o}{\frac{1}{h_1} + \frac{\delta_1}{\lambda_1} + \frac{\delta_2}{\lambda_2} + \frac{1}{h_2}}$$

$$900 = \frac{1350 - 30}{\frac{1}{9} + \frac{0.12}{0.18} + \frac{\delta_2}{0.033} + \frac{1}{18}}$$

$$\delta_2 = 0.02\,\text{m}\,(20\text{ mm})$$

The interlayer temperature t_2 will be

$$t_2 = t_1 - \frac{q \times \delta_1}{\lambda}$$

$$= 1250 - \frac{900 \times 0.12}{0.18} = 650°\text{C}$$

Hence, rock wool is within safe temperature limits. In fact we can reduce the fiber board layer to about 0.1 m and still be within limits.

Conductivity of steel is 50 w/m°C.

The overall thermal resistance of the wall without a steel jacket is

$$k = \frac{1}{\frac{1}{h_1} + \frac{\delta_1}{\lambda_1} + \frac{\delta_2}{\lambda_2} + \frac{1}{h_2}} = \frac{1}{\frac{1}{9} + \frac{0.12}{0.18} + \frac{0.02}{0.033} + \frac{1}{18}}$$

$$= 0.695 \text{ w/m}^2\text{°C}$$

Hence, addition of a 5-mm steel jacket has no effect on the overall heat transfer.

EXAMPLE 3.3

An indirectly heated cylindrical furnace with an inner diameter of 230 mm is 1000 mm long. An electric heater is embedded in the lining at 250 mm diameter. The refractory used for insulation has a conductivity of 0.2 w/m°C.

If the heating element operates at 1200°C determine:

1. The power delivered inside the furnace, assuming inner surface at 900°C.
2. The outer diameter of the furnace, if the heat loss is to be limited to about 30% of the useful power.

Neglect end losses. Heat transfer coefficient from the outer cover to atmosphere is 18 w/m^2°C.

Solution

The furnace is shown in Figure 3.5. For simplicity the lining can be divided into two parts inner ① and outer ② with the heating element ③ acting as the partition.

The heat transferred from the element to the inner lining q_1 is

$$q_1 = \frac{t_1 - t_2}{\frac{2.3}{2\pi\lambda} \log \frac{d_2}{d_1}} \text{ w/m}$$

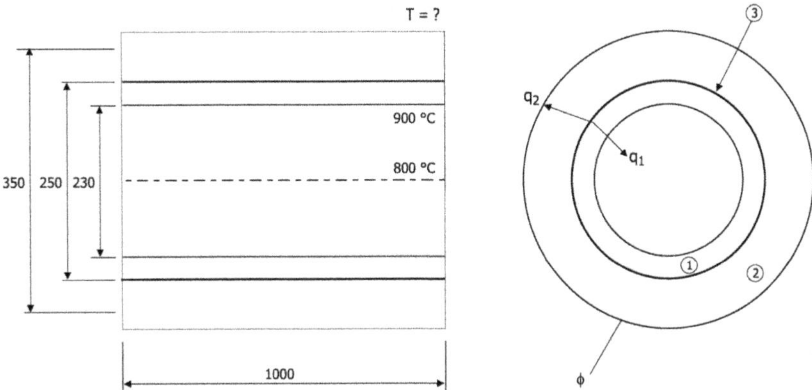

Figure 3.5 Cylindrical furnace (Example 3.3).

Substituting

$$q_1 = \frac{1200 - 900}{\frac{2.3}{2\pi \times 0.2} \log \frac{0.25}{0.23}} = 4.3 \text{ kW}$$

30% of 4.3 kW, i.e., 1.43 kW is to be dissipated from outer surface q_2 = 1.43 kW

The heat dissipated at 18 w/m^2°C (h_2).

Assume an outer diameter 450 mm. The outer surface area is

$A_2 = \pi \times 0.45 = 1.41$ m^2/m length

Heat dissipated through the outer surface is

$A_2 \times h_2 \times (t_3 - t_o) = 1.43$ kW

where t_3 = outer surface temperature.

$1.41 \times 18 \times (t_3 - 30) = 1.43 \times 10^3$

$t_3 = 86$°C

This temperature and attendant heat loss can be reduced by slightly increasing the outer diameter (~460 mm) and the permissible loss (~2.8%). Some trial calculations will be required.

The actual heat loss will be higher if we consider the end losses.

Total power required is 4.3 + 1.43 + End losses
= 5.73 kW + End losses
≃ 7.8 kW

The heat loss can also be reduced by using a double layer outer lining (e.g., ceramic fiber or rock wool).

3.3 THE SHAPE FACTOR

The usual basic two-dimensional differential equation under steady state conduction and the assumption of constant conductivity is

$$\frac{\partial^2 t}{\partial x^2} + \frac{\partial^2 t}{\partial y^2} = 0 \tag{3.21}$$

The solution of this equation gives the two-dimensional (x, y) temperature in the given body.

There are many solutions for the above equation under given boundary conditions. One of the solutions applicable to finite bodies having a finite inner and outer temperatures is

$$Q = S\lambda(t_f - t_o) \times \tau \tag{3.22}$$

where λ = Thermal conductivity w/m°C
$t_f - t_o$ = Temperature difference between the inner and outer sides(°C)
τ = Time sec.
Q = Total heat conducted in τ
S = Shape factor

As the name suggests, the shape factor depends on the geometry (shape) and the dimensions of the conducting body. It is determined by analytical and computational techniques.

Some shape factors for situations related to furnaces are shown in Figure 3.6. A more extensive collection can be found in references {3, 4}. The solved problems will make the use of the shape factor clear.

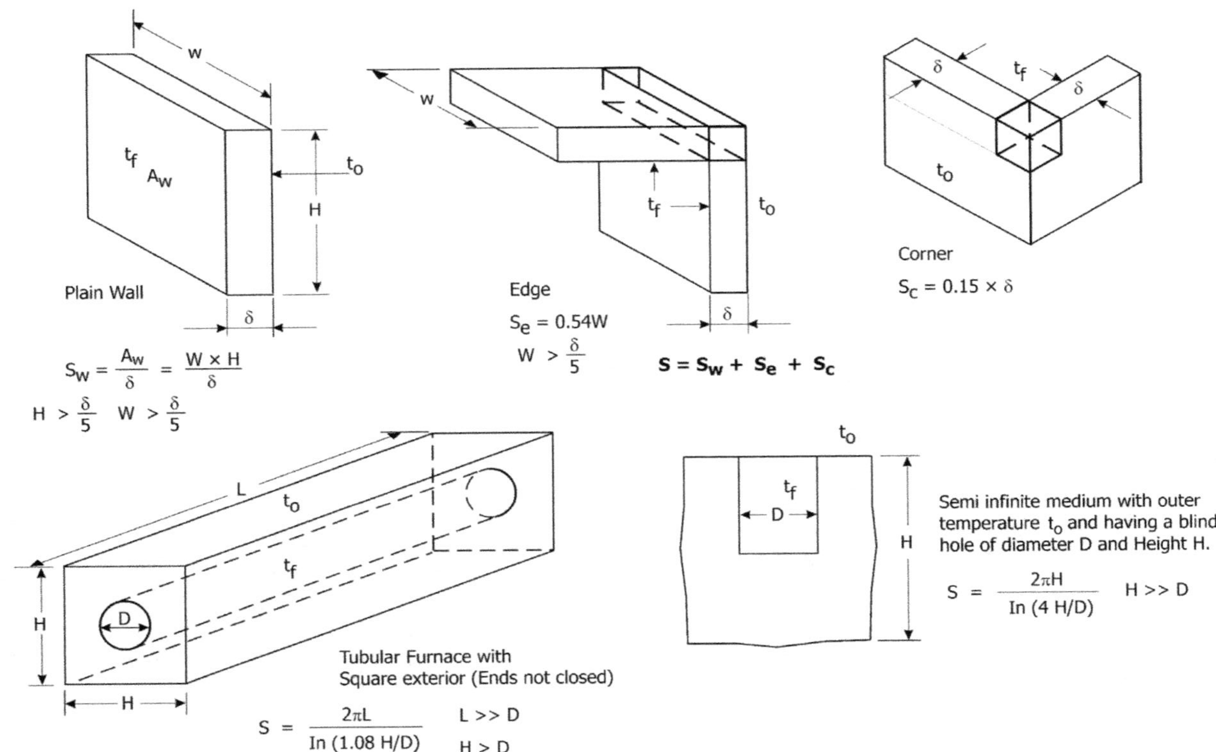

Plain Wall
$S_W = \frac{A_W}{\delta} = \frac{W \times H}{\delta}$
$H > \frac{\delta}{5}$ $W > \frac{\delta}{5}$
Edge
$S_e = 0.54W$
$W > \frac{\delta}{5}$
Corner
$S_C = 0.15 \times \delta$
$S = S_W + S_e + S_C$
Tubular Furnace with Square exterior (Ends not closed)
$S = \frac{2\pi L}{\ln (1.08\ H/D)}$ $L >> D$ $H > D$
Semi infinite medium with outer temperature t_o and having a blind hole of diameter D and Height H.
$S = \frac{2\pi H}{\ln (4\ H/D)}$ $H >> D$

Figure 3.6 Some selected shape factors.

With a little ingenuity the factor can be used for multiple layer walls.

EXAMPLE 3.4

The dimensions of a continuous duty furnace are 1500 × 800 × 700 mm. The inside temperature is 800°C and the outer wall temperature is 65°C. The walls of the furnace are made of fire-clay and have a thickness of 450 mm. The thermal conductivity of the wall λ is 0.8 w/m°C.

Determine the heat lost by conduction in a period of 24 h.

Solution

The furnace has a rectangular shape and hence has 3 × 2 = 6 walls, 12 edges, and 8 corners.

For walls, the shape factor is

$$S_w = \frac{A}{\delta} \quad \text{where } A = \text{the area}, \quad \delta = \text{the thickness.}$$

For 1500 × 800 mm, and 450 mm thickness (there are 2 such walls)

$$A = 1.5 \times 0.8 \times 2 = 2.4$$

$$S_{w1} = \frac{2.4}{0.45} = 4.8$$

Similarly

$$S_{w2} = \frac{2 \times 0.8 \times 0.7}{0.45} = 2.5$$

and

$$S_{w3} = \frac{2 \times 0.8 \times 0.7}{0.45} = 4.6$$

Hence

$$S_w = 4.8 + 2.5 + 4.6$$
$$\simeq 20$$

For edges, the shape factor S_e is

$= 0.54L$ where L is the edge length

$= 0.54\ (4(1.5 + 0.8 + 0.7))$

$= 6.48$

For corners, the shape factor S_c is

$= 0.15 \times \delta \times 8$

$= 8 \times 0.15 \times 0.45 = 0.54$

Hence, the total shape factor is

$$S = 20 + 6.48 + 0.54$$
$$= 27.02$$

The total heat loss in 24 h is

$= S\ \lambda(t_f - t_o) \times \tau$

$= 27.02 \times 0.8 \times (800 - 65) \times 24 \times 3600$

$= 1373 \times 10^6$ J

Note: If we consider the shape factor

$$S = 27.02 = 100\% \text{ then}$$
$$S_w = 74\%,\ S_e = 24\% \text{ and } S_c = 2\%$$

Hence, contribution of the corners to the heat loss is quite negligible.

3.4 GRAPHICAL METHOD FOR WALL HEAT TRANSFER AND DESIGN

In section 3.3, we have seen that the rate of heat flow q, (w/m^2) for a plain wall is given by

$$q = \frac{(t_f - t_o)}{\frac{1}{h_1} + \frac{\delta}{\lambda} + \frac{1}{h_2}}$$
$$= k(t_f - t_o)$$

where k is the overall heat transfer coefficient.

Therefore,

$$k = \frac{1}{\frac{1}{h_1} + \frac{\delta}{\lambda} + \frac{1}{h_2}}$$

or $\frac{1}{k} = \lambda$ overall thermal resistance.

For a multiple-layered wall, the middle term δ/λ in the above equation will be replaced by $\frac{\delta_1}{\lambda_1} + \frac{\delta_2}{\lambda_2} + \cdots$ for the individual layers having thickness δ_n and conductivity λ_n.

The above equation also shows that there is a linear relationship between $(t_f - t_o)$ and $1/k$, namely

$$t_f - t_o = q \times k \text{ or for any layer}$$

$$t_{wt} - t_{w(i-1)} = q\frac{1}{ki} = q\frac{\delta i}{\lambda i}$$

These linear relationships can be used for graphical construction to determine the layer temperatures and the rate of heat flow q.

The following data are assumed to be available:

- Furnace source (flame) temperature t_f
- Outside temperature t_o
- Heat transfer coefficients h_1 and h_2
- Thickness δ_n and the conductivities of different layers λ_n

First, the thermal resistances $\frac{1}{ki}$ for the wall are calculated. For this purpose the flame to wall (inside) and from wall to atmosphere (outside) heat transfer coefficients are also regarded as ficticious resistances.

Let the calculated resistances be

$$\frac{1}{h_1}, \frac{\delta_1}{\lambda_1}, \frac{\delta_2}{\lambda_2}, \ldots, \frac{1}{h_2}$$

These resistances are plotted on the x axis (Figure 3.7). On the y axis, temperature is plotted on any convenient scale.

The temperature point B inside is connected by a straight line to the temperature point C outside. Point A marks the initial temperature of the wall.

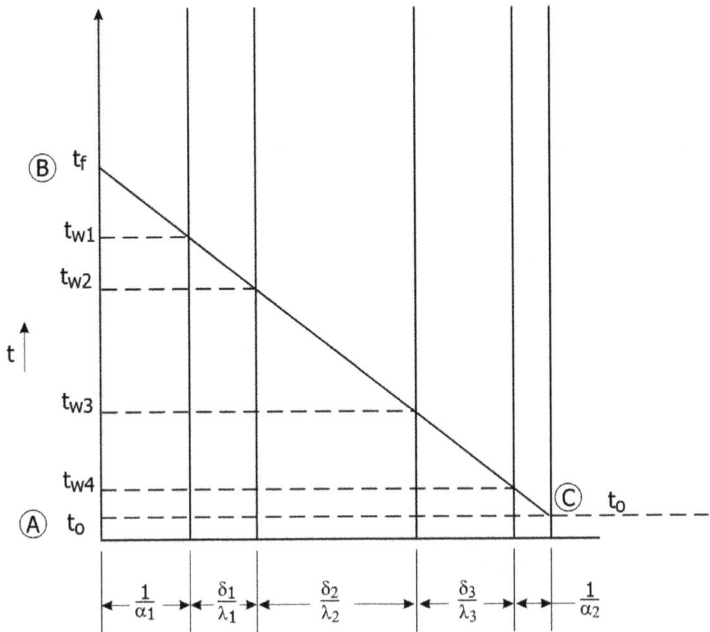

Figure 3.7 Steady state heat flow through a three-layered wall (graphical method).

The various wall temperatures $t_{w1}\ldots t_{wn}$ are read on the temperature axis at the layer boundary and the temperature line *BC*.

The rate of heat flow q will be

$$q = k(t_f - t_o)$$

$$= (t_f - t_o)\frac{1}{\frac{1}{h_1}+\sum\frac{\delta n}{\lambda n}-\frac{1}{h_2}}$$

$$= AB \times AC$$

EXAMPLE 3.5

A three-layered wall is made as follows:

- I layer (hot side) fire clay, 240 mm, $\lambda = 1.2$ w/m°C
- II layer (cera wool), 150 mm, $\lambda = 0.12$ w/m°C
- III layer (Rock wool), 200 mm, $\lambda = 0.03$ w/m°C

The furnace temperature is 1300°C and the outside temperature is 30°C. The heat transfer coefficient on the hot side (h_1) is 0.2 kw/m^{2}°C and that on the outside (h_2) is 12 w/m^{2}°C.

Determine the wall temperatures graphically.

Solution

First we will calculate the thermal resistances.

The total resistance (k)

$$k = \frac{1}{\frac{1}{h_1} + \frac{\delta_1}{\lambda_1} + \frac{\delta_2}{\lambda_2} + \frac{\delta_3}{\lambda_3} + \frac{1}{h_2}}$$

$$\frac{1}{h_1} = \frac{1}{200} = 0.005$$

$$\frac{\delta_1}{\lambda_1} = \frac{0.24}{1.2} = 0.200$$

$$\frac{\delta_2}{\lambda_2} = \frac{0.15}{0.12} = 1.250$$

$$\frac{\delta_3}{\lambda_3} = \frac{0.02}{0.03} = 0.660$$

$$\frac{1}{h_2} = \frac{1}{12} = 0.080$$

Therefore,

$$k = \frac{1}{0.005 + 0.200 + 1.250 + 0.660 + 0.080} = \frac{1}{2.195} = 0.456 \frac{\text{w}}{\text{m}^{2}\,°\text{C}}$$

The various resistances are plotted against temperature in Figure 3.8. Note that the resistance offered by the heat transfer coefficient h_1 is very small and cannot be plotted on a convenient scale.

From the graph:

1. Wall inner temperature ($t_f = t_{w1}$) $\simeq$ 1300°C
2. Fire clay and cera wool boundary temperature $t_{w2} = 119$°C

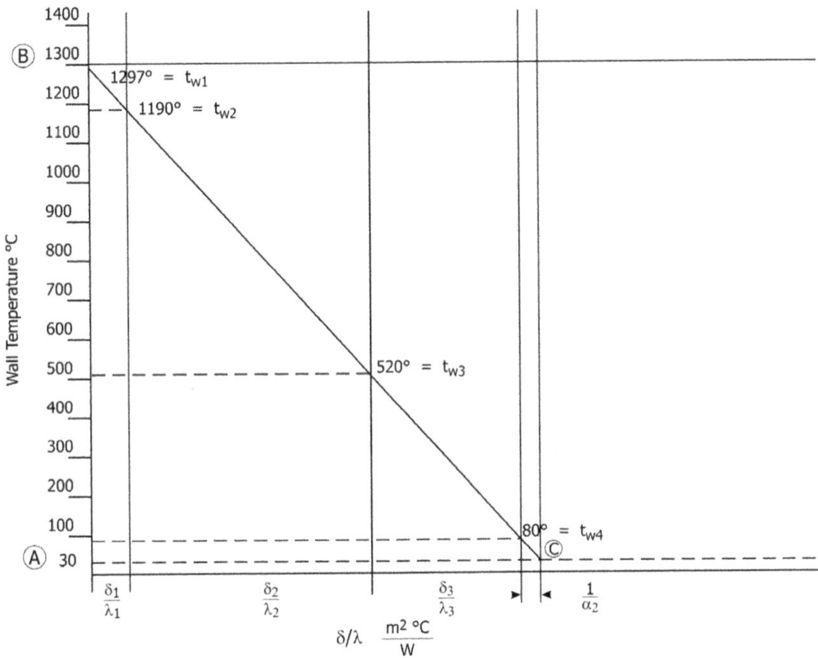

Figure 3.8 Graphical solution (conduction through multilayered wall).

3. Cerawool and rock wool boundary temperature $t_{w3} = 520°C$
4. Outer wall temperature $t_{w3} = 90°C$

Solution of Example 3.5 by analytical method

The overall conductivity of the wall (k) is

$$k = \frac{1}{\frac{1}{h_1} + \frac{\delta_1}{\lambda_1} + \frac{\delta_2}{\lambda_2} + \frac{\delta_3}{\lambda_3} + \frac{1}{h_2}}$$

$$= 0.456$$

Hence, heat flow rate q is

$$= 0.456\ (1300 - 30)$$

$$= 580\ \text{w/m}^2$$

For h_1, i.e., the first resistance

$$t_{w1} = t_f - q\left(\frac{1}{h_1}\right)$$

$$= 1300 - 580\,(0.005)$$

$$= 1297°\text{C} \simeq 1300°\text{C}$$

For h_1 and δ_1 resistances

$$t_{w2} = t_f - q\left(\frac{1}{h_1} + \frac{\delta_1}{\lambda_1}\right)$$

$$= 1300 - 580(0.005 + 0.2)$$

$$= 1300 - 120$$

$$= 1180°\text{C}$$

For h_1, δ_1, and δ_2 resistances

$$t_{w3} = t_f - q\left(\frac{1}{h_1} + \frac{\delta_1}{\lambda_1} + \frac{\delta_2}{\lambda_2}\right)$$

$$= 1300 - 580(0.005 + 0.2 + 1.25)$$

$$= 456°\text{C}$$

For h_1, δ_1, δ_2, and δ_3 resistances

$$t_{w4} = t_f - q\left(\frac{1}{h_1} + \frac{\delta_1}{\lambda_1} + \frac{\delta_2}{\lambda_2} + \frac{\delta_3}{\lambda_3}\right)$$

$$= 1300 - 580(0.005 + 0.2 + 1.25 + 0.66)$$

$$= 74°\text{C}$$

3.5 CONVECTION

In convection heat transfer occurs across the boundary between a solid and a fluid (gas or liquid) in contact with it. The fluid is in motion. Heat transfer will be from the fluid to

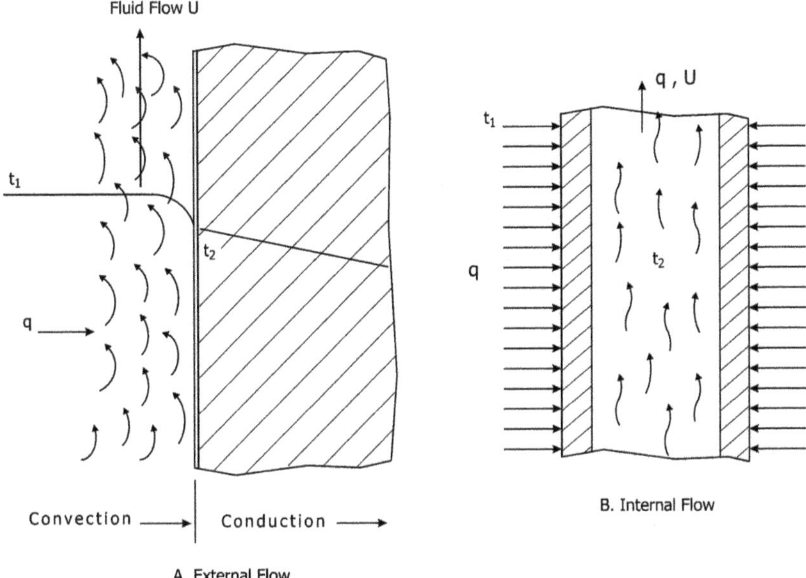

Figure 3.9 Convective heat transfer.

the solid such as hot combustion gases circulating along the object in the furnace or the furnace walls. It may also take place from a hot object or wall to the gases or liquid circulating around it. If the temperature of the hot medium is t_1 and that of the cold medium t_2 then the heat transferred across the contact surface of area A (m^2) is (see Figure 3.9(A))

$$Q = h_c(t_1 - t_2)A \text{ w} \tag{3.23}$$

where h_c is the convective heat transfer coefficient (w/m$^{2\circ}$C). The heat transferred across unit area will be

$$q = \frac{Q}{A} = h_c(t_1 - t_2)\,(\text{w/m}^2) \tag{3.24}$$

It can be seen that the above equations are similar to that for conduction. However, the only similarity between conduction and convection is the fact that both depend directly on the temperature difference $(t_1 - t_2)$.

The heat transfer coefficient is not a single property of either solid or fluid medium but depends on the thermal and other physical properties of both mediums.

Some of the properties related to the heat transfer coefficient are density, velocity, viscosity, temperature, conductivity, specific heat, etc. The involvement of a large number of properties makes it impossible to develop a single, universal formulation for calculating the heat transfer coefficient. It depends on the situation and conditions of the surfaces and fluids involved in transfer. Thus, h_c is a variable and not a physical constant like thermal conductivity. For the same pair of fluid and solid, there will be different values of convective coefficients, each applicable to a particular situation only. A partial list of factors that influence the coefficient is as follows:

1. Nature of fluid flow—This may be laminar or turbulent, or mixed. It is determined from the Reynolds number (Re) which is given by

   $$\mathrm{Re} = \frac{U\ell}{\upsilon}$$

   For $\mathrm{Re} < 2320$ the flow is laminar and for $\mathrm{Re} > 10^4$ it is turbulent. For $2320 < \mathrm{Re} < 10^4$ the flow is mixed (see Chapter 2).
2. Physical properties of the fluid such as density, conductivity, viscosity, specific heat, and pressure.
3. Shape of contact surfaces such as flat plate, tube, and coiled tube, and their dimensions.
4. Roughness of the solid surface which determines the friction.
5. Velocity of the fluid, which may be determined from conditions of flow such as natural and forced. Fluids expand on heating and their density lowers. Hence, a density difference naturally occurs between a cold and a hot fluid taking part in convection. This creates a motion that is natural. Convection under this natural

force is called "natural convection," which is influenced by the coefficient of expansion and gravity.

In many instances the fluid is "forced" to move along the surface. This is done by using a pump, or pan, etc. Convection under such mechanically induced flow is called "forced convection." Besides the gravity and expansion forces, natural convection also occurs under a concentration difference that is accompanied by mass transfer.

6. Temperature of fluid which affects all physical properties.
7. The flow of fluid may be "external," such as air flowing over a furnace outer wall, or "internal," such as fluid flowing through a tube (Figure 3.9(B)).

   The heat transfer coefficients for "external" and "internal" flow situations are different.

Determination of Convective Heat Transfer Coefficient

Considering the difficulties in calculating the heat transfer coefficient from basic principles, it is determined by experimental methods based on the theory of similarities.

Experiments are conducted on small-scale models in which heat transfer takes place in a manner similar to the large-scale phenomena of interest. The theory of similarity requires that the model and the real process must have the same determining criteria and the same mathematical relations between the criteria.

Critical relationships are experimentally established for a particular situation of interest (e.g., cooling fluid flowing through a hot pipe). These relationships are used to estimate the heat transfer coefficients applicable to a real large-scale situation.

Theory of models and similarity criteria make it easy and economical to estimate the transfer coefficients. It also saves considerable time. For details of these methods see [1, 2].

The various parameters and properties that are involved in a given heat transfer situation are grouped in dimensionless numbers. There are many such numbers, some of which are of interest and are given below. These numbers are called "similarity criteria."

$$\text{Nusselt number Nu} = \frac{h\ell}{\lambda} \tag{3.25}$$

$$\text{Grashof number Gr} = \frac{g\beta(t_s - t_{f\ell})\ell^3}{\upsilon} \tag{3.26}$$

$$\text{Peclet number Pe} = \frac{U\ell}{\alpha} \tag{3.27}$$

$$\text{Reynolds number Re} = \frac{U\ell}{\upsilon} \tag{3.28}$$

$$\text{Prandtl number Pr} = \frac{\upsilon}{\alpha} \tag{3.29}$$

$$\text{Stanton number St} = \frac{h}{\rho c U} = \frac{\text{Nu}}{\text{Re} \cdot \text{Pr}} \tag{3.30}$$

where

h = Heat transfer coefficient (w/m°C)
U = Fluid velocity (m/sec)
λ = Thermal conductivity of fluid (w/m°C)
g = Gravitational attraction (m/sec)
β = Coefficient of volumetric expansion (1/°C)
ℓ = Any suitable dimension of the system, m (e.g., diameter of pipe)
υ = Kinematic viscosity (N m²/sec)
α = Thermal diffusivity (m²/sec)
ρ = Density (kg/m³)
c = Specific heat (J/kg°C)
t_s = Temperature of solid (°C)
$t_{f\ell}$ = Temperature of fluid (°C)

The various dimensionless groups given above are related to various aspects of the fluid flow at the solid-fluid boundary. Their significance is as follows:

The Reynolds number relates the forces of inertia and viscosity in the fluid. It is thus a criteria to distinguish between laminar and turbulent flow.

The Peclet number shows the ratio between heat transfer by convection and conduction.

The Grashof number considers the ratio between gravitation force and viscous force, hence applicable to natural convection.

The Prandtl number is a ratio of two physical properties—viscosity and thermal diffusivity. It can be obtained from handbooks. It relates the viscous and thermal fields in the fluid.

For determining the convective heat transfer coefficient, the Nusselt number (and Stanton number which contains Nu) is the most important group as it contains the unknown coefficient h.

Relationships between the dimensionless numbers are constructed to fit the data obtained from model experiments based on the similarity principle. They generally have a form such as

$$\text{Nu} = A(\text{Re})^n\,(\text{Pr})^m \text{ — Forced Convection}$$

or

$$\text{Nu} = K(\text{Gr})^x\,(\text{Pr})^y, \text{ etc. — Natural Convection}$$

where A and K are constants and m, n, x, y, and so on are the powers.

From such relationships, the Nusselt number is derived and the heat transfer coefficient is calculated. It is again pointed out that the given relationships apply only to a particular situation. We will consider several examples to make this point clear.

3.6 FORCED CONVECTION

There are two main situations where we will come across forced convection.

In many furnaces we use fans to establish a gas flow over the work and surrounding walls, roof, etc. The velocities are generally low (1–10 m/sec). This is modeled by flow over flat plates.

In some situations we used forced cooling of furnace parts by circulating cool water under pressure through cooling tubes or coils. These are modeled as forced circulation through (generally) circular tubes.

We will examine these two situations with an aim to estimate the heat transfer coefficients. We will begin with a discussion of the boundary layer phenomenon, which is a decisive factor in forced circulation.

3.6.1 Boundary Layer and Convection

In Chapter 2, we have discussed the boundary layer phenomenon that appears when a fluid flows over a flat surface.

If the bulk fluid velocity before the contact with the surface is U_f, it develops a profile (see Figure 3.10(A)) near the surface. At the surface, a thin layer of fluid comes to rest (velocity zero). The velocity then gradually increases to U_f and the thickness of the layer from the surface to the point where $U = U_f$ is called the boundary layer δ_U. The boundary layer increases to a maximum stable thickness at a distance X_U over the surface. The above discussion assumes that the fluid flow is laminar. Inside the boundary layer the flow is viscous.

A similar phenomenon is observed when there is a temperature difference between the solid surface and the fluid flowing over it. Figure 3.10(A) shows a fluid with a uniform bulk temperature t_f flowing over a surface with a lesser temperature t_s. The flow is laminar. It is observed that the temperature between the bulk fluid and the surface develops a profile similar (but not equal) to the velocity profile, and a thermal boundary layer δ_T appears over the surface. This layer also increases and stabilizes at a certain distance X_T. The heat transfer from the bulk fluid to the surface takes place mainly by conduction through the distance δ_T in which the fluid has very little velocity.

This shows that convection, i.e., heat transfer by fluid circulation, is always accompanied by conduction. Only at the entrance, i.e., at $X = 0$, the transfer is by pure convection, and

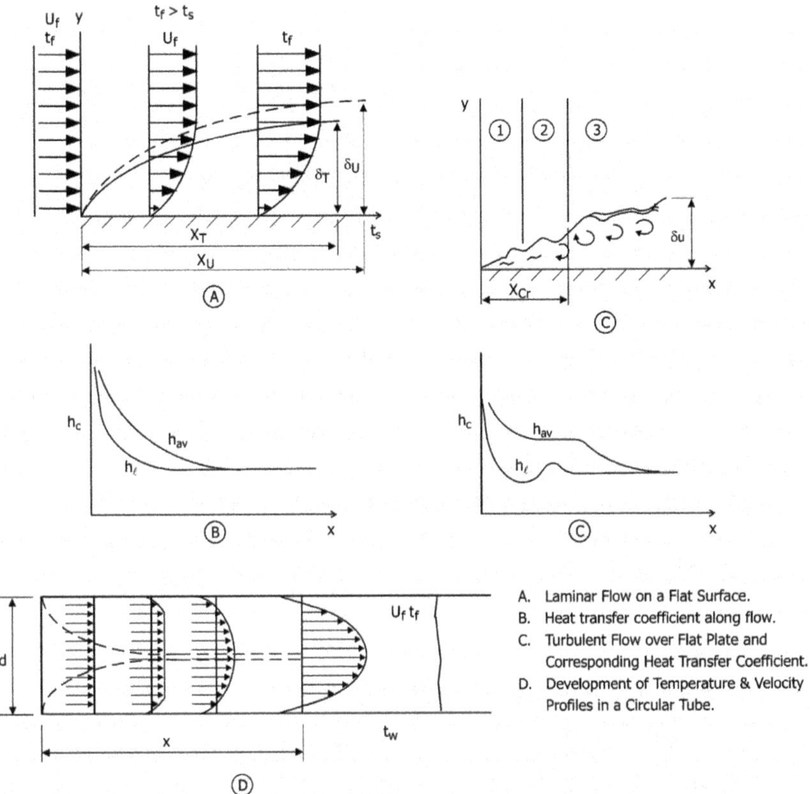

Figure 3.10 Boundary layer and temperature profiles in convection.

the convection coefficient is very high. Further downstream ($X > 0$), conduction becomes a limiting factor. This reduces the transfer coefficient h_c to a limit, as shown in Figure 3.10(B).

The convection heat transfer coefficient is, therefore, not constant but depends on the location of the spot along the stream. We have to consider the "local" transfer coefficient and an "overall" or "average" transfer coefficient. The "average" coefficient is usually determined by taking a logarithmic average in the region of interest.

The above discussion points out the importance of the Nusselt number (Nu), which is a ratio of the convection coefficient and the thermal conductivity.

When the fluid flow is turbulent (Figure 3.10(C)), there are circulation and eddies in the boundary layer and the effective thickness of the layer δ is reduced. The heat transfer coefficient is high but the concept of local and average transfer coefficients still apply.

Similar situations arise in the flow through a tube. Mechanical (flow-based) and thermal (heat-based) boundary layers are created around the circumference of the tube (see Figure 3.10(D)). This gives rise to velocity and temperature profiles through the tube section. At some point (X) along the tube the boundary layers meet and the velocity and temperature stabilize. The profiles are usually parabolic. Since at the tube entrance the transfer coefficient is high, it reduces and stabilizes the velocity and temperature further.

To conclude, the convection coefficient depends on the plane of interest. It is also higher for a turbulent flow. We will assume that the thermal and velocity boundary layers coincide so that X_T and X_U coincide.

3.6.2 Forced Convection over Flat Plate

Heat transfer from or to a horizontal plate in a longitudinal flow of air is a very interesting phenomenon which will be useful in design modeling.

Consider a flat horizontal plate of length L and width W over which air is flowing at a bulk velocity U_f and temperature t_f in the x direction (Figure 3.10(A)). Let the surface temperature be t_s with $t_f > t_s$ so that heat is transferred from the air to the plate.

As discussed previously, velocity and thermal boundary layers will form on the surface. The thickness of the layers δ_u and the δ_t will gradually increase and stabilize in the flow direction. Here we are concerned with the velocity boundary layer δ_u.

The Reynolds number for the flow will be given by

$$\mathrm{Re} = \frac{U_f \times W}{\upsilon} \tag{3.31}$$

where υ is the kinematic viscosity (m/sec) at t_f.

The flow will be laminar, transitional, or turbulent depending on Re, so that

$$\mathrm{Re} < 2000 \text{ — laminar}$$

$$2000 < \mathrm{Re} < 10 \text{ — transitional}$$

$$\mathrm{Re} < 10 \text{ — turbulent}$$

The thickness of the boundary layer for $\mathrm{Re} < 10^5$ at a position x is given by

$$\delta_{x\ell} = \frac{5.83}{\sqrt{\mathrm{Re}_x}} \tag{3.32}$$

$$\text{where} \quad \mathrm{Re}_x = \frac{U_f x}{\upsilon} \tag{3.33}$$

is the local Reynolds number.

The Nusselt number at x is given by

$$\mathrm{Nu}_x = 0.335\,\mathrm{Re}_x^{1/2}\,\mathrm{Pr}^{1/3} \tag{3.34}$$

where Pr is the Prandtl number at t_f. The heat transfer coefficient at x will be

$$h_x = \mathrm{Nu}_x \times \frac{\lambda}{X} \tag{3.35}$$

λ being the thermal conductivity of air (W/m °C) at t_f.

The mean heat transfer coefficient along the length can be obtained from

$$\mathrm{Nu} = 0.67\,\mathrm{Re}^{1/2}\,\mathrm{Pr}^{1/3} \tag{3.36}$$

and

$$h_m = \mathrm{Nu}_m \times \frac{\lambda}{L} \tag{3.37}$$

If the flow is turbulent, $\mathrm{Re} > 10^5$, the thickness of the turbulent boundary layer at a distance x is given by

$$\delta_{xT} = \frac{0.37x}{\sqrt[5]{\mathrm{Re}_x}} \tag{3.38}$$

where Re_x is given by Eq. (3.33).

The heat transfer coefficient at x for turbulent flow can be obtained from the following correlation

$$Nu_x = 0.0255\, Re_x^{0.8}\, Pr^{0.3} \tag{3.39}$$

The mean heat transfer coefficient for turbulent flow is given by

$$Nu_m = 0.032\, Re^{0.8}\, Pr^{0.3} \tag{3.40}$$

If the critical Reynolds number is assumed to be ~10^5, the critical length x_{cr} and the maximum thickness of laminar boundary layer δ_{cr} can be obtained from Equation (3.41) and Equation (3.42).

$$x_{cr} = 10^5 \times \frac{\upsilon}{U_f} \tag{3.41}$$

$$\delta_{cr} = 1.8 \times 10^3 \times \frac{\upsilon}{U_f} \tag{3.42}$$

For air and for a small temperature difference between the bulk air and temperatures the factor $(Pr)^{0.3}$ can be neglected.

Depending on the air velocity U_f, the flow pattern will be generally laminar at the front edge and change gradually to transitional and turbulent. Hence, there will be a wide variation in the heat transfer coefficient near the front edge. The flow can be made entirely turbulent by using a turbulizer grid. For a mixed pattern we will have to calculate the coefficient for both laminar ($x \leq x_{cr}$) and turbulent ($x \geq x_{cr}$) conditions.

Solved examples that follow will make the calculating procedure clear.

EXAMPLE 3.6

A plate having a length 1.0 m and a width 0.5 m is subjected to air flow parallel to its surface. The surface temperature of the plate is 30°C.

Calculate the mean heat transfer coefficient from the air to the plate if the bulk temperature of air is 400°C and its velocity is 1.0 m/sec.

Also determine the heat transferred to the plate.

Solution

At 400°C the properties of air are

$$\upsilon = 63.09 \times 10^{-6}\ \text{m}^2/\text{sec},\ \text{Pr} = 0.678,\ \lambda = 5.21 \times 10^{-2}\ \text{w/m°C}$$

The critical dimension $\ell_o = 1.0$ m, $U = 1.0$ m/sec

$$\text{Re} = \frac{U \times \ell_o}{\upsilon} = \frac{1 \times 1}{63.09 \times 10^{-6}} = 1.59 \times 10^4$$

Critical Re is 5×10^5. As the above calculated Re $< 5 \times 10^5$, the flow is laminar.

The correlation to be used for average Nu is

$$\begin{aligned}\text{Nu} &= 0.67 \times (\text{Re})^{0.5}(\text{Pr})^{0.3}\\ &= 0.67 \times (1.59 \times 10^4)^{0.5} \times (0.678)^{0.3}\\ &= 75.2\end{aligned}$$

$$\text{Nu} = \frac{\ell_o h_c}{\lambda} \text{ hence } h_c = \text{Nu} \times \frac{\lambda}{\ell_o}$$

where h_c is the (mean) convective heat transfer coefficient.

$$\begin{aligned}h_c &= 75.2 \times (5.21 \times 10^{-2}/1)\\ &= 3.92\ \text{w/m}^2\text{°C}\end{aligned}$$

Assuming that only one (upper) side of the plate is exposed to the flow, heat transferred to the plate is

$$\begin{aligned}Q &= h_c \times (t_f - t_p) \times \text{Area}\\ &= 3.92 \times (400 - 30) \times (1 \times 0.5)\\ &= 725\ \text{w}\end{aligned}$$

EXAMPLE 3.7

For the plate in the previous example, determine the local boundary layer thickness (δ) and the local heat transfer coefficient along the length. All other conditions remain unchanged.

Solution

We will determine the local parameters at point x so that $x/\ell_o = 0.1, 0.2, \ldots, 1.0$ from the front edge of the plate.

The boundary layer thickness at a distance x from the front edge is given by

$$\delta = \frac{4.64x}{\sqrt{\mathrm{Re}_x}}$$

$$\mathrm{Re}_x = \text{Local Reynolds number} = \frac{U \times x}{\upsilon}$$

For $x = 0.1\ \ell_o = 0.1 \times 1.0 = 0.10$

$$\mathrm{Re}_{x1} = \frac{1 \times 0.10}{63.09 \times 10^{-6}} = 1.59 \times 10^3$$

Similarly

$$\mathrm{Re}_{x2} \times 10^3 = 3.17,\ \mathrm{Re}_{x3} = 4.77,$$
$$\mathrm{Re}_{x5} = 7.95,\ \mathrm{Re}_{x8} = 12.72,\ \mathrm{Re}\ell_o = 15.9$$

The local Nusselt number (Nu_x) is given by

$$\mathrm{Nu}_x = 0.332(\mathrm{Re}_x)^{0.5} \times (\mathrm{Pr})^{0.3}$$

$$= 0.332 \times (1.59 \times 10^3)^{0.5} \times (0.678)^{0.3} = 11.8$$

$$h_{cx1} = 11.8 \times \frac{5.21 \times 10^{-2}}{0.1}$$

$$= 6.14\ \mathrm{w/m^2\,{}^\circ C}$$

Similarly

$$\left.\begin{matrix} h_{cx2} = 4.33 & h_{cx8} = 2.17 & \\ h_{cx3} = 3.54 & h_{cx\ell} = 1.94 & h_{cx5} = 2.75 \end{matrix}\right\} \mathrm{w/m^2\,{}^\circ C}$$

Boundary layer thickness δ_x

$$\delta_{x1} = \frac{4.64 \times 0.1}{\sqrt{1.59 \times 10^3}} = 0.011\ \mathrm{m}$$

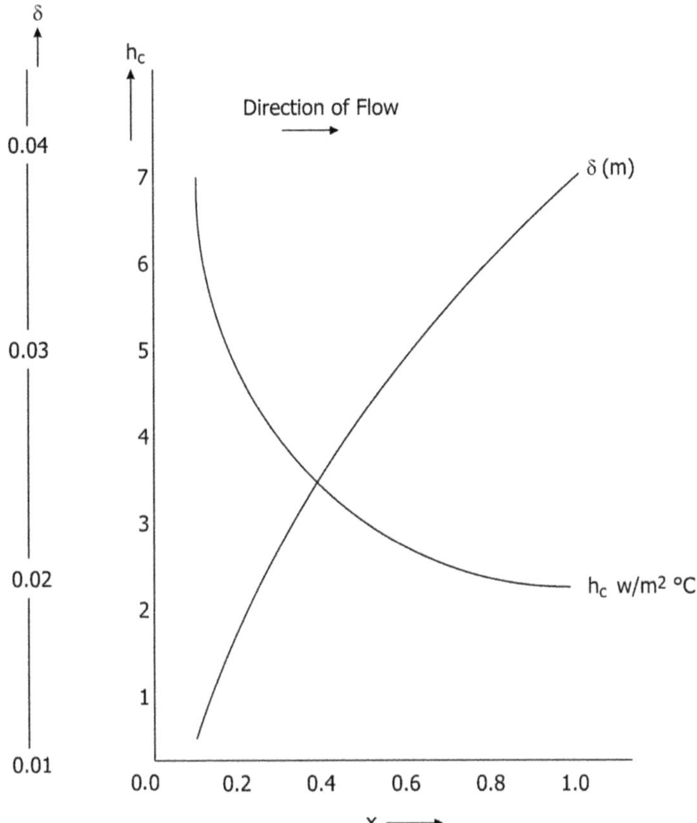

Figure 3.11 Variation of local transfer coefficient and boundary layer thickness along a plate in forced convection (Example 3.7).

Similarly

$$\left.\begin{array}{lll} \delta_{x2} = 0.016 & \delta_{x8} = 0.033 & \\ \delta_{x3} = 0.020 & \delta_{x\ell} = 0.037 & \delta_{x5} = 0.026 \end{array}\right\} \mathrm{m}$$

The above results are shown graphically in Figure 3.11.

EXAMPLE 3.8

For the flat plate in Example 3.6, determine the air velocity at which the flow over the plate is turbulent. All other conditions remain the same.

Also, determine the mean heat transfer coefficient under turbulent conditions and the heat transferred to the plate.

Solution

The flow will become turbulent when the Reynolds number over the length exceeds 5×10^5

$$\text{Re}_L = \frac{U \times L}{\upsilon} \quad \text{or} \quad U = \frac{\upsilon \times \text{Re}_L}{L}$$

Kinetic viscosity υ at 400°C is 63.09×10^{-6} cm²/sec

$$U = \frac{5 \times 10^5 \times 63.09 \times 10^{-6}}{1.0}$$

$$= 32 \text{ m/sec}$$

Hence, the velocity of air must be greater than 32 m/sec to achieve turbulence. Prandtl number at 400°C = 0.678 and the conductivity of air $\lambda = 5.21 \times 10^{-2}$ w/m°C.

The correlation for the average Nusselt number for turbulent flow is

$$\text{Nu}_L = 0.036 \ \text{Pr}^{0.3} \ (\text{Re}_L{}^{0.8} - 23{,}200)$$

for $\text{Re}_L > 5 \times 10^5$, $\text{Pr} > 0.5$

This correlation includes the transition between laminar and turbulent flow at the beginning.

Substituting the above values and taking $U = 40$ m/sec (> 32 m/sec critical value)

$$\text{Re}_L = \frac{40 \times 1.0}{63.09 \times 10^{-6}} = 6.35 \times 10^5$$

$$\text{Nu} = 0.036(0.678)^{0.3}[(6.35 \times 10^5)^{0.8} - 23{,}200]$$

$$= 662$$

$$h_c = \text{Nu}\frac{\lambda}{\ell_o} = 662 \times \frac{5.21 \times 10^{-2}}{1.0}$$

$$= 34.5 \text{ w/m}^2\text{°C}$$

If the flow is turbulent from the beginning the applicable correlation is

$$= 0.036(\text{Pr})^{0.3} \times (\text{Re})^{0.8}$$

$$= 0.036 \times (0.678)^{0.3} \times (6.35 \times 10^5)^{0.8}$$

$$= 14.0$$

$$h_c = 1410 \times \frac{5.21 \times 10^{-2}}{1.0}$$

$$= 73.2 \text{ w/m}^2\,{}^\circ\text{C}$$

To satisfy the condition of no transition, the flow must be modified by using a turbulator grid or barrier at the front edge of the plate.

For $h_c = 34.5$, the heat transferred is

$$Q = 34.5\ (400 - 30) \times (1 \times 0.5)$$
$$= 6.38 \text{ kW}$$

For $h_c = 73.2$, the heat transferred is

$$Q = 73.2(400 - 30) \times (1 \times 0.5)$$
$$= 13.5 \text{ kW}$$

EXAMPLE 3.9

For the same plate in previous examples determine the boundary layer thickness and the local heat transfer coefficient along the length. The bulk air speed is 40 m/sec.

Solution

We have seen that at $U = 40$ m/sec the flow is turbulent with or without a transition zone.

For convenience, the plate length (1.0 m) is divided into 10 equal sections, each of length 0.1 m.

The local Reynolds number along the plate will be

$$\text{Re}_{x1} = \frac{40 \times 1.0}{63.09 \times 10^{-6}} = 6.34 \times 10^4$$

Similarly

$$Re_{x2} = 1.27 \times 10^5, \quad Re_{x3} = 1.902 \times 10^5,$$

$$Re_{x5} = 3.70 \times 10^5$$

$$Re_{x7} = 4.438 \times 10^5, \quad Re_{x8} = 5.72 \times 10^5,$$

$$Re_L = 6.34 \times 10^5$$

The local thickness of turbulent boundary layer (δ_x) is given by

$$\delta_x = \frac{0.37x}{\sqrt[5]{Re_x}}$$

Substituting the above calculated values of Re_x

$\delta_{x1} = 0.034$ m $\quad \delta_{x7} = 0.0192$ m

$\delta_{x2} = 0.007$ m $\quad \delta_{x8} = 0.021$ m

$\delta_{x3} = 0.00976$ m $\quad \delta_{xL} = 0.0256$ m

$\delta_{x5} = 0.0135$ m

The local Nusselt number is given by

$$Nu = 0.0288(Re_x)^{0.8} \times (Pr)^{0.3}$$

$$Nu_{x1} = 0.0288 \times (6.34 \times 10^4)^{0.8} \times (0.678)^{0.3}$$

$$= 177$$

$$h_{cx1} = 177 \times \frac{5.21 \times 10^{-2}}{0.1}$$

$$= 92 \text{ w/m}^2\text{°C}$$

Similarly

$Nu_{x2} = 80.8 \quad Nu_{x7} = 62$

$Nu_{x3} = 74.5 \quad Nu_{x8} = 61.2$

$Nu_{x5} = 67.2 \quad Nu_{xL} = 58.5$

The results of these calculations are shown graphically in Figure 3.12.

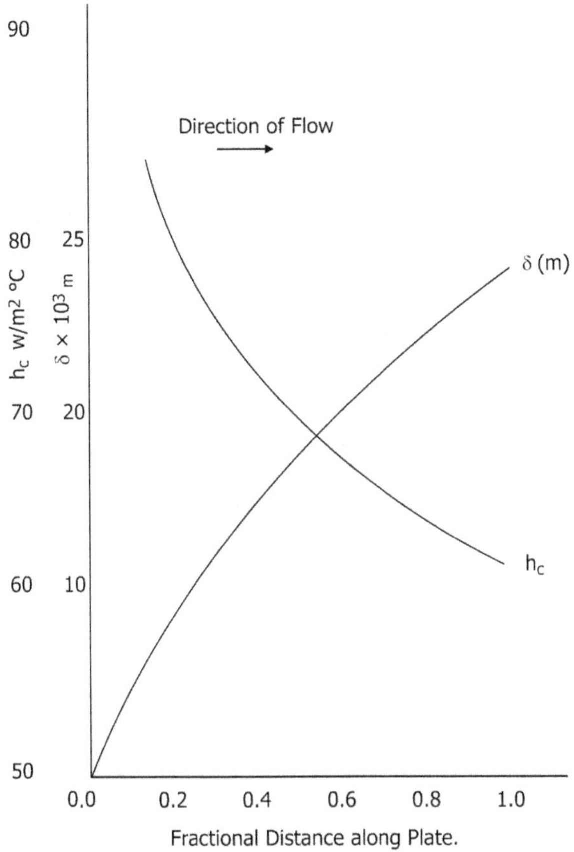

Figure 3.12 Variation of boundary layer thickness (*d*) and heat transfer coefficient for turbulant flow over a flat plate (Example 3.9).

3.6.3 Forced Convection inside Tubes

As discussed earlier, forced convection takes place when the fluid flow is caused by an external agency such as a pump. Forced flow of fluids in tubes is used in many devices as a method of heating or cooling. In this book we are mainly concerned about the flow of water used to cool furnace walls, seats, etc. Circular tubes are more commonly used, though square tubes are used—mainly for induction coils.

The use of water as the cooling (or occasionally heating) medium puts a restriction on temperatures encountered in the process. As water boils at 100°C the upper temperature is restricted to about 80°C. The inlet temperature will be the atmospheric temperature, between 10 to 30°C depending on the season and location. The outlet temperature of water will have to be above 30° and below 80°C, i.e., about 60 to 70°C.

Depending on the bulk speed at the inlet (U m/sec) and the dimension and temperature, the flow inside the tube may be laminar, transient, or turbulent. This will be decided by the Reynolds number ($\mathrm{Re}_f = Ud/\upsilon$).

$\mathrm{Re} < 2000$	Flow is laminar
$\mathrm{Re} > 10^4$	Flow is fully turbulent,
$2000 < \mathrm{Re} < 10^4$	Flow is transitional

3.6.3A Laminar Flow

At $\mathrm{Re} < 2000$ the fluid has a uniform velocity U_∞ and temperature T_∞ entering the tube. As it flows forward the fluid experiences a friction drag (due to wall roughness) and a viscous drag due to shear (adjacent to wall). This leads to the formation of a boundary layer δ_h. The boundary layer thickens progressively and the velocity distribution changes gradually from uniform to parabolic. At some distance from the entry the boundary layers from the opposite sides meet and the flow stabilizes, acquiring a uniform velocity profile which continues further, without change. Thus, there are two sections in the flow direction. The first section extends from the inlet to a length x_h at which the flow pattern stabilizes. In the second section, which extends from x_h onward (to $x = L$), the flow profile remains constant.

Similar transitions take place in the temperature profile of the flow. The temperature, which is uniform (T_∞) at the entry, gradually changes to a parabolic distribution due to the formation of the thermal boundary layer δ_t. At a certain distance x_t it stabilizes and continues further without change.

Thus there are two thermal sections. First, from entry to x_t where the uniform profile changes leading to stabilization at x_t. Then, from x_t to $x = L$, the thermal profile remains essentially constant.

The distances x_h and x_t can be calculated from the relations

$$x_h = 0.05 \text{ Re}_f \times D \tag{3.43}$$

$$x_t = 0.05(\text{Re}_f \text{ Pr}_f) \times D \tag{3.44}$$

The inlet regions 0 to x_h or 0 to x_t can extend for a distance 50–100 diameters from the inlet. The coefficient of heat transfer h_c is the highest at the entrance and decreases further. We have, therefore, to consider two transfer coefficients — the "local" coefficient at a given distance x and a mean coefficient for a given range (e.g., $x = 0$ to $x = x_t$, etc.)

The correlation used for circular tubes with a well-established laminar flow is

$$\text{Nu} = 4.36 \tag{3.45}$$

For a well-developed laminar flow and constant wall temperature

$$\text{Nu} = 3.66 \tag{3.46}$$

There are many other correlations. They are not considered here because we will rarely come across laminar flow in tubes in our designs.

3.6.3B Turbulent Flow

The laminar profile of the fluid flow in a pipe changes when the Reynolds number exceeds the critical value $\text{Re}_{cr} = 2000$. It is transient for Re between 2000 and 10^4. In this range the flow is partly laminar and partly turbulent. The later component increases as Re approaches 10^4. At higher values of Re the flow is fully turbulent.

The development of turbulencet δ_t and the hydrodynamic δ_H boundary layers takes place in a manner similar to that in laminar flow. However, the stabilization to a fully developed hydrodynamic and thermal profile takes place over a shorter distance (~ 10δ) from the entrance.

There are many correlations available for determining the Nusselt number and the heat transfer coefficient h_c for turbulent flow in pipes. Only two correlations which are useful for our purpose are given below. They are based on the Reynolds number Re_f $\frac{Ud}{\gamma}$ and Prandtl number Pr ($= \gamma/\alpha = \gamma c\rho/\lambda$).

1. Dietus-Boelter Correlation

$$Nu = 0.023\ Re^{0.8}\ Pr^{n} \tag{3.47}$$

where

$n = 0.4$ for heating and $n = 0.3$ for cooling.

This correlation is applicable in the range

$6000 < Re < 10^7$ and $0.5 < Pr < 120$.

2. Sider-Tate Correlation

$$Nu = 0.027\ Re^{0.8}\ Pr^{0.3}\left[\frac{Pr_f}{Pr_w}\right]^{0.25} \tag{3.48}$$

The Reynolds number in both correlations is calculated for the internal diameter of the tube. Pr_f and Pr_w refer to Prandtl numbers calculated for bulk fluid average temperature and the tube wall temperatures, respectively. The term Pr_f/Pr_w takes care of the change of properties with temperature. If the temperature range is small, this term can be neglected.

The Sieder-Tate correlation is applicable in the range

$6000 < Re < 10^7$ and $0.7 < Pr < 10^4$.

The correlation used in solved problems that follow uses a similar correlation with a slight change in the constants.

$$Nu = 0.027\ Re^{0.8}\ Pr^{0.3}\left[\frac{Pr_f}{Pr_w}\right]^{0.25} \tag{3.49}$$

There are two types of transfer conditions encountered in heat transfer. In the first condition the tube wall is at a constant temperature. The fluid temperature changes from inlet (t_1 or t_{in}) to outlet (t_2 or t_{out}).

In the second condition the thermal flux q(w/m^2) incident on the tube is constant so that the wall temperature also changes from the inlet to the outlet point, (Figure 3.9(B) and Figure 3.10(D)). As discussed before, the fluid temperature has a "profile" at any cross section. Hence, we consider a "bulk" fluid temperature T_b which is the bulk mean, average or actually measured temperature at that cross section.

The term "average temperature" (without the word bulk) will mean the arithmetic average. For example, if the inlet temperature is t_1 and the outlet temperature is t_2 the average temperature is $t_1 + t_2/2$.

However, when the temperature profile is not linear we will have to use log mean temperature difference, given by

$$\Delta t_{LMTD} = \frac{t_1 - t_2}{2.3 \log t_2/t_1} \tag{3.50}$$

In most applications we use circular tubes and the control dimension (ℓ_o) is the diameter d. If a rectangular tube of (internal) size a and b is used, the equivalent diameter is given by

$$d = 4\frac{\text{Area}}{\text{Perimeter}} = 4\frac{ab}{2a + 2b} = \frac{2ab}{a + b} \tag{3.51}$$

EXAMPLE 3.10

A 15-mm diameter pipe is cooled by water flowing through it. The inlet temperature of water is 25°C and the outlet temperature is 65°C. The velocity of water is 3.0 m/sec. The lengthwise mean temperature of the tube wall is 80°C.

Now determine:

1. The heat transfer coefficient (h_c)
2. Amount of heat transferred (Q)
3. Length of the tube (L)

Solution

The arithmetic mean temperature of water is

$$t_f = \frac{25+65}{2} = 45°C$$

At 45°C the properties of water are

$\rho = 988.8$ kg/m^3	$\upsilon = 0.657 \times 10^{-6}$ m^2/sec
$C = 4.174$ kJ/kg°C	$\lambda = 62.6 \times 10^{-2}$ w/m°C
Pr = 3.97	$\alpha = 15.5 \times 10^{-8}$ m^2/sec

The Reynolds number at 45°C

$$\text{Re} = \frac{Ud}{\upsilon} = \frac{3 \times 15 \times 10^{-3}}{0.657 \times 10^{-6}} = 6.85 \times 10^4$$

$\text{Re} = 6.85 \times 10^4 > 10^4$, hence, the flow is marginally turbulent. The Prandtl number at 65°C is 3.76

Hence $\frac{\text{Pr}_f}{\text{Pr}_s} = \frac{3.97}{3.76} = 1.06$ and so can be neglected.

The correlation for the Nusselt number is (for $0.5 < \text{Pr} < 120$ and $6000 < \text{Re} < 10^7$)

$$\text{Nu} = 0.023 \times \text{Re}^{0.8}\ \text{Pr}^{0.4}$$

The correlation Pr_f/Pr_s need not be used, as the temperature difference $t_s - t_f$ is small.

$$\text{Nu} = 0.023 \times (6.85 \times 10^4)^{0.8} \times 3.97^{0.4}$$

$$= 296$$

The length is not known. It is assumed that $L/d > 50$, therefore no correction for the inlet length is required (this will be checked later). The heat transfer coefficient h_c will be

$$h_c = \text{Nu}\frac{\lambda}{d} = 296 \times \frac{62.6 \times 10^{-2}}{15 \times 10^{-3}}$$

$$= 1.24 \times 10^4 \text{ w/m}^2\text{°C}$$

The mass rate of water flow m

$= \text{velocity} \times \text{cross section area} \times \text{density}$

$$= U \times \frac{\pi d^2}{4} \times \rho$$

$$= 3 \times \frac{\pi(15 \times 10^{-3})^2}{4} \times 988.8 = 0.524 \text{ kg/sec}$$

The amount of heat transferred to water/sec

Q = mass flow rate × sp. heat × temperature difference

$= m \times c \times (t_2 - t_1)$

$= 0.524 \times 4.174 \times (65 - 25)$

$Q_1 = 87.5$ kW

As the wall temperature gradient is not known we will use the log mean temperature difference $\Delta t_{\log}$ given by

$$= \frac{t_1 - t_2}{2.3 \log \frac{t_s - t_1}{t_s - t_2}}$$

$$= \frac{65 - 25}{2.3 \log \frac{80 - 25}{80 - 65}}$$

$= 30.8°\text{C}$

As the heat transfer coefficient h_c is known, heat taken up by water through the wall is

$Q_2 = h_c \times \text{wall area} \times \Delta t_{\log}$

$= h_c \times \pi d L \times \Delta t_{\log}$

$= 1.24 \times 10^4 \times \pi \times 15 \times 10^{-3} \times L \times 30.8$

$= 1.80 \times 10^4 L$ w

Equating Q_1 and Q_2

$$L = \frac{87.5 \times 10^3}{1.80 \times 10^4}$$

$= 4.86$ m

EXAMPLE 3.11

A steel pipe having 15 mm diameter and 3.0 m length is cooled by water flowing through it. The wall temperature is 90°C. The inlet temperature of water is 30°C and the outlet temperature is 60°C. The velocity of water is 3.0 m/sec.

Determine the heat transfer coefficient and the total heat transferred to water.

Solution

The average temperature of water is

$$T_{av} = \frac{30+60}{2} = 45°\text{C}$$

At this temperature the properties of water are

$\rho = 984$ kg/m^3 $\qquad$ $\upsilon = 651 \times 10^{-6}$

$C = 4.174$ kJ/kg°C $\qquad$ $\lambda = 64.1 \times 10^{-2}$ w/m°C

Hydraulic dimension $\ell_o = d = 0.015$ m.
Thermal diffusivity α of water at 45°C

$$= \frac{\lambda}{\rho c}$$

$$= \frac{64.1\times 10^{-2}}{984\times 4.174} = 1.56\times 10^{-4}\ \text{m}^2/\text{sec}$$

$$\text{Pr} = \frac{\upsilon}{\alpha} = \frac{651\times 10^{-6}}{1.56\times 10^{-4}} = 4.17$$

The diameter of pipe is 0.015 m = ℓ_o

$$\text{Re} = \frac{\ell_o U}{\upsilon} = \frac{0.015\times 3}{0.651\times 10^{-6}}$$

$$= 6.91\times 10^4$$

Re > 2000. Hence, the flow is turbulent. The ℓ/d ratio for the tube is $\frac{3}{0.015} = 200 > 50$.

Hence, using the Dittus-Boetter correlations

$$= 0.023\ \mathrm{Re}^{0.8} \times \mathrm{Pr}^{0.4}$$

$$= 0.023(6.91 \times 10^4)^{0.8} \times (4.17)^{0.4} = 3.03 \times 10^2$$

$$h_c = \mathrm{Nu}\frac{\lambda}{\ell_o} = 3.03 \times 10^2 \times \frac{6.41 \times 10^{-2}}{0.015}$$

$$= 1.29 \times 10^4\ \mathrm{w/m^2\,°C}$$

The tube internal diameter is 0.015 and length given is 3.0 m. The surface area of the tube is

$$A = \pi dL$$

$$= \pi \times 0.015 \times 3.0 = 0.141\ \mathrm{m}^2$$

The total heat transferred to water is

$$= A \times h_c \times (t_1 - t_2)$$

$$= 0.141 \times 1.29 \times 10^4 \times (60 - 30)$$

$$= 5.46 \times 10^4\ \mathrm{w}$$

EXAMPLE 3.12

Determine the change in heat transfer coefficient for problem 3.6 if the velocity of water is increased to 5 m/sec. All other parameters are the same.

Solution

If the water velocity (U) is 5 m/sec and tube diameter is 15 mm

$$\mathrm{Re} = \frac{Ud}{\upsilon} = \frac{5 \times 15 \times 10^{-3}}{0.657 \times 10^{-6}}$$

$$= 1.14 \times 10^5$$

The flow is turbulent. Using the same correlation

$$\mathrm{Nu} = 0.023\ \mathrm{Re}^{0.8} \times \mathrm{Pr}^{0.4}$$

$$= 0.023(1.14 \times 10^5)^{0.8} \times (3.97)^{0.4} = 443$$

The heat transfer coefficient is

$$h_c = \text{Nu}\frac{\lambda}{d} = 4.43 \times 10^2 \times \frac{62.6 \times 10^{-2}}{15 \times 10^{-3}}$$

$$= 1.85 \times 10^4$$

The mass flow rate of water is

$$m = U \times \frac{\pi d^2}{4} \times \rho = \frac{5 \times \pi (15 \times 10^{-3})^2 \times 988.8}{4}$$

$$= 0.874 \text{ kg/sec}$$

Heat transferred to water

$$Q_1 = m \times c \times (t_2 - t_1)$$
$$= 0.874 \times 4.174 \times (65 - 25)$$
$$= 146 \text{ kW}$$

The log mean temperature difference $\Delta t_{\log} = 30.8°\text{C}$
Heat transferred from wall to water

$$Q_2 = h_c \times \pi d L \times \Delta t_{\log}$$

$$= 1.85 \times 10^6 \times \pi \times 15 \times 10^{-3} L \times 30.8$$
$$= 2.69 \times 10^4 L \text{ w}$$

Equating Q_1 and Q_2

$$L = \frac{146 \times 10^3}{2.69 \times 10^4}$$

$$= 5.43 \text{ m}$$

$$\frac{h_{c_{3U}}}{h_{c_{5U}}} = \frac{1.24 \times 10^4}{1.85 \times 10^4} = 67\%$$

EXAMPLE 3.13

The outer wall temperature of a furnace is 100°C. The ambient temperature is 30°C. Wall height is 0.5 m.

Calculate

1. The heat transfer coefficient h_c for natural convection from the wall.
2. The heat transfer coefficient in the wall temperature range 200–50°C.

Solution

We first calculate Gr, Pr, and their product Gr · Pr

$$\text{Gr} = \frac{g\beta_f \ell_o^3 \Delta t}{\upsilon^2}$$

$$g = 9.81\,\text{m/sec}^2 \quad \beta_f = \frac{1}{30+273} = 2.68\times10^{-3}\,{}^\circ\text{K}^{-1} \quad t_o = 0.5$$

$\Delta t = 100 - 30 = 70$°C, $\upsilon = 23.13 \times 10^{-3}$ m/sec^2

Substituting

$$\text{Gr} = \frac{9.81\times 2.68\times 10^{-3}\times 0.5^3\times 70}{(23.13\times 10^{-6})^2}$$

$$= 0.43\times 10^9$$

$$\text{Pr} = 0.685 \text{ (from table*)}$$

$$\text{Gr}\cdot\text{Pr} = 0.295\times 10^9$$

The correlation for Gr · Pr. < 10^9 is

$$\text{Nu} = 0.6(\text{Gr}\cdot\text{Pr})^{1/4}$$

$$= 0.6(0.295\times 10^9)^{1/4}$$

$$= 78.6$$

$$h_c = \text{Nu}\frac{\lambda}{\ell_o}$$

Average thermal conductivity is 2.9×10^{-2} w/m°C

Substituting

$$h_c = \frac{78.6\times 2.9\times 10^{-2}}{0.5}$$

$$= 4.56 \text{ w/m°C}$$

---

* Standard tables found in handbooks listed in the *Bibliography*.

Repeating the calculations for 200, 150, and 50°C the following results are obtained

T°C	h_c w/m^2 °C
200	4.97
150	4.69
100	4.56
50	3.68

This shows that the convection heat transfer coefficient changes very little with temperature and can be regarded as constant at about 4.5 w/m°C.

3.6.4 Heat Transfer in Coils

There are many instances when parts of furnaces are required to be cooled by water circulation. Typical examples are metal walls of a vacuum furnace, heating (insulation) coils of an induction heating set-up, doors and windows. Cooling is achieved by flowing water through coils welded to the walls (Figure 3.13). The heat transfer takes place by forced convection between the tube walls and water.

The flow along the tube wall experiences a lengthwise component and a centrifugal component (due to continuously bending path). These two components produce a screwlike or helical flow pattern. The centrifugal or secondary flow around the wall appears only at the bend and depends on the bend radius R.

It is experimentally found that the mixed flow pattern occurs if the Reynolds number of the fluid Re_f is between two critical values Re_{cr1} and Re_{cr2}. For a helical coil of radius r and tube diameter d.

$$\mathrm{Re}_{cr1} = \frac{16.4}{\sqrt{d/R}} \qquad \left(\text{For } \frac{d}{R} \geq 8 \times 10.4\right) \tag{3.52}$$

and

$$\mathrm{Re}_{cr2} = 18{,}500\left[\frac{d}{2R}\right]^{0.28} \tag{3.53}$$

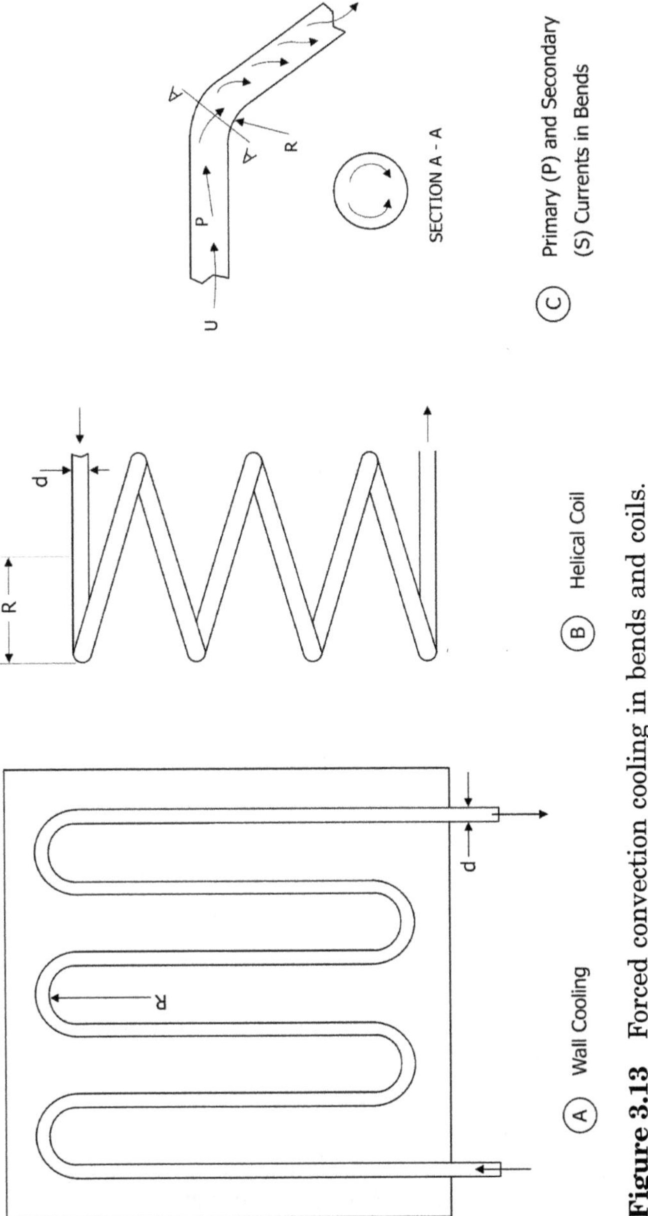

Figure 3.13 Forced convection cooling in bends and coils.

For $Re_f < Re_{cr1}$ the flow is laminar and there is no secondary circulation.

The Nusselt number for laminar flow is

$$Nu = 3.66 \text{ for } Pr > 6 \tag{3.54}$$

If Re_f is between Re_{cr1} and Re_{cr2}, i.e.,

$$Re_{cr1} < Re_f < Re_{cr2}$$

there is secondary circulation and the Nusselt number is obtained by the usual correlation

$$Nu = 0.021\, Re_f^{0.8} Pr_f^{0.4} \left[\frac{Pr_f}{Pr_w}\right]^{0.25} \tag{3.55}$$

the flow is fully turbulent.

If $Re_f > Re_{cr2}$, the Nusselt and the heat transfer coefficient h_c is first determined by the above correlation and a correction ϵ is applied so that

$$hc_{cor} = \epsilon h_c \tag{3.56}$$

where $$\epsilon = 1 + 1.8\frac{d}{R}$$

For simple bends the enhanced transfer coefficient $h_{c_{cor}}$ appears only at a little distance after the bend (downstream) and disappears further.

The effect of bends and coiling is more pronounced on the friction factor and the head requirements.

EXAMPLE 3.14

Cooling water is circulated through an induction heating coil made of copper tube having 6 mm external and 4.5 mm internal diameters. The coil diameter is 100 mm.

The internal wall temperature of the coil is 80°C. The inlet temperature of water is 20°C and the outlet temperature is 65°C. The velocity of water is 1.5 m/sec.

Determine the heat transfer coefficient from tube wall to water and the rate of heat transfer per meter length of pipe.

If the coil has 10 turns, calculate the total heat removed by the water.

Solution

Average water temperature is

$$t_{av} = \frac{t_{\text{in}} + t_{\text{out}}}{2} = \frac{20 + 65}{2} = 425°\text{C}$$

Properties of water at 42.5°C

$\upsilon = 0.634 \times 10^{-6}$ m²/sec $\qquad Pr = 4.11$

$\lambda = 63.1 \times 10^{-2}$ w/m°C $\qquad C = 4.174$ kJ/kg°C

$\rho = 993$ kg/m³ $\qquad \alpha = 15.4$ m²/sec

$d/R = 4.5/50 = 9 \times 10^{-2} > 8 \times 10^{-4}$

$$\text{Re}_f = \frac{Ud}{\upsilon} = \frac{1.5 \times 4.5 \times 10^{-3}}{0.634 \times 10^{-6}}$$

$$= 1.06 \times 10^4$$

First critical Reynolds number

$$\text{Re}_{cr1} = \frac{16.4}{\sqrt{dR}} = \frac{16.4}{\sqrt{4.5 \times 10^{-3} \times 50 \times 10^{-3}}} = 1.09 \times 10^3$$

Second critical Reynolds number

$$\text{Re}_{cr2} = 18{,}500 \left[\frac{d}{2r}\right]^{0.28}$$

$$= 18{,}500 \left[\frac{4.5}{100}\right]^{0.28} = 7{,}760$$

$\text{Re}_f > \text{Re}_{cr2}$, hence, the heat transfer coefficient can be calculated by the usual correlation,

$\text{Nu} = 0.023\ (\text{Re}_f)^{0.8}\ (\text{Pr})^{0.4}$

and a correction applied.

We first calculate h_c for a straight pipe

$$\text{Nu} = 0.023(1.06 \times 10^4)^{0.8}(4.11)^{0.4}$$
$$= 67.2$$
$$h_c = \text{Nu}\frac{\lambda}{d}$$
$$= \frac{67.2 \times 63.1 \times 10^{-2}}{45 \times 10^{-3}} = 9,420 \text{ w/m}^2\,°\text{C}$$

Correction ϵ is given by

$$\epsilon = 171.8\frac{d}{R} = 1 + 1.8\frac{4.5}{50} = 1.162$$
$$h_c \text{ for coil} = \epsilon h_c = 9,420 \times 1.162$$
$$= 10,900 \text{ w/m}_2\,°\text{C}$$

Surface area of 1 m length of pipe

$$= \pi\, d\, L$$
$$= \pi \times 4.5 \times 10^{-3} \times 1$$
$$= 14.1 \times 10^{-3} \text{ m}^2$$

Heat transferred through 1 m length

$$= 10,900 \times 14.1 \times 10^{-3}$$
$$= 154 \text{ w/°C}$$

The coil has 10 turns of 100 mm diameter

$$\text{Length of pipe} = \pi \times 100 \times 10^{-3} \times 10$$
$$= 3.14 \text{ m}$$
$$\text{Heat transferred} = 154 \times 3.14$$
$$= 484 \text{ w/°C}$$

The coil wall temperature is 80°C.
Mean water temperature is 45°C.
Mean temperature difference between the wall and water is

$$\Delta t = 80 - 45 = 35°\text{C}$$

Hence, the heat removed by water

$$= 484 \times 35 = 17 \text{ kW}$$

3.7 NATURAL CONVECTION (FLAT WALLS)

Natural or free convection occurs due to the movement of a fluid by the force of gravity. If a cold fluid (e.g., air) comes in contact with a hot surface it gets heated, resulting in expansion and a density lower than that in the cold state. The hot fluid rises due to lesser gravity. This results in an upward motion of the hot fluid. No external flow inducer (e.g., fan) is required. Hence, the resultant convection is natural or free.

Depending on the extent (critical length) of the solid surface, the fluid motion may be laminar or transient.

Such a situation occurs in many practical cases. The external walls of a furnace are warm. They are in contact with the cold atmosphere. This induces a motion of air over the walls resulting in heat transfer by free convection from the wall to the surrounding. A similar situation arises inside a furnace (e.g., electrically heated furnace). The internal (trapped) air circulates along the walls and over the charge and transfers heat from or to it.

The correlations for free convection are generally of the form

$$\mathrm{Nu} = A(\mathrm{Gr} \cdot \mathrm{Pr})^n \tag{3.57}$$

where Gr = Grashoff number

$= \dfrac{g\beta \ell_o^3 \Delta t}{\upsilon^2}$ relating volumetric expansion (β), gravity (g) kinematic viscosity, and reference dimension (ℓ_o)

Pr = Prandtl number

$= \dfrac{\upsilon}{\alpha}$ relating kinematic viscosity (υ), thermal diffusivity ($\alpha = \lambda/\rho c$)

Nu = Nusselt number

$= \dfrac{h_e \ell_o}{\lambda}$ relating the heat transfer coefficient (h_c), reference dimension (ℓ_o), and thermal conductivity (λ)

The choice of reference dimension depends on the situation. The applicable constants A and n are derived from experimental data and are tabulated for the situation under consideration.

The decisive factor in such correlations is the numerical value of Gr · Pr.

There may be some variation in the values of A and n that appear in the correlations developed by different workers but the general form of the equation remains the same.

It is assumed that the physical properties (λ, α, υ, etc.) are constant, there is no pressure gradient, and the inertia forces are negligible compared to buoyancy (expansion, gravity, and viscosity) forces. The volumetric expansion coefficient ρ is the reciprocal of the absolute wall temperature (1/K or K^{-1}).

The decisive product (Gr · Pr) is also a dimensionless number called a Rayleigh number (Ra).

Free Convection in Flow over Vertical Plate or Wall

When air or some other fluid flows over a hot vertical plate or wall, the flow may be laminar or transient as shown in Figure 3.14. The critical dimension ℓ_o in this case is the height of the wall.

There are different correlations developed for different ranges of (Gr · Pr) values as given below. The general form is $\mathrm{Nu} = A\,(\mathrm{Gr} \cdot \mathrm{Pr})^n$

$$\left.\begin{array}{l} \text{1. For } 10^{-3} < \mathrm{Gr} \cdot \mathrm{Pr} < 5 \times 10^{-2} \\ \quad A = 1.18, \quad n = 1/8 \end{array}\right\} \tag{3.58}$$

$$\left.\begin{array}{l} \text{2. For } 5 \times 10^{2} < \mathrm{Gr} \cdot \mathrm{Pr} < 2 \times 10^{9} \\ \quad A = 0.54 - 0.6, \quad n = 1/4 \end{array}\right\} \tag{3.59}$$

$$\left.\begin{array}{l} \text{3. For } \mathrm{Gr} \cdot \mathrm{Pr} > 2 \times 10^{9} \\ \quad A = 0.135, \quad n = 1/3 \end{array}\right\} \tag{3.60}$$

$$\left.\begin{array}{l} \text{4. For } \mathrm{Gr} \cdot \mathrm{Pr} < 10^{-3} \\ \quad \mathrm{Nu} = 0.45 \end{array}\right\} \tag{3.61}$$

For horizontal plates similar correlations are used.

For a horizontal plate where the heated surface faces up (e.g., furnace top), the heat transfer coefficient h_c calculated using conditions given in Equation (3.58) to Equation (3.61) is increased by 30%.

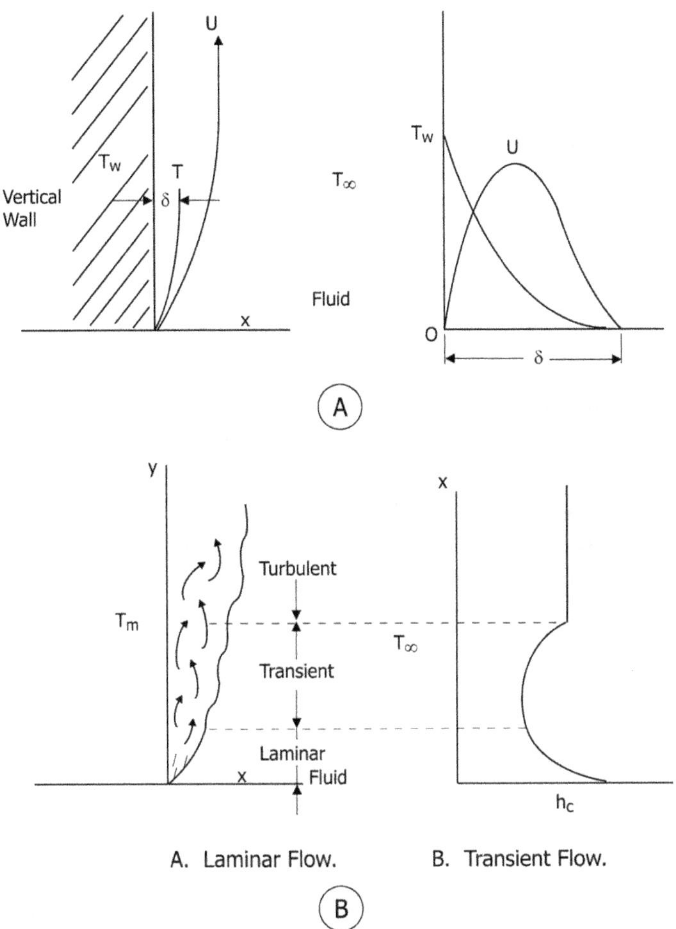

Figure 3.14 Natural (free) convection over a vertical wall.

For a horizontal plate with the hot surface facing down (e.g., furnace bottom) the transfer coefficient as calculated is reduced by 30%.

The critical dimension ℓ_o for both the horizontal surfaces is taken as the width or the lower dimension.

The general procedure for calculating the heat transfer coefficient h_c is as follows:

1. From the given data the numbers Gr and Pr are calculated.

2. The product Gr · Pr is determined from which the applicable correlation is chosen.
3. From the correlation the Nusselt number Nu is determined.
4. The heat transfer coefficient is given by

$$h_c = \text{Nu}\,\frac{\lambda}{\ell_o} \tag{3.62}$$

3.7.1 Free Convection over Horizontal Pipes

Air circulation around horizontal, inclined, or cylindrical hot objects gives rise to natural convection. Typical circulation patterns are shown in Figure 3.15. The experimental correlations are similar to that for vertical objects in natural convection, i.e., they relate the product Gr · Pr to the Nusselt number. The critical dimension in this case is the external diameter d of the cylinder.

Figure 3.15 shows that the air stream coming from the lower side moves along the circumference and disintegrates into a vortex in the upper part, giving a turbulent pattern.

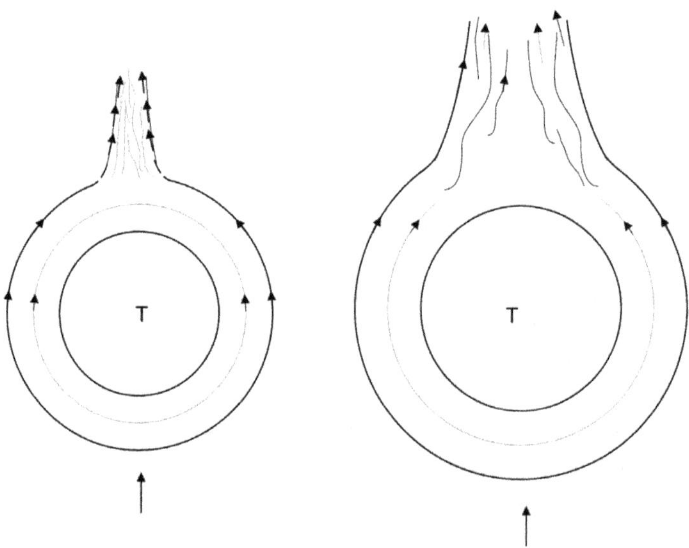

Figure 3.15 Free convection over horizontal tubes.

Consequently the heat transfer coefficient h_c is not uniform all over. From the bottom it changes as shown by the dotted line. The breakup of the laminar pattern depends on the diameter, as shown. The correlation given below is for the overall (average) Nusselt number for cylindrical objects.

$$\text{Nu} = A(\text{Gr}\cdot\text{Pr})^n \tag{3.63}$$

where Pr_f and Pr_w are the Prandtl numbers for fluid (air) at a distance and at the wall, respectively.

$$\text{For Gr}\cdot\text{Pr from } 10^3 \text{ to } 10^9,\quad A = 0.3 \text{ to } 0.5\ n = 0.25 \tag{3.64}$$

$$\text{For Gr}\cdot\text{Pr} > 10^9\ A = 0.18,\quad n = 0.3 \tag{3.65}$$

EXAMPLE 3.15

The external temperature of a horizontal cylindrical pipe of 0.03 m diameter is 100°C. The bulk air temperature around the pipe is 30°C.

Determine the convective heat transfer coefficient for the pipe.

How does the transfer coefficient change if the pipe diameter changes from 0.2 to 0.03 m?

All other conditions remaining the same, how does the transfer coefficient change if the pipe temperature varies between 1000 and 100°C?

Solution

At an average temperature $\frac{100+30}{2} = 65°\text{C}$, the properties of air are

$$\beta = \frac{1}{65+273} = 2.96\times 10^{-3}\,{}^\circ\text{K}^{-1}$$

$$\upsilon = 19.5\times 10^{-6}\ \text{m}^2/\text{sec},\quad \ell_{av} = 2.93\times 10^{-2}\ \text{w/m°C}$$

$$\text{Pr}_{av} = 0.695,\quad \text{Pr}_f(\text{at } 30°\text{C}) = 0.701, \text{Pr}_w\ (\text{at } 100°\text{C}) = 0.698$$

also $\ell_o = d = 0.03$ m, $g = 9.81$ m²/sec $\Delta t = 100 - 30 = 70$°C

$$\text{Gr} = \frac{g\beta\ell_0{}^3\Delta t}{\upsilon^2}$$

$$= \frac{9.81 \times 2.96 \times 10^{-3} \times 0.03^3 \times 70}{(19.5 \times 10^{-6})^2} = 1.43 \times 10^5$$

$$\text{Gr} \cdot \text{Pr} = 1.43 \times 10^5 \times 0.698 = 1.01 \times 10^5$$

$$\text{Nu} = 0.5\,(1.01 \times 10^5)^{0.25} = 18.91$$

$$h_c = \text{Nu} \times \frac{\lambda}{\ell_o} = 8.98 \times \frac{2.93 \times 10^{-2}}{3 \times 10^{-2}}$$

$$= 8.71 \text{ w/m}^2\text{°C}$$

We will now calculate h_c for 0.2, 0.5, 0.15, and 0.05 m pipe diameters. Repeating the above calculations ℓ_o (d) values are as follows:

d(m)	Gr · Pr	Nu	h_c w/m²°C
0.20	0.296×10^8	36.9	5.40
0.15	1.250×10^7	29.7	5.81
0.10	0.369×10^7	21.9	6.42
0.05	0.462×10^6	13.0	7.02
0.03	0.998×10^5	17.8	8.65

The results show that h_c increases as the pipe diameter decreases, but the rate of increase is not much.

If the pipe (0.03 m diameter) temperature is 1000°C, various parametric values are

$$\beta = \frac{1}{1000 + 273} = 7.86 \times 10^{-4}\ \text{°K}^{-1},$$

$$\upsilon = 177.1 \times 10^{-6} \text{ m}^2/\text{s} \qquad \lambda = 8.07 \times 10^{-2} \text{ w/m°C}$$

$$Pr_{av} = 0.703,\ \ Pr_w\ (1000\text{°C}) \cdot 0.719,\ \ Pr_f(30\text{°C}) \cdot 0.704$$

$$\ell_o = d = 0.03 \text{ m},\ \ g = 9.81 \text{ m/sec}^2, \Delta t = 1000 - 30 = 970\text{°C}$$

Substituting

$$\mathrm{Gr} = \frac{9.81 \times 7.86 \times 10^{-4} \times 0.03^3 \times 970}{\left(177.1 \times 10^{-6}\right)^2}$$

$$= 6.04 \times 10^7$$

$$\mathrm{Gr.Pr} = 6.04 \times 10^7 \times 0.703$$

$$= 4.24 \times 10^7$$

$$\mathrm{Nu} = 0.5\left(4.2 \times 10^7\right)0.25$$

$$= 22.5$$

$$h_c = 22.5 \frac{8.07 \times 10^{-2}}{3 \times 10^{-2}}$$

$$= \underline{60.5 \text{ w/m°C}}$$

Repeating the calculations for 800, 600, 400, and 200°C with appropriate values of β, υ, λ, etc. the results are shown below

T °C	h_c w/m² °C
1000	8.71
800	11.10
600	11.10
400	11.90
200	8.75
100	8.71

The results show that the free convection coefficient h_c is practically independent of the hot body temperature.

EXAMPLE 3.16

A steel bar of diameter d = 20 mm is cooled in a cross flow of dry nitrogen flowing at a velocity 1.0 m/sec. The temperature of nitrogen and steel are 25 and 950°C, respectively.

Calculate

1. The heat transfer coefficient for the bar surface
2. The heat transfer coefficient for the velocity range 1.0 to 100 m/sec

All other conditions remaining same.

Solution

Properties of nitrogen at 25°C are

$\rho = 1.166$ kg/m^3 $\qquad \upsilon = 15.5 \times 10^{-6}$ m^2/sec

$C = 1.031$ kJ/kg °C $\qquad \lambda = 26.1 \times 10^{-3}$ w/m°C

Pr = 0.698 $\qquad d = 0.020$ m

$$\text{Re}_f = \frac{Ud}{\upsilon} = \frac{1 \times 0.020}{15.5 \times 10^{-6}} = 1290$$

The correlation for the Nusselt number is

$$\text{Nu} = 0.26(\text{Re})^{0.6}\,(\text{Pr})^{0.37}\left[\frac{\text{Pr}}{\text{Pr}_s}\right]^{0.25}$$

Here Pr_s is at the surface temperature 900°C

$$\text{Pr}_s(900°\text{C}) = 0.765$$

$$\frac{\text{Pr}}{\text{Pr}_s} = \frac{0.698}{0.765}$$

$$= 0.912$$

$$\text{Nu} = 0.26(1290)^{0.6}(0.698)^{0.37}(0.912)^{0.25}$$

$$= 0.26 \times 73.5 \times 0.875 \times 0.977$$

$$= 16.3$$

Note: As Pr is practically constant for gases the constant (0.26) and the terms (Pr)0.37, (Pr/Pr.)0.25 can be combined to obtain a correlation of the form,

$$\text{Nu} = 0.22\ (\text{Pr})^{0.6}$$

$$h_c = \text{Nu}\frac{\lambda}{d} = 16.3 \times \frac{26.1 \times 10^{-3}}{20 \times 10^{-3}}$$

$$= 21.3\ \text{w/m}^{2}\,°\text{C}$$

Repeating the calculations for $U = 3, 5, 10, 100$ m/sec

U	Re	Nu	h_c
1	1,290	16.3	21.3
3	3,870	31.3	40.8
5	6,450	42.5	55.6
10	12,900	64.4	84.3
100	1.29×10^5	256.0	336.0

3.7.2 Free Convection inside Enclosures

Many furnaces (enclosures) have no gas or air circulation intentionally taking place inside. Indirectly, a heated electric furnace is a typical example of this type.

The heat transfer inside such furnaces is a very complicated process and it is difficult to afford a rigorous theoretical analysis (see Section 3.14). Without suffering large errors, it is safe to assume that the enclosed air gets heated and transfers the heat to the charge (on bottom) and to the walls by convection. Radiation from the heater to the walls and the charge also takes place simultaneously.

Natural convection currents are set up around the charge and the wall which are at a temperature lower than the heater (see Figure 3.16).

Hot air rises along the walls causing convectional heat transfer from the gas to the walls.

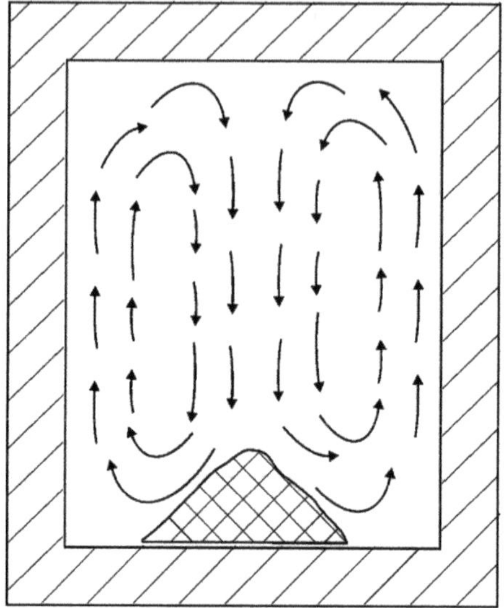

Figure 3.16 Free convection inside a furnace.

The convection heat transfer coefficient h_c can be determined by a correlation similar to that used earlier, i.e.,

$$\left.\begin{aligned} \mathrm{Nu} &= A(\mathrm{Gr}\cdot\mathrm{Pr})^n \quad \text{and} \\ h_c &= \mathrm{Nu} \end{aligned}\right\} \tag{3.66}$$

However, the conductivity λ used in such cases is the effective conductivity λ_{ef} given by

$$\lambda_{ef} = \epsilon\lambda_o \tag{3.67}$$

where λ_o is the tabulated or normal conductivity.

The coefficient ϵ has a value 1.0 to 20.0 (linear) in the range $10^3 < \mathrm{Gr}\cdot\mathrm{Pr} < 10^9$ (Figure 3.17). This arises because free convection inside the furnace occurs in a limited space. The air moving along the wall is in layers and heat transfer in layers is mainly by conduction. Physical properties such as λ, β, υ, Pr

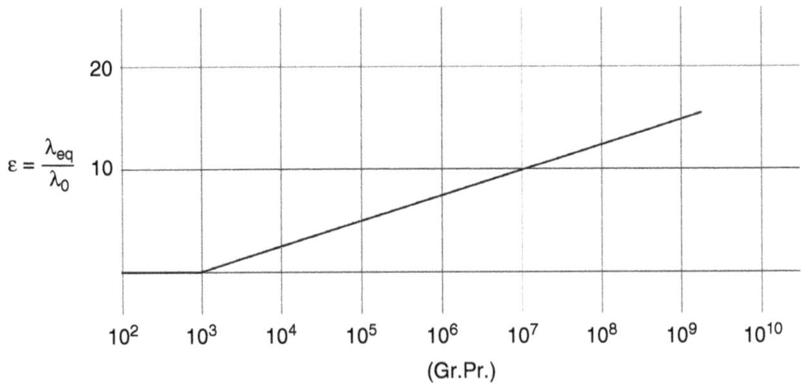

Figure 3.17 Relation between Gr · Pr and equivalent conductivity λ_{eq} for free convection inside furnace.

are those at a mean temperature

$$t = \frac{t_f + t_w}{2} \tag{3.68}$$

EXAMPLE 3.17

An indirectly heated furnace has a wall height 0.5 m. The temperatures inside the furnace (steady state) are

Heater	1300°C
Air	1200°C
Wall	1100°C

Determine the convection heat transfer coefficient for the wall.

Solution

Average temperature for air-wall is

$$t_{av} = \frac{1200 + 1100}{2} = 1150°\text{C}$$

The air properties at this temperature are

$$\beta = \frac{1}{1150 + 273} = 7.03 \times 10^{-4}\ \mathrm{K}^{-1}$$

$$\upsilon = 216 \times 10^{-6}\ \mathrm{m^2/sec}$$

$$\mathrm{Pr} = 0.723 \quad \lambda_o = 8.28 \times 10^{-2}\ \mathrm{w/m°C}$$

Also $\Delta t = 1200 - 1100 = 100°\mathrm{C},\ \lambda_o = 0.5\ \mathrm{m}$

$$\mathrm{Gr} = \frac{9.81 \times 7.03 \times 10^{-4} \times 0.5^3 \times 100}{(216 \times 10^{-6})^2}$$

$$= 1.85 \times 10^6$$

$$\mathrm{Gr} \cdot \mathrm{Pr} = 1.85 \times 10^6 \times 0.723$$

$$\frac{\lambda_{ef}}{\lambda_o} = 1.33 \times 10^6$$

$$= \epsilon = 6.0 \quad \text{for Gr} \cdot \text{Pr} \sim 10^6$$

$$\mathrm{Nu} = 0.5\,(1.33 \times 10^6)^{0.25} = 17.0$$

$$h_c = \mathrm{Nu}\frac{\lambda_{ef}}{\ell_o}$$

$$= 17 \times \frac{8.28 \times 10^{-2} \times 6}{0.5}$$

$$= 16.9 \sim 17\ \ \mathrm{w/m^2\ °C}$$

3.8 RADIATIVE HEAT TRANSFER

Radiation is a process in which heat transfer takes place without physical contact. All bodies radiate or emit electromagnetic waves in the space around them. The radiations are mainly concentrated in the ultraviolet, visible, and infrared parts of the spectrum and cover a wavelength range λ from 0.1 to 100 μm. Radiation occurs due to the internal atomic or molecular vibrations and takes place at all temperatures except zero Kelvin.

The intensity or power (w/m^2) of these radiations is not uniform over the entire wavelength range (see Figure 3.18). It shows a maxima and tapers off toward long and short wavelengths. Radiation power increases with temperature and the maxima shifts toward shorter wavelengths, e.g., at 800 K, λ_{max} ~ 3.5 μm and at 1200 K, λ_{max} ~ 2.5 μm. The area under the intensity-wavelength curve represents the total radiated power. The thermal radiation emitted by real bodies is directional, i.e., it is not radiated uniformly in the space around the object. A more detailed discussion on the nature of thermal radiation is given in Chapter 14.

When radiant energy is incident on a body, some energy is reflected, some absorbed, and some transmitted through

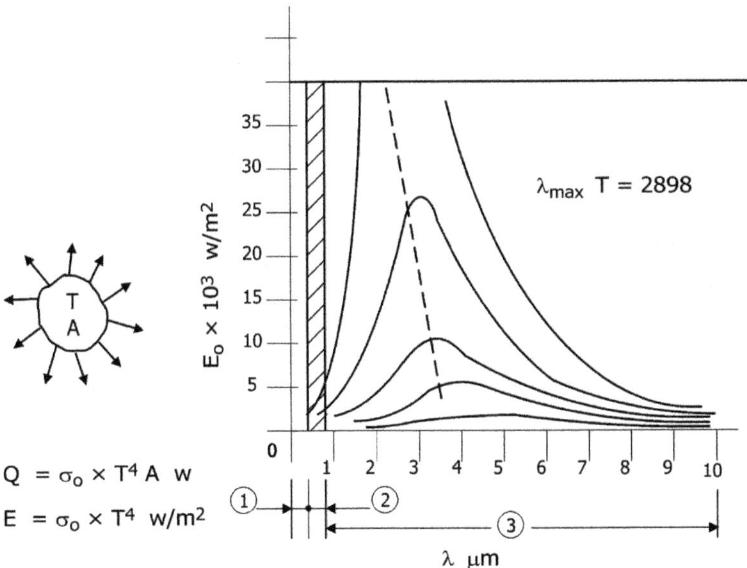

Figure 3.18 Radiating black body and its spectrum.

so that,

$$Q_{\text{in}} = Q_R + Q_A + Q_T \tag{3.69}$$

where Q_{in} = the incident radiation energy (w), and Q_R, Q_A, and Q_T are respectively the reflected, absorbed, and transmitted energies. On dividing both sides of the above equation by Q_{in}, we get

$$\frac{Q_R}{Q_{\text{in}}} + \frac{Q_A}{Q_{\text{in}}} + \frac{Q_I}{Q_{\text{in}}} = 1 \tag{3.70}$$

$$\text{or} \quad R + A + T = 1 \tag{3.71}$$

where R, A, and T are the fractions of incident energy that are reflected, absorbed, and transmitted. These fractions represent the characteristic properties of the material receiving radiation so that

$$\left.\begin{aligned} \frac{Q_R}{Q_{\text{in}}} &= \text{reflectivity} \\ \frac{Q_A}{Q_{\text{in}}} &= \text{absorptivity} \\ \frac{Q_I}{Q_{\text{in}}} &= \text{transmissivity} \end{aligned}\right\} \tag{3.72}$$

We will generally come across opaque bodies so that $T = 0$ and $R + A = 1$.

An ideal absorbing body will absorb all the incident energy. This body is called a "black body" for which R and T are zero and $A = 1$. Recalling that thermal radiation consists of all electromagnetic waves in the range 0.1 to 100 μm, a black body must be capable of absorption throughout this wavelength region. A black body by definition is an ideal absorber, which also indicates that it is a good emitter or radiator (assuming a constant temperature).

The radiating power of a black body depends on the fourth power of its absolute temperature. This is called the Stefan-Boltzmann law.

$$E_o = \sigma_o T^4 \text{ w/m}^2 \tag{3.73}$$

where

E_o = Radiated power or thermal flux w/m^2

T = Absolute temperature of the body K.

σ_o = Proportionality constant and is called the Stefan-Boltzmann constant. This is a universal constant and has a value 5.67×10^{-8} w/m^2.

The total energy radiated by a black body (Q_o) of area A (m^2) is

$$Q_o = E_o A = \sigma_o T^4 A \text{ w} \tag{3.74}$$

As the Stefan-Boltzmann constant is very small we will modify the above equation for engineering calculations.

$$Q = E_o A = \sigma_o \left[\frac{T}{100}\right]^4 A \text{ w} \quad \text{where } \sigma_o = 5.67$$

This will take care of the factor 10^{-8} appearing in the original Stefan-Boltzmann constant. We will follow this modification throughout.

Black body is a very useful and ideal concept. Real bodies absorb some of the incident energy. This absorption is selective, i.e., restricted to some particular wavelength regions or bands. These bands are characteristic of the material. Thus the energy radiated by real bodies is less than E_o in Equation (3.73).

To overcome the problem of selective absorption of real bodies the concept of "gray body" is evolved. A gray body radiates and absorbs energy all over the spectrum like a black body but the intensity E is less than E_o. Gray bodies thus have an absorptivity independent of wavelength.

The ratio of radiant energy of a gray body (E) and that of a black body (E_o) is called the emissivity ϵ.

$$\epsilon = \frac{E}{E_o} \tag{3.75}$$

From Equation (3.73)

$$E = \sigma_o \, \epsilon \, T^4 \ \ \text{w/m}^2 \tag{3.76}$$

Emissivity has a value less than 1.0 and depends on the material, surface preparation, and temperature. Values of emissivities of many materials are available in reference books. Actual value displayed in practice may be different. In this book we will assume all bodies as gray.

Later in this chapter we will come across gases that take part in radiative transfer. For these, the reflectivity is zero and $A + T = 1$.

We have seen that all bodies (black, gray, or otherwise) absorb radiation and radiate or emit it. The ratio of absorbed radiation power A and emitted radiation power E for a black body is 1 as by definition it absorbs all the incident power and emits it back.

For other bodies, the radiated power E is less than the absorbed power A. Kirchoffs law states that the ratio of absorbed power to radiated power is the same for all bodies. It depends on the temperature and is equal to the power radiated by a black body at that temperature.

$$\frac{E_T}{A_T} = E_o = \sigma_o T^4 \tag{3.77}$$

or $$E_T = \sigma_o A_T T^4 \tag{3.78}$$

where

E_T = power emitted at temperature T by the body
A_T = power absorbed at temperature T

Comparing Eqation (3.76) and equation (3.78) shows that absorptivity A is numerically equal to emissivity, i.e., $A = \epsilon$.

3.9 RADIATION EXCHANGE BETWEEN BODIES

In the previous sections we have reviewed the radiating properties and nature of thermal radiation. In practice we seldom come across single radiating bodies. There are (at least) two bodies (or their surfaces) that form a radiating pair exchanging radiation or heat.

If the two bodies are black bodies with temperature T_1 and T_2 (°K), where $T_1 > T_2$, the net heat exchanged between them is

$$Q = Q_1 - Q_2$$

The above statement assumes that all the heat radiated by one body is received by the other and vice versa. This is a tall order. Such a condition is possible only in two cases—two radiating parallel plates and a body totally enveloped by a large body (Figure 3.19 A). In both cases all the nonradiating surfaces must be perfectly insulated.

In all other cases of radiation exchange only a part of radiation from one body will incident on the other. The actual surfaces taking part in exchange will depend on the geometric arrangement of the two bodies.

This gives rise to the concept of "view factor" which is discussed later.

3.9.1 Radiative Exchange between Two Parallel Surfaces

Figure 3.19(B) shows two parallel plates 1 and 2 radiating toward each other. The plates are of equal size and area F(m^2) and are separated by a small distance d (m).

A_1, R_1, ϵ_1 are the absorptivity, reflectivity, and emissivity of plate 1, and A_2, R_2, ϵ_2 that of the second plate. T_1 and T_2 are the steady state temperatures and $T_1 > T_2$.

Consider a ray starting from 1 having a power E_1 (w/m^2). This ray will strike 2 where part of the radiation will be absorbed and part will be reflected. The radiation exchange will be as follows:

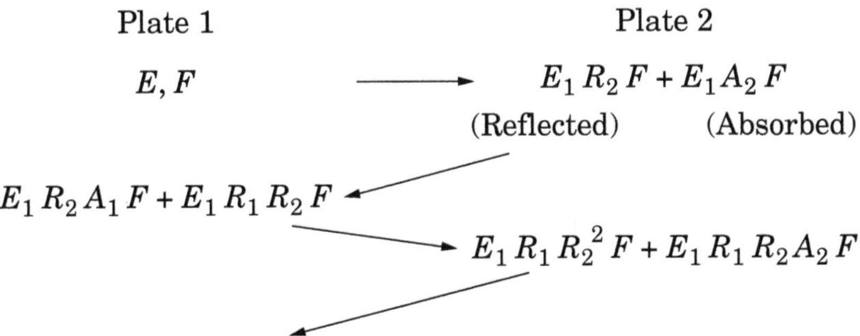

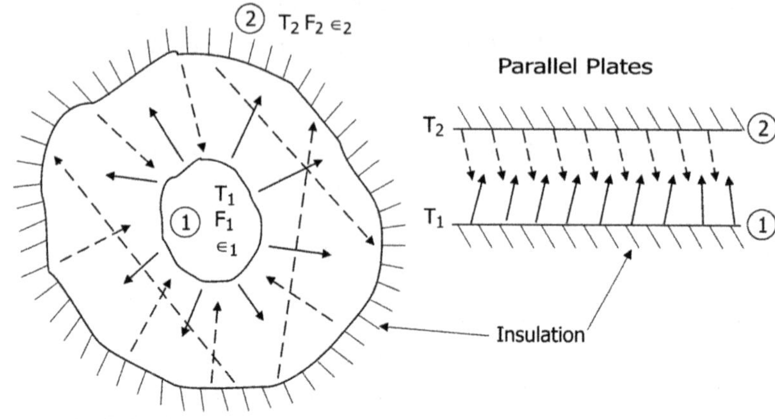

A. Total Radiation Exchange

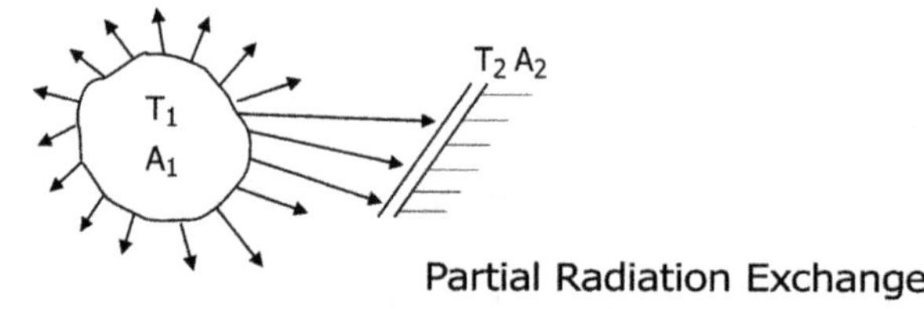

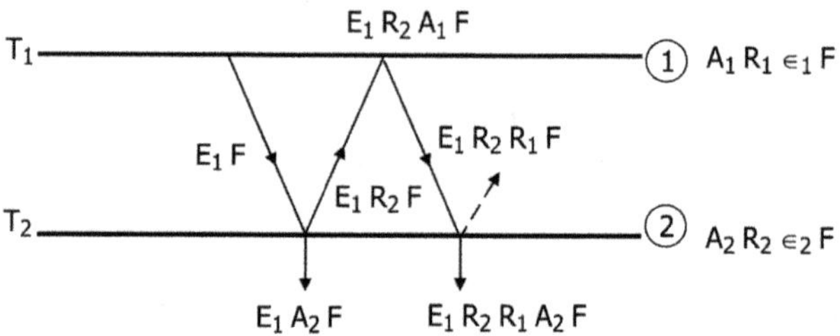

B. Radiation Exchange between Parallel Plates

Figure 3.19 Radiation exchange between parallel plates.

There will be infinite exchanges. The total energy absorbed by plate 2 will be

$$Q_1 = E_1A_2F[1 + R_1R_2 + (R_1R_2)2 + (R_1\,R_2)3 + \ldots]$$

$$= E_1A_2F\sum_{n=0}^{\infty}(R_1R_2)^n \tag{3.79}$$

Similarly, the total energy absorbed by plate 1 will be

$$Q_2 = E_2 A_1 F \sum_{n=0}^{\infty} (R_1 R_2)^n$$

The net radiation exchange will be

$$Q = Q_1 - Q_2 = (E_1 A_2 F - E_2 A_1 F)(R_1 R_2)^n \tag{3.80}$$

For an opaque material we know that

$A + R = 1$

Hence $R_1 = (1 - A_1)$ and $R_2 = (1 - A_2)$

Also, the summation in brackets approaches or converges to $1/(1 - R_1 R_2)$, i.e.,

$$(R_1 R_2)^n = 1/[1 - (1 - A_1)(1 - A_2)]$$

From Equation (3.76)

$E_1 = \sigma_o \epsilon_1 T_1^{\,4}$ and $E_2 = \sigma_o \epsilon_2 T_2^{\,4}$

Also from Equation (3.78) $A_1 = \epsilon_1$ and $A_2 = \epsilon_2$.
Making these substitutions in Equation (3.80) we get

$$\underset{1\to2}{Q} = \sigma_o \frac{1}{\frac{1}{\epsilon_1} + \frac{1}{\epsilon_2} - 1}\left(T_1^4 - T_1^4\right) \tag{3.81}$$

If we write

$$\epsilon_{\text{cor}} = \frac{1}{\frac{1}{\epsilon_1} + \frac{1}{\epsilon_2} - 1} \tag{3.82}$$

the preceding equation becomes

$$\underset{1\to2}{Q} = \sigma_o \epsilon_{\text{cor}}\left(T_1^4 - T_2^4\right) \tag{3.83}$$

Comparing Equation (3.83) with that for a single gray, radiating body, i.e.,

$$Q = \sigma_o \, \epsilon \, T^4 F \tag{3.84}$$

We see that they are similar except that the emissivity ϵ in Equation (3.84) is replaced by ϵ_{cor} in Equation (3.83). Hence, the modified emissivity is called "corrected" emissivity.

The distance d between the plates is assumed to be small as compared to the width. If this condition is not satisfied the expression for ϵ_{cor} becomes quite complicated.

Similarly, area(s) of the two plates do not appear in exception for ϵ_{cor}. This is because both plates completely view or see each other.

3.9.2 Radiative Exchange between Article and Enclosure

In furnace design we commonly come across a situation where one radiating surface completely encloses the other. For example the object kept in a furnace (enclosure) is completely surrounded by the walls (Figure 3.19(A)).

In this case the radiation energy exchanged is given by

$$Q = \sigma_o \frac{1}{\frac{1}{\epsilon_1} + \left[\frac{1}{\epsilon_2} - 1\right]\frac{F_1}{F_2}} \left(T_1^4 - T_2^4\right) F_1 \text{ w} \tag{3.85}$$

so that

$$\epsilon_{cor} = \frac{1}{\frac{1}{\epsilon_1} + \left[\frac{1}{\epsilon_2} - 1\right]\frac{F_1}{F_2}} \tag{3.86}$$

In the above equation we see that the areas F_1 and F_2 of two bodies appear in ϵ_{cor}. This is because the article radiation is fully received by the enclosure, but all the radiation from the enclosure is not received by the article. Some of the wall radiation falls on the walls, as shown in Figure 3.19. This gives rise to the definition of "view factor." The fraction of the total radiation from one body that is received by the second body is the view factor ϕ.

In the above case of an article within an enclosure

$$\phi_{1\to 2} = 1 \quad \text{but} \quad \phi = F_1/F_2$$

In the case of the two parallel surfaces discussed earlier

$$\underset{1\to 2}{\phi} = \underset{2\to 1}{\phi} = 1$$

View factor depends on the geometry and placement of the radiating bodies. The formulae for calculating view factors for many situations are available in handbooks.

In many cases the calculation of view factor becomes impossible and we have to resort to approximation. In this book we will generally require the use of only the two cases reviewed above.

Note: In the case of article-enclosure situation, it is not necessary for both to be circular. What is required is that the article be of convex shape and the enclosure concave. Similarly, the article need not be concentric with the enclosure. It can be anywhere in the enclosure.

Again consider Equation (3.86) for ϵ_{cor}. If the article is very large, occupying most of the cavity, $F_1/F_2 \to 1$ and the expression reduces to ϵ_{cor} for parallel surfaces (Equation (3.82)).

If the article is very small, $F_1/F_2 \to 0$ because most of the enclosure radiates to itself. ϵ_{cor} then becomes ϵ_1.

3.10 RADIATION SCREENS

Radiative heat exchange between two bodies can be appreciably reduced or modified by placing a screen between the two (Figure 3.20(A)).

The screen is made from a thin sheet of a suitable metal. It is highly reflective, opaque to the radiation, and has a low thermal mass. For high reflectivity its surface is polished.

Consider a screen placed between two parallel plates 1 and 2 at temperatures T_1 and T_2, where $T_1 > T_2$. The screen temperature is T_{Sc}.

The screen will receive, reflect, and absorb radiation coming from plates 1 and 2 so that

$$T_1 > T_{Sc} > T_2$$

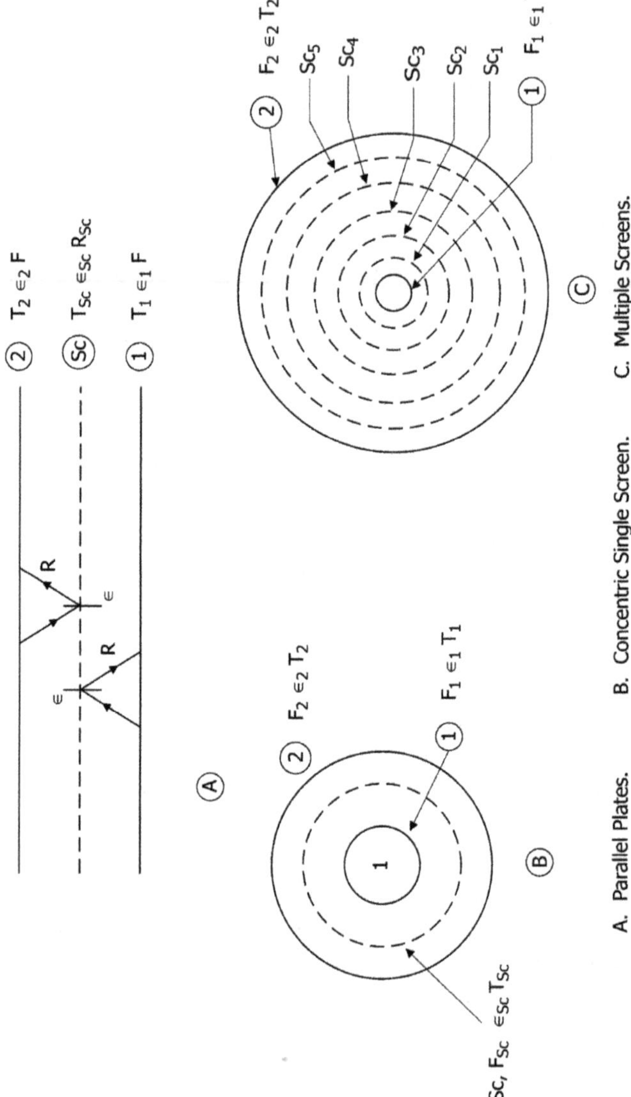

Figure 3.20 Radiation with screens.

Plate 1 and the screen will form one pair of radiating bodies. The screen and plate 2 will form the other pair. Applying Equation (3.86) to the two pairs

$$q_{1\to \mathrm{Sc}} = \sigma_o \underset{1\to \mathrm{Sc}}{\epsilon_{\mathrm{cor}}} \left((T_1)^4 - (T_{\mathrm{Sc}})^4\right) \mathrm{w/m^2} \tag{3.87}$$

and

$$q_{\mathrm{Sc}\to 2} = \sigma_o \underset{1\to \mathrm{Sc}}{\epsilon_{\mathrm{cor}}} \left((T_1)^4 - (T_{\mathrm{Sc}})^4\right) \mathrm{w/m^2} \tag{3.88}$$

where ϵ_{cor} is the corrected emissivity and ϵ_1, ϵ_{Sc}, and ϵ_2 are the emissivities of plate 1, screen, and plate 2, respectively. As the plates are of equal size the view factor $F_1/F_2 = 1$.

Hence according to Equation (3.86)

$$\underset{1\to \mathrm{Sc}}{\epsilon_{\mathrm{cor}}} = \frac{1}{\frac{1}{\epsilon_1} + \left[\frac{1}{\epsilon_{\mathrm{Sc}}} - 1\right]} \tag{3.89}$$

$$\underset{\mathrm{Sc}\to 2}{\epsilon_{\mathrm{cor}}} = \frac{1}{\frac{1}{\epsilon_{\mathrm{Sc}}} + \left[\frac{1}{\epsilon_2} - 1\right]} \tag{3.90}$$

$$\underset{1\to 2}{\epsilon_{\mathrm{cor}}} = \frac{1}{\frac{1}{\epsilon_1} + \left[\frac{1}{\epsilon_2} - 1\right]} \tag{3.91}$$

Under steady state conditions

$$q_{1\to \mathrm{Sc}} = q_{\mathrm{Sc}\to 2}$$

We can use Equation (3.87) and Equation (3.88) to determine the unknown temperature T_{Sc} and the radiation flux $q_{(1\to 2)\mathrm{Sc}}$.

If the two radiating bodies form an enclosure-object pair with a screen placed between them (Figure 3.20(B)) similar equations can be developed in which the view factor will now appear in ϵ_{cor} calculation. Corrected emissivities are

$$\underset{1\to\mathrm{Sc}}{\epsilon_{cor}} = \frac{1}{\frac{1}{\epsilon_1} + \frac{F_1}{F_{Sc}}\left[\frac{1}{\epsilon_{Sc}} - 1\right]} \tag{3.92}$$

$$\underset{\mathrm{Sc}\to 2}{\epsilon_{cor}} = \frac{1}{\frac{1}{\epsilon_1} + \frac{F_{Sc}}{F_2}\left[\frac{1}{\epsilon_2} - 1\right]} \tag{3.93}$$

$$\underset{(1\to 2)\mathrm{Sc}}{\epsilon_{cor}} = \frac{1}{\frac{1}{\epsilon_1} + \frac{F_1}{F_2}\left[\frac{1}{\epsilon_2} - 1\right]} \tag{3.94}$$

where F_1, F_{Sc}, and F_2 are the surface areas. The above treatment can be applied to any number of screens between the radiating surfaces. At each screen the radiating flux is successively reduced.

Let there be n number of screens between 1 and 2 with absorptivities (emissivities) $\epsilon_{Sc1}, \epsilon_{Sc2}, \ldots, \epsilon_{Sc}$, etc. The radiation from 1 will be partially reflected and absorbed by Sc_1 which will radiate toward Sc_2, etc. (Figure 3.20(C)).

If 1, 2, and screens are concentric

$$q_{(1\to 2)\mathrm{Sc}} = \sigma_o \underset{(1\to 2)\mathrm{Sc}}{\varepsilon_{cor}} \left((T_1)^4 - (T_2)^4\right) \mathrm{w/m^2} \tag{3.95}$$

where

$$\underset{(1\to 2)\mathrm{Sc}}{\epsilon_{cor}} = \frac{1}{\frac{1}{\epsilon_1} + \sum_{n=1}^{n}\left[\frac{2}{\mathrm{Sc}i} - 1\right]\frac{F}{\mathrm{Sc}i}} \tag{3.96}$$

The total flux Q(w) will be

$$Q = \sigma_o \underset{(1\to 2)\mathrm{Sc}}{\epsilon_{cor}} \left((T_1)^4 - (T_2)^4\right)F_1 \tag{3.97}$$

Under steady state conditions

$$q_{1\to Sc1} = q_{Sc1\to Sc2} = \cdots = q_{Sc(n-1)\to Scn} = q_{1\to 2}$$

If all the screens are of equal area

$$F_1 = F_{Sc1} = F_{Sc2} = \ldots F_{Scn} = F_2 \text{ and}$$

$$\epsilon_{cor(1\to 2)Sc} = \frac{1}{\frac{1}{\epsilon_{1,2}} + \sum_{n=1}^{n}\left[\frac{2}{\epsilon_{Sci}} - 1\right]}$$

Additionally if all emissivities are equal

$$\epsilon_1 = \epsilon_1 = \epsilon_{Sc1} = \epsilon_{Sc2} = \cdots \epsilon_{Scn} = \epsilon_2 \text{ and}$$

$$\epsilon_{cor} = \frac{1}{\frac{1}{\epsilon} + n\left[\frac{2}{\epsilon} - 1\right]}$$

Addition of one screen will reduce the radiation from $1 \to 2$ by about 50%. Successive additional screens will further reduce the flux progressively. This will be clear from Example 3.21 through Example 3.23. For economic and constructional reasons the maximum number of screens does not exceed 4–5.

Radiation screens are very useful in vacuum furnaces as ordinary insulating materials cannot be used there. The material of screens must be capable of withstanding the expected temperature. Generally materials used are tungstan, molybdenum, and tantalum for high temperatures (> 1200°C) and stainless steels for lower temperatures.

3.11 RADIATION EXCHANGE INSIDE AND OUTSIDE FURNACES

In furnace design we frequently come across a situation in which a charge is on the furnace bottom and the furnace is heated by one of the following modes (Figure 3.21(A)):

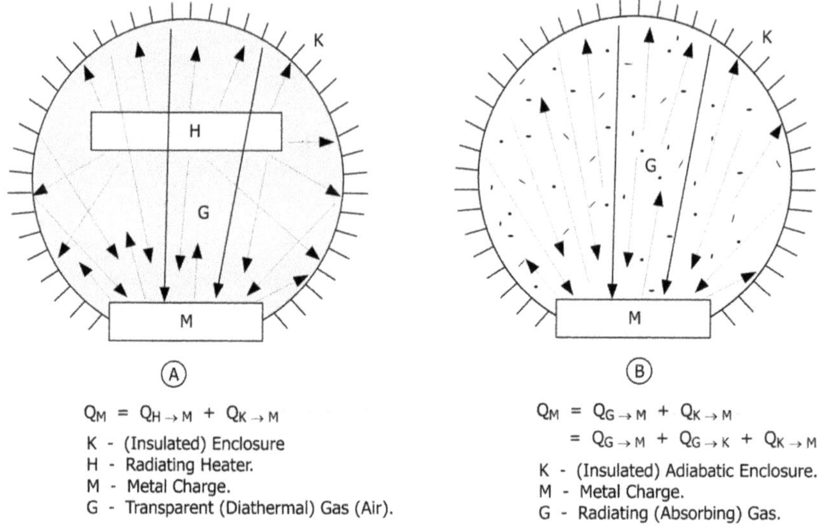

Figure 3.21 Radiation exchange inside furnace.

1. Heating is indirect so that either the (electric) heater or the flame is situated outside the wall. The wall is heated and radiates toward the charge.

   The enclosure space between the wall and the charge may be occupied by a transparent (diathermal) gas such as air or hydrogen.

   Alternatively the space may be occupied by a protective gas, i.e., a radiation-absorbing gas containing CO_2, H_2O, etc.

2. The furnace may be heated by burning a fuel in the enclosure so that the intermediate space is filled by combustion gases containing radiation-absorbing (attenuating) components such as CO_2, H_2O, SO_2, etc.

   The heat transmission to the charge is now by two paths.

   Gas → Charge

   Gas → Wall → Charge

   Both paths will be affected whenever radiation passes through gas.

3. The heater is in the enclosure but it is passive so that the enclosure is filled with air which is transparent.

   This mode may employ an electric heater or a radiating tube.

   The heat transfer to the charge will be by two paths.

   Heater → Charge

   Heater → Wall → Charge

In the first mode where the enclosure contains transparent gas the radiant flux to the charge can be calculated as

$$q = \sigma_o \epsilon_{cor}\left[(T_k)^4 - (T_m)^4\right] \text{w/m}^2 \tag{3.98}$$

where T_k and T_m are the wall and charge temperatures (K), and

ϵ_{cor} = the corrected emissivity

$$\left.\begin{aligned} &= \frac{\epsilon_k \epsilon_m}{\epsilon_k + \phi_{km}\epsilon_m(1-\epsilon_k)} \\ &= \frac{\epsilon_k \epsilon_m}{1-(1-\epsilon_k)(1-\phi_{km}\epsilon_m)} \end{aligned}\right\} \tag{3.99}$$

where ϵ_k and ϵ_m are wall and charge emissivities, and

ϕ_{km} = the view factor = F_m/F_k

When the radiation path is through an absorbing gas the flux is attenuated by absorption. Many parameter such as emissivities of the wall, gas, and charge, their mutual view factors, temperatures, etc., come into play, thus making it virtually impossible to develop an expression for the calculation of the applicable overall emissivity ϵ_{Cor}. One Of the expressions available is given below. It assumes that the charge is radiated from the wall only, and the radiation passes through an absorbing gas (Figure 3.21(B)).

$$\epsilon_{gkm} = \frac{\epsilon_g \epsilon_m [1+\phi_{km}(1-\epsilon_g)]}{\epsilon_g + \phi_{km}(1-\epsilon_g)[1-(1-\epsilon_g)(1-\epsilon_m)]} \tag{3.100}$$

where ϵ_{gkm} is the corrected emissivity for the assumed radiation path gas → wall → metal. ϵ_g, ϵ_m are gas and work emissivities, respectively, and ϕ_{km} is the view factor (wall → work) = Fk/Fm. The work is convex so that $\phi_{mm} = 0$ and $\phi_{km} = 1$.

The flux on the work will be

$$q_m = \sigma_o \epsilon_{gkm}\left[(T_g)^4 - (T_m)^4\right] \text{w/m}^2 \tag{3.101}$$

The wall temperature T_k which is unknown can be calculated by the expression given below.

$$T_k^4 = T_m^4 + (T_k - T_m)^4 \frac{1}{1 + \frac{\phi_{km}\epsilon_m(1-\epsilon_g)}{\epsilon_g[1+\phi_{km}(1-\epsilon_g)(1-\epsilon_m)]}} \tag{3.102}$$

3.12 RADIATION IN ABSORBING MEDIA

The discussion on radiant heating until now assumed that the medium between the two surfaces is completely transparent to the whole radiation spectrum.

In practice we come across many situations where the medium participates and modifies the radiation passing through it. Some examples are:

1. When a fuel is burnt (combusted) in a furnace the volume of the furnace is occupied by the combustion products, which contain CO_2 and H_2O besides other gases. CO_2 and H_2O have selective absorption or emissivity and modify the heat transmission.
2. In some furnaces we introduce special protective gases to protect the charge from oxidation. These special atmospheres also contain CO_2 and H_2O.
3. In some processes like baking bread, cakes, etc., CO_2 and H_2O evolve from the dough. Here again the heat transfer is affected.

In general, monoatomic and biatomic gases like CO_2, O_2, N_2, H_2, Ar, and He are transparent to radiation, hence, ordinary air can be considered reasonably transparent.

Tri- and polyatomic gases such as CO_2, H_2O, SO_2, and hydrocarbons have selective absorption. This arises because of electron energy states in their molecules.

In effect, the emissivity of gas mixtures containing tri-atomic gases is considerably different from that of transparent gases (unity). The following procedure is adopted for calculating the emissivity of participating gas mixtures ϵ_g.

1. The emissivity depends on the amount of absorbing gas in the mixture. Hence, from the given composition, the partial pressure of the absorbing component p is determined in kPa or atmosphere (100 kPa) units.

   $$p = 100 \times \% \text{ absorbing component kPa} \qquad (3.103)$$

2. The absorption also depends on the thickness of the gas layer. The thicker the beam, the more the absorption. The thickness of the beam depends on the geometry of the enclosure (furnace). The standardized measurement of beam length is called the effective or mean beam length (L_e). For rectangular space (muffle)

   $$L_e = 3.6\frac{V}{A}\text{ m} \qquad (3.104)$$

   where V is the volume and A the area of the enclosure.

   Tables for values of L_e for different geometries are available in the literature.
3. After the determination of p and L_e their product pL_e is determined. This product has dimensions m·atm, or m·kPa, or m·kN/m^2.
4. Emissivity of a gas also depends on its temperature. This dependancy is not linear and the mathematical relationship is quite complicated. For common tri-atomic gases graphs have been prepared for emissivity against temperature at different pL_e values.
5. Referring to Figure 3.22 and Figure 3.23 the emissivity of the required gas (CO_2, H_2O, etc.) is determined at the given temperature and pL_e is calculated.

6. If more than one absorbing gas (e.g., CO_2 and H_2O) is present in the gas, their individual emissivities are added to obtain total emissivities. However, in most instances the absorption bands of these gases overlap. Also the bands are restricted to certain wave bands. To correct these two factors, the gases are assumed to be "Gray" and a correction factor (β) is applied.
7. Graphs have been presented for correction factors giving the factor for the determined partial pressure and the product pL_e. Graphs for CO_2, H_2O, and correction factor for H_2O are shown in the Figure 3.22, Figure 3.23, and Figure 3.24.
8. Thus, in a furnace gas if CO_2 and H_2O are both present, the emissivity of gas ϵ_g is given by

$$\epsilon_g = p_{CO_2}L_e + \beta p_{H_2O}L_e \tag{3.105}$$

If only one absorbing gas is present (e.g., H_2O)

$$\epsilon_g = p_{H_2O}L_e$$

In furnace calculation the factor β and other corrections are often neglected as they are negligible.

9. The radiation emitted by the gas will be

$$E_q = \epsilon_g \times 5.67\left\{\frac{T_g}{100}\right\}^4 \text{ w/m}^2 \tag{3.106}$$

where T_g is the absolute temperature of the gas.

10. Another method of determining ϵ_g when both CO_2 and H_2O are present is by using graphs as in Figure 3.22, Figure 3.23, and Figure 3.24. Here a correction Δe is determined from the ratio $p_{H_2O}/(p_{H_2O} + p_{CO_2})$ where the denominator is the sum of individual emissivities. Thus

$$\epsilon_g = p_{CO_2}L_e + p_{CO_2}L_e - \Delta e$$

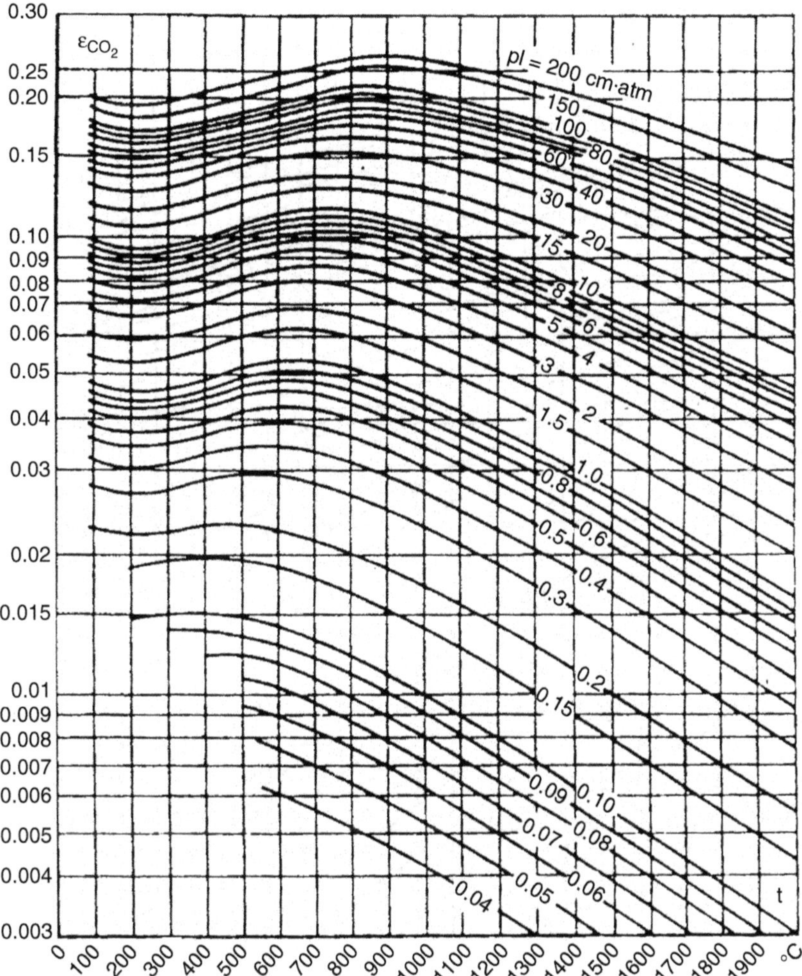

Figure 3.22 Emissivity of CO_2 as a function of P and ℓ.

11. If the gas is at temperature T_g (abs) and the enclosing surface is at T_s (abs) the net radiation exchange will be

$$Q = 5.67 \times A\left\{\epsilon_g\left[\frac{T_q}{100}\right]^4 - \epsilon_s\left[\frac{T_s}{100}\right]^4\right\} \text{ w} \tag{3.107}$$

where ϵ_s is the absorptivity (emissivity) and A the area of enclosure.

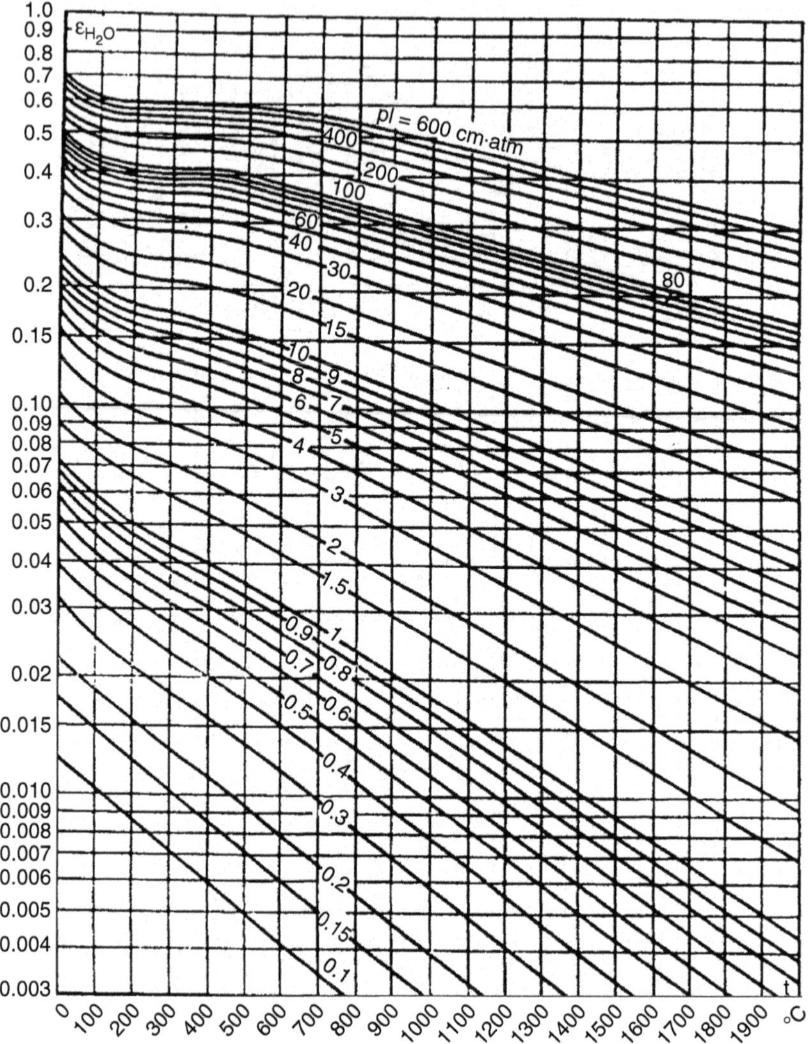

Figure 3.23 Emissivity of water vapor as a function of P and ℓ.

Heat exchange between gray bodies in a closed space filled by absorbing gas

A situation that arises frequently in furnace design is as follows:

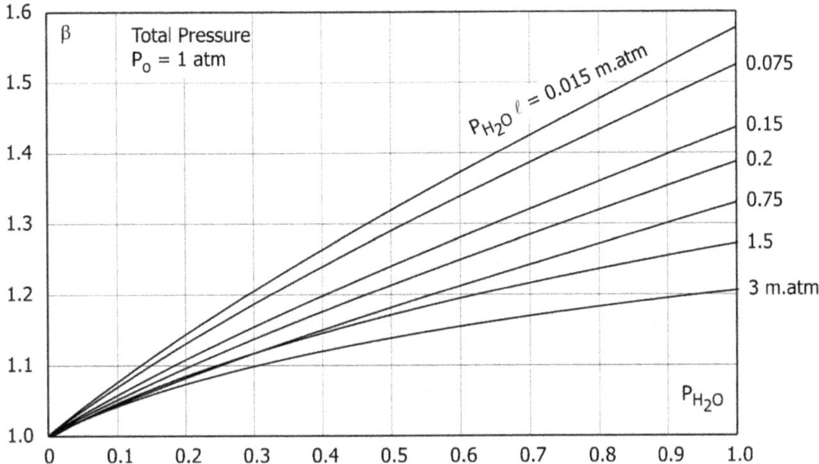

Figure 3.24 Correction factor β for partial pressure of water vapor.

The furnace enclosure (muffle) is filled with hot absorbing gases at temperature t_g.

A charge (metal) is on the hearth and is at a known temperature t_m. The temperature of the walls t_ℓ is known. Then, what is the heat flux that strikes the charge?

If we analyze the heat flow, some heat from the gases strikes the walls. The walls then emit heat to the charge through the gas. In this transmission the flux is attennuated due to selective absorbitivity of the gas.

Some heat is radiated directly from the gas to the charge.

The enclosure is concave and radiates some heat to itself.

The mathematical solution of this problem depends on calculating the applicable value of the gas → wall → charge and gas charge emissivity $\epsilon_{g\ell m}$. Without going into the details of the solution we quote the formula for $\epsilon_{g\ell m}$ applicable in this situation.

$$\epsilon_{g\ell m} = \epsilon_g \epsilon_\ell \frac{\phi_{21}(1-\epsilon_g)+1}{\phi_{21}(1-\epsilon_g)[\epsilon_\ell + \epsilon_g(1-\epsilon_\ell)] + \epsilon_g} \tag{3.108}$$

where

ϵ_g = emissivity of gas (as calculated by the method described in the previous section)

ϵ_ℓ = emissivity of the lining (walls)

ϕ_{21} = the view factor between the walls and the metal = F_m/F_ℓ

The heat flux is then

$$Q_{g\ell m} = \sigma_o \, \epsilon_{g\ell m} \, F_1 \left(T_g^4 - T_\ell^4\right)$$

The above formula assumes that the walls are adiabatic, i.e., they do not lose any incident heat.

The example that follows will clarify the application of the above method.

3.13 RADIATION LOSS FROM FURNACE OPENINGS

A furnace or hot enclosure may have openings such as windows or doors. There may be unintentional openings such as imperfectly fitting doors, burner tiles, etc. All these openings radiate considerable heat to the surroundings and represent a loss.

If the walls with such openings are relatively thin (~ few mm) the loss can be calculated by the usual formula

$$Q = \sigma_o \, (T_F)^4 A \quad \text{w} \tag{3.109}$$

where A is the opening area (m^2) and T_F is the furnace temperature (K).

Furnaces usually have thick walls and the radiation emanating from the inner opening are absorbed or reflected by the lining width (Figure 3.25). In such cases, the radiation loss is determined from the size and depth of the opening with the help of Figure 3.25. A coefficient ϕ is determined and the loss is given by

$$Q = \sigma_o (T_F)^4 \, \phi A \quad \text{w} \tag{3.110}$$

EXAMPLE 3.18

A steel slab measuring 200 × 300 × 20 mm is kept in a furnace having a 300 × 350 mm bottom and 120 mm height. Emissivity of steel is 0.8 (Figure 3.26).

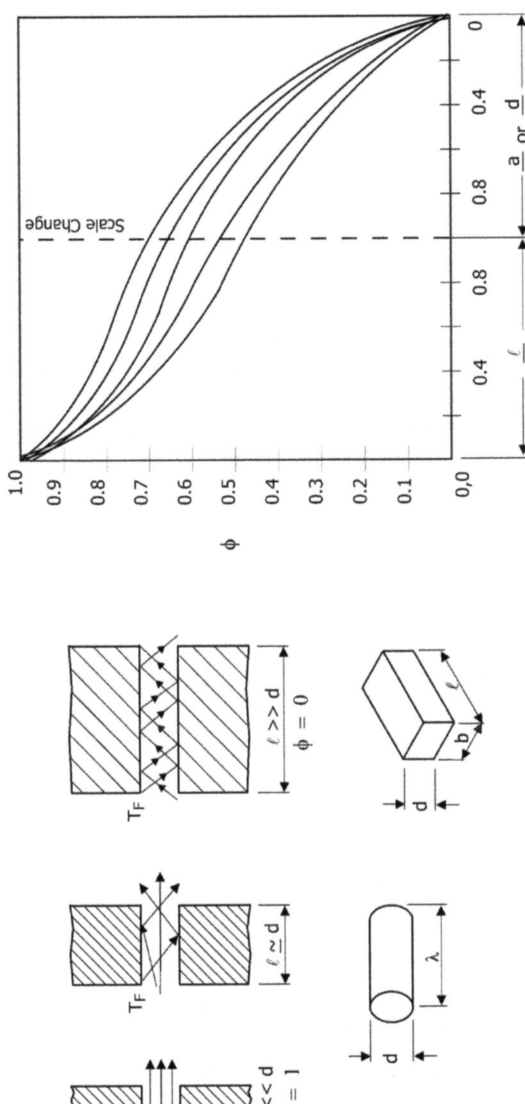

Figure 3.25 Radiation through wall openings d.

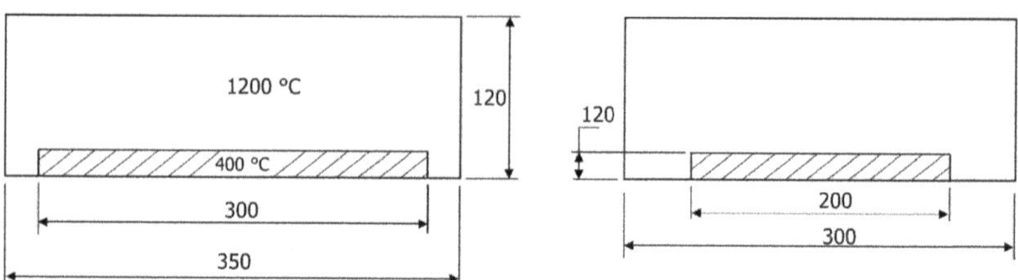

Figure 3.26 [Example 3.18].

The furnace contains combustion gases at 1200°C. The gases contain 15% CO_2 and 8% H_2O.

The walls of the furnace are at 1100°C and the combustion gases can be considered gray.

If the slab is 400°C determine the radiant flux density it receives from the gases and the walls.

Solution 1

Area of slab resting on furnace hearth is $0.3 \times 0.2 = 0.06$ m^2

Area of furnace bottom is $0.35 \times 0.3 = 0.105$ m^2

Hearth area not covered by slab is $0.105 - 0.06 = 0.045$ m^2

$$\text{The slab occupies } \frac{0.06}{0.105} \times 100 = 57\% \text{ hearth area}$$

The volume of furnace excluding that occupied by the slab is

$$(0.35 \times 0.3 \times 0.12) - (0.3 \times 0.2 \times 0.02)\ V = 0.0114 \text{ m}^3$$

Surface area of walls (excluding area occupied by the charge) is

Top	$0.35 \times 0.30 \times 0.3 = 0.105$
Hearth	$0.35 \times 0.30 - 0.06 = 0.045$
Sides	$2.00 \times 0.30 \times 0.12 = 0.072$
Sides	$2.00 \times 0.35 \times 0.12 = 0.084$
	Total wall area F_ℓ=0.306 m^2

$$\text{Mean beam length} = 3.6\frac{V}{F} = 3.6\frac{0.0114}{0.306} = 0.134 \text{ m} = Le$$

Assuming atmospheric pressure = 100 kPa

$$pCO_2 = 0.15 \times 100 = 15 \text{ kPa}$$

$$Le \times pCO_2 = 0.134 \times 15 = 2.01 \text{ mkPa}$$

$$pH_2 = 0.08 \times 100 = 8 \text{ kPa}$$

$$Le \times pH_2 = 0.134 \times 8 = 1.072 \text{ mkPa}$$

From Figure 3.22 and Figure 3.23, for wall temperature 1100°C

$$\epsilon_{CO_2} = 0.062$$

$$\epsilon_{H_2O} = 0.042$$

Correction factor $\beta = 1.08$
Hence, the emissivity of the gas is

$$\epsilon_g = 0.062 + 1.08 \times 0.042 = 0.1074$$

The angle factor between the slab and the walls is

$$\phi = \frac{\text{Surface area (exposed) of slab}}{\text{Surface area of walls}} = \frac{F_m}{F_\ell}$$

$$= \frac{0.3 \times 0.2 + 2 \times 0.02 \times 0.3 + 2 \times 0.02 \times 0.2}{0.306}$$

$$= 0.26$$

Corrected emissivity for slab/wall heat exchange is

$$\epsilon_{cor} = \epsilon_g \epsilon_m \frac{\phi(1-\epsilon_g)+1}{\phi[(1-\epsilon_g)(\epsilon_m + \epsilon_g(1-\epsilon_m))] + \epsilon_g}$$

$$= 0.104 \times 0.8 \frac{0.26(1-0.104)+1}{0.26[(1-0.104)(0.8+0.104(1-0.8))] + 0.104}$$

$$= 0.347$$

The flux density radiated by the gas and the wall to the slab is

$$q = \epsilon_{cor}\sigma_o \left((T_g)^4 - (T_m)\right)^4$$

$$T_g = 1{,}200 + 273 = 1{,}473 \text{ K} \qquad T_m = 400 + 273 = 673 \text{ K}$$

$$= 0.347 \times 5.7\left\{\left[\frac{1473}{100}\right]^4 - \left[\frac{673}{100}\right]^4\right\}$$

$$= 89{,}151 \text{ w/m}^2$$

EXAMPLE 3.19

Combustion gases in a furnace contain 13% CO_2 and 11% H_2O. The total pressure of the gases is 98.1 kPa (1 atm) and the gas temperature is 1350°C.

Furnace volume is 18 m^3 and the surface area is 58 m^2.

Calculate the emissivity of the gas mixture and the radiant flux emitted by it.

Solution

For convenience we can assume 1 atm = 100 kPa without introducing significant error. Then

$$P_{CO_2} = 100 \times 0.13 = 13 \text{ kPa and}$$

$$P_{H_2O} = 100 \times 0.11 = 11 \text{ kPa}$$

$$\text{Mean beam length} = 3.6 \times \frac{V}{F} = Le$$

$$= 3.6 \times \frac{18}{58} = 1.12 \text{ m}$$

$$P_{CO_2} Le = 13 \times 1.12 = 14.56 \text{ m} \cdot \text{kPa and}$$

$$P_{H_2O} Le = 11 \times 1.12 = 12.32 \text{ m} \cdot \text{kPa}$$

From Figure 3.22 and Figure 3.23 at 1350°C

$$\epsilon_{CO_2} = 0.07$$

$$\epsilon_{H_2O} = 0.08$$

From graph (3.21) the correction factor for H_2O is $\beta = 1.1$
Hence the total emissivity ϵ_g is

$$\epsilon_g = \epsilon_{CO_2} + \beta \times \epsilon_{H_2O}$$

$$= 0.07 + 1.1 \times 0.08$$

$$= 0.158$$

The radiation flux density of the gas is

$$q = 5.7 \times 0.158 \left[\frac{1{,}350 + 273}{100}\right]^4$$

$$= 11{,}572 \text{ w/m}^2$$

EXAMPLE 3.20

An oil-fired furnace operates at 1200°C. The furnace gases contain 15% CO_2 and 13% H_2O. The dimensions of the furnace are 1.5 × 1.5 × 1 m. The charge occupies 45% of hearth area.

If the charge is at 400°C determine the radiant flux density it receives from the gases and the walls.

The height of the charge above the hearth is 250 mm.
Emissivity of the charge and the walls is 0.8.

Solution

Volume (V) of the furnace except that occupied by the charge is

$$1.5 \times 1.5 \times (1 - 0.25) = 1.69 \text{ m}^3$$

Surface area (F) participating in radiative exchange

$$\begin{aligned} 1.5 \times 1.5 &= 2.25 \text{ (top)} \\ 2 \times 1.5 \times 1 &= 3.00 \text{ (sides)} \\ 2 \times 1.5 \times 1 &= 3.00 \text{ (sides)} \\ 0.55(1.5 \times 1.5) &= 1.24 \text{ (hearth)} \\ \text{Total} &= 9.49 \text{ m}^2 \end{aligned}$$

Mean beam length Le

$$= 3.6\frac{V}{F} = 3.6 \times \frac{1.69}{9.49} = 0.6411\,\text{m}$$

Assuming atmospheric pressure = 100 kPa

$$P_{CO_2} = 100 \times 0.15 = 15\ \text{kPa}$$

$$P_{H_2O} = 100 \times 0.13 = 13\ \text{kPa}$$

$$Le \times P_{CO_2} = 0.6570 \times 15 = 9.855\ \text{m} \cdot \text{kPa}$$

$$Le \times P_{H_2O} = 0.6570 \times 13 = 8.341\ \text{m} \cdot \text{kPa}$$

From Figure 3.22 and Figure 3.23 at 1200°C

$$\epsilon_{CO_2} = 0.085$$

$$\epsilon_{H_2O} = 0.08$$

$$\epsilon_g = \epsilon_{CO_2} + \beta \times \epsilon_{H_2O}$$

$$= 0.085 + 1.1 \times 0.08 = 0.173$$

The angle factor (view factor) for the charge and the walls is

$$\Phi = \frac{\text{Surface area (exposed) of the charge}}{\text{Surface area of the walls}}$$

$$= \frac{F_c}{F_w} = \frac{1.01}{9.49} = 0.106$$

By Equation (3.108) the corrected emissivity for the gas and the lining to metal (charge) is

$$\epsilon_{\text{cor } g,\ell \to m} = \epsilon_g \epsilon_m \frac{\phi(1-\epsilon_g)+1}{\phi(1-\epsilon_g)(\epsilon_\ell + \epsilon_g(1-\epsilon_\ell)) + \epsilon_g}$$

$$= 0.173 \times 0.8 \frac{0.106(1-0.173)+1}{0.106(1-0.173)(0.8+0.173(1-0.8))+0.173} \quad \mathrm{T}$$

$$= 0.612$$

he flux density of radiation from the gas and the wall

$$q = \sigma_o \epsilon_{cor} \left\{ \left[\frac{T_1}{100} \right]^4 - \left[\frac{T_2}{100} \right]^4 \right\}$$

$$= 5.7 \times 0.612 \left\{ \left[\frac{1200 + 273}{100} \right]^4 - \left[\frac{400 + 273}{100} \right]^4 \right\}$$

$$= 157{,}068 \text{ w/m}^2$$

EXAMPLE 3.21

A vacuum electric furnace has a cylindrical heater having 300 mm diameter and 450 mm height. Maximum heater temperature is 1500°C. The heater is concentric with a 12-mm thick metal vessel having a 600 mm diameter and 1000 mm height. The atmospheric temperature is 30°C. The emissivities of the heater and the vessel wall are 0.8 and 0.7, respectively.

Determine the radiation flux from the heater to the wall.

Solution

Figure 3.27(A) shows the furnace section. For simplicity we consider only the radiant flux. The corrected emissivity (ϵ_{cor}) will be

$$\epsilon_{cor} = \frac{1}{\frac{1}{\epsilon_1} + \frac{F_1}{F_2}\left[\frac{1}{\epsilon_2} - 1\right]}$$

F_1 = The heater area

$$= \pi \times 0.3 \times 0.45 = 0.42 \text{ m}^2$$

F_2 = The wall area

$$= \pi \times 0.6 \times 1.0 = 1.88 \text{ m}^2$$

Also $\epsilon_1 = 0.8$ and $\epsilon_2 = 0.7$

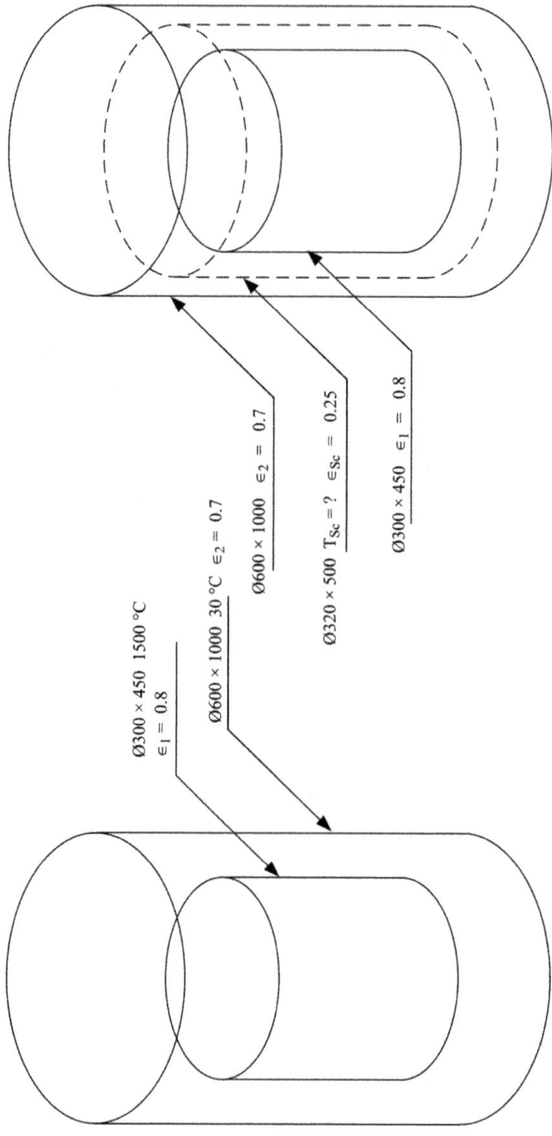

Figure 3.27 [Example 3.21 and 3.22].

substituting

$$\epsilon_{cor} = \frac{1}{\frac{1}{0.8} + \frac{0.42}{1.88}\left[\frac{1}{0.7} - 1\right]}$$

$$= 0.743$$

The heat radiated from the heater to the wall is given by

$$Q = \sigma_o \epsilon_{cor} \left\{ \left[\frac{T_1}{100}\right]^4 - \left[\frac{T_2}{100}\right]^4 \right\} \times F_1$$

$$Q = 5.67 \times 0.743 \left\{ \left[\frac{1500 + 273}{100}\right]^4 - \left[\frac{30 + 273}{100}\right]^4 \right\} \times 0.42$$

$$\frac{1.75}{0.4} = 175 \text{ kW}$$

or $= 416 \text{ kW/m}^2$

The above calculations show that a large flux is incident on the wall. This flux will raise the wall temperature to an unacceptable level. Hence, the wall will have to be cooled by providing water cooling. This cooling can be achieved by water circulation through tubes. This is a convective heat transfer problem and is discussed in the convection section of this chapter.

EXAMPLE 3.22

In the furnace in the previous example, a cylindrical radiation shield of tantalum is inserted between the heater and the wall.

The shield diameter is 320 mm and the height is 500 mm. Emissivity of the shield is 0.25. The furnace and wall temperatures are the same (1500 and 30°C, respectively).

Determine

1. The heat flux between the heater and the wall
2. The screen temperature

Solution

The arrangement of the heater, screen, and walls is shown in Figure 3.27(B).

Let the screen temperature be T_{Sc} (unknown). Under steady state conditions

$$q_{1\to 2} = q_{1\to Sc} = q_{Sc\to 2}$$

Also

F_1 = The heater area = 0.42 m^2

F_{Sc} = The screen area = $\pi \times 0.320 \times 0.5$ = 0.5 m^2

F_2 = The wall area = 1.88 m^2

Also

$$\underset{1\to 2}{\epsilon_{cor}} = \frac{1}{\frac{1}{\epsilon_1} + \frac{F_1}{F_2}\left[\frac{1}{\epsilon_2} - 1\right]} = 0.743 \text{ (previous example)}$$

$$\underset{1\to Sc}{\epsilon_{cor}} = \frac{1}{\frac{1}{\epsilon_1} + \frac{F_1}{F_{Sc}}\left[\frac{1}{\epsilon_{Sc}} - 1\right]}$$

$$= \frac{1}{\frac{1}{0.8} + \frac{0.42}{0.5}\left[\frac{1}{0.25} - 1\right]} = 0.265$$

$$\underset{Sc\to 2}{\epsilon_{cor}} = \frac{1}{\frac{1}{\epsilon_{Sc}} + \frac{F_{Sc}}{F_2}\left[\frac{1}{\epsilon_2} - 1\right]}$$

$$= \frac{1}{\frac{1}{0.25} + \frac{0.5}{1.88}\left[\frac{1}{0.7} - 1\right]} = 0.243$$

As $q_{1\to Sc} = q_{Sc\to 2}$

$$F_1 \underset{1\to Sc}{\epsilon_{cor}} \left\{\left[\frac{T_1}{100}\right]^4 - \left[\frac{T_{Sc}}{100}\right]^4\right\} = F_{Sc} \underset{Sc\to 2}{\epsilon_{cor}} \left\{\left[\frac{T_{Sc}}{100}\right]^4 - \left[\frac{T_2}{100}\right]^4\right\}$$

Substituting the above calculated values

$$0.42 \times 0.265 \left\{ \left[\frac{1500+273}{100} \right]^4 - \left[\frac{T_{Sc}}{100} \right]^4 \right\}$$

$$= 0.5 \times 0.243 \left\{ \left[\frac{T_{Sc}}{100} \right]^4 - \left[\frac{30+273}{100} \right]^4 \right\}$$

$T_{Sc} = 1480°K$ or $1207°C$

Heat flux from the heater to the screen

$$q_{1 \to Sc} = \sigma_o \, \epsilon_{\substack{cor \\ 1 \to Sc}} \left\{ \left[\frac{T_1}{100} \right]^4 - \left[\frac{T_{Sc}}{100} \right]^4 \right\} F_1$$

$$= 5.67 \times 0.265 \left\{ \left[\frac{1500+273}{100} \right]^4 - \left[\frac{1207+273}{100} \right]^4 \right\} \times 0.42$$

$$= 32.2 \text{ kW or } 76 \text{ kW/m}^2$$

Comparing this with the flux $1 \to 2$ without screen

$$\frac{32}{175} = 0.18$$

This shows that the addition of one screen has reduced the flux by 82%.

Note: That this reduction of flux is applicable only to a screen having the given emissivity and area.

To maintain the wall temperature a little above 30°C, water cooling of the wall may be required.

EXAMPLE 3.23

In the furnace in Example 3.21, five screens are used to reduce the radiant flux to the wall (Figure 3.28(A)). The dimensions and emissivities of the screens are as follows:

	Sc.1	Sc.2	Sc.3	Sc.4	Sc.5
Diameter (mm)	340	360	400	450	500
Height (mm)	480	500	540	600	650
Emissivity	0.25	0.25	0.7	0.8	0.6

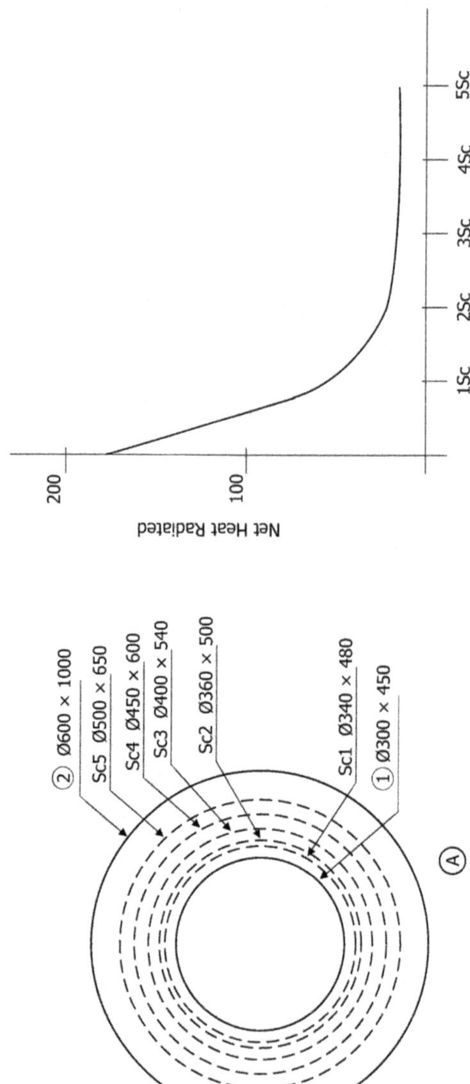

Figure 3.28 Effect of multiple screens on radiation (Example 3.23).

The dimensions and temperature of the heater and the wall are same as in example.

Determine the radiant flux on the wall after insertion of each screen.

Solution

From the above data the areas of screens and their emissivities are as given below.

	Heater	Sc.1	Sc.2	Sc.3	Sc.4	Sc.5	Wall
Area (m^2)	0.42	0.513	0.565	0.679	0.848	1.02	1.88
ϵ	0.8	0.25	0.25	0.7	0.8	0.6	0.7

The flux from the heater to the wall as calculated previously is 416 kW/m²

$$\underset{1\to 2}{\epsilon_{cor}} = \frac{1}{\frac{1}{0.8} + \frac{0.42}{1.88}\left[\frac{1}{0.7} - 1\right]} = 0.743$$

$$\underset{1\to Sc1}{\epsilon_{cor}} = \frac{1}{\frac{1}{0.8} + \frac{0.42}{0.513}\left[\frac{1}{0.25} - 1\right]} = 0.276$$

$$\underset{Sc1\to Sc2}{\epsilon_{cor}} = \frac{1}{\frac{1}{0.25} + \frac{0.513}{0.565}\left[\frac{1}{0.25} - 1\right]} = 0.149$$

$$\underset{Sc2\to Sc3}{\epsilon_{cor}} = \frac{1}{\frac{1}{0.25} + \frac{0.565}{0.679}\left[\frac{1}{0.7} - 1\right]} = 0.230$$

$$\underset{Sc3\to Sc4}{\epsilon_{cor}} = \frac{1}{\frac{1}{0.7} + \frac{0.679}{0.848}\left[\frac{1}{0.8} - 1\right]} = 0.613$$

$$\underset{Sc4\to Sc5}{\epsilon_{cor}} = \frac{1}{\frac{1}{0.8} + \frac{0.848}{1.02}\left[\frac{1}{0.6} - 1\right]} = 0.553$$

$$\underset{Sc5\to 2}{\epsilon_{cor}} = \frac{1}{\frac{1}{0.6} + \frac{1.02}{1.88}\left[\frac{1}{0.7} - 1\right]} = 0.525$$

Overall corrected emissivity

$$\underset{(1\to 2)\mathrm{Sc}}{\epsilon_{\mathrm{cor}}} = \frac{1}{\underset{1\to 2}{\epsilon_{\mathrm{cor}}} + \sum_{n=1}^{5} \underset{\mathrm{Sc}}{\epsilon_{\mathrm{cor}}}}$$

For 5 screens

$$\underset{(1\to 2)5\mathrm{Sc}}{\epsilon_{\mathrm{cor}}} = \frac{1}{\frac{1}{0.743} + \frac{1}{0.276} + \frac{1}{0.149} + \frac{1}{0.230} + \frac{1}{0.613} + \frac{1}{0.553}}$$

$$= 0.0532$$

For 4 screens

$$= \frac{1}{\frac{1}{0.743} + \frac{1}{0.276} + \frac{1}{0.149} + \frac{1}{0.230} + \frac{1}{0.613}}$$

$$= 0.057$$

Similarly
For 3 screens

$$\underset{(1\to 2)3\mathrm{Sc}}{\epsilon_{\mathrm{cor}}} = 0.062$$

For 2 screens

$$\underset{(1\to 2)2\mathrm{Sc}}{\epsilon_{\mathrm{cor}}} = 0.0857$$

For 1 screen

$$\underset{(1\to 2)1\mathrm{Sc}}{\epsilon_{\mathrm{cor}}} = 0.201$$

For no screen

$$\underset{1\to 2}{\epsilon_{\mathrm{cor}}} = 0.743$$

The net radiation in each case will be given by

$$Q_{(1\to 2)}\mathrm{Sc} = \sigma_o \underset{(1\to 2)\mathrm{Sc}}{\varepsilon_{\mathrm{cor}}} \left\{ \left[\frac{T_1}{100}\right]^4 - \left[\frac{T_2}{100}\right]^4 \right\} F$$

Here, $T_1 = 1500°C$, $T_2 = 30°C$, $F_1 = 0.42\ m^2$, $\sigma_o = 5.67\ w/m^2$, and ϵ_{cor} as calculated above.

Substituting these we get

$$Q_{(1\rightarrow 2)}\ 0sc. = 175.0\ kW$$
$$Q_{(1\rightarrow 2)}\ 1sc. = 47.0\ kW$$
$$Q_{(1\rightarrow 2)}\ 2sc. = 19.0\ kW$$
$$Q_{(1\rightarrow 2)}\ 3sc. = 14.6\ kW$$
$$Q_{(1\rightarrow 2)}\ 4sc. = 13.4\ kW$$
$$Q_{(1\rightarrow 2)}\ 5sc. = 12.5\ kW$$

These results are plotted in Figure 3.28(B). This shows that no substantial reduction in radiation occurs beyond the third screen. However, the fourth and fifth screens may be necessary for safety.

3.14 EXTENDED SURFACES

Consider the end part of a typical process furnace shown in Figure 3.29. The furnace has an inner muffle in the form of a ceramic tube. The inner muffle extends beyond the furnace to a certain length L. The part of the tube inside the furnace is at a higher temperature t_f while the part which extends beyond is exposed to the atmospheric temperature t_o.

Heat will be conducted out of the furnace by the extension and transferred to the surrounding by convection and radiation. Thus, a temperature gradient will be set up in the extension as shown.

If the extension (ℓ) is sufficiently long, the tip (at $x = \ell$) will be at atmospheric temperature and there will be no transfer of heat at the end. At the base ($x = 0$) the temperature is the highest and so the heat transfer at base will be maximum. For an extension of limited length, heat will be lost from the tip.

There are many such instances involving extended surfaces that we will come across. These extensions, when incorporated intentionally, are called "fins." They are used to dissipate the heat generated in a limited volume such as internal combustion engines and gear boxes.

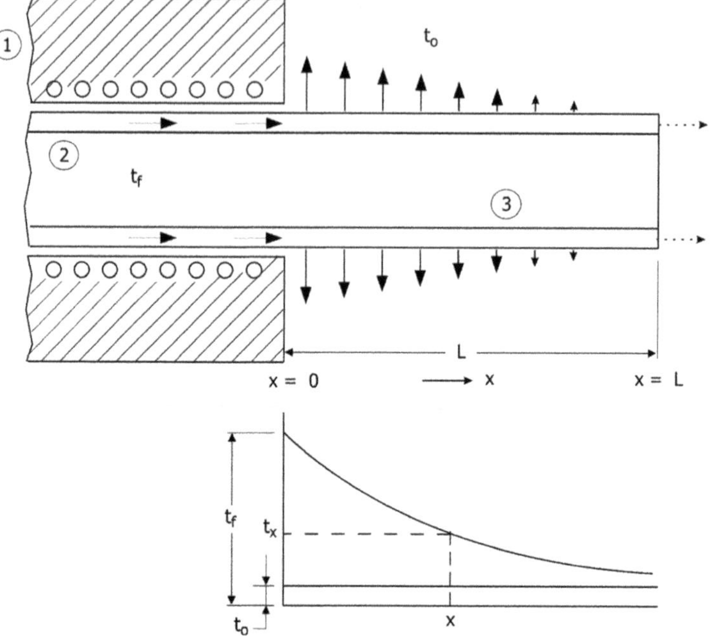

Figure 3.29 Cooling (heat loss) from extended surfaces.

There are two problems associated with extensions that are of our interest.

1. What should be the minimum length of the extension to assure a desired temperature at the tip?
2. How much heat will be lost through such an extension (end losses)?

The differential equation for temperature distribution in the extension is based on heat balance. Heat conducted by the extension cross section will be dissipated to the atmosphere by transfer (convection + radiation) through the surface.

Assume that the heat transfer coefficient h is constant and the conductivity of the extension be λ (also constant).

$$\frac{d^2 \nu}{dx^2} m^2 - \nu = 0 \tag{3.111}$$

where v is the temperature differance $t_x - t_o$ and

$$m = +\sqrt{\frac{h\rho}{\lambda A}} \tag{3.112}$$

where ρ is the perimeter (m) of the extension and A is the cross sectional area (m^2). The dimensions of m are

$$m = \sqrt{\frac{\text{w}}{\text{m}^2\,{}^\circ\text{C}} \times \text{m} \Big/ \frac{\text{w}}{\text{m}^2\,{}^\circ\text{C}} \times \text{m}^2} = \frac{1}{\text{m}} \tag{3.113}$$

A simplified solution to Equation (3.111) above is

$$v_x = v_f \frac{\cosh\,(m(\ell - x))}{\cosh\, m\ell} \tag{3.114}$$

Here, $v_x = t_x - t_o$ and $v_f = t_f - t_o$

The heat lost to the surroundings (i.e., lost from the furnace as end loss) is

$$Q = v_f \sqrt{h\rho\lambda A}\,\tanh\,(m\ell) \tag{3.115}$$

If ℓ is very large $\cosh m\ell \to \infty$ and $\tanh m\ell \to 1$

$v_x = \ell = 0$

If ℓ is very large so that there is no heat loss at the tip, we can also use the expression

$$t_x = t_f e^{-mx}\,{}^\circ\text{C}$$

and $Q = t_f\sqrt{h\rho\lambda A}$ w

EXAMPLE 3.24

A ceramic tubular muffle has outer diameter 100 mm and inner diameter 80 mm. A length 0.1 m of the muffle extends out of the furnace which is at 1200°C. The surrounding temperature is 30°C

Thermal conductivity of muffle $\lambda = 1.0$ w/m°C

Heat transfer coefficient (average) $h = 200$ w/m²°C

Determine the temperatures along the extension and the heat lost.

Solution

Cross sectional area of tube A

$$= \frac{\pi}{4}\left(100^2 - 80^2\right) \times 10^{-6} = 2.83 \times 10^{-3}\ \text{m}^2$$

Perimeter (outer) of the tube

$$\rho = \pi \times 100 \times 10^{-3} = 0.314\ \text{m}$$

$$m = \sqrt{\frac{h\rho}{\lambda A}} = \sqrt{\frac{200 \times 0.314}{1.6 \times 2.83 \times 10^{-3}}}$$

$$= 118\ \text{m}^{-1}$$

As the muffle has very low conductivity the heat transfer will be mainly near the base. Consider distances of 0.01 m from the base at

$x = 0$ $\quad t = 1200°\text{C}$

$x = 0.01$ $\quad t = 1200 \times e^{-118 \times 0.01} = 369°\text{C}$

$x = 0.02$ $\quad t = 1200 \times e^{-118 \times 0.02} = 113°\text{C}$

$x = 0.03$ $\quad t = 1200 \times e^{-118 \times 0.03} = 34°\text{C}$

Thus, heat will be lost through the first 3 cm length of the muffle.

Heat lost will be

$$Q = t_f \sqrt{h\rho\lambda A}\ \text{w}$$

$$= 1200 \times \sqrt{200 \times 0.314 \times 1.6 \times 2.83 \times 10^{-3}}$$

$$= 640\ \text{w}$$

EXAMPLE 3.25

Calculate the temperatures along an extension tube made of Fe-Cr-Al alloy. The furnace temperature is 1000°C. The dimensions and other conditions are similar to those in Example 3.24.

Thermal conductivity of alloy tube (λ) = 15 w/m°C

Solution

Cross sectional area $A = 2.83 \times 10^{-3}$ m^2

Perimeter = 0.314 m

$\lambda = 15$ w/m°C

$h = 200$ w/m^2°C

$$m = \sqrt{\frac{200 \times 0.314}{15 \times 2.83 \times 10^{-3}}} = 38.5$$

As the conductivity is much higher and the tube is of finite length, we will use the expression

$$t = t_f \frac{\cosh\,(m(\ell - x))}{\cosh m\ell}$$

here $m = 38.5$ m^{-1}, $\ell = 0.3$ m (30 cm)
let $x = 0.05$ m

$$t = 1000 \frac{\cosh\,(38.5\,(0.3 - 0.05))}{\cosh 38.5(0.3)}$$

$$= 175°\text{C}$$

Similar calculations give the following results

x		t°C
0.005	~	1000
0.01	~	816
0.05	~	175
0.1	~	30

Showing that at x ~ 10 to 12 cm, the temperature reaches the surrounding temperature.

The heat dissipated is

$$Q = t_f \sqrt{h\rho\lambda A}\ \tanh\,(m\ell)\ \text{w}$$

$$= 1000\sqrt{(200 \times 0.314 \times 15 \times 2.83 \times 10^{-3})}\ \tanh\,(38.5 \times 0.3)$$

$$= 1.63\ \text{kW}$$

Comparing these results it can be seen that the hot length for metallic muffle is about twice that of ceramic muffle. The heat dissipation is also about 2.5 times more.

EXAMPLE 3.26

A furnace enclosure has a surface temperature varying between 200 and 50°C. The atmospheric average temperature is 30°C.

If the enclosure has an emissivity 0.8, determine the heat radiated by the enclosure.

What will be the change in radiation if the surface is given aluminium paint having an emissivity 0.3?

Solution

The thermal radiation flux from the wall to the atmosphere will be

$$q = \sigma_0 \epsilon_w \left\{ \left[\frac{T_w}{100} \right]^4 - \left[\frac{30+273}{100} \right]^4 \right\} \text{ W/m}^2$$

$$\epsilon_w = \text{The wall emissivity} = 0.8 \text{ (unpainted)}$$

$$q_{200°C} = 5.67 \times 0.8 \left\{ \left[\frac{200+273}{100} \right]^4 - \left[\frac{30+273}{100} \right]^4 \right\}$$

$$= 5.67 \times 0.8\,[500.5 - 84.3]$$

$$= 1.9 \text{ kW/m}^2$$

Similarly

$$q_{150°C} = 5.67 \times 0.8 \left\{ \left[\frac{150+273}{100} \right]^4 - \left[\frac{30+273}{100} \right]^4 \right\}$$

$$= 1.07 \text{ kW/m}^2$$

$$q_{100°C} = 0.5 \text{ kW/m}^2$$

$$q_{50°C} = 0.113 \text{ kW/m}^2$$

For the aluminium painted wall $\epsilon = 0.3$

$$q_{200°C} = 0.708 \text{ kW/m}^2$$

$$q_{150°C} = 0.401 \text{ kW/m}^2$$

$$q_{100°C} = 0.185 \text{ kW/m}^2$$

$$q_{50°C} = 0.041 \text{ kW/m}^2$$

These results are plotted in Figure 3.30 which shows that the radiation loss decreases rapidly with the decrease in the wall temperature. Also, the loss from a wall painted with low emissivity paint is much lower (35%) than the bare wall.

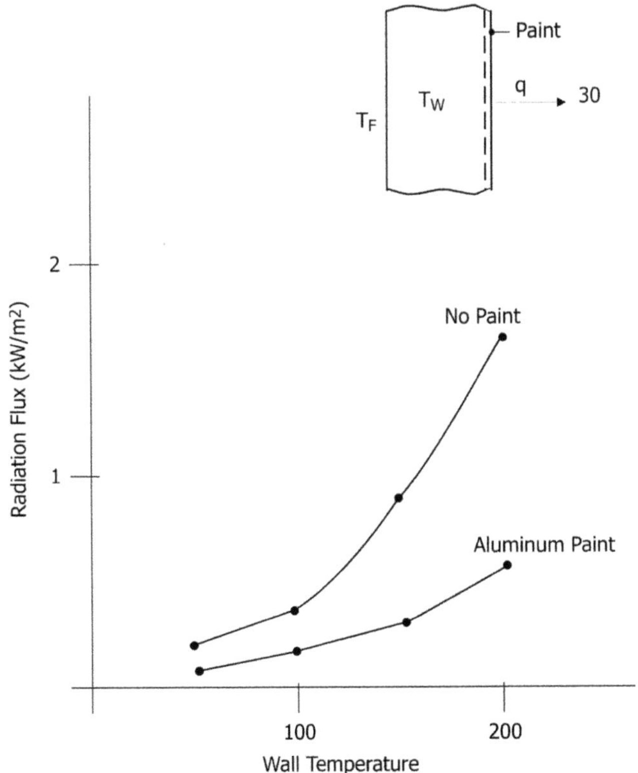

Figure 3.30 [Example (3.26)]

The radiative heat transfer coefficient will be obtained by dividing the calculated flux values by the temperature difference.

$$\text{e.g.,} \quad hr_{100°C} = \frac{1.07}{100-30} = 15 \text{ W/m}^2\text{°C} \quad \text{unpainted}$$

$$hr_{100°C} = \frac{0.401}{100-30} = 5.73 \text{ W/m}^2\text{°C} \quad \text{painted}$$

If the furnace temperature T_F and the wall thickness, etc. are the same, a bare wall will conduct heat faster than a painted wall. This fact is useful in the design of continuous and batch-type furnaces.

Chapter 4

Transient Conduction

CONTENTS

4.1 INTRODUCTION

In this mode of conduction heating the temperature of a body varies with both time and the coordinates. Basically there are three variables to deal with. They are time (τ), temperature (t), and the coordinates of the point (x, y, z) in the body.

The differential equation connecting time, temperature, and coordinates is

$$\frac{\partial t}{\partial \tau} = \alpha\left[\frac{\partial^2 t}{\partial x^2} + \frac{\partial^2 t}{\partial y^2} + \frac{\partial^2 t}{\partial z^2}\right] \tag{4.1}$$

or $$\frac{\partial t}{\partial \tau} = \alpha \nabla^2 t$$

A large number of physical properties of the body are involved in the heat transfer. In addition, certain properties of the surrounding medium also take part in the heating of the body.

Hence, a short list of such involved parameters becomes

1. Density (ρ), specific heat (c), and thermal conductivity (λ)
2. Time (τ)
3. Position (x, y, z)
4. Dimensions of the body ($\ell_1, \ell_2, \ell_3, \ldots$)
5. Temperature of the surrounding medium (t_o)
6. Heat transfer coefficient (h), etc.

Thus Equation (4.1) will have a general solution of the form

$$t = f(x, y, z, t, \alpha, h, c, \ell_1, \ell_2, \ldots)$$

All physical properties vary with the temperature; the actual solution of Equation (4.1) poses many problems, making it impossible to have one general solution.

To simplify the problem we make an assumption that in the temperature range involved all the physical properties are constant. Even this simplification makes it possible to have a general solution only for geometrically simple (or well-defined) bodies such as slabs or plates, cylinders, and spheres. The reason being that such regular shaped bodies can be described by a coordinate system, such as cartesian and cylindrical. It should be noted that actual bodies involved in practice are rarely geometrically regular. Hence, the solutions available from theory are useful only as guides.

In numerical solutions of equations we generally use average physical properties.

A three-dimensional numerical solution becomes unwieldy and its solution, even in simple cases, requires the help of computers.

In practice we reduce a three-dimensional problem to a single-dimensional one. This is achieved by making the body an infinite (very large length) cylinder or an infinite slab (with finite thickness and very large surface, etc.).

With these simplifications Equation (4.1) in one dimension becomes

$$\frac{\partial t}{\partial \tau} = \alpha \frac{\partial^2 t}{\partial x^2} \tag{4.2}$$

4.2 SOLUTION BY USING CHARTS

There are many methods of solving the above equation. All solutions are particular solutions for a given geometry of the heated body. The number of variables is reduced by grouping them into dimensionless numbers. The following dimensionless numbers and groups are frequently used.

$$\text{Dimensionless temperature } \Theta = \frac{t_f - t}{t_f - t_i} \tag{4.3}$$

where

t_f = Source (furnace) temperature (°C)

t = Temperature of the body at time t (°C)

t_i = Initial temperature of the body, i.e., the outside temperature (°C)

$$\text{Biot number (Bi)} = \frac{h\delta}{\lambda} \tag{4.4}$$

where

h = Heat transfer coefficient w/m^2°C

δ = Characteristic dimension of the body m

= 1/2 thickness of a plate

= Radius of a cylinder

= Radius of a sphere, etc.

λ = Thermal conductivity (w/m°C)

$$\text{Fourier number (Fo)} = \frac{\alpha\tau}{\delta^2} \tag{4.5}$$

where

α = Thermal diffusivity of the body m^2/sec

$= \frac{\lambda}{c\rho}$

λ = Thermal conductivity (w/m°C)

τ = Time (sec)

c = Specific heat (J/kg°C)

ρ = Density (kg/m^3)

The solution of Equation (4.2) then takes the form

$$\Theta = \Psi\,(\text{Bi, Fo}, x/\delta) \tag{4.6}$$

where

x = Coordinate of the point in the body at which the temperature or time is to be determined.

Note: The solutions and methods described here are applicable to both heating and cooling.

Mathematical details of the solutions of these equations are of no interest to the furnace designer and, hence, are not discussed here.

As noted earlier, solutions are possible only for regular bodies such as infinite plates, infinite cylinders, and spheres. These solutions are functions of the dimensionless temperature Θ, the Biot (Bi), and the Fourier (Fo) numbers.

The designer has, in general, two problems for which solutions are desired.

1. Given a certain temperature at a certain location in the body, what is the time required to achieve this temperature if the body is kept under a known thermal flux?
2. Given the time, what will be the temperature at a desired point (or location) in the body when the body is kept in a known thermal flux?

The usual assumption is that during the process of heating, the concerned physical properties (density, thermal diffusivity, etc.) are constant.

A number of charts and tables are published containing the solutions. These graphs make it easy to obtain quick and reasonably reliable solutions to the two problems posed above.

One such method of using the charts is outlined below.

Figure 4.1 to Figure 4.6 contain the charts prepared for the center and surface dimensionless temperature Θ for a plate, cylinder, and sphere. The temperature is correlated for various Biot and Fourier numbers of the body. When any two variables from Θ, Bi, and Fo are known the third can be found out from the chart.

For example, suppose that an infinite cylinder (very long length) is to be heated for a certain time τ in a known thermal flux h. The temperatures at the center (axis) and the surface are to be determined after τ.

First determine the Fo and Bi numbers using the given or known physical properties of the body (cylinder).

From Figure 4.1 to Figure 4.6 read the temperature Θ for these Bi and Fo numbers.

Note: The figures are correlation of Fo, Bi, and Θ for the center (axis) of the cylinder.

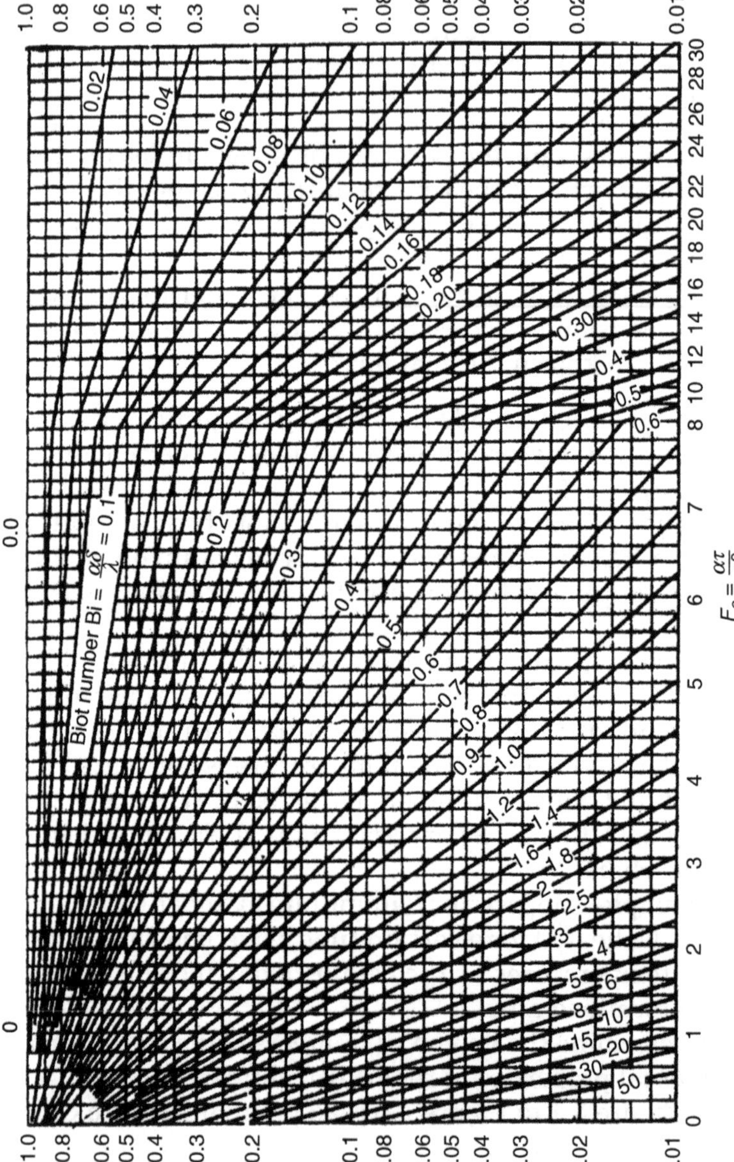

Figure 4.1 Charts showing relation between Θ, Fo, and Bi for the surface of a plate.
Note: The vertical axis represents dimensionless temperature $\Theta = (t_f - t)/(t_f - t_o)$.

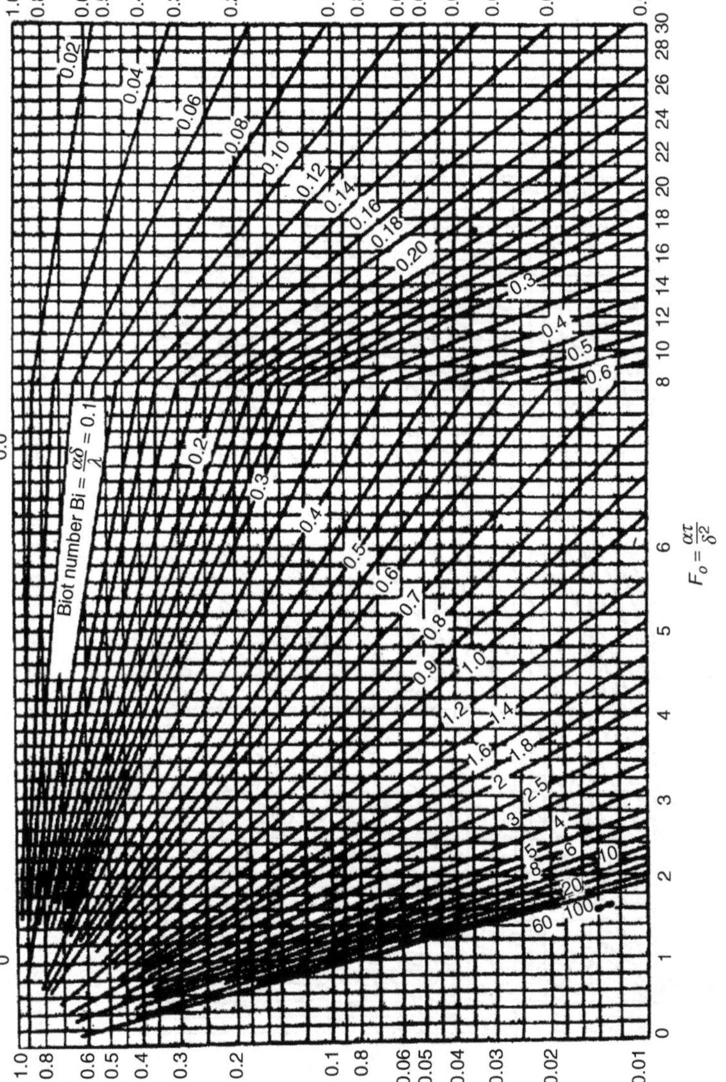

Figure 4.2 Charts showing relation between Θ, Fo, and Bi for the middle plane of a plate.
Note: The vertical axis represents dimensionless temperature $\Theta = (t_f - t)/(t_f - t_o)$.

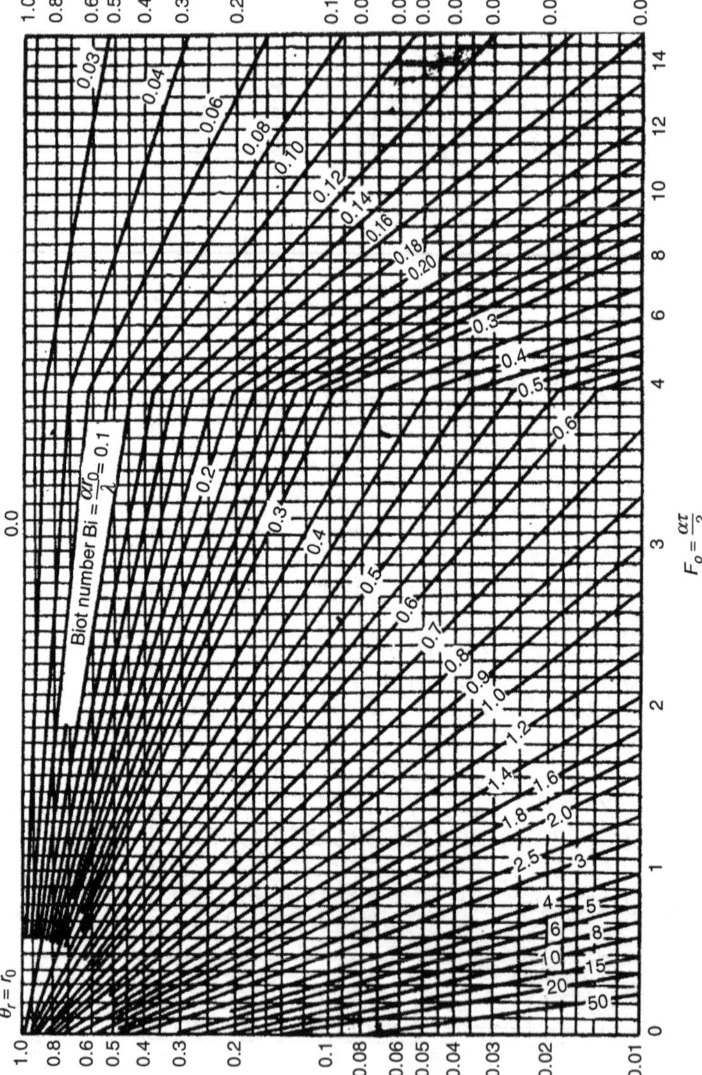

Figure 4.3 Charts showing relation between Θ, Fo, and Bi for the surface of a cylinder. *Note:* The vertical axis represents dimensionless temperature $\Theta = (t_f - t)/(t_f - t_o)$.

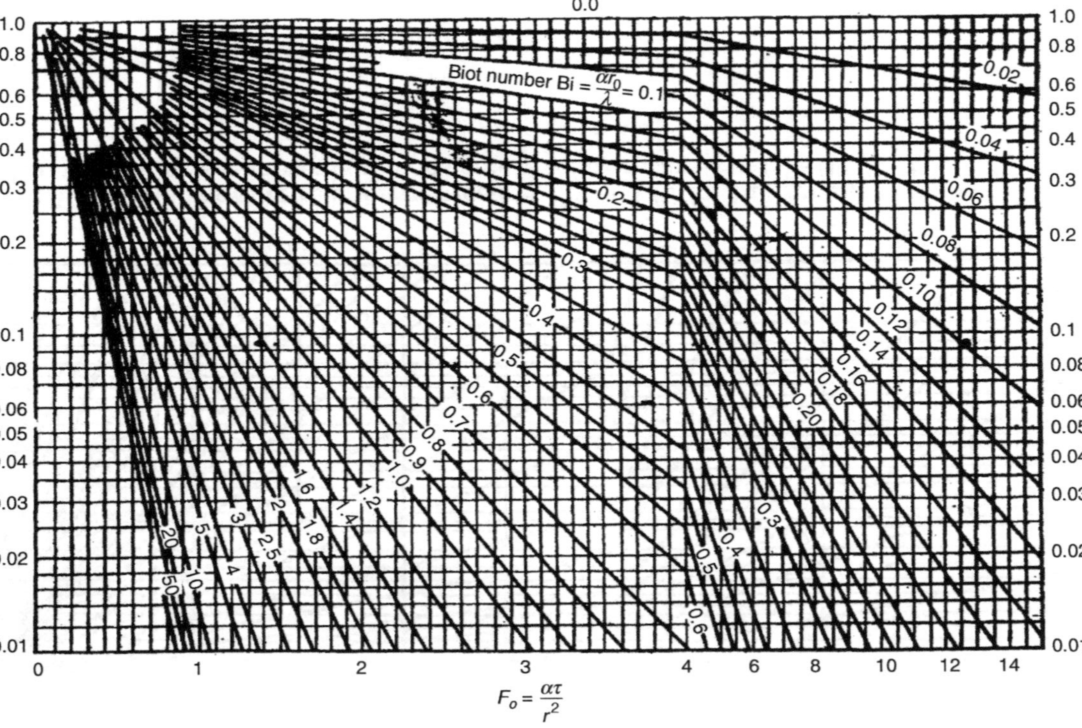

Figure 4.4 Charts showing relation between Θ, Fo, and Bi for the axis of a cylinder.
Note: The vertical axis represents dimensionless temperature $\Theta = (t_f - t)/(t_f - t_o)$.

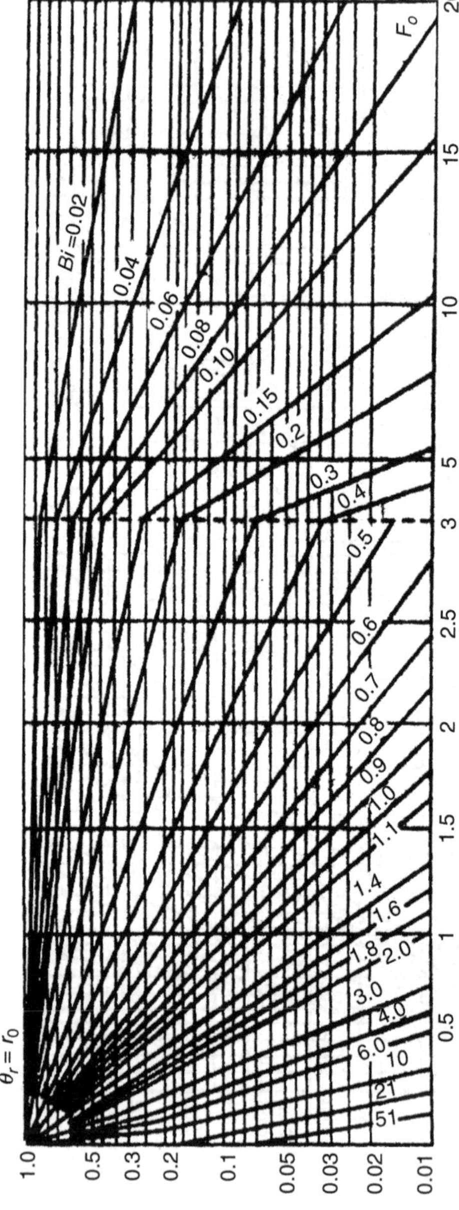

Figure 4.5 Charts showing relation between Θ, Fo, and Bi for the surface of a sphere. *Note:* The vertical axis represents dimensionless temperature $\Theta = (t_f - t)/(t_f - t_o)$.

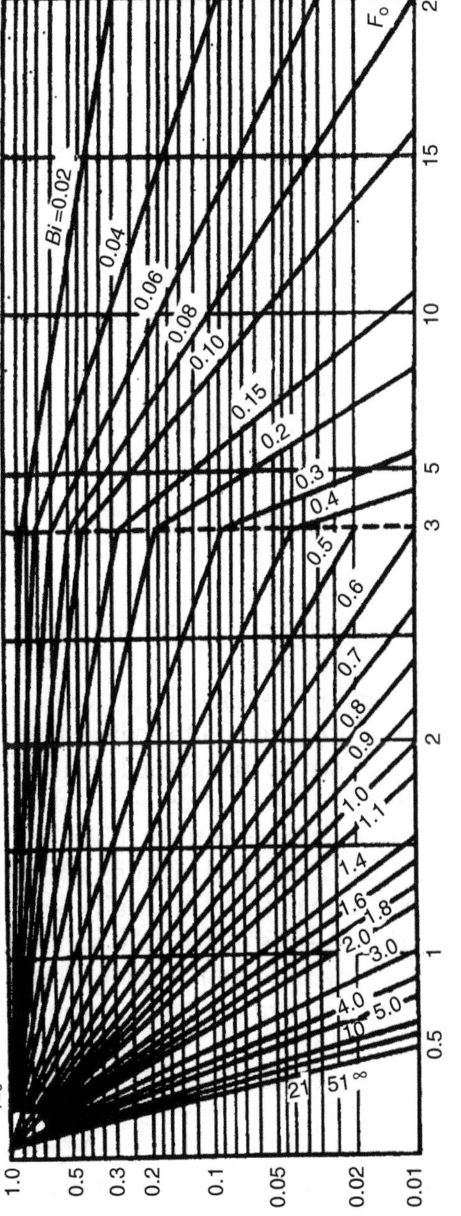

Figure 4.6 Charts showing relation between Θ, Fo, and Bi for the center of a sphere.
Note: The vertical axis represents dimensionless temperature $\Theta = (t_f - t)/(t_f - t_o)$.

Then determine the central temperature from the relation.

$$\Theta = \frac{t_f - t_{x=0}}{t_f - t_o}$$

where

t_f = Furnace temperature (°C)
$t_x = 0$ = Temperature of axis (°C)
t_o = Initial temperature (°C)

In a similar manner determine the surface temperature $t_{x=1}$ by using the appropriate figure.

The inverse problem of determining the time τ required to achieve a certain temperature at the center or axis can be solved by using the same charts.

Figure 4.3 to Figure 4.6 will be useful for solving similar problems regarding a plate or a sphere.

4.3 HEATING OF BODIES OF FINITE SIZE

The analytical method described in the previous section considered bodies of infinite (very large) length or surface area compared to other dimensions.

In practice, the bodies are finite, i.e., have a limited length or surface area. The heating of such finite bodies of regular shape can be done by a method known as the method of multiplied solutions.

The finite body of interest is considered to be formed by the intersection of two or more infinite bodies (Figure 4.7).

A rectangular parallelopiped of sides $2\delta x$, $2\delta y$, and $2\delta z$ is formed by the intersection of three mutually perpendicular infinite plates having thicknesses of $2\delta x$, $2\delta y$, and $2\delta z$, respectively.

Similarly, a finite cylinder can be considered as formed by the intersection of an infinite cylinder (radius r) and an infinite plate of thickness 2δ.

For each component, infinite body, the Biot and Fourier numbers (Bi, Fo) are calculated from the given conditions. By using the appropriate chart (Figure 4.1 to Figure 4.6) for each component body, the dimensionless temperature Θ at the required point (usually the center or axis of the finite body) is determined.

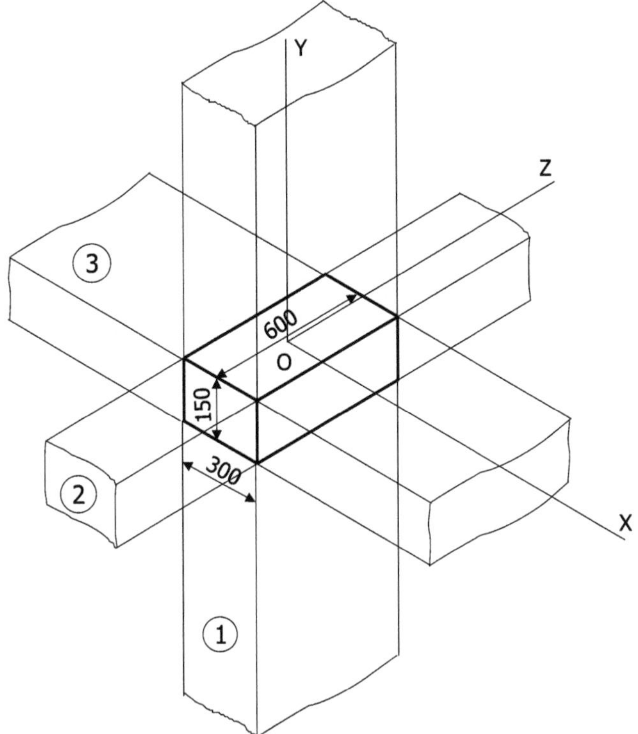

Figure 4.7 Finite body by intersection of three infinite bodies (Example 4.4).

Suppose these temperatures are Θx, Θy, and Θz; then the dimensionless temperature Θ of the finite body, at the point of interest is

$$\Theta = \Theta x \times \Theta y \times \Theta z \tag{4.7}$$

From Θ the actual temperature t is determined by the usual way, i.e.,

$$t = t_f - (t_f - t_o) \times \Theta \tag{4.8}$$

where t_f is the final (furnace) temperature and t_o is the initial temperature °C.

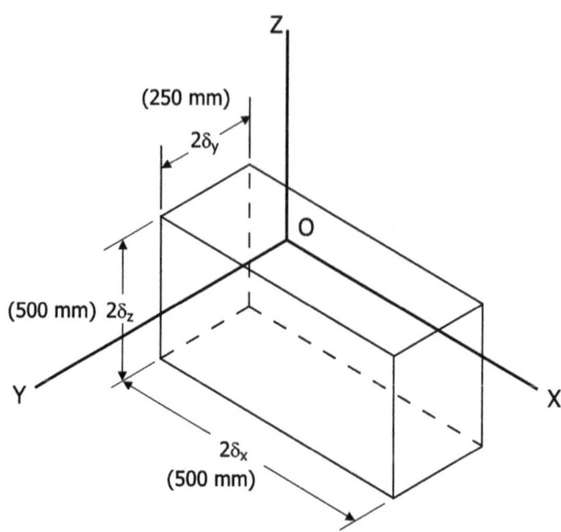

Figure 4.8 Finite rectangular box.

A long rectangular bar having its edges $2\delta x$ and $2\delta y$ can be treated in a similar way. The bar can be considered to be formed by the intersection (Figure 4.8) of two plates having thicknesses $2\delta x$ and $2\delta y$ each.

We first determine Θx and Θy in the usual way by calculating Bi and Fo numbers and consulting the appropriate chart. Then

$$\Theta = \Theta x \times \Theta y \tag{4.9}$$

For a short cylinder formed by a plate of thickness $2\delta z$ and a long cylinder of diameter $2r$, we determine Θz and Θr and find Θ by using the relation

$$\Theta = \Theta z \times \Theta r \tag{4.10}$$

In practice we rarely come across perfectly regular shapes of finite size. The shapes are of a large geometrical variety. These shapes cannot be treated rigorously by analytical methods. We usually approximate the given shape to the nearest regular body and estimate the heating or cooling time. In such cases the estimate has to be reasonable and backed by experience.

The preceding method of multiplied solutions can be used for determining the temperature at any point in the finite body. This application is not discussed here.

EXAMPLE 4.1

A steel shaft having a diameter of 120 mm is initially at a temperature of 26°C and is placed in a furnace with a temperature of 900°C.

Determine the time required to heat the shaft when the axis acquires a temperature of 820°C.

Also determine the surface temperature at the above-determined time.

$\lambda = 21$ w/m°C, $a = 6.2 \times 10^{-6}$ m²/sec, $h = 180$ w/m²°C

Solution

The Biot number (Bi) is

$$\text{Bi} = \frac{h\gamma_o}{\lambda} = \frac{180 \times 0.06}{21}$$

$$= 0.51$$

$$t_f = 900°\text{C}, \quad t_o = 26°\text{C}, \quad t_{\gamma o} = 820°\text{C}$$

$$\Theta = \frac{t_f - t_{\gamma_o}}{t_f - t_o} = \frac{900 - 820}{900 - 26} = \frac{80}{874}$$

$$= 0.092$$

Referring to Θ, Fourier number Fo, from the graph for cylinders, at Bi = 0.51

$$\text{Fo} = 2.9$$

$$\tau = \frac{\gamma_o^2 - \text{Fo}}{a} = \frac{0.06^2 \times 2.9}{6.2 \times 10^{-6}} = \frac{3.6 \times 2.9 \times 10^3}{6.2}$$

$$= 1683 \text{ sec} = 28 \text{ min} = \text{say } 1/2 \text{ h}$$

For Bi = 0.051 and Fo = 2.9, the surface value of the dimensionless temperature Θ is obtained from Bi, Θ, and Fo graph for surface. Hence

$$\Theta = \frac{t_f - t_{\gamma=\gamma_o}}{t_f - t_o} = \frac{900 - t_{\gamma=\gamma_o}}{900 - 26} = 0.07$$

$$\therefore\ t_{\gamma=\gamma o} = 838°\text{C}$$

EXAMPLE 4.2

A rectangular billet of steel measures 150 × 300 × 600 mm with an initial temperature of 25°C. It is placed in a furnace at 1400°C.

What will be the temperature of the center of ingot after 2 hours?

Coefficient of heat transmission $h = 150$ w/m^2°C
Coefficient of thermal diffusivity $a = 7 \times 10^{-6}$ m^2/sec
Coefficient of thermal conductivity $\lambda = 28$ w/m°C

Solution

The billet can be considered as having formed by the intersection of three infinite plates—1, 2, and 3.

The dimensionless temperature Q at any point in a finite rectangular billet is equal to the product of dimensionless temperatures of the three infinite plates forming the given rectangle.

Let t_c be the temperature at the center (located at origin O)

$$\frac{t_f - t_c}{t_f - t_i} = \frac{t_f - t_{x=0}}{t_f - t_i} \times \frac{t_f - t_{y=0}}{t_f - t_i} \times \frac{t_f - t_{z=0}}{t_f - t_i} \qquad \text{(i)}$$

t_i = Initial temperature = 25°C

t_f = Furnace temperature = 1400°C

For a plate having thickness $(2\delta x) = 600$ mm

$$\text{Bi} = \frac{h\delta x}{\lambda} = \frac{150 \times 0.3}{28} = 1.6$$

$$\text{Fo} = \frac{a\tau}{\delta x^2} = \frac{7 \times 10^{-6} \times 2 \times 3600}{0.3^2} = 0.56$$

From chart for Bi = 1.6 and Fo = 0.5

$\Theta x = 0.66$ (ii)

For a plate of thickness $(2\delta y) = 300$ mm

$$\text{Bi} = \frac{150 \times 0.15}{28} = 0.8$$

$$\text{Fo} = \frac{7 \times 10^{-6} \times 2 \times 3600}{0.15^2} = 2.24$$

From chart for Bi = 0.8 and Fo = 2.24

$\Theta y = 0.28$ (iii)

For a plate of thickness $(2\delta z) = 150$ mm

$$\text{Bi} = \frac{150 \times 0.075}{28} = 0.4$$

$$\text{Fo} = \frac{7 \times 10^{-6} \times 2 \times 3600}{0.075^2} = 8.842$$

From chart for Bi = 0.4 and Fo = 8.842

$\Theta z = 0.055$ (iv)

$$\frac{t_f - t_c}{t_f - t_i} = \frac{1400 - t_c}{1400 - 25} = 0.66 \times 0.28 \times 0.055 = 0.0101$$

$$t_c = 1400 - 12 = 1386°\text{C}$$

EXAMPLE 4.3

For the billet in Example 4.2, what will be the temperature at the center (t_c) after one hour of heating under the same conditions?

Solution

The Bi will remain unchanged.

$$\text{For Bi} = 1.6 \text{ Fo} = \frac{7 \times 10^{6} \times 3600}{0.3^2} = 0.28$$

$$\Theta x = 0.8$$

Similarly

For Bi = 0.8 Fo = 1.12

$$\Theta y = 0.65$$

and

For Bi = 0.4 Fo = 4.480

$$\Theta z = 0.22$$

$$\frac{t_f - t_c}{t_f - t_i} = \frac{1400 - t_c}{1400 - 25} = \Theta x \times \Theta y \times \Theta z$$

$$= 0.8 \times 0.65 \times 0.22$$

$$t_c = 1400 - 157 = 1243°\text{C}$$

What will be the temperature at the center after heating for 30 and 15 min under the same conditions?

Again, Bi will remain unchanged.

Bi = 1.6	Fo = 0.14	$\Theta x = 0.95$
Bi = 0.8	Fo = 0.56	$\Theta y = 0.76$
Bi = 0.4	Fo = 2.24	$\Theta z = 0.48$

$$\frac{t_f - t_c}{t_f - t_i} = \frac{1400 - t_c}{1400 - 25} = \Theta x \times \Theta y \times \Theta z$$

$$= 0.95 \times 0.76 \times 0.48 = 0.615$$

$$t_c = 1400 - 1357 \times 0.3466 = 899°\text{C}$$

For	Bi = 1.6	Fo = 0.07	$\Theta x = 1.0$
	Bi = 0.8	Fo = 0.28	$\Theta y = 0.82$
	Bi = 0.4	Fo = 1.12	$\Theta z = 0.75$

$$\frac{t_f - t_c}{t_f - t_i} = \frac{1400 - t_c}{1400 - 25} = \Theta x \times \Theta y \times \Theta z$$

$$= 1.0 \times 0.82 \times 0.75 = 0.615$$

$$t_c = 1400 - 1357 \times 0.615 = 554°\text{C}$$

EXAMPLE 4.4

A steel ingot measuring 100 × 250 × 500 mm is at an initial temperature of 30°C. It is placed in a furnace at a temperature of 1250°C for a period of 1 hr.

Determine the temperature at the center and at the center of faces of the ingot at the end of the heating period.

For steel $\lambda = 26$ w/m°C $a = 6.5 \times 10^{-6}$ m²/sec, $\rho = 7.83 \times 10^3$ kg/m³

Heat transfer coefficient to the ingot $\alpha = 200$ w/m²°C.

Solution

The ingot is shown in Figure 4.7. The origin is chosen at the center.

The dimensionless temperature at any point Θ in the ingot will be equal to the product of the dimensionless temperatures Θ*x*, Θ*y*, and Θ*z* of the constituent slabs at that point, i.e., (for center $x = y = z = 0$),

$$\Theta o = \Theta_{x=0} \times \Theta_{y=0} \times \Theta_{z=0}$$

Here, the thicknesses of the constituent plates are 100, 250, and 500 mm.

First calculate Bi, Fo, and Θ for the center and surface of the three constituent slabs.

$$\delta x = 0.250 \text{ m} \qquad T = 60 \text{ min} = 3600 \text{ sec}$$

$$a = 6.5 \times 10^{-6} \text{m}^2/\text{sec} \quad \lambda = 26 \text{ w/m°C}$$

$$h = 200 \text{ w/m°C}$$

$$\text{Fo} = \frac{a\tau}{\delta x^2} = \frac{6.5 \times 10^{-6} \times 3600}{0.25^2} = 0.375$$

$$\text{Bi} = \frac{h\delta x}{\lambda} = \frac{200 \times 0.25}{28} = 1.92$$

Similarly

For $\delta y = 0.05$ m

$$\text{Fo} = 9.36 \quad \text{Bi} = 0.3846$$

And

For $\delta z = 0.125$ m

Fo = 1.5, Bi = 0.96

Referring to Figure 4.3 and Figure 4.4, the center and surface dimensionless temperatures Θ are

δm	Θcenter	Θsurface
$\delta x = 0.250$	0.64	0.34
$\delta y = 0.05$	0.05	0.035
$\delta z = 0.125$	0.42	0.48

The dimensionless temperature at the center is

$$\Theta c = \Theta_x \times \Theta_y \times \Theta_z$$

$$= 0.64 \times 0.05 \times 0.42 = 0.01344$$

$$t_c = t_f - (t_f - t_a) \times \Theta c$$

$$= 1250 - (1250 - 30) \times 0.01344$$

$$= 1086°\text{C}$$

For the same Fo and Bi numbers the values of $\Theta x = dx$, $\Theta y = dy$, and $\Theta z = dz$ are obtained by referring to the appropriate figure.

In Figure 4.7 P is a point at the center of the face, $x = y = 0$ and $z = dz$. It can be seen that there will be six such points, one on the center of each face, and these six points will form three pairs.

The center of each face, e.g., point P is on the face of one slab and on the axis of the other two slabs. Hence, Θ for each such point will be obtained by the multiplication of

e.g., $\Theta_{\text{center } x}$, $\Theta_{\text{center } y}$ and $Q_{\text{surface } z}$

i.e., $\Theta_P = \Theta_{\text{center } x} \times \Theta_{\text{center } y} \times \Theta_{\text{surface } z}$

$$\Theta_{Px} = 0.64 \times 0.05 \times 0.48 = 0.01536$$

$$t_P = 1250 - 1220 \times 0.01536$$

$$= 1231°\text{C}$$

Similarly

$$\Theta_{Py} = 1241°C$$

$$\Theta_{Pz} = 1238°C$$

4.4 TRANSIENT HEATING (COOLING) OF A SEMIINFINITE SOLID

The concept of a semiinfinite solid is very useful in tackling many practical problems. In such a solid the x direction extends from 0 to ∞ (i.e., there is no $-x$) and the solid extends to $\pm\infty$ in y and z directions. Thus, we have one surface, i.e., the $x = 0$ plane. This surface can be subjected to a variety of heating or cooling conditions so as to obtain the temperature history in the x direction. There are many solutions applicable under different conditions. We will consider a few of interest to us. For a comprehensive discussion of the various available solutions refer to Carslaw and Jaeger {3}.

An excellent discussion of the limitations and areas of applications of these methods is given by Bejan {2}.

4.4.1 Instantaneous Temperature Change at Surface

In the first case we consider a semiinfinite solid at a constant temperature T_i through-out. At a certain instant ($t = 0$) the plane $x = 0$ is suddenly brought to a temperature $T_s > T_i$ for heating and $T_s < T_i$ for cooling. Over time the heat penetrates unidirectionally along x direction and the temperature rises progressively. At a sufficiently long distance from $x = 0$ the temperature is still T_i (or T_∞). The rise in temperature with time t is shown in Figure 4.9.

The assumption of an instantaneous rise in the surface temperature from T_s to T_i is difficult to realize practically. Some kind of heat transfer process will apply to each real situation and there will be some time until the temperature increases to T_s.

Some practical situations approach the assumption. These are the processes in which the article is dipped in a

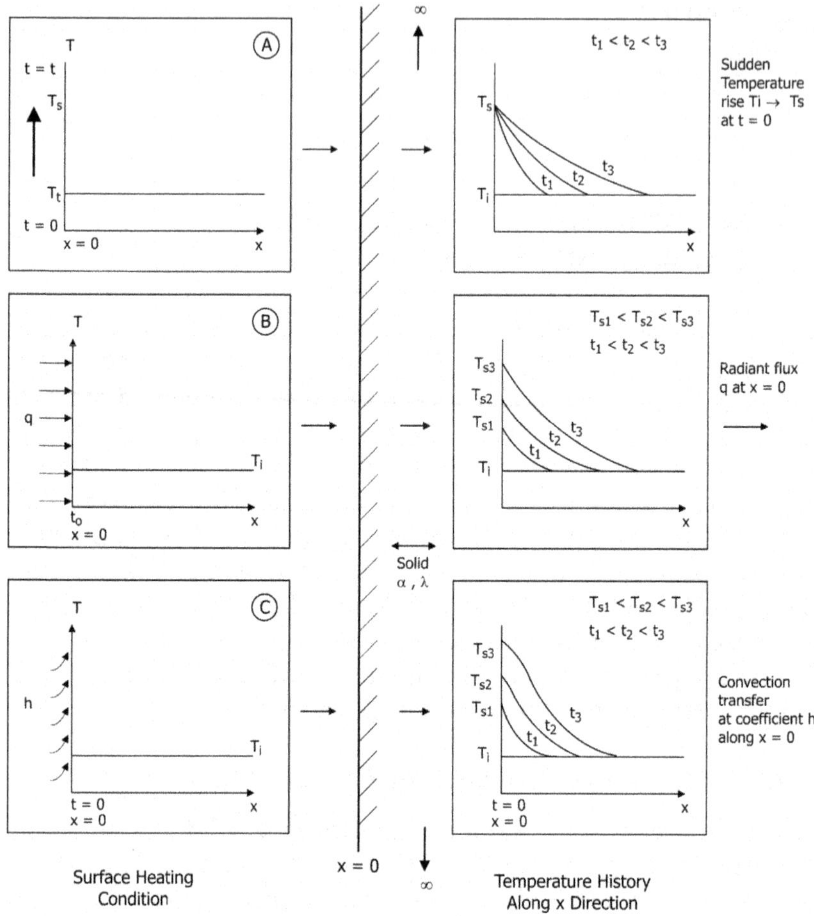

Figure 4.9 Unidirectional heating (cooling) of a semiinfinite solid.

large, quiescent fluid bath. Quenching of a hot steel article in oil or heating of metal in a salt bath are some examples. The fluid surrounds the surface from all sides and the temperature can be assumed to rise instantaneously.

Consider such an article dipped in a fluid bath at temperature T_∞. Let the temperature of the article be T_i ($T_\infty < T_i$).

Consider various points below the surface at different depths. Let the radius (or thickness) of the article be r_o.

The surface will come to temperature T_i according to our assumption, but the interior will be mostly at T_i at $t = 0$. As time progresses the temperature transient will progress toward the center so that there will be a transient depth δ which will approach r_o with time. This time in which $\delta \ll r_o$ is called the "early regime". The progress of δ will depend on diffusivity α (assumed constant).

Thus, for $t \ll \dfrac{r_o^2}{\delta}$ the transient temperature is in an early stage. (4.11)

It can be shown that $\delta \sim (\alpha t)^{1/2}$

At $t \simeq \dfrac{r_o^2}{\alpha} t_c$ the transient will have reached the center. This is called the critical or transient time. (4.12)

For $t \gg \dfrac{r_o^2}{\alpha}$ the internal temperature will level out and will be T_∞ throughout after a very long time (late regime) (4.13)

It can be seen that the transient or internal temperature will be governed by the thermal diffusivity α, depth r_o, and process time t. The assumption of instantaneous temperature rise (or fall) from $T_i \rightarrow T_\infty$ at the surface automatically sets the surface heat transfer coefficient to infinity.

For the early regime, i.e., when $t \ll$ the temperature at a depth x from the surface at time t, is given by

$$\frac{T_{xt} - T_\infty}{T_i - T_\infty} = erf \frac{x}{2\sqrt{\alpha t}} \tag{4.14}$$

where

$T_{x,t}$ = Temperature at x and time t

T_∞ = Fluid temperature

T_i = Initial temperature

The surface flux at time t is given by

$$q = \lambda \frac{T_i - T_\infty}{\sqrt{\pi \alpha t}} \tag{4.15}$$

where λ is the thermal conductivity (w/m°C).

We have discussed the condition for the validity of the basic assumption of instantaneous temperature change. A more practical condition is the existence of some transfer process at the surface. Two important cases are

1. Surface subjected to constant radiation flux q (w/m²)
2. Surface heat transfer by convection current of fluid with a transfer coefficient h (w/m°C)

4.4.2 Constant Radiation Flux

When the surface is irradiated by radiation flux q, its temperature gradually increases from T_i to T_{s1}, T_{s2}, T_{s3} etc. as shown in Figure 4.9(B). Heat will penetrate deeper with time as the surface temperature increases. The equation for finding the temperature at point x after time t is

$$T_{x,t} - T_i = \frac{2q}{\lambda}\left[\frac{\alpha t}{\pi}\right]^{1/2} exp\left[-\frac{x}{4\alpha t}\right] - \frac{qx}{\lambda} erfc \frac{x}{2\sqrt{\alpha t}} \tag{4.16}$$

where "*erfc*" is the complementary error function = (1 – "*erf* ")

For surface, i.e., $x = 0$

$$T_{0,t} - T_i = \frac{2q}{\lambda}\left[\frac{\alpha t}{\pi}\right]^{1/2} \tag{4.17}$$

4.4.3 Surface Heating by Convection

The temperature at a point x at time t when the surface is in contact with a well-stirred fluid is given by

$$\frac{T_{x,t} - T_\infty}{T_i - T_\infty} = erf \frac{x}{2\sqrt{\alpha t}} + exp\left[\frac{hx}{\lambda} + \frac{h^2 \alpha t}{\lambda^2}\right] erfc\left[\frac{x}{2\sqrt{\alpha t}} + \frac{h}{\lambda}\sqrt{\alpha t}\right] \tag{4.18}$$

where h is the heat transfer coefficient.

This equation is rather complicated. The solution depends on two-dimensionless groups x/δ (dimensionless distance) and the thermal penetration depth $h\delta/\lambda$. (Remember that $2\sqrt{\alpha t} \simeq \delta$ and h/λ is the ratio of convective/conductive transfer)

Equation (4.18) is a more general and refined form of Equation (4.14) which assumes an infinite heat transfer coefficient. (If we put $h \to \infty$ in Equation (4.18) the second term vanishes and we get the Equation (4.14).

Solution of Equation (4.18) for inner temperatures is shown in Figure 4.9(C) in which

$\dfrac{T_{xt} - T_\infty}{T_i - T_\infty}$ is plotted against $\dfrac{x}{2\sqrt{\alpha t}}$ for various values of $\dfrac{h}{\lambda}\sqrt{\alpha t}$

Note: The three terms are dimensionless.

Until now we have surveyed two methods for determining the transient conditions in conductive heat transfer. The first method (see Section 4.2) is based on charts (Figure 4.1 to Figure 4.6) prepared from the exact solutions of Equation 4.1 for simple geometric shapes. The dimensionless temperature is determined in relation to the two dimensionless numbers, Biot number (Bi) and Fourier number (Fo). These numbers were defined in Equation 4.4 and Equation 4.5.

$$\text{Bi} = \frac{h\delta}{\lambda} \qquad \text{and} \qquad \text{Fo} = \frac{\alpha\tau}{\delta^2}$$

The Biot number relates internal resistance to conduction (δ/λ), external resistance ($1/h$), and the thickness or the radius δ.

If Bi 0 (practically for Bi < 0.1) and the object is this thin, $\Theta \to 1$ at all locations. Thus the charts are not useful for Bi < 0.1.

On the other hand, if Bi $\rightarrow \infty$, (practically Bi > 100) for given δ and λ, h must be very high. This means that the surface instantly attains the temperature of the medium. This is not practical. There must be some temperature gradient at and near the surface.

In a process such as quenching, a vapor blanket of cooling medium forms around the object and momentarily slows down cooling of the surface.

Thus the charts are useful for Bi in the mid range i.e., $0.1 \leq \text{Bi} \leq 100$.

The Fourier number relates diffusivity α and object dimension δ to time τ hence it is called as dimensionless time. For small time τ or a small combination of τ/δ^2 (practically < 100) the charts show no change in. However transients do occur at or mean the surface in small τ or Fo.

The Charts are useful only for relatively large Bi and Fo.

There are many practical applications in which either the objects are thin or we have interest in the temperature history of surface layers.

The transient analysis of second type discussed in Section 4.4.1 and Section 4.4.2 are useful when Bi and / or Fo are small. This is because of the assumption of semi-infinite solid. This makes are surface ($x = 0$) and near interior available for detailed analysis. There is also no restriction of shape.

There is a third method of analysis in which we can investigate the transient behavior in situations arising in a combination of small Bi and large Fo. This method is discussed in the next section.

4.4.4 The Late Regime

The time period after the transient has reached the critical dimension (r_o) is the "late regime." At all $t >> r_o^2/\alpha$, the temperature of the solid levels out until the surface and center are practically at equal temperature (Figure 4.10).

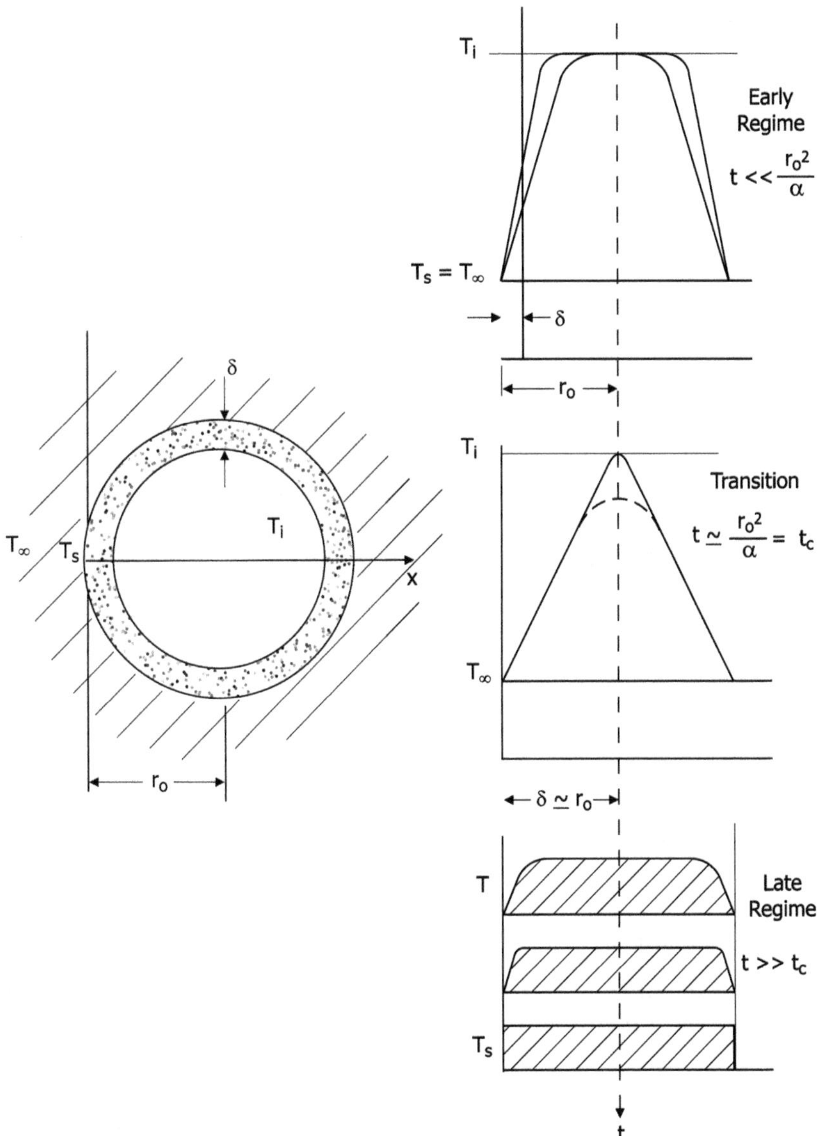

Figure 4.10 Three stages of transitional heating.

The temperature of a point x in the body then depends on the rate of removal from the surface (hA/°C) and the heat stored (ρcV). The method used for the analysis is called the "lumped parameter" concept. The solution is given as follows.

$$\frac{T - T_\infty}{T_1 - T_\infty} = exp - \left[\frac{hA}{\rho cV}(t - t_c) \right] \tag{4.19}$$

where $t - t_c$ is the time period after the transition, i.e., after $t >> r_o^2/\alpha$. A is the surface area and V the volume of the solid.

T = Temperature at the time t

T_∞ = Temperature of the fluid

T_1 = Temperature of the beginning of the late regime, i.e.,

T when $t = r_o^2/\alpha$

This method is applicable when hr_o/λ, i.e., the Biot number is much less than 1, therefore

$$\left. \begin{aligned} \text{Bi} = \frac{hr_o}{\lambda} << 1 \\ \text{and} \quad t >> r_o^2/\alpha \end{aligned} \right\} \tag{4.20}$$

EXAMPLE 4.5

A steel ball having radius 4.0 mm is heated to 900°C throughout and then dropped in an oil bath at 50°C. For the ball

Thermal diffusivity α = 0.053 cm²/sec

Specific heat c = 0.5 J/gm°C Density ρ = 7.8 gm/cm³

Heat transfer coefficient (average, long term) h = 0.2 w/cm²°C.

Determine

1. The time regimes of cooling
2. The temperature gradient in each regime

Solution

The cooling regimes in the ball will depend on the ratio

$$\tau = \frac{r_o^{\,2}}{\alpha} = \frac{0.4^2}{0.053} \sim 3.0 \text{ sec}$$

The early regime will be when

$$\tau \ll \frac{r_o^{\,2}}{\alpha} \text{ i.e.,} \quad \tau \ll 3.0 \text{ sec}$$

Let $\tau = 0.1$ sec

In early regime it is presumed that the surface temperature of the ball becomes equal to the oil temperature, i.e., 50°C immediately on submersion. The center temperature is 900°C.

At a depth of 0.1 cm from the surface the temperature will be

$$\frac{t_{0.1} - t_\infty}{t_i - t_\infty} = erf \frac{x}{2\sqrt{\alpha\tau}}$$

$t_\infty = 50°\text{C}, \quad t_i = 900°\text{C}, \quad x = 0.1 \text{ cm}, \quad \tau = 0.1 \text{ sec}$

substituting

$$\frac{t_{0.1} - 50}{900 - 50} = erf \frac{0.1}{2\sqrt{0.053 \times 0.1}} = erf\ 0.687$$

$$= 0.65$$

$$t_{0.1} = 0.65(850) + 50 \sim 600°\text{C}$$

Similarly $t_{0.2} = 850°\text{C}$

$$t_{0.3} = 900°\text{C}$$

$$t_{0.4} = 900°\text{C}$$

This shows that the cooling transient has not reached the center.

The transient regime is defined as

$$\tau_c \simeq \frac{r_o^2}{\alpha} \simeq 3.0 \text{ sec}$$

Let $\tau = 2.5$ sec

Similar calculation shows that in the transient regime

$$t_{0.1} = 173°\text{C}$$

$$t_{0.2} = 297°\text{C}$$

$$t_{0.3} = 400°\text{C}$$

$$t_{0.4} = 526°\text{C}$$

This shows that the center has started cooling.

For long term, i.e., steady state

$$\tau >> \frac{r_o^2}{\alpha}, \quad \text{i.e.,} \quad \tau >> 3.0 \text{ sec}$$

Let $\tau = 15$ sec

Volume of the ball

$$V = \frac{4}{3}\pi r^3 = 0.267 \text{ cm}^3$$

Surface area $A = 4\,\pi\, r^2 = 2.01$ cm^2

We will calculate the temperature at the center after 15 sec, i.e., (18 – 3) sec after the transient period of 3 sec when the center has reached about 500°C.

$t_\infty = 50°\text{C}$ $t_1 = 500°\text{C}$

The equation to be applied for longtime cooling is

$$\frac{t_{0.1} - t_{\infty}}{t_i - t_{\infty}} = exp - \left[\frac{hA}{\rho c V}(\tau - \tau_c) \right]$$

$$\frac{t_{0.4} - 50}{500 - 50} = exp - \left[\frac{0.2 \times 2.01}{7.8 \times 0.5 \times 0.267} 15 \right] = exp - 5.80$$

$$= 3.03 \times 10^{-3}$$

$$t_{0.4} = 3.03 \times 10^{-3} \times 450 + 50$$

$$= 51.3°\text{C}$$

showing that the center has cooled.
If we consider

$\tau = 10$ sec $\quad t = 60°\text{C}$ and

$\tau = 5$ sec $\quad t = 110°\text{C}$

4.5 TRANSIENT CONDUCTION — FINITE DIFFERENCES METHOD

The basic differential equation for unidirectional transient conduction is

$$\frac{\partial t}{\partial \tau} = a \frac{\partial^2 x}{\partial t^2}$$

In the previous section we have seen the analytical method of solving this equation, subject to certain boundary conditions and with the assumption that the concerned physical properties remain essentially constant.

Another convenient method of solving the above equation is called the finite differences method. This is a computational method in which the required derivative $\partial t/\partial \tau$ is replaced by a simple algebraic expression.

The body (in x direction) is fictitiously divided into n small segments of equal width Δx. The heating time is divided into k equal intervals of $\Delta\tau$, As unidirectional conduction is assumed,

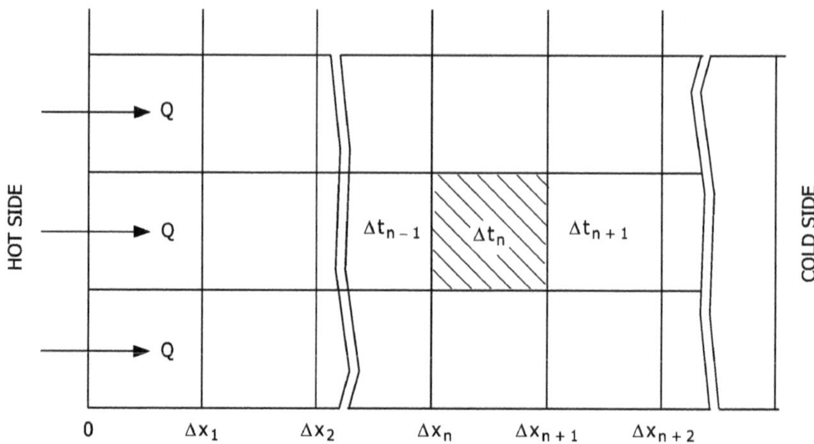

Figure 4.11 Transient conduction by finite difference method.

the temperature of any segment $t_{\Delta x,n}$ will be dependant on the temperature of the previous segment $t_{\Delta x,n-1}$ on one side, and the temperature $t_{\Delta x,n+1}$ of the next segment on the other side (Figure 4.11). This will hold for all segments at all time intervals k.

Thus the heat Q_1 received by a segment Δx_n will depend on the instantaneous values of $t_{k,\,n-1}$ and $t_{k,\,n+1}$.

Similarly the heat Q_2 leaving Δx_n and entering Δx_{n+1} will depend on $t_{k,n}$ and $t_{k,n+1}$ for the chosen interval k.

Assuming that the physical properties such as conductivity λ, specific heat c, density ρ, etc., remain constant, we can write

$$Q_1 = \frac{\lambda}{\Delta x}(t_{k,n-1} - t_{k,n}) F\Delta\tau \tag{4.21}$$

$$Q_2 = \frac{\lambda}{\Delta x}(t_{k,n} - t_{k,n+1}) F\Delta\tau \tag{4.22}$$

where F is the surface area of element Δx.

The element Δx will not transmit all the heat (Q_1) it receives but will retain some to increase its internal heat. Hence $Q_2 < Q_1$.

$$\begin{aligned} Q_1 - Q_2 &= \frac{\lambda}{\Delta x}[t_{k,n-1} - 2t_{k,n} + t_{k,n+1}] F\Delta\tau \\ &= Mc\Delta t \end{aligned} \tag{4.23}$$

where M is the mass of the element. Noting that $M = F\Delta x\rho c$ and $\Delta t = t_{k+1,n} - t_{k,n}$,

$$F\Delta x\rho c\Delta t = \frac{\lambda}{\Delta x}[t_{k,n-1} - 2t_{k,n} + t_{k,n+1}]F\Delta\tau$$

$$\Delta t = t_{k+1,n} - t_{k,n} = \frac{\lambda\Delta\tau}{\rho c\Delta x^2}[t_{k,n-1} - 2t_{k,n} + t_{k,n+1}] \tag{4.24}$$

But $\dfrac{\lambda}{\rho c} = \alpha$ (diffusivity), and hence,

$$t_{k+1,n} = \frac{\alpha\Delta\tau}{\Delta x^2}[t_{k,n-1} - 2t_{k,n} + t_{k,n+1}] + t_{k,n} \tag{4.25}$$

If we choose Δx so that

$$\frac{\alpha\tau}{\Delta x^2} = \frac{1}{2} \tag{4.26}$$

we get,

$$t_{k+1,n} = \frac{t_{h,n-1} + t_{k,n+1}}{2} \tag{4.27}$$

Equation 4.27 is the basis of finite difference method. It states that the temperature $t_{k,n}$ of the nth segment at interval $\Delta\tau_k$ is equal to the arithmetic mean of the temperatures of segments immediately before (Δx_{n-1}) and immediately after (Δx_{n+1}) in the previous time interval ($\Delta\tau_{k-1}$).

For the successful application of this method it is necessary to correctly choose Δx and $\Delta\tau$. Usually the x dimension of the body is given. This x is divided into a number of segments having equal width Δx. If Δx chosen is too large, the number of segments becomes small and the results obtained are not correct. On the other hand, if the width is too small, the number of segments is large. This increases accuracy but takes a long time for computations. Generally, 6 to 10 segments of x are convenient and give reliable results. Once Δx is chosen, $\Delta\tau$ is determined from Equation (4.26). The total heating time t then gets divided into equal time intervals $\Delta\tau$. The next solved example will make the application of this method clear.

EXAMPLE 4.6

A furnace wall made of fire clay bricks has a thickness of 200 mm and is initially at a temperature 60°C. The temperature of the furnace is suddenly increased to 1400°C. The outer surface of the wall is at an ambient temperature 30°C. Find the temperature distribution of the wall 2 h after the heating starts and the heat absorbed by the wall in this time.

Density of bricks	ρ	1.9×10^3 kg/m^3
Thermal conductivity	λ	0.85 kg/m°C
Specific heat	c	0.9 kJ/kg°C
Heat transfer coefficient inside the furnace	h_i	0.32 kw/m^2°C
Heat transfer coefficient outside the wall	h_o	12 w/m^2°C

Solution

The thermal diffusivity will be

$$a = \frac{\lambda}{c\rho} = \frac{0.85}{1.9 \times 10^3 \times 0.9 \times 10^3}$$

$$= 0.5 \times 10^{-6} \text{ m}^2/\text{sec}$$

Divide the wall into eight segments to obtain Δx

$$\Delta x = \frac{200}{8} = 25 \text{ mm}$$

$$\Delta \tau = \frac{\Delta x^2}{2a}$$

$$= 625 \text{ sec} = 10.4 \text{ min}$$

The total time is 2 h, hence,

$$\text{time intervals are } \frac{120}{10.4} = 11.6, \text{ say } 12$$

The furnace at the beginning of heating, i.e., at $\Delta\tau < 0$ is at 60°C. Hence, the whole wall, throughout its thickness is at 60°C.

At $\Delta\tau = 0$ when the heating is started the inner wall will attain 1400°C very quickly. The initial temperature of the inner side of the wall will be the mean of source and ambient temperatures

$\therefore$ Δt at $0\Delta x$, $0\Delta\tau$

$$= \frac{1400 + 60}{2} = 730°\text{C}$$

All other segments are at 60°C at $\Delta\tau = 0$.

In the next time increment, $1\Delta\tau$, the inner temperature of the wall will attain 1400°C and will remain constant.

We assume that the temperatures of various segments of the wall, i.e., $1\Delta x$, $2\Delta x, \ldots, n\Delta x$ will increase progressively by one step at a time, i.e., at $\Delta\tau$. Only the first segment will get heated. At the next instant the first and second segments will get heated, and so on.

For calculations the expression used will be

$$t_{k,n} = \frac{t_{k-1,n-1} + t_{k-1,n+1}}{2}$$

The results are best presented in a tabular form as shown in the Table 4.1. The time segments form rows and the wall segments form columns.

For example:

If the temperature at $4\Delta t$ and $2\Delta x$ is to be calculated, note that at

$t_{3\Delta\tau}$, $1\Delta x$ is 814°C

and $t_{3\Delta\tau}$, $3\Delta x$ is 144°C

Table 4.1 [Example 4.6]

Time	← Furnace Inner Side				Segment Number			Furnace outer side →			
$\Delta\tau$	$0\Delta x$	$1\Delta x$	$2\Delta x$	$3\Delta x$	$4\Delta x$	$5\Delta x$	$6\Delta x$	$7\Delta x$	$8\Delta x$	$9\Delta x$	$10\Delta x$
$0\Delta\tau$	730	60	60	60	60	60	60	60	60	60	60
$1\Delta\tau$	1400	395	60	60	60	60	60	60	60	60	60
$2\Delta\tau$	1400	730	228	60	60	60	60	60	60	60	60
$3\Delta\tau$	1400	814	395	144	60	60	60	60	60	60	60
$4\Delta\tau$	1400	898	479	227	102	60	60	60	60	60	60
$5\Delta\tau$	1400	940	562	290	143	81	60	60	60	60	60
$6\Delta\tau$	1400	981	615	352	185	101	70	60	60	60	60
$7\Delta\tau$	1400	1008	666	400	226	127	80	65	60	60	60
$8\Delta\tau$	1400	1033	704	443	263	153	96	70	63	60	60
$9\Delta\tau$	1400	1052	738	484	298	180	111	80	65	61.5	60
$10\Delta\tau$	1400	1069	768	518	332	204	130	88	71	63	60
$11\Delta\tau$	1400	1084	794	550	361	231	146	100	76	65	60
$12\Delta\tau$	1400	1097	817	577	390	253	165	111	82	68	60

Hence, $t_{4\Delta\tau}$, $2\Delta x$

$$= \frac{814 + 144}{2} = 479°\text{C}$$

In a similar manner all the temperatures for various time and wall segments are calculated.

Note: The temperature of the outer wall is practically constant for all the heating time, i.e., 2 h.

The problem states that the outer temperature is held at 60°C and the atmospheric temperature is 30°C. Heat will be lost from the wall to the environment by convection and radiation at the given transfer rate of 12 w/m^2°C. This loss is not calculated here.

The problem also states that the wall temperature (inside) is raised to 1400°C. The heat source (flame or electric element) will have to be at a slightly higher temperature (say 1450–1500°C) to have a positive transfer coefficient of 0.32 kW to the wall.

The average temperature of the wall at the end of the 2 h heating period is

$$\frac{\frac{1400+1097}{2}+\frac{1097+817}{2}+\frac{817+577}{2}+\frac{577+390}{2}+\frac{390+253}{2}+\frac{253+165}{2}+\frac{165+111}{2}+\frac{111+82}{2}+\frac{82+68}{2}+\frac{68+60}{2}}{10}$$

$$= \frac{4291}{10} \sim 430°\text{C} = t_a$$

The thickness of the wall is 200 mm. Let us calculate the heat stored for the 1 m^2 area

$$\text{Volume of wall } V = 1 \times 0.2 = 0.2 \text{ m}^3$$

$$\text{Specific heat } c = 0.9 \text{ kJ/kg°C}$$

$$\text{Density } \rho = 1.9 \times 10^3 \text{ kg/m}^3$$

$$\text{Time } \tau = 2 \text{ h}$$

Heat stored per 2 h period

$$= V\rho c t_a$$

$$= 0.2 \times 1.9 \times 10^3 \times 0.9 \times 430$$

$$= 147 \times 10^3 \text{ kJ/2 h}$$

4.6 APPLICATION OF THE FINITE DIFFERENCE METHOD TO A MULTILAYERED WALL

The finite differences method in the previous section can also be applied to a two- or more layered wall (i.e., a wall constructed of two or more materials). Firstly, only the intermediate boundary between the layers needs to be separately calculated.

Consider a two-layer wall as shown in Figure 4.12. Let Y,Y′ be the intermediate boundary between the material's X (on the hot side) and X′ (on the cold side). Let the temperature of X and X′ immediately before and after the boundary be t_1 and t_2 and let λx and $\lambda x'$ be their conductivities, and Δx and $\Delta x'$ their thicknesses.

The heat flux will be the same on the left and right side of the boundary. The thermal resistance of each layer is $\frac{\Delta x}{\lambda}$ and the heat flow is under a potential $t_1 - t$ on the left and $t - t_2$ on the right of the boundary. (The whole arrangement can be considered as two resistances in series with a potential difference (P.D.) of V volts)

$$\frac{t_1 - t}{\frac{\Delta x}{\lambda x}} = \frac{t - t_2}{\frac{\Delta x'}{\lambda x'}} \tag{4.28}$$

which simplifies to

$$t = \frac{\frac{\Delta x'}{\lambda x'} t_1 + \frac{\Delta x}{\lambda x} t_2}{\frac{\Delta x'}{\lambda x'} + \frac{\Delta x}{\lambda x}} \tag{4.29}$$

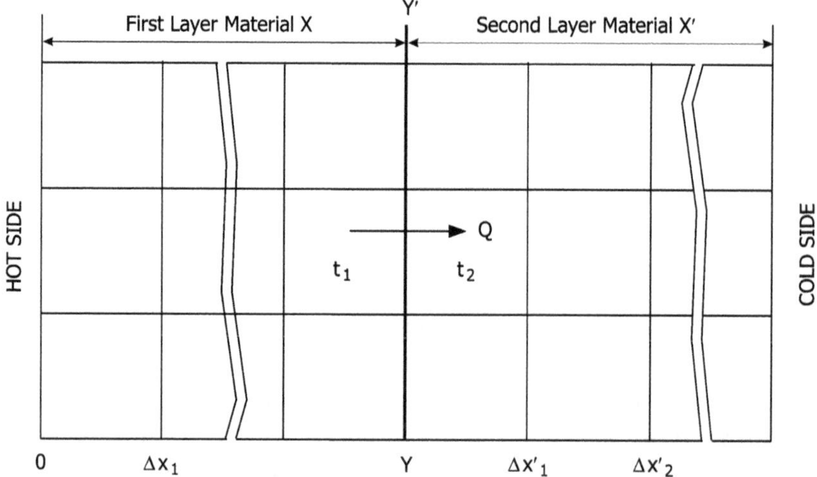

Figure 4.12 Transient conduction by finite difference method for two-layered wall.

Secondly, it should be remembered that although the process is taking place in the same time τ or time element $\Delta\tau$, the two layers have different conductivities and thicknesses. Hence, the line or length segments Δx will be different.

If $\Delta\tau$ is the common time segment by using Equation (4.26)

$$\Delta\tau = \frac{\Delta x'^2}{\alpha'} = \frac{\Delta x^2}{\alpha} \tag{4.30}$$

$$\therefore\ \Delta x' = \Delta x\sqrt{\alpha'/\alpha} \tag{4.31}$$

where α' and α are the thermal diffusivities of $\Delta x'$ and Δx.

This method can also be used to determine the minimum thickness of the wall. For this it is necessary to know the maximum service temperature or the refractoriness of the wall material.

EXAMPLE 4.7

A two-layer furnace wall consists of a 60-mm ceramic wool material on the hot side and an 80-mm rock wool layer on the cold (outer) side.

The initial temperature of the furnace and the atmosphere is 30°C. The heat is then turned on to obtain a temperature of 1250°C on the inside.

Find the heat absorbed per square meter by the wall in a heating period of 90 min and the distribution of the wall temperature at the end of heating.

For ceramic wool

$$\text{Density } \rho_1 = 300 \text{ kg/m}^3$$

$$\text{Thermal conductivity } \lambda_1 = 0.110 \text{ w/m°C}$$

$$\text{Specific heat } c_1 = 1.07 \text{ kJ/kg°C}$$

For rock wool

$$\text{Density } \rho_2 = 100 \text{ kg/m}^3$$

$$\text{Thermal conductivity } \lambda_2 = 0.032 \text{ w/m°C}$$

$$\text{Specific heat } c_2 = 0.75 \text{ kJ/kg°C}$$

Heat transfer coefficient inside the furnace = 0.25 kw/m^{2}°C.

Heat transfer coefficient between outer wall and atmosphere = 10 w/m^{2}°C.

Solution

For ceramic wool

$$\text{Thermal diffusivity } \alpha_1 = \frac{\lambda_1}{c_1\rho_1}$$

$$= \frac{0.110}{1.07\times 10^3 \times 300}$$

$$= 0.342\times 10^{-6}\ \text{m}^2/\text{sec}$$

For rock wool

$$\text{Thermal diffusivity } \alpha_2 = \frac{\lambda_2}{c_2\rho_2}$$

$$= \frac{0.032}{0.75\times 10^3 \times 10^2}$$

$$= 0.4\times 10^{-6}\ \text{m}^2/\text{sec}$$

Let us divide the ceramic wool layer into three layers of 20 mm each. Then

$$\Delta\tau = \frac{0.02^2}{2\times 0.342\times 10^{-6}}$$

$$= 9.6\,\text{mins} \simeq 10\ \text{min}$$

and $$2\Delta x = 0.02\sqrt{\frac{0.4}{0.342}}$$

$$= 21\ \text{mm} \simeq 20\ \text{mm}$$

Similarly, the thermal resistances of the $1\Delta x$ and $2\Delta x$ segments are

$$r_1 = \frac{0.1020}{0.110} = 0.18\ \frac{\text{m}^2\,°\text{C}}{\text{w}}$$

$$r_2 = \frac{0.020}{0.32} = 0.625\ \frac{\text{m}^2\,°\text{C}}{\text{w}}$$

The temperature distribution is shown in Table 4.2. The first temperature at $0\Delta\tau$, $1\Delta x_o$ is shown as the average of 1250 and 30°C. The inner face then attains 1250°C and retains it throughout the 90-min period.

Except for, this, the whole lining, inner and outer, is at 30°C at the beginning and then gradually gets warmed up.

The temperatures are calculated by the usual formula

$$t_n = \frac{t_{n-1\Delta n-1} + t_{n+1\Delta n+1}}{2}$$

TABLE 4.2 [Example 4.7]

Time	Central Layer							
$\Delta\tau$	$1\Delta x_0$	$1\Delta x_1$	$1\Delta x_2$	$1\Delta x_3$	$2\Delta x_0$	$2\Delta x_1$	$2\Delta x_2$	$2\Delta x_3$
$0\Delta\tau$	640	30	30	30	30	30	30	30
$1\Delta\tau$	1250	335	30	30	30	30	30	30
$2\Delta\tau$	1250	640	182	30	30	30	30	30
$3\Delta\tau$	1250	716	335	60	30	30	30	30
$4\Delta\tau$	1250	792	338	266	45	30	30	30
$5\Delta\tau$	1250	830	549	312	148	38	30	30
$6\Delta\tau$	1250	900	571	460	171	89	34	30
$7\Delta\tau$	1250	910	680	482	274	102	60	32
$8\Delta\tau$	1250	965	696	590	292	167	67	45
$9\Delta\tau$	1250	973	778	606	378	180	106	48
			60mm		80 mm			Totaling 140 mm

The temperature at the contact layer, i.e., $1\Delta x_3$ is calculated by the formula,

$$t_1\Delta x_3 = \frac{t_{1\Delta x_2} X r_2 + t_{2\Delta x_0} X r_1}{r_1 + r_2}$$

where r_1 and r_2 are the thermal resistances of the two materials.

$$r_1 = \frac{\Delta x_1}{\lambda_1} \quad \text{and} \quad r_2 = \frac{\Delta x_2}{\lambda_2}$$

After 90 min the average temperature of the outer wall is

$$\frac{32 + 45 + 48}{3} = 42°\text{C}$$

Average temperature of the inner wall, i.e., ceramic fiber lining is

$$\frac{1250 + 2 \times 778 + 2 \times 973 + 1 \times 606}{6} = 893°\text{C}$$

Average temperature of rock wool lining is

$$\frac{378 + 2 \times 180 + 2 \times 106 + 48}{6} = 166°\text{C}$$

The amount of heat stored by 1 m^2 of the linings:
For ceramic wool

$$\text{Thickness} = 3 \times 20 = 600 \text{ mm} = 0.06 \text{ m}$$

$$\text{Specific heat} = 1.07 \text{ kJ/kg°C}$$

$$\text{density} = 300 \text{ kg/m}^3$$

$$\text{Heat absorbed/m}^3 = 0.06 \times 1 \times 300 \times 1.07 \times 10^3 \times 893$$

$$= 17{,}200 \text{ kJ}$$

For rockwool:

$$\text{Thickness, } 4 \times 20 = 0.800 \text{ mm} = 0.08 \text{ m}$$

$$\text{Specific heat} = 0.75 \text{ kJ/kg°C}$$

$$\text{density} = 100 \text{ kg/m}^3$$

$$\text{Heat absorbed/m}^3 = 0.08 \times 1 \times 100 \times 0.75 \times 10^3 \times 166$$

$$= 1000 \text{ kJ}$$

Total heat obsorbed by the lining in 90 min

$$= 17200 + 1000 = 18,200 \text{ kJ/m}^2$$

The temperature of the outer wall is > 30°C from $7\Delta t$ and has an average value of 42°C. Heat lost to the environment is

$$= 10 \times (42 - 30)(90 - 60) \times 60$$

$$= 216 \text{ kJ/m}^2, 90 \text{ min}$$

Suppose that for design purposes, the following restrictions apply:

1. Hot face temperature < 1400°C
2. Rock wool temperature < 750°C
3. Furnace outer wall temperature < 60°C

A glance at the table will show that restriction (1) is fixed by the data and selection of proper ceramic fiber.

The interface temperature (at $2\Delta x_0$) is well below 750°C as required by restriction (2).

The outer wall temperature is below 60°C, satisfying restriction (3). However, if this restriction could be loosened, we can cut down the last outer layer, $(2\Delta x_3)$, thus reducing the cost.

4.7 CONCENTRATED HEAT SOURCES

Laser and electron beams are two important heating sources that are now commercially available and are well established in some specialized fields such as welding and drilling. These are radiation sources that can be concentrated (focused) at very small radius spots (r_f= 0.1 to 1.0 mm) and have a very high energy density (~ 10^4–10^8 w/cm^2).

The interaction of such beams and target material is a very interesting topic in heat transfer. These beams can be pulsed (switched) at a frequency, or can be continuous. In case of a pulsed beam, each pulse deposits a certain "dose" of energy J on the target spot. A continuous beam delivers energy as a function of time (w).

Depending on the characteristics of the beam and the properties of the target material, the beam is absorbed at the surface or penetrates well inside. The interaction of these beams with the target is discussed in detail in Chapter 11.

We will review the available theoretical models that attempt to describe the temperature field created by such concentrated sources in the target area and volume.

It is assumed that the exposure time of the beam on the target is small (~ 0.1–10^{-4} sec) and the heat spread outside the target area is negligible. It is further assumed that the physical properties of the target (conductivity λ, diffusivity α, specific heat c, etc.) are constant.

4.7.1 Instantaneous Point Source

In this model, energy J is deposited over a vanishingly small area (a point) in time $\tau_i << \tau$. Considering the point as a small sphere situated at the center, the temperature t at a point r is given by

$$t_{r,\tau} = \frac{Q}{8\rho c\,(\pi\alpha\tau)^{3/2}}\,exp - \left[\frac{r^2}{4\alpha t}\right] \tag{4.32}$$

where Q is the energy deposited (J) at the beginning (t = 0) and r is the radial distance from the center.

Just after the momentary heat liberation, (τ = 0) temperature at the center (r = 0) is infinite and will decrease at every point as $t^{-3/2}$. At large values of τ ($\rightarrow \infty$), $t \rightarrow t_{0.1}$.

The time duration of the process and the infinitesimal size of the hotspot are the key features of this model.

Similar models are available for instantaneous line- and plane-heat sources. They are not described here. The instantaneous point source model is useful for pulsed radiation sources.

If the heat source is in the form of a disk of radius a, the temperature along the disk axis, i.e., the z coordinate, is given by

$$t(z,\tau) = \frac{q}{2\sqrt{\pi\alpha\tau}}\left(1 - e^{-\frac{a^2}{4\alpha\tau}}\right)e^{-\frac{z^2}{4\alpha\tau}} \tag{4.33}$$

This model can be applied to a laser spot where the energy is (assumed) absorbed in a thin cylinder (disk-like) region on the surface.

4.7.2 Continuous Sources

An uninterrupted or continuous source will deliver a power q (w) for the whole period $o \rightarrow \tau$. The temperature of the target will continuously increase. Assuming an infinitesimal spot at the origin, the temperature at a distance (radius) r is given by

$$t_{(r,\tau)} = \frac{q}{4\pi r} erfc\left[\frac{r}{2\sqrt{\alpha\tau}}\right] \tag{4.34}$$

at $\tau \rightarrow \infty$, $erfc\ \alpha/\sqrt{\alpha t} \rightarrow 0$, so after a long time the temperature will assume a steady state value

$$t_{(r,\infty)} = \frac{q}{4\pi r} \tag{4.35}$$

All the heat supplied will spread out uniformly and be absorbed by the target, and the temperature will change by $1/r$.

This model is applicable to continuous radiation sources for which the distance r from the source and time τ of irradiation are important to achieve the necessary temperature at the desired location.

In many instances where we apply a train of pulses for a certain duration we get a quasi-continuous source. An analytical solution exists for this situation but the calculations are cumbersome. Usually we know the number of pulses delivered in a given time and average energy (J) per pulse. This can be converted to an average power (w) delivered to the target in given time.

If heat is supplied to the target surface ($z = 0$) over a circle of radius a, the temperature change along the axis with time τ is given by

$$t(z,\tau) = \frac{2q\sqrt{\alpha t}}{\lambda}\left[ierfc\ \frac{z}{2\sqrt{\alpha\tau}} - ierfc\frac{\sqrt{z^2+a^2}}{2\sqrt{\alpha\tau}}\right] \tag{4.36}$$

where λ is the conductivity (w/m°C).

On the surface at the center ($z = 0$) the temperature is given by

$$t(o,\tau) = \frac{2q\sqrt{\alpha t}}{\lambda}\left[1.1284 - ierfc\ \frac{a}{2\sqrt{\alpha\tau}}\right] \tag{4.37}$$

since $2ierfc\ o = 1.1284$. The second term in the bracket is usually small and can be neglected. Hence,

$$t(o,\tau) = \frac{1.1284q\sqrt{\alpha\tau}}{\lambda} \tag{4.38}$$

or $$q = \frac{0.886\lambda t}{\sqrt{\alpha\tau}} \tag{4.39}$$

The above equation can be used to calculate the power density for obtaining a certain temperature at the center of the spot.

Equation (4.36) to Equation (4.39) are very useful for the estimation of power densities or temperatures in the case of concentrated sources. They, however, do not consider the change of state (melting, evaporation) and the latent heat absorbed during heating.

Concentrated sources are increasingly used for surface-heat treating in which the rates of heating and cooling at the focal spot are important.

The rate of heating for a given power density q_o and irradiation time τ is given by

$$\frac{dt}{d\tau} = \frac{q_o}{\sqrt{\lambda c \rho \pi \tau}}$$

where c and ρ are the specific heat and density respectively.

The rate of cooling is given by

$$\frac{dt}{d\tau} = \frac{q_o}{\sqrt{\pi \lambda c \rho}} \left[\frac{1}{\sqrt{\tau}} - \frac{1}{\sqrt{\tau - \tau_i}} \right] \tag{4.40}$$

where τ is the time after the irradiation is stopped (i.e., the cooling time) and τ_i is the pulse duration or irradiation time.

Mathematical models for a moving source are also available. These are useful for welding operations and are not discussed here.

EXAMPLE 4.8

A pulse with an energy 10 J is incident on a thin steel plate. The focal spot is small and can be treated as a point.

Density of steel ρ = 7.86 gm/cm^3

Specific heat c = 0.5 J/gm°C

Thermal diffusivity α = 0.15 cm^2/sec

If the plate is to be joined to a small spring at about 1000 °C, determine the time within which the process has to be carried out to obtain a suitable temperature at a depth 0.1 mm.

Solution

For an instantaneous point source the expression for temperature is

$$t_{z,\tau} = \frac{Q}{8\rho c\,(\pi\alpha\tau)^{3/2}}\, exp - \frac{z^2}{4\alpha\tau}$$

Given $Q = 10$ J, $\tau = 0.05$ s, $z = 0.01$ cm

$$t_{(0.01,0.05)} = \frac{10}{8\times 7.86\times 0.5}\times\frac{1}{(\pi\times 0.15\times 0.05)^{3/2}}$$

$$\times\; exp - \frac{0.01^2}{4\times 0.15\times 0.05}$$

$$= \frac{10}{31.4}\times\frac{1}{3.62\times 10^{-3}}\times 0.997$$

$$= 87°\text{C}$$

For the same spot at $\tau = 0.025$

$$t = \frac{10}{31.4}\times\frac{1}{1.28\times 10^{-3}}\; exp - \frac{0.01^2}{4\times 0.15\times 0.05}$$

$$= 250°\text{C}$$

Similarly

For $\tau = 0.012$ sec $t = 740°$C

and for $\tau = 0.005$ sec

$$t = \frac{10}{31.4}\times\frac{1}{(\pi\times 0.15\times 0.005)^{3/2}}\; exp\; \frac{0.01^2}{4\times 0.15\times 0.005}$$

$$= \frac{10}{31.4}\times\frac{1}{1.14\times 10^{-4}}\times 0.967$$

$$= 2700°\text{C}$$

Thus the joining has to be carried out within about 10 m/sec after the pulse.

EXAMPLE 4.9

A continuous laser beam having 600 W power is focused on the surface of a semiinfinite steel plate. Radius of the focal spot is 150 μm.

Diffusivity of steel (α) = 0.08 cm²/sec
Thermal conductivity (λ) = 0.51 W/m°C
Initial temperature throughout the slab = 30°C

Assume that all the incident power is absorbed by the target.

Determine the temperature along the beam axis at a spot located 1.0 mm from the surface after 0.5, 1.0, 1.5, and 2.0 sec of exposure.

Solution

The equation to be used is

$$T_{(z,\tau)} = \frac{2q_o\sqrt{\alpha\tau}}{\lambda}\left[ierfc\,\frac{z}{2\sqrt{\alpha\tau}} - ierfc\,\frac{\sqrt{z^2 + r_f^2}}{2\sqrt{\alpha t}}\right]$$

$$q_o = \text{power density} = \text{power/spot area}$$

$$= \frac{600}{\pi \times 0.015^2} = 8.5 \times 10^5 \text{ w/cm}^2$$

$$\alpha = 0.08 \text{ cm}^2\text{/sec } \lambda = 0.51 \text{ w/m°C}, \quad z = 0.1 \text{ cm}$$

For $\tau = 0.5$ sec substituting in the above equation

$$T(0.1, 0.5) = \frac{2 \times 8.5 \times 10^5}{0.51}\sqrt{0.08 \times 0.5}$$

$$\times\left[ierfc\frac{0.1}{2\sqrt{0.08 \times 0.5}} - ierfc\,\frac{\sqrt{0.015^2 + 0.1^2}}{2\sqrt{0.08 \times 0.5}}\right]$$

$$= 6.67 \times 10^5\,(ierfc\,0.25 - ierfc\,0.253)$$

$$= 6.67 \times 10^5\,(0.6982 - 0.6565)$$

$$= 6.67 \times 10^5 \times 0.007$$

$$= 667°\text{C}$$

Note: The term in brackets gives a small, dimensionless multiplier.

Similarly

$$t(0.1, 1.0) = 954°\text{C} \quad (t = 1.05 \text{ sec})$$

$$t(0.1, 1.5) = 1150°\text{C} \quad (t = 1.5 \text{ sec})$$

$$t(0.1, 2) = 1300°\text{C} \quad (t = 2.0 \text{ sec})$$

If we change the point location, i.e., the z coordinate, the temperature history at that point can be calculated in a similar way.

Note: The surface (z = 0) assumes a very high temperature in a very short time and there will be some evaporation loss.

EXAMPLE 4.10

The plate in Example 4.9 is to be surface hardened by scanning or rastering the beam over the surface. The temperature at the surface is not to exceed 1300°C after 1.0 m/sec.

Determine the power density required. Also determine the temperature along the beam axis at a depth of 200 μm after 10^{-3} sec on the application of the power density calculated above.

Solution

The equation for surface temperature can be obtained by putting $z = 0$ in the expression for $T_{(z,\,t)}$ in Example 4.8.

$$T_{(0,t)} = \frac{2q_o\sqrt{\alpha t}}{\lambda}\left(ierfc\, o - ierfc\frac{r_f}{2\sqrt{\alpha t}}\right)$$

$$= \frac{2q_o\sqrt{\alpha t}}{\lambda}\left(1.1284 - ierfc\frac{r_f}{2\sqrt{\alpha t}}\right)$$

Given $\tau = 10^{-3}$ sec, $\alpha = 0.08$ cm²/sec, $\lambda = 0.51$ w/cm°C, $r_f = 0.015$ cm substituting

$$1300 = \frac{2q_o\sqrt{0.08\times 10^{-3}}}{0.51}\left[1.1284 - ierfc\frac{0.015}{2\sqrt{0.08\times 10^{-3}}}\right]$$

$$= q_o\frac{17.88\times 10^{-3}\times 0.9611}{0.51} = 3.37\times 10^{-2}$$

$$q_o = \frac{1300\times 10^2}{3.37}$$

$$= 3.86\times 10^4 \text{ w/cm}^2$$

The temperature at $z = 200\ \mu$m or 0.02 cm

$$t(0.02, 10^{-3}) = \frac{2\times 3.86\times 10^4\times 8.94}{0.51}$$

$$\times\left[ierfc\frac{0.02}{17.88\times 10^{-3}} - ierfc\frac{\sqrt{0.02^2+0.15^2}}{17.88\times 10^{-3}}\right]$$

$$\simeq 60°\text{C}$$

The beam power required is

$$Q = \text{power density}\times\text{spot area}$$

$$= 3.86\times 10^4\times\pi\times 0.015^2 = 27.3 \text{ w}$$

If the time t is reduced to an order of 10^{-4} to 10^{-6} sec, the power required will increase.

4.8 TRANSIENT CONDUCTION GRAPHICAL METHOD (SCHMIDT'S METHOD)

Consider a single-layer wall extending in the x direction as shown in Figure 4.13. The wall is at a uniform temperature t_o at the beginning of heating, i.e., when $\tau = 0$.

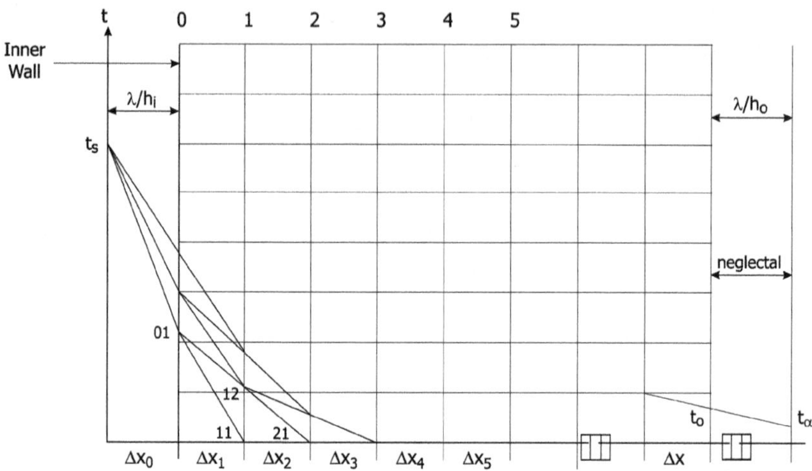

Figure 4.13 Schmidt's chart considering the source to wall heat transfer.

After the heating starts, the temperature profile changes along with time. After a very long time ($\tau = \infty$) the temperature profile will be an inclined straight line indicating steady state conditions.

Starting from $\tau = 0$ to $\tau = \tau$ we can gradually develop the temperature profile by the graphic (Schmidt's) method described here.

It is assumed that the physical properties of the wall (conductivity λ, diffusivity α, etc.) remain constant. We may choose mean values to improve accuracy.

First, the thickness x of the wall is divided into a number of equal segments Δx. The number of segments should not be too small or too large.

The appropriate time interval $\Delta\tau$ is obtained in a manner similar to the numerical method in the previous section, i.e.,

$$\Delta\tau = \frac{\Delta x^2}{2\alpha} \tag{4.42}$$

On graph paper we fix the wall on the x axis and mark the division of segments Δx_1, Δx_2, etc.

On the vertical (y) axis we plot the temperatures to any convenient scales.

On the left hand of the y axis ($-x$, y) the source (heater or flame) temperature is fixed at a distance λ/h from the axis. λ is the thermal conductivity of the wall and h is the heat transfer coefficient from the source to the wall. Hence, λ/h has dimension m.

This may not always be necessary. The wall can be fixed at the required temperature. For positive heat transfer from the source to the wall, it is necessary that the source is at a slightly higher temperature (50–100°C) than the given wall temperature.

After fixing wall segments (Δx) and the temperature, the maximum furnace temperature is marked on the y axis (t_f). Name the segments Δx_0 (the inner wall), Δx_1, Δx_2, etc., and mark the initial temperature t_o.

The t_f point at Δx_0 is joined to the t_0 point on Δx_2. This shows that in the first interval $\Delta\tau$ the heat has penetrated the segment Δx_1. Note that this line will intersect Δx_1 at t_1'.

In the next step t_1' is connected to Δx_3 to show that heat has now penetrated to Δx_3. This line is intersected by Δx_2 and t_2' showing rise in temperature of Δx_2 at $2\Delta\tau$. This point is joined to the point t_f on the y axis.

This procedure is successively repeated to cover all the wall segments.

This is shown schematically in Figure 4.14 for the first few wall segments.

For this figure, the scales are arbitrary. t_f is the wall temperature and t_o is the initial or outside temperature. t_o is constant throughout the wall at $\tau < 0$.

The development of the chart progresses as follows:

At $\tau = 0$ the wall (inner) attains a temperature t_f.

At $\tau = \Delta\tau_1$ only the first segment gets heated and all other segments are at t_o.

Hence, we join t_f and t_{21} by a straight line which intersects Δx_1 at 11, which is its new temperature.

In the next instant, i.e., $\tau = \Delta\tau_2$, the heat progresses to the segment Δx_2 and all other segments are at t_o. So points 11

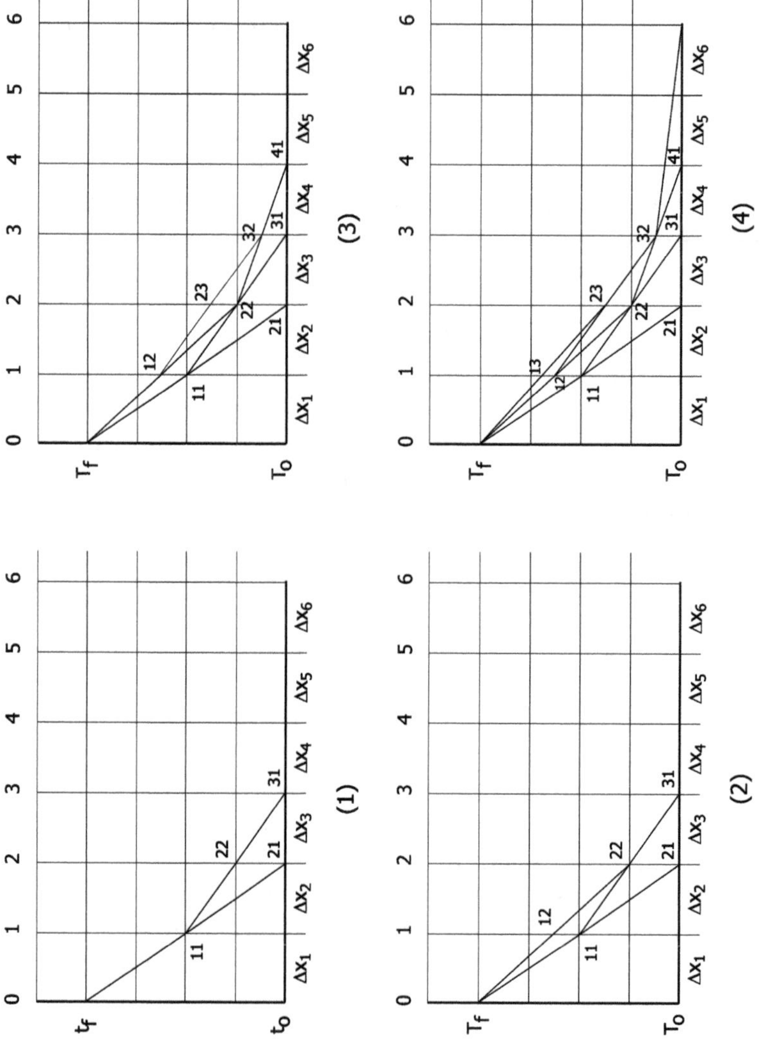

Figure 4.14 Development of temperature profile in the wall.

and 31 are joined. They intersect Δx_2 at 22 which is its new temperature.

In this time $\Delta\tau_2$ the temperature of the segment Δx_1 is also increasing. This is determined by joining 22 and t_f. The new temperature of Δx_1 is given by the point 12.

In a similar manner the chart is developed for one segment at a time. Remember that the temperature of a segment is always fixed by the temperatures of two adjoining segments (left and right) at the immediately previous time, i.e., again,

$$t_{\Delta_{xn}\Delta\tau_{n-1}} = \frac{t_{\Delta x_{n-1}\Delta\tau_{n-1}} + t_{\Delta x_{n-1}\Delta\tau_{n-1}}}{2} \tag{4.43}$$

This is similar to the finite differences method of the previous section. However, in this case we solve the above equation graphically, therefore much more quickly.

As the assumption that the inner wall instantaneously attains the source temperature is not exactly valid, we can include the source-to-wall heat transfer in the chart.

The heat is incident on the wall from the source, which has a transfer coefficient h and a conductivity λ, hence the thermal resistance is λ / h_i m. We have to include this as a fictitious wall segment on the left of the wall as shown in Figure 4.15.

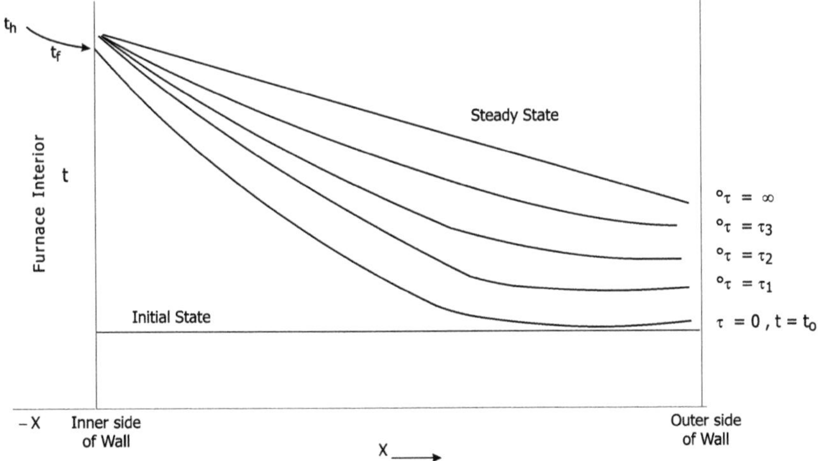

Figure 4.15 Change in temperature of wall with time.

A similar fictitious resistance exists on the outer wall and can be taken into consideration in a similar manner. This resistance is usually very small and is neglected.

Schmidt's method can also be used for a multiple-layered wall. The interface temperature will have to be calculated as in previous method.

The chart can also be used for cooling and heating of large ingots, plates, etc. For design purposes we can use the chart for

1. Determining the wall thickness for keeping the outer wall temperature to a desired level in given time
2. To determine the outer wall temperature in a given period and to obtain heat lost to the surroundings

Change in temperature profile of the wall with time is shown in Figure 4.15.

Chapter 5

Fuels and Their Properties

CONTENTS

5.1 INTRODUCTION

Fuels are sources of heat. These can be divided into two major categories. All combustible fuels belong to the first category. Electricity as a fuel does not require combustion and thus belongs to the second category.

In this chapter we will review combustible fuels. These can be subdivided into three types:

1. Solid fuels—Coal, coke, anthracite, wood, charcoal, etc.
2. Liquid fuels—Distillates such as kerosene, diesel, furnace oils, tars, etc.
3. Gaseous fuels—Natural gas, biogas, liquified petroleum gases such as butane, propane, etc.

Small and medium ovens and furnaces do not use solid fuels, hence they are not discussed in this book except in waste incineration.

The liquid fuels used in the heating of furnaces and ovens are distillates and furnace oils which are treated at length.

There are many fuel gases made from coal and oils. The use of such derived gases is now dwindling. The gases mainly used and universally available are natural gas or its blends, LPG, such as butane or propane, or their mixtures. Biogas or sewage gas are becoming increasingly popular because of availability, economics, and social and ecological need. Hence, only these gases which are mainly used are discussed in detail.

5.2 PROPERTIES OF FUELS

Many properties of gaseous and liquid fuels are peculiar to their type. For oils the properties are given per kg, and for gases, in m^3.

The heat obtained by burning a unit quantity is called its heating or calorific value. This is common property to all fuels. It is determined by using an appropriate calorimeter. It can also be calculated theoretically. These calculations are discussed in detail in subsequent sections. Faced with an unknown or new fuel the designer can get a reasonable estimate of its heating value from these calculations.

For determining the calorific value and many other combustion parameters, the chemical analysis of fuels is the next important property.

Fuel oils contain a very large number of hydrocarbons and organic compounds. These contain elements like carbon, hydrogen, and sulfur. In the combustion process these elements burn with exothermic reactions, give off heat, and form gaseous oxides. These together form the combustible mass. Hence, the chemical analysis giving their content is the important factor, not the way they are chemically combined to form the "oil."

Other minor constituents of oil occasionally occurring are ash and moisture. There may be some oxygen also.

Some fuel oils contain traces of waxes, which are heavy hydrocarbons and are difficult to burn. They may clog the fine passages in burners.

Viscosity of oils is the next important property. It is related to flow and atomization of oils. It is also used to classify the various commercial grades of oils.

The method of classification is not universally standardized. The designations may be by letters A, B, C, ... or by numbers. The units of reporting viscosity also differ from country to country. Thus, though the absolute units are m^2/sec, there are other units such as Engler (°E), Seybolt (sec), and Redwood (sec). The viscosity decreases with increasing temperature. Thus, highly viscous (and cheap) oils can be made to flow by heating. Relations between temperature, viscosity (various units), and grades of fuel oil is shown in Figure 5.1.

Fuel gases contain one of the following hydrocarbons as main combustible components—methane (CH_4), propane (C_3H_8), or butane (C_4H_{10}).

There may be minor contents of other hydrocarbons and some hydrogen also. Natural gas contains N_2 and CO_2 as noncombustible gases in small amounts. Fuel gases do not contain ash. Propane and butane are generally supplied as liquids under pressure. Bulk natural gas is supplied through pipe lines or in bottles.

The density or specific gravity of oils and gases are also reported.

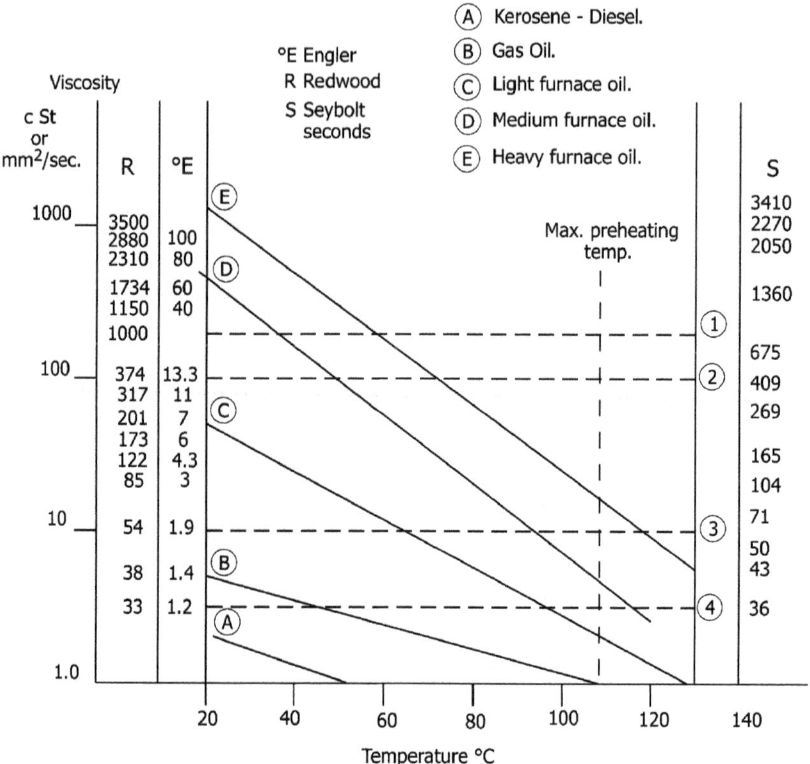

Figure 5.1 Temperature-viscosity relation for some common furnace oils.

From the analysis of combustible contents we can calculate the air required for combustion, volume, and contents of combustion products, and the theoretical maximum temperature of the flame. These parameters form the basis of furnace design and fuel selection.

Experimental methods for determining the fuel properties are not discussed as the designer does not require these methods.

Subsequent sections and solved examples will clarify the use of these properties in design.

Table 5.1 and Table 5.2 list the important properties of selected fuels.

TABLE 5.1 Properties of Selected Gaseous Fuels

Property	Units	Natural Gas	LPG Propane	LPG Butane	Biogas
Chemical composition	% vol.	CH_3 - 95–98 N_2 - 2–4 H_2S < 1.0	C_3H_8 > 98	C_4H_{10} > 93	CH_4 - 65–70 CO_2 - 25–30
Density (gas)	kg/m^3	0.73–0.78	2.0	2.5	—
Density (liquid)	kg/m^3	—	510	580	—
Cal. Value (higher)	kJ/m^3	38	93	119	—
Cal. Value (lower)	kJ/m^3	37.5	86	112	—
Theoretical combustion air.	m^3/m^3 gas	9.4	24	30	6.3
Volume of combustion products.	m^3/m^3 gas	10.4	25.8	33.5	6.6
Composition of combustion products % vol. at theoretical air	CO_2	9.6	11.6	12	15
	H_2O	21	15.4	15	9.8
	N_2	69.74	73	73	76
Flame temperature	°C	1950–2100	2100	2170	2150
Limits of flammability	% Vol. Gas in air	3.0–12	2.3–10	1.7–9	—
Burning velocity	cm/sec	30	38	38	—

Note: The properties given above are for design purposes only. There will be some variation according to the source.

TABLE 5.2 Properties of Some Liquid Fuel Oils

Property	Units	Petrol, Diesel, Kerosene	Kerosene Gas Oil	Light Oil	Medium Oil	Heavy Oil
Kinematic viscosity						
40°C	cSt or mm²/sec	0.75	1.6–5	50	1200	1500
100°C		—	0.6–1.2	3.5	20	20
Specific gravity	—	0.73	0.83	0.93	0.95	0.95
Flash point	°C	43	38	76	86	110
Flame temperature	°C	2022	2028	2100	2100	2022
Calorific value	MJ/kg	43–45	45–46	43	43	43
Air required for combustion	m^3/kg	10.8	10.7	11.0	11.5	11.5
Min. temperature for pumping	°C	Atmospheric	> –10	> 8	> 38	> 40
Min. temperature for atomization	°C	–10	7.0–10	> 55	> 90	> 100
C/H ratio by mass	—	6.5	6.5	7.3	7.4	7.4
Residue and ash %	—	—	0.02	0.05	0.05	0.08

Note: The properties given above are typical but not standard. There can be some variation. Consult the supplier for applicable properties of his product.

5.3 LIQUID FUELS

All liquid fuels are derived from the distillation of crude oil and coal tars. Distillation gives various fractions at different temperatures. A very large number of other organic primary chemicals are also obtained from distillation.

The liquid fractions used as industrial fuels vary from low boiling point—light fractions, such as kerosene—to high boiling point known as heavy fractions, which are also much thicker (viscous).

At the lower end of distillation we get thick (almost solid) fractions such as tars, paraffin, pitch, waxes, etc. At the other end we get refinery gases which are mainly gaseous hydrocarbons such as propane (C_3H_8) and butane (C_4H_{10}).

The oils obtained from distillation are not composed of single chemicals. They are mixtures of many hydrocarbons, i.e., compounds of carbon C, and hydrogen H having a general formula C_xH_y that distill in a particular temperature range.

Depending on the source of crude oil and the practice adopted in distillation, a number of oils are obtained. The variety is further increased by blending different oils. Hence, for a given furnace application many "brand name" oils are offered by different manufacturers. The manufacturer supplies the necessary data for use of their products. By comparing these data from different suppliers it can be seen that there is little variation in the useful physical properties, such as heating (calorific) value, viscosity, etc.

Liquid fuels require burners for their efficient combustion. Some liquid fuels are very viscous, making their flow through pipes and small openings in burners very difficult. Such fuels require preheating to reduce their viscosity. Burners and preheating arrangements are discussed in subsequent sections.

Fuel prices vary widely. Gasoline or petrol is the costliest and most volatile. It is rarely used for industrial heating. Diesel and kerosene come next in price. Kerosene is used in some special furnaces (kitchen, bakery, etc.) as it is cheaper than diesel and safer to handle and transport.

Furnace oils come next in price. Many grades such as heavy, medium, and light are offered. The choice depends on the application and availability.

Thick oils and tar are the cheapest industrial fuels but they are difficult to burn efficiently. They require preheating machinery.

As the viscosity of oils depends on temperature, there can be considerable variation depending on season.

An external but highly influencing factor in the cost of oils is the global fluctuation in crude oil prices.

Important properties of liquid fuels of interest are given in Table 5.2.

Variations in viscosity of these oils is shown in Figure 5.1. The standardization and classification of fuel oils varies from country to country. Hence, the data given should be taken for guidance only.

5.4 GASEOUS FUELS

There is a large variety of gaseous fuels. Coke oven and blast furnace gases are by-products of iron and steelmaking processes. Producer gas is obtained from coal, and water gas is obtained by reacting coal with steam.

Natural gas is tapped from the earth. Refinery gases are by-products of distillation of crude oil. Depending on availability, gases are also blended to obtain a suitable fuel.

Decomposition of organic waste under anaerobic conditions produces biogas, which is produced in considerable quantities in rural areas.

Gases can be transported over great distances by pipelines. Some gases are compressed and supplied in cylinders. Refinery gases such as butane and propane are liquified under pressure (liquified petrolium gas (LPG) and are transported as liquids in cylinders or tankers. Natural gas (NG) is compressed and transported as gas by pipelines or cylinders. In some processes, NG is fractionated and reformed to obtain lean gas and other by-products.

On a small scale (for chemicals laboratories, etc.) fuel gas is produced by evaporation of light liquid fuels such as kerosene and diesel.

Thus, the number of gaseous fuels available is very large. The regional availability decides the type of gas supplied to the industries. Thus, coke oven and blast furnace gases or their blends are available around steelmaking plants. Natural gas is generally mined from the sea bottom and is supplied in coastal regions. Many countries have a grid or network of underground pipelines throughout the whole country.

Besides the availability and transportability, gaseous fuels have other advantages. They burn cleanly, i.e., do not produce environmentally dangerous products of combustion, have no ash, and can be burned in relatively simple burners. Gases are also cheaper than liquid fuels.

Except refinery gases (LPG) all gases have methane (CH_4) as the main burning component. LPG consists of propane (C_3H_8) or butane (C_4H_{10}), or their (generally) 50:50 mixture. Very small amounts of other combustible hydrocarbons, sulfur (H_2S), and noncombusting constituents such as nitrogen and oxygen are also present.

Only LPG fuels are presently available in the market of india an other developing countries. However, it appears that natural gas will soon be distributed in coastal regions.

Important properties of fuel gases are given in Table 5.1. The gases locally available will have slightly different properties. It is recommended that the furnace designer consult the supplier for actual properties. The solved examples will be helpful in estimating working properties.

5.5 BIOGAS

The oils and gaseous fuels discussed in previous sections were basically natural fuels refined and processed for industrial applications.

Biogas is perhaps the only man-made gaseous fuel and is rapidly acquiring importance in homes and some industries. Hence, its production and applications are described in some detail.

Sewage waste from urban and rural communities, farms, dairies, food processing plants, hospitals and the like contain a large amount of organic wastes and water. They form a sludge or muddy aggregate. If this sludge is kept out of contact of air (i.e., in anaerobic conditions) it is digested by inherent microorganisms and converted to methane. This gas is called biogas.

Besides 65–70% methane as the combustible component, biogas contains small amounts of H_2, H_2S, N_2, and the rest H_2O. Its calorific value is 23–25 MJ/m^3. Natural gas contains 92–97% methane and has a calorific value of 35–40 MJ/m^3. Thus, compared to natural gas, biogas is a lean fuel. It is used mainly as a domestic gas and in food processing and water heating applications where energy demand is not intensive.

Combustion of biogas is covered in Example 5.3. Properties of biogas are given in Table 5.1.

Though technically a "lean" fuel, biogas has several advantages.

1. Raw material for biogas manufacturing is available in plenty locally at any large or small sewage collection facility. The sludge has no cost. In fact, its disposal by gas-making is a boon to the waste disposal problem.
2. Raw material is (redundant) available year-round.
3. After making the gas, the sludge volume is greatly reduced. The digested sludge can be used as manure or for land fill. It is relatively inert and poses no disposal problems.
4. As will be seen in the subsequent section the gas manufacturing method is very simple and requires neither technical personnel for maintenance nor complex machinery. Hence, the capital investment and running costs are also very low.

In very few cities does collected sewage receive any treatment. It is freely discharged to rivers, ponds, and oceans. Due to its biologically harmful contents, this practice gives rise to extensive pollution and health problems. Villages and small communities, farms, and food industries have no collection and disposal facilities. Biogas production minimizes all such problems.

Nonrenewable fuel supplies such as crude oil and natural gas are dwindling at a fast rate. Their cost is also ever increasing. Biogas offers an ideal solution (at least to a significant level) to this problem.

The process of converting organic sewage to gas is a biochemical process depending on the life and regeneration of bacteria. Industrial wastes contain a number of chemicals, oils, grease, and the like. These chemicals are poisonous to bacteria. Hence, gas generation from such sewage is difficult. Some pretreatment to remove these poisons is necessary. This increases the cost and thus, the simplicity of biogas generation is lost.

The gas generation is a slow and temperature-dependent process. It requires 4–16 h to complete the digestion. The conversion rate depends on the content of digestible matter (volatile matter) in the sludge. Digestion is optimum between 20 to 40°C. Thus, the gas generation slows down in winter. To overcome this problem some generators have an external (electric or steam) heating facility. If the temperature is very high the combustible component (methane) decreases.

There are two basic processors used for gas generation. Both types are briefly discussed in Section 5.5.1 and Section 5.5.2.

5.5.1 Single Stage Generation

This is a very simple generator and is shown schematically in Figure 5.2.

It consists of two parts, both made of Mild Steel. The lower part is the generator and sludge container. It has an opening(s) to admit dewatered and thickened sludge. The charging may be done by gravity flow or by a sludge pump. The bottom has a conical shape with an outlet to discharge spent sludge.

The top part consists of a dome-shaped floating gas collector. During digestion the generated gas bubbles collect in the dome, raising it as the gas accumulates. The gas is taken off through a flexible connection near the top of the dome. The gas may be compressed and stored in a separate gas tank.

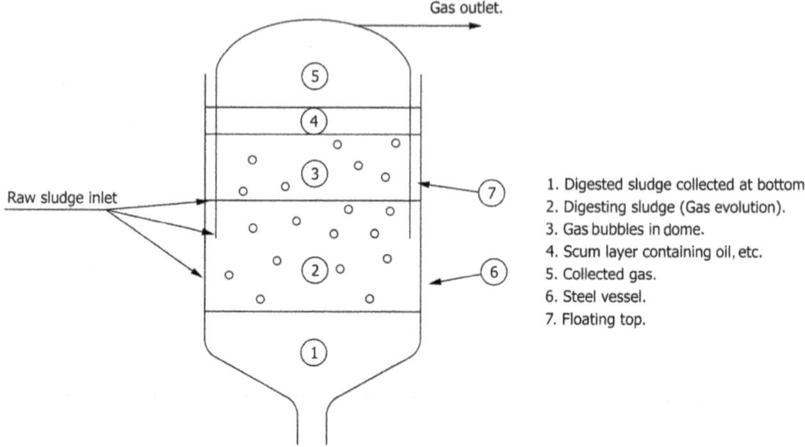

Figure 5.2 Single stage biogas generator.

As the gas is taken off more sludge is periodically charged. In this generator the digesting sludge is quiescent and generation is slow. This can be remedied by including some stirring arrangement which will expose sludge for digestion.

Single-stage generation is extensively used in rural communities.

5.5.2 Two Stage Generator

A two-stage generator is shown in Figure 5.3. There are two vessels. The first stage is a steel vessel with a fixed top, otherwise it is similar to a single-stage generator. It has a mechanically-driven stirrer and heating panels as shown.

The second stage is also a steel tank but has a floating cover. The two stages are interconnected at different levels. Well-mixed and partially-digested sludge is admitted from I to II stage by these interconnections.

The digestion started in the first stage is completed in the second stage.

The accumulated gas in the floating cover is taken off from a connection near the top.

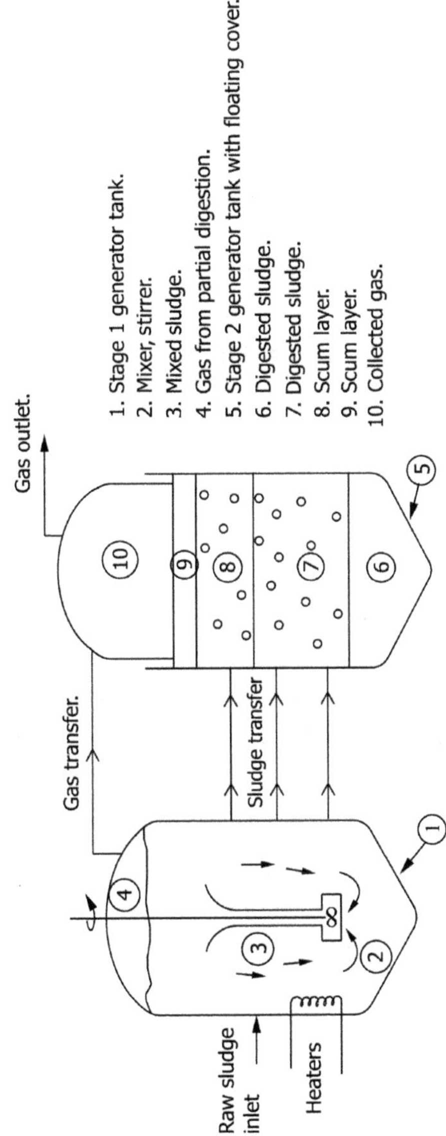

Figure 5.3 Two-stage biogas generator.

This generator gives more complete and faster gas evolution. It is suitable for large plants based on urban sewage or food industries.

Biogas generators are designed by the following criteria:

1. The population to be served. This may be human beings, dairy animals, etc.
2. Production and availability of dewatered and toxin-free sludge.
3. Composition of sludge, pH, Volatile Matter (VM), etc.
4. Expected digestion rate, temperature variation, availability of auxilliary power.

5.6 HEATING (CALORIFIC) VALUE

The amount of heat (Joules) available from the combustion of a unit quantity of a fuel is its "heating value." For liquid fuels it is generally given on weight basis (J/kg) and for gaseous fuels on volume basis (J/m^3).

A measured quantity of fuel is burned with oxygen in a calorimeter. The heat imparted by the fuel to the calorimeter and the water in calorimeter jacket is measured.

Most fuels contain hydrogen in combined or free form. Combustion of hydrogen produces water, which is in liquid or vapor form depending on the temperature of measurement. Generally, in a laboratory the products of combustion are cooled to room temperature (~20°C).

During combustion the water is evaporated, because the temperature in the calorimeter is quite high. The water vapor absorbs a considerable amount of heat (in both sensible and latent form) when it rises from room temperature to that of combustion.

Thus

$$Q^w = Q^{ws} + Q^{wl} + Q^{vs}$$

Where

Q^w = Total heat in water

Q^{ws} = Sensible heat in water from room temperature to 100°C

$Q^{w\ell}$ = Latent heat of vaporization

Q^{vs} = Latent heat of vapor (steam) from 100°C to the combustion temperature

During calorimetric measurement, the temperature comes down to room temperature. Hence, all the heat (Q_W) absorbed is returned and appears in the measurement. Hence,

$$Q^h = Q^c + Q^w$$

where Q^h is the measured heat and Q^c is the heat of combustion.

In industrial heating the component Q_w is never used as the products of combustion do not cool below 100°C in the furnace. Hence, this component is always lost. The heating value of the fuel which is actually available is lower than Q^h by an amount Q^w. Hence, $Q^\ell = Q^h - Q^w$, where Q^ℓ is the lower heating value and Q^h is the higher heating value.

Thus, two heating values, Q^h and Q^ℓ, are usually quoted.

When only one value is quoted without qualification (h or ℓ) the higher value or true value is presumed.

If a fuel does not contain hydrogen, its higher and lower heating value is the same.

The heating value can be calculated from basic combustion data for constituent elements such as carbon, hydrogen, sulfur, etc. There are semi-empirical formulae for determining the heating value. These calculations are discussed in subsequent sections.

Once the heating value of a fuel is known (either experimental or calculated) the following data can be determined:

1. Air required for combustion
2. Volume of combustion products
3. Theoretical combustion temperature
4. Composition of combustion products
5. Heat content of combustion products

These calculations will be discussed in subsequent sections. In many cases these data are supplied by the manufacturer.

The published data or manufacturers' data show some variations. These are due to the source, processing, and blending which vary from country to country.

The data given in Table 5.1 and Table 5.2 are sufficient for design purposes.

5.7 CALCULATION OF CALORIFIC VALUE

All gaseous and liquid fuels contain carbon and hydrogen as combustion elements. They are generally in the form of complex hydrocarbons or other organic molecules. The exact formulae of these molecules are difficult to list and moreover such information is not required for combustion calculations. Some fuels also contain small amounts of sulfur (~1%) which makes a minor contribution to combustion.

The basic combustion reaction is that of oxidation with atmospheric oxygen and the evolution of heat, as these reactions are exothermic. For carbon, the reaction for complete combustion is

$$C + O_2 \rightarrow CO_2 + 408{,}860 \text{ kJ/k mole} \tag{5.1}$$

For hydrogen, though the reaction is similar, the water that is formed may be in a vapor or liquid condition depending on the temperature. Thus

$$H_2 + \tfrac{1}{2}O_2 \rightarrow (H_2O)_{liq} + 286{,}223 \text{ kJ/k mole} \tag{5.2}$$

or

$$H_2 + \tfrac{1}{2}O_2 \rightarrow (H_2O)_{vap} + 241{,}800 \text{ kJ/k mole} \tag{5.3}$$

The difference 286,223 – 241,800 = 44,423 kJ/mole is due to the release of sensible and latent heat of water during condensation. This gives rise to two calorific values. The higher calorific value of Q^h presumes all water (H_2o) in liquid or condensed form. The lower calorific value Q^l presumes all H_2o in vapon (steam) form. For fuels that do not contain hydrocarbon the higher and lower calorific values are the same.

Methane (CH_4) is the simplest hydrocarbon and is a constituent of natural gas and biogas. Combustion of methane is

$$CH_4 + 2O_2 \rightarrow CO_2 + 2(H_2O)_{vap} + 805,560 \text{ kJ/mole} \qquad (5.4)$$

Combustion of higher hydrocarbons is not simple and practically realized heat values do not agree with the theoretical values.

Sulfur in fuels is rarely in free form. The simplest compound is H_2S which is present in many gaseous fuels. Its combustion is according to the equation

$$H_2S + 1.5\ O_2 \rightarrow SO_2 + (H_2O)_{vap} + 519,188 \text{ kJ/mole} \qquad (5.5)$$

If sulfur is assumed to be in elemental form, the oxidation reaction is

$$S + O_2 \rightarrow SO_2 + 296,927 \text{ kJ/mole} \qquad (5.6)$$

The heat evolved in the above given reactions is kJ/mole of the combusting (oxidizing) element or compound. It can be converted to kJ/kg or kJ/m^3 by using the appropriate molecular weight. A list of such important oxidation reactions and molecular weights is given in Table 5.3 and Table 5.4.

Table 5.3 Heat generated in Some Common Combustion Reactions

Reaction		Heat Evolved		
		kJ/mol	kJ/kg	kJ/m^3
$C + O_2$	$\rightarrow CO_2$	408,841	34,068	—
$CO + 1/2\ O_2$	$\rightarrow CO_2$	282,986	—	12,627
$C + 1/2\ O_2$	$\rightarrow CO$	110,389	—	—
$S + O_2$	$\rightarrow SO_2$	296,927	9,131	—
$H_2S + 1.5O_2$	$\rightarrow SO_2 + H_2O$	518,996	15,265	23,170
$H_2 + 1/2\ O_2$	$\rightarrow H_2O_{vap}$	241,746	119,910	10,785
$H_2 + 1/2\ O_2$	$\rightarrow H_2O_{liq}$	286,223	143,112	12,770
$CH_4 + 2O_2$	$\rightarrow CO_2 + 2H_2O$	803,112	50,062	35,818
$C_4H_{10} + 6.5O_2$	$\rightarrow 4CO_2 + 5H_2O$	2,660,083	45,770	118,646
$C_3H_8 + 5O_2$	$\rightarrow 3CO_2 + 4H_2O$	2,046,047	46,402	91,255

TABLE 5.4 Some Important Atomic and Molecular Weights

Element or Compound	Chemical Symbol and Atomic Weight	Molecular Formula	Molecular Weight
Hydrogen	H = 1	H_2	2
Oxygen	O = 16	O_2	32
Nitrogen	N = 14	N_2	28
Carbon	C = 12	—	—
Sulfur	S = 32	—	—
Water	—	H_2O	18
Carbon dioxide	—	CO_2	44
Carbon monoxide	—	CO	28
Methane	—	CH_4	16
Propane	—	C_3H_8	44
Butane	—	C_4H_{10}	58
Sulfur dioxide	—	SO_2	64
Hydrogen sulfide	—	H_2S	34

The calorific or the heating value of a fuel is the heat generated by a unit quantity by the combustion of that fuel. For solid and liquid fuels the unit quantity is 1 kg, for gases it is 1 m^3.

If we know from the analysis the amount of combustible elements per unit present in the fuel, the calorific value can be determined by using the above oxidation reactions.

We have seen in the previous sections that fuels, especially liquid and solid fuels, contain noncombustible components such as extraneous (mechanical) moisture, ash, and nitrogen. These have to be taken into account while calculating the calorific value as they absorb heat. Similarly, the combustible elements, C, H, and S are not in their elemental form but as complex organic compounds. On heating, these compounds break up or dissociate. In doing so, they absorb heat which is not apparent in calorific value. Lastly, the moisture — extraneous or generated in combustion — may be as vapor or condensed as liquid. These two forms have different heat contents as seen earlier.

The previeusly discussed factors make it impossible to calculate the exact calorific value by theory. Such a theoretical value is always higher than the practically realizable heat.

To obviate this problem we use empirical formulae for estimating the calorific value. For solid and liquid fuels Mendeleev's formulae is used for higher heating value Q^h.

$$Q^h = 4.187(81C + 300\,H - 26(O - S))\ \text{kJ/kg}. \tag{5.7}$$

Where C, H, O, and S are the percentage amounts of these elements in 1kg fuel. The lower heating value Q^ℓ is given by

$$Q^\ell = 4.187(81C + 300H - 26(O - S) - 6(W + 9H))\ \text{kJ/kg}. \tag{5.8}$$

where W is the moisture content.

Gaseous fuels (at least those considered in this book) contain one predominant combustible hydrocarbon. It is either methane (CH_4), propane (C_3H_8), or butane (C_4H_{10}). There are still small amounts of other hydrocarbons, CO_2, and H_2O. Thus the theoretical calorific values for gaseous fuels are also on the higher side. The following empirical formula gives sufficiently accurate practical values for gaseous fuels

$$\begin{aligned} Q^\ell &= 0.108H + 0.126CO + 234H_2S + 0.358CH_4 \\ &\quad + 0.913C_3H_8 + 1.187C_4H_{10}\ \text{MJ/m}^3 \end{aligned} \tag{5.9}$$

where H, CO, H_2S, C_3H_8, C_4H_{10} are the volume percentages of these components in the gas.

If the composition of a gaseous fuel is known, it is always quicker and easier to use the value supplied by the manufacturer or in the published tables such as in Table 5.1.

For liquid fuels, it can be seen from Table 5.2 that the calorific value does not change much from oil to oil. The viscosity of the oil is a good indicator of practical calorific value.

5.8 COMBUSTION AIR REQUIREMENTS AND PRODUCTS

We can use Reaction (5.1) to Reaction (5.6) given before for calculating the air required and the volume of combustion products. The reactions can be balanced quantity-wise by considering the molecular weights or volumes. Avogadro's law, which states that 1 k mole of all gases at equal pressure and temperature occupy equal volume and at 0°C and 1 bar (1 atmosphere (ATM)) conditions, this volume is 22.4 m³. The solids have zero molecular volume. Thus, for combustion of carbon

$$\left.\begin{array}{lll} & C+O_2 & CO_2 \\ \text{Mol wt} & 12+32 & 44 \\ \text{Volume} & 0+1 & 1 \\ \text{Thus} & 1\text{ kg C}+2.67\text{ kgO}_2 & 3.67\text{ kg CO}_2 \\ \text{or} & 0\text{ vol C}+1\text{ vol O}_2 & 1\text{ vol CO}_2 \end{array}\right\} \tag{5.10}$$

for combustion of hydrogen

$$\left.\begin{array}{lll} & H_2+\frac{1}{2}O_2 & H_2O \\ \text{Mol wt} & 2+16 & 18 \\ \text{Volume} & 1+\frac{1}{2} & 1 \\ \text{or by wt} & 1+8 & 9 \end{array}\right\} \tag{5.11}$$

for combustion of methane

$$\left.\begin{array}{lll} & CH_4+2O_2 & CO_2+2H_2O \\ \text{Mol wt} & 16+64 & 44+36 \\ \text{Volume} & 1+2 & 1+2 \end{array}\right\} \tag{5.12}$$

We can write similar equations for propane, butane, sulfur and the like.

The left hand side (L.H.S.) of Equation (5.10), Equation (5.11), and Equation (5.12) gives the oxygen required for combustion of 1 kg or 1 m³ of fuel. Note that air (from which the oxygen is taken) contains nitrogen to oxygen 3.31:1 by mass and 3.762 : 1 by volume. Hence, nitrogen comes in with oxygen; therefore the air for combustion of unit fuel can be.

The R.H.S. of these equations gives the quantities of combustion products. These are predominantly CO_2, H_2O, and traces of SO_2. The products are in molecular volumes. Nitrogen coming in on the L.H.S. will appear unaffected on the R.H.S. The sum of

CO_2, H_2O, and N_2 will give the molecular volume of combustion products. Any moisture (in mol) coming in with the fuel will also appear on the R.H.S. and should be added to the combined volume of N_2, O_2, and CO_2. Multiplying the molecular volume by 22.4 will give the volume in m^3.

The percentage composition of the combustion products can be determined from the above data.

If we assume that all the heat derived from combustion is taken up by the products of combustion, the temperature "t" to which these products will be heated can be calculated from the relation

$$t = \frac{Q^h}{C_c \times V_c} + t_o \tag{5.13}$$

where C_c and V_c are the specific heat (J/kg°C), V_c the volume of combustion products, Q^h the calorific value, and t_o the atmospheric temperature in °C. Generally t_o is very small compared to t and is neglected.

The temperature so derived is called the "adiabatic or theoretical flame temperature." This is the maximum possible temperature. In practice, the measured temperature is lower by 200–400°C. This is because not all the heat is absorbed by the gases. Some heat is invariably lost.

A number of solved problems following this section will clarify the process of calculation of the various combustion parameters.

5.8.1 Combustion Air and Practical Requirements

The amount of air as calculated above is just sufficient, according to the chemical equation, to bring about complete combustion (stoichiometric air).

Details of combustion dynamics are discussed in the next chapter. It is sufficient here to state that there are mechanical problems involved in bringing together the fuel and air at all combustion sites. Hence, it is necessary to provide a certain amount of air in addition to stoichiometric air to assure satisfactory combustion. The excess air used is about 1–10%. It will increase the volume V_c of combustion air. Uncombusted oxygen and accompanying nitrogen will appear in the combustion products.

Some of the heat evolved will be absorbed by the excess air and will escape through the flue gases, representing loss of fuel.

The increased volume will lower the theoretical temperature of the flame as can be seen from Equation (5.13). The lowered temperature will decrease the heat transfer inside the furnace. Excess air, therefore, lowers the fuel efficiency and furnace efficiency but in most cases, loss of fuel is economically more harmful. It may also give rise to ecological problems due to release of fuel to the atmosphere.

Similar problems will arise if the air supply is deficient.

Up to about 1–3% excess air does not produce severe economic or operational problems.

5.8.2 Preheating of Air

In steady state operation, the combustion products leave the furnace at a temperature 200–300% °C less than the furnace temperature. They still have appreciable sensible heat content which is lost if unused.

This heat can be partially used to heat the fresh air supplied for combustion by using a recuperator. The outgoing hot gases transfer heat to incoming cold air. It is possible to heat air to 300–600°C in a properly designed heat exchanger.

Assuming that the initial temperature of air is 25°C and it is heated to a temperature t (>100°C), the heat content of the air will be

$$H = 1.01 \times t \text{ kJ/kg} \tag{5.14}$$

where 1.01 is the specific heat of dry air (kJ/kg).

If X is the specific humidity (i.e., actual moisture content) of moist air (g/kg), the heat content is

$$H = 1.01 \times t + X(2688 + 2.008t) \tag{5.15}$$

where 2688 is the heat content of steam.

$$= \text{water } 0° \rightarrow 100°\text{C} + 2270 \text{ latent heat}$$

It is rarely necessary to consider atmospheric humidity in combustion calculations.

5.9 SOLID WASTE AND GARBAGE

Solid waste produced as an unavoidable part of living is posing a formidable problem of safe and economical disposal. Increasing population, ways of modern living with packaged and processed goods, and nonavailability of disposal sites are some of the reasons.

Incineration or burning the waste/garbage is one of the solutions available to minimize the disposal problem. The volume of waste is reduced to about 10% by incineration. Industrial wastes frequently contain hazardous chemicals. By properly handling their combustion process, the toxicity of these chemicals can be substantially reduced.

Considering the social and economic importance of the waste disposal problem we will discuss here the combustion process involved, and later the principle of design of incinerators.

The solid waste placed on the hearth of the incineration furnace will behave both as a "charge" receiving heat and as a fuel giving off heat by combustion.

Waste or garbage is an extremely heterogeneous aggregate. It consists of organic waste containing C, N, H, and S; metals, mainly aluminum and steel; all types of plastics; glass ceramic; simple clay; textiles; and many other materials. The distribution of these materials in a garbage mass is also nonuniform.

The composition of the garbage is ever-changing. It depends on the social and economical structure of the population, locality, season, and many other factors. It also contains a lot of uncombined (i.e., mechanically mixed) moisture.

Establishments such as hospitals; biological and other laboratories; dairies and milk processing units; fish and meat production and processing units; hotels; restaurants; vegetables; and other markets produce waste peculiar to their products. Almost all industries and establishments produce paper and other packaging waste. Human beings produce garbage containing unusable or unwanted products such as glass, plastics, ceramics, and metals, in addition to kitchen waste.

Daily collected garbage from large cities has now reached a few thousand tons. Though incineration is one of the feasible disposal methods, the combustion process of such a heterogeneous material is extremely complicated. All the constituents of garbage show wide variations in their combustion characteristics.

Garbage is collected and stored in an open space. It gets mixed with rain water. This will depend on season and duration of exposure before incineration. This water has to be driven off in the incineration. Only after such “drying” does the combustible mass catch fire and burn on its own, i.e., behaves as a fuel.

Garbage, even when dry, contains a lot of organic matter. Its burning produces many hazardous gases and pyrolized chemicals. Pyrolysis means incomplete combustion due to oxygen deficiency or low temperature. It is necessary to carry out the combustion to completion and clean the combustion products before they can be safely released into the atmosphere. It is therefore necessary to carry out the combustion in two stages. In the first stage the garbage is dried, i.e., the moisture is removed at about 600°C. The dry mass catches fire with the evolution of gases. Ash and other noncombustible materials separate out.

In the next stage the gases are heated to about 1000°C in excess air. This brings about complete combustion, oxidizing hazardous chemicals and objectionable odors.

In practical incinerators the two stages are carried out in a single or double chamber depending on the size of charge and its chemical nature.

In Chapter 6 we will discuss their combustion process and apparatus in some detail. Design of incinerators will be dealt with separately.

Due to the reasons discussed earlier it is not possible to present the composition and heating value of garbage as a fuel. Some typical compositions of urban wastes are given below.

Household Garbage

Proximate Analysis →		
Moisture	—	10%
Volatile matter	—	60%
Fixed carbon	—	8%
Ash, metals, glass	—	22%

Ultimate Analysis % (Dry basis)		
Total carbon	—	50.0
Hydrogen	—	7.0
Oxygen	—	42.0
Nitrogen	—	0.7
Sulfur	—	0.3

Municipal Refuse

Proximate Analysis %		
Moisture	—	65–73
Volatile matter	—	15–20
Fixed carbon	—	4–5
Ash, metals, glass	—	5–15
Ultimate Analysis % (Dry basis)		
Carbon	—	50–54
Hydrogen	—	6–8
Nitrogen	—	1–1.5
Oxygen	—	30–35
Ash	—	1–2

The heating value of garbage is about 65–110 MJ/kg. Density is 160–700 kg/m^3.

It is again pointed out that the above quoted data are typical and not standard. Each incineration problem will have to take into account the composition of locally available garbage. We will take up an example on calculating the combustion requirements.

5.10 INCOMPLETE COMBUSTION

Liquid and gaseous fuels require atomizing and mixing for their efficient combustion. Both are mechanical processes and have inherent limitations. To overcome these we use excess air so that more than enough oxygen is available in the fuel-air mixture.

In spite of burner design and excess air provision, some fuel (1–2%) usually escapes with the combustion gases. Thus combustion is about 98–99% efficient.

During starting and turn down, the fuel/air ratio changes and frequently the air supply is deficient—resulting in incomplete combustion.

Incomplete combustion represents loss of fuel and heat. It also reduces the flame temperature.

In this section we will investigate the phenomenon of incomplete combustion.

There are instances where fuels (generally gaseous) are intentionally combusted incompletely. Protective atmospheres are generated by this technique. However, the process of combustion is drastically modified to achieve this. We will review such atmosphere production considering the combustion of methane. The reaction ordinarily carried out in excess air will be written as

$$CH_4 + x(2O_2) + N_2 \rightarrow CO_2 + 2H_2O + (N_2 + O_2)$$
$$(+805,560 \text{ kJ/mole}) \qquad (5.16)$$

where $2O_2$ is the stoichiometric oxygen requirement and x represents the excess fraction. If the air supplied provides less than stoichiometric oxygen the reaction will be

$$CH_4 + y(2O_2) + N_2 \rightarrow CO_2 + 2H_2O + (\overline{CH_4}) + N_2 + \Delta H \qquad (5.17)$$

where y represents deficiency fraction.

$(\overline{CH_4})$ represents unburnt methane and ΔH, i.e., the heat evolved is < 35,800 kJ/mole.

35,800 – ΔH will represent the heat lost and $(\overline{CH_4})$ the fuel lost due to oxygen deficiency.

However, if the deficiency ratio is about 5–10%, the heat evolved will be substantial. The theoretical temperature for correct oxygen is 2040°C which will be reduced due to less oxygen. Let us assume a flame temperature of 1800–1900°C. The ignition temperature of methane is 720–850°C. Hence, it is unlikely that the unburnt methane $(\overline{CH_4})$ will escape unchanged.

It is more likely that (CH_4) will be decomposed as follows:

$$(CH_4) \rightarrow C + 2H_2 - 59{,}218 \text{ kJ/mole}$$

The carbon particles will thus form smoke or soot. The dissociation will absorb heat.

Alternatively, all the hydrogen will be oxidized with available oxygen.

Considering the temperature, concentration, and the presence of catalyzing surfaces, it is more likely that the methane will decompose on the lines of many possible and complex reactions such as

$$CH_4 \rightarrow C + 2H_2 - 59{,}218 \text{ kJ/mole}$$

or

$$CH_4 + H_2O_{steam} \rightarrow CO + 3H_2 - \Delta H$$

and so on.

These reactions are endothermic and will absorb heat, thus increasing the heat loss. Thus, the reaction will now be

$$CH_4 + x(2O_2) + [N_2] \rightarrow CO_2 + H_2O + H_2 + CO + C + [N_2] + \Delta H$$

The reactions with unburned CH_4 are not fully predictable, hence the presence and quantity of products like C, H_2, and CO and even to some extent CH_4 can be found only by the analysis of the combustion products.

Free carbon can be detected by the flame luminosity or appearance of soot and smoke. Unburned CH_4 may impart a typical odor. If the fuel is directly burned over food or other sensitive material it will impart a disagreeable smell to the product.

It is observed that the quantity of CO and H_2 in the fuel gases is approximately equal. Example 5.4 will show a method of calculating the heat and fuel loss under oxygen deficiency conditions.

However, actual prevailing conditions may change and produce different combustion products.

5.11 COMBUSTION AND POLLUTION

The combustion products contain a number of major and minor constituents. The major constituents are carbon dioxide, carbon monoxide, nitrogen, oxygen, and water vapor.

Minor constituents are hydrogen sulfide (H_2S), sulfur di- and trioxides (SO_2, SO_3), nitrogen oxides (NO_x), unburned fuel (C_xH_y), carbon, etc.

Carbon monoxide and all the minor constituents listed above are a source of pollution. There are two main reasons for their occurrence:

1. Fuel impurities which originate from the fuel processing, grade, and mainly from the source, e.g., some crude oils contain significant sulfur and other polluting components.
2. Inefficient combustion arising from incorrect air/fuel ratio due to burner design or operation, or wrong settings of the regulating valves. Incorrect choice of flame length, atomization process, and mismatch of burner placement with respect to furnace size make air-fuel mixing inefficient and cause incomplete combustion.

A well-controlled burner system operating under steady state conditions emits the polluting agents at a much lower level. It is mainly transient conditions such as start, stop, and turn down that give rise to objectionably high polluting levels. There is practically no control over the inherent impurities in the first reason given above. A fuel can be processed to reduce the level of impurities, but that increases the cost.

Thus, there are three ways to lower the pollution from combustion products.

1. Using a better combustion system (burners) and controls to assure complete combustion at all times. This will also lower fuel consumption.
2. If economically possible, change the fuel or its grade or source.
3. Remove the pollutants before the combustion products are released in the atmosphere.

Besides combustion there are other sources of pollutants. During hardening, when a hot object is immersed in quenching oil, a considerable amount of smoke and fumes arise. Cleaning and descaling plants, food processing plants, and so on also give rise to gaseous pollutants and objectionable odors.

Incineration is another example where the pollution arises indirectly by the combustion of charge. Incineration can produce a large number of pollutants due to the heterogeneous nature of waste matter being burned.

Growing industrialization has increased fuel consumption. All kinds of pollutants are appearing in the atmosphere at an alarming rate. The environmental and ecological damages arising due to their presence have reached dangerous levels.

For example, combustion of furnace oils introduces the following amounts of pollutants (kg/10^3 l of oil) into the atmosphere.

Dust -1.6, C_x H_y -0.154, CO -0.56, SO_2 -0.195, NO_x -8

Remembering that the use of furnace oil is in 10^9–10^{12} l/day, we can get some idea about the stupendous quantity that is released in the air. Fuel oil is not the only industrial fuel; there are other oils, all types of gases, and distillate fuels (petroleum, diesel) that are used in automotive and aviation industries. All these fuels add their own share of pollutants to the air.

The biological effects of these pollutants on life, vegetation, and ecology are listed further to give an idea about how dangerous the problem really is to life on this planet. As time and pollution go on, more and more harmful effects are becoming apparent.

1. **Dust**—This arises from ash, uncombusted or partially combusted solid charge (waste incineration), carbon, or soot from incomplete combustion of fuels and the like. The dust particles have about 0.1–5.0 micron diameters. Very fine dust is inhaled during respiration and can clog fine air passages in lungs. Ash also contains elements like lead, zinc, vanadium, arsenic, mercury, cadmium, nickel, etc. These are deposited in the lungs and produce respiratory or carcinogenic problems.

Fine dust (< 3μm) remains suspended in the air for a long time and acts as nucleant to fog and smog.

2. **Sulfur and nitrogen oxides**— Sulfur and nitrogen oxides like SO_2, SO_3, and NO_x are gaseous. They dissolve and react with water to form sulfuric or nitric acid. When combustion products containing these oxides come in contact with rain, they are converted to acids. The resulting rain is called "acid rain." The acids formed are highly corrosive. When such rain falls on metals, it attacks electric lines, transformers, switch gear, metal roofs, railway tracks and so on.

   The acids are also extremely dangerous to all kinds of vegetation and life. These acids severely affect fish, forests, masonry (stones), and statues. They also attack lungs and respiratory systems and irritate eyes.

   The large scale destruction of marine and plant life and occurrence of smog in many countries is linked to acid rain.

   Oil combustion produces about 500 mg of these oxides for every m^3 of combustion products, while the safe level is about 100–150 mg/m^3.
3. **Carbon Monoxide**—Incomplete combustion produces CO, which is highly toxic. It reduces hemoglobin's capacity to transport oxygen to tissues and attacks the nervous system.

   At 0.005% in atmosphere CO is not very dangerous, but at higher concentrations it can become fatal. As CO is slightly lighter than air it spreads very rapidly. Many industrial casualities are proven to be caused by CO.

   The only fortunate property of CO is that it is highly unstable. If temperature and other conditions are proper, it is converted to CO_2.
4. **Carbon Dioxide**—This is a product of complete combustion and a major constituent (~10–15%) of combustion gases. The only natural process of absorbing CO_2 is by plant respiration where it is split and gives oxygen. All "life" and combustion on earth depends on atmospheric oxygen.

> Industries produce large concentrations of CO_2. Extensive deforestation diminishes the planet's natural capacity to convert CO_2 into O_2. This increases the CO_2 concentration in the atmosphere.
>
> CO_2 interferes with transmission of heat from the earth resulting in increased temperature. This gives rise to the so-called "green house" effect.
>
> Thus, high CO_2 affects respiration and comfort. In very high amounts, CO_2 attacks the nervous system, and can even be fatal.

This is only a very short list of the major effects of combustion pollutants. The list of pollutants and that of ecological damages is ever-growing.

Almost all developed nations have rigid standards of permissible emissions and are trying to clean the atmosphere. Developing and undeveloped countries have either no standards or have very loose standards, and practically no enforcement.

Another interesting fact about combustion pollution is that it is not a local phenomenon. Most industries use chimneys (which are effectively static pumps) to disperse the pollutants to a particular height. If a certain chimney is not effective, it is replaced by a taller one.

Wind picks up the gases from a chimney and dumps them at a further place. The wind velocity increases with altitude. Hence, a taller chimney will carry the pollutants still further.

The only positive method of reducing or removing the issuing pollutants is to clean the gas after combustion and before its release to the atmosphere. There are various techniques applied for cleaning. No one method can clean the gas of all pollutants. Here, a few relevant methods are discussed further.

Large particles are removed by bag filters. The gas passes through cloth bags. The dust is trapped by the small openings in the cloth weaving.

Fine dust, soot and smoke can be removed in an electrostatic precipitator over a high-voltage charged grid. Fine particles have an electric charge and they are attracted by an oppositely charged grid.

Sulfur oxides are absorbed by subjecting gases to water or lime milk spray.

Nitrogen oxides are best controlled by adjusting flame temperature (usually lowering) and by recirculating some of the gases through the flame. Specially designed low NO_x burners are available.

Carbon monoxide is similarly lowered by combustion control. It can also be lowered by post-combustion heating. This heating also oxidizes pyrolized organic gases from industrial waste incineration.

Carbon dioxide cannot be lowered as it is an unavoidable product of combustion. It can be dispersed by a chimney and undue concentration can be avoided by extensive tree-planting programs.

It is not possible to discuss all the pollutants and details of cleaning processes adopted in this chapter. Reference should be made to the extensive literature available and the advice of consultants should be sought.

Most of the furnaces and ovens considered in this book are unlikely to produce extensive pollution problems. However, global views and effects of damages to all kinds of life are going to tighten antipollution legislation. The designer should be fully aware of these problems.

Addition of cleaning equipment or change to more sophisticated burning systems, considerably increases the cost.

This chapter has considered only the pollution from fuel consumption in furnaces. There are far greater problems involving internal combustion engines, power plants, etc., which are beyond the scope of this book.

EXAMPLE 5.1

A fuel oil has following properties:

Analysis W+%	
Carbon	– 86.20
Hydrogen	– 12.30
Sulfur	– 1.50
Calorific value	– 43.40 MJ/kg
Theoretical air for combustion	– 10.71 m³/kg oil
Volume of combustion products	– 11.40 m³/kg

Composition of combustion products, % volume

CO_2 – 13.5%
H_2O – 12.3%
N_2 – 74.2%

Theoretical flame temperature is 2240°C. If the oil is supplied with 10% deficit air, now determine:

1. Volume and composition of combustion products
2. Amount of heat and fuel lost
3. Theoretical flame temperature

Solution

The combustion reaction for oil can be written as

$$\left.\begin{array}{lll} C + O_2 & \longrightarrow & CO_2 \\ H_2 + \tfrac{1}{2}O_2 & \longrightarrow & H_2O \\ S + O_2 & \longrightarrow & SO_2 \end{array}\right\} + 43.4\ \text{MJ/kg oil}$$

$O_2 + N_2$: 10.71 m^3/kg = 100% combustion air → 2.25 m^3 O_2 and 8.46 m^3 N_2

Only 90% air is made available

$\therefore 0.9 \times 10.71 = 9.64\ m^3$ available air

This air will contain

$O_2 - 2.025\ m^3$
$N_2 - 7.615\ m^3$

1 kg oil requires 2.25 $m^3 O_2$ for complete combustion. Hence, 2.025 $m^3 O_2$ will combust.

$$\frac{2.025}{2.25} = 0.9\ \text{kg oil}$$

Hence, unburned oil is 0.1 kg.

It will contain 0.0862 C

(in kg) 0.0123 H_2

0.015 S

Let us assume that the carbon in unburned fuel reacts with water vapor.

$$C + H_2O \rightarrow CO + H_2$$

$$\text{Mole } 1 + 1 \rightarrow 1 + 1$$

Unburned oil contains 0.0862 kg C, i.e.,

$$\frac{0.0862}{12} = 0.00718 \text{ mole C}$$

This requires 0.00718 mole of H_2O and gives 0.00718 mole each of CO and H_2.

$$\text{Volume of } H_2O \text{ required} = 0.00718 \times 22.4$$
$$= 0.1608 \text{ m}^3$$

Volumes of CO and H_2 formed are the same, i.e., 0.160 m^3 each.

Volume of combustion products is 11.40 m^3/kg.

For 0.9 kg oil, the products of combustion will be 10.26 m^3. It will contain (vol %)

$$CO_2 - 13.5 = 1.385 \text{ m}^3$$

$$H_2O - 12.3 = 1.262 \text{ m}^3$$

$$N_2 \quad - 74.2 = 7.613 \text{ m}^3$$

To this we have to add CO and N_2 and subtract H_2O from incomplete combustion.

$$CO_2 = 1.385 \text{ m}^3$$

$$H_2O = 1.262 - 0.1608 = 1.1012 \text{ m}^3$$

$$N_2 = 7.613 \text{ m}^3$$

$$CO = 0.1608 \text{ m}^3$$

$$H_2 = 0.1608 + 0.0123 \text{ (from fuel)} = 0.1731$$

Total volume of incomplete combustion products is 10.433. The composition (vol %) is

CO_2 – 13.28
H_2O – 10.55
N_2 – 72.97
CO – 1.54
H_2 – 1.66

The calorific value of oil is 43.4 MJ/kg.
Out of 1 kg only 0.9 kg is burned. Hence,

$43.4 \times 0.9 = 39.06$ MJ evolved.

For the reaction heat absorbed is 9.902 MJ/kg C.
Unburned fuel is 0.1 kg and contains 0.0862 kg C, hence, heat absorbed is

$9.902 \times 0.0862 = 0.853$ MJ

Net heat available

$$= 39.06 - 0.853$$
$$= 38.207 \text{ MJ}$$

hence,

$$\frac{38.2}{43.4} = 8.8\% \text{ heat is lost}$$

Theoretical flame temperature is

$$t = \frac{Q^h(\text{available})}{C_c \times V_c}$$

Assuming $C_c = 1.65$ kJ/m^3°C

$$t = \frac{38{,}207}{1.65 \times 10.433} = 2{,}219°\text{C}$$

EXAMPLE 5.2

The composition of a sample of waste material to be incinerated is as follows (in wt %)

Moisture	–	24.0
Carbon	–	28.0
Hydrogen	–	3.5
Oxygen	–	22.0
Nitrogen	–	0.33
Sulfur	–	0.16

Noncombustible matter (ash, metals, glass, etc.) –23.10

Determine (on 1 kg basis)

1. Heat required to be externally supplied for removal of moisture
2. Heat taken up by noncombustible matter
3. Theoretical and empirical calorific value
4. Air required for combustion
5. Volume and composition of combustion products
6. Theoretical flame temperature
7. Overall heat balance
8. Net heat available for external use

Solution

The sample contains 25% moisture. In incineration the moisture will first be expelled.

Assuming initial temperature 25°C, on 1 kg basis, the heat required to raise the temperature of moisture to 100°C

$$= W \times C_w \times (t_2 - t_1)$$

where W = weight (=1 kg)

C_w = specific heat of water (J/kg) ≈ 4.2 kJ/kg

$t_2 - t_1 = 100 - 25 = 75$

substituting

$$\text{Heat in water} = 1 \times 4.2 \times 75$$
$$= 315 \text{ kJ/kg}$$

Latent heat of evaporation is 2270 kJ/kg. The vapor will absorb heat to reach combustion temperature, which is unknown. Assuming the combustion temperature as 1500°C, the heat absorbed will be (specific heat being 2.0 kJ/kg °C)

$$= W \times C_s(t_3 - 100)$$
$$= 1 \times 2(1500 - 100)$$
$$= 2800 \text{ kJ/kg}$$

Total heat absorbed by the moisture for heating from 25 to 1500°C is

$$2270 + 315 + 2800 = 5385 \text{ kJ/kg}$$

1 kg of sample will contain 0.25 kg moisture which will absorb

$$5385 \times 0.25 = 1346.25 \text{ kJ} = He$$

Calculating the calorific value from Mendeleev's formula

$$Q^h = 4.187[81 \times \%\text{C} + 300 \times \%\text{H} - 26(\%\text{O} - \%\text{S})]$$

substituting

$$= 4.187[81 \times 0.28 - 300 \times 0.035 - 26(22 - 0.16)]$$

we get

$$Q^h = 10{,}911 \text{ kJ/kg}$$

As expected, this value is lower and more realistic. We will use this value of Q^h for further calculations.

Air required for combustion, volume, and contents of combustion products is calculated and given in the following Table (5.5).

TABLE 5.5 Calculations for Air Required, Quantity of Combustion Products and its Constituents for Waste Incineration (on 100 kg waste basis)

Waste					Air				Combustion Products						
Component	Content %	Mass kg	Molecular mass	Qty k mole	O_2 k mole	N_2 k mole	Total (Air) k mole	Total (Air) m^3	CO_2 k moles	H_2O k moles	SO_2 k moles	O_2 k moles	N_2 k moles	Total k mole	Total m^3
C	28	28	12	2.333	2.333	3.213	9.503		2.333				12.1	17.533	17.533
H	3.5	3.5	2	1.75	0.875	−0.687				1.75					× 22.4
S	0.16	0.16	32	0.005	0.005	2.526	+2.526	12.029			0.005				$393m^3$
O	22	22	32	0.687	0.687	× 3.762									%
N	0.33	0.33	28	0.0118	—	9.5028	12.029	× 22.4	—	—	—	—	—		$CO_2 = 13.4$
H_2O	24	24	18	1.333	—			270		1.333	0.0118				$H_2O = 17.6$
Ash	23.1	23.1	—	—	—										SO_2 = —
															O_2 = —
															$N_2 = 69.0$
0% Excess Air					2.526	9.503	12.029	270	2.333	3.083	0.005	—	12.111		20.874
50% Excess Air					3.8	14.254	18.054	404.4	2.333	3.083	0.005	1.274	14.254	20.874	× 22.4
															$468\ m^3$
															%
															$CO_2 = 11.16$
															$H_2O = 14.74$
															$O_2 = 6.1$
															$N_2 = 68.28$

This shows that
For 0% excess air

Air required – 2.70 m^3/kg

Combustion products – 3.93 m^3/kg

Contents of combustion products (%)

CO_2 – 13.3

H_2O – 17.6

SO_2 – Trace

N_2 – 69.1

Q^h – calorific value – 10911 kJ

Assuming specific heat of combustion products $C_c = 1.6$

$$Q_h = C_c V_c t$$

$$t = \frac{10,911}{3.93 \times 1.6}$$

$$= 1,735°C$$

For 50% excess air

Air required – 4.04 m^3/kg

Combustion products – 4.68 m^3/kg

Contents of combustion products (%)

CO_2 – 11.12

H_2O – 14.55

O_2 – 6.10

N_2 – 68.23

SO_2 – Trace

The sample contains 23% noncombustible matter and ash. Assuming specific heat of this matter as 0.8 kJ/kg°C and an average temperature of 600°C the heat absorbed by 1 kg of noncombustible is

$$= 1 \times 0.8 \times (600 - 25)$$

$$= 460 \text{ kJ/kg}$$

At 23% ash the heat absorbed by ash in 1 kg of garbage is

$$460 \times 0.23 = 106 \text{ kJ}$$

Heat evolved by carbon (28%) in the waste,

$$C + O_2 \rightarrow CO_2 + 34070 \text{ kJ/kg °C}$$

Thus for 0.28 carbon the heat evolved is

$$34,070 \times 0.28 = 9,540 \text{ kJ}$$

Heat evolved by hydrogen (3.5%)

$$H_2 + \tfrac{1}{2}O_2 \rightarrow H_2O + 121,025 \text{ kJ/kg.}$$

For 0.035 kg hydrogen the heat is

$$121,025 \times 0.035 = 4,236 \text{ kJ}$$

Sample contains 0.16% sulfur.
Assuming that it is in elemental form

$$S + O_2 \rightarrow SO_2 + 9,131 \text{ kJ}$$

For 0.0016 kg sulfur the heat evolved is

$$9,131 \times 0.0016 = 14.6 \text{ kJ}$$

The total heat evolved on combustion of 1 kg waste is

$$9,540 + 4,236 + 14.6 \cong 13,790 \text{ kJ} = Q^h$$

The actual combustion process is not likely to be simple elemental oxidation. The heat evolved may be less than the above calculated value.

The combustion temperature is

$$= \frac{10,911}{4.68 \times 1.65}$$

$$= 1,457°\text{C}$$

Assume that the combustion products leave the incineration chamber at 600°C. To bring about complete combustion and remove odors, this gas will have to be heated to about 1000°C in the mixing chamber or the flue.

Quantity of combustion products is about 4–5 m^3/kg and has a specific heat 1.6 kJ/m^3 °C.

Hence, additional heat required

$$H_A = 5 \times 1.6(100 - 600)$$
$$= 3200 \text{ kJ}$$

We can now calculate heat balance.

$$\text{Heat required externally} = H_e + H_A$$
$$= 1346 + 3200$$
$$= 4546 \text{ kJ/kg}$$

Heat available from combustion

$$Q^h = 10911 \text{ kJ/kg}$$

Theoretical available heat

$$= 10911 - 4546 = 6365 \text{kJ/kg}$$

Assume 10% loss to walls, etc., and 106 kJ as heat taken by ash.

Net heat available ≅ 5600 kJ/kg waste. This heat can be utilized for water heating and other purposes.

EXAMPLE 5.3

A sewage (bio) gas has the following composition (volume basis)

CH_4 – 65%

CO_2 – 33%

N_2 – 1%

H_2S – 1%

Now determine:

1. The heating value
2. Air required for combustion
3. Amount of flue gases with no excess air
4. Composition of flue gases
5. Theoretical flame temperature

Solution

The given composition of the gas shows that methane is the only major combustible component. H_2S is minor.

The methane reaction is

$$CH_4 + 2O_2 \rightarrow CO_2 + 2H_2O + 35,800 \text{ kJ/m}^3$$

At 65% CH_4 by volume the heat evolved by combustion of methane from 1 m^3 gas is

$$35,800 \times 0.65 = 23,270 \text{ kJ}$$

The H_2S reaction is

$$H_2S + 1.5O_2 \rightarrow SO_2 + H_2O + 23,170 \text{ kJ/m}^3$$

At 1% H_2S in the gas, the heat contribution from H_2S is

$$23,170 \times 0.01 = 231 \text{ kJ}$$

The total heat available from 1 m^3 gas is

$$23,270 + 231 = 23,501$$

$$Q^h = 23,500 \text{ kJ/m}^3$$

The methane reaction shows that 1 m^3 methane requires 2 m^3 oxygen.

Hence, 0.65 m^3 methane will require

$$2 \times 0.65 = 1.30 \text{ m}^3 \text{ O}_2$$

The H_2S reaction shows that 1 m^3 H_2S requires 1.5 m^3 oxygen.

Hence, 0.01 m^3 will require

$1.5 \times 0.01 = 0.015$ m^3 O_2

Total oxygen required is

$1.30 + 0.015 = 1.315$ m^3

At N_2:O_2 volume ratio 3.762 nitrogen corresponding to the above oxygen is

$1.315 \times 3.762 = 4.947$ m^3

The air required for combustion is

$$1.315 + 4.947 = 6.262\ \text{m}^3/\text{m}^3\ \text{gas}$$

On combustion, 1 m^3 methane produces 1 m^3 each of CO_2 and H_2O, hence 0.65 m^3 will produce

0.65 m^3 CO_2
0.65 m^3 H_2O

Similarly, 1 m^3 H_2S produces 1 m^3 each of SO_2 and CO_2
Hence, 0.01 m^3 H_2S will produce

0.01 m^3 SO_2
0.01 m^3 H_2O

Total gases produced by combustion are

$$\left.\begin{array}{l} 0.65\ \text{m}^3\ \text{CO}_2 \\ 0.65 + 0.01 = 0.66\ \text{m}^3\text{H}_2\text{O} \\ 0.01\ \text{m}^3\ \text{SO}_2 \end{array}\right\}$$

As 1 m^3 of gas already contains 0.33 m^3 CO_2 and 0.01 m^3 N_2, the combustion products will contain

$0.65 + 0.33 = 0.98$ m^3 CO_2
0.65 m^3 H_2O

0.01 m^3 SO_2

4.947 + 0.01 = 4.957 m^3 N_2

Total amount of flue gases *Vcp* is 6.597 m^3, say 6.6 m^3

The % composition of combustion products is

$$\left.\begin{array}{l} CO_2 - 14.8\% \\ H_2O - 9.8\% \\ SO_2 - 0.16\% \\ N_2 - 76.0\% \end{array}\right\}$$

Let us assume a flame temperature of 1800°C. At this temperature the specific heats of constituent gases are (kJ/m^3°C)

CO_2 – 2.4226

H_2O – 1.9055

N_2 – 1.4705

Since the percentage of SO_2 is very small, it can be neglected. The specific heat contribution of these gases to the combustion gas mixture in Equation (5.26) is in proportion to their amount. Hence,

CO_2 – 2.4226 × 0.148 = 0.3585

H_2O – 1.9055 × 0.098 = 0.1867

N_2 – 1.4705 × 0.76 = 1.1176

Total 1.6628 kJ/m^3 °C

$$C_{cp} = 1.6628 \text{ kJ/m}^3 \text{ °C}$$

$$Q^h = V_{cp} \times C_{cp} \times t_c$$

$$23{,}500 \times 10^3 = 6.6 \times 1.6628 \times 10^3 \times t_c$$

$$\therefore\ t_c = 2141 \text{ °C}$$

Due to excess air and incomplete burning, this temperature will be about 1900°C.

EXAMPLE 5.4

A fuel oil has the following composition (%)

Carbon	– 86.5
Hydrogen	– 11.5
Oxygen	– 1.0
N, S and ash	– 1.0

Now determine:

1. The heating (calorific) value per kg
2. Air required for combustion
3. Volume and composition of combustion products
4. Theoretical combustion temperature
5. Composition and amount of combustion product at 20% excess air
6. Theoretical combustion temperature at 20% excess air

Solution

Calorific value as per Mendeleev formula

$$Q^h = 4.187(81 \times \text{C} + 300 \times \text{H} - 26(\text{O} - \text{S}))\ \text{kJ/kg}$$

Neglecting the last term, as % O and S are small, and substituting

$$\begin{aligned} Q^h &= 4.187(81 \times 86.5 + 300 \times 11.5) \\ &= 43.781\ \text{MJ/kg} \\ &\cong 44\ \text{MJ/kg} \end{aligned}$$

We can determine Q^h from heat given by the C and H reactions.

$$1\ \text{kg C} \rightarrow 34{,}070\ \text{kJ/kg}$$

$$\text{Hence, } 0.865\ \text{C} \rightarrow 34{,}070 \times 0.865 = 29{,}470.55\ \text{kJ}$$

Similarly, 1 kg $H_2 \rightarrow$ 143,112 kJ/kg

Hence, 0.115 $H_2 \rightarrow 143{,}112 \times 0.115$

$= 16{,}457.88$ kJ

Adding Equation (5.28) and Equation (5.29) and neglecting O, S, and N

$Q^h = 45.928 \sim 46.0$ MJ/kg

This value of Q^h is higher than that obtained in Equation (5.27). This is expected because the second method assumes that oil contains elemental C and H, which is not true, hence, we will take $Q^h = 44.0$ MJ/kg for further calculations.

To calculate air required (theoretical) for combustion of 1 kg oil consider the reactions for C and O

$C + O_2 \rightarrow CO_2$

1 kg C will require 2.666 kg O_2

$\therefore$ 0.865 kg $\rightarrow 2.66 \times 0.865$

$= 2.306$ kg O_2

Similarly, $H_2 + \frac{1}{2}O_2 \rightarrow H_2O$

1 kg H_2 will require 8 kg O_2

$\therefore$ 0.115 kg $\rightarrow 0.115 \times 8$

$= 0.92$ kg O_2

Hence, total O_2 required for 1 kg of oil

$= 2.306 + 0.92$

$= 3.226$ kg

On weight basis $N_2 : O_2$ for air is 3.31

$\therefore$ N_2 corresponding to 3.226 kg O_2 is

$3.31 \times 3.226 = 10.678$ kg

Air required for combustion of 1 kg oil is

$3.226 + 10.678 = 13.904$ kg

On volume basis taking density of air as 1.293 kg/m^3

Air required = 10.75 m^3

From above combustion equations

1 kg C will produce 3.666 kg CO_2

$\therefore$ 0.865 kg $\rightarrow$ 3.17 kg CO_2

Similarly,
1 kg will produce 9 kg H_2O

$\therefore$ 0.115 $\rightarrow$ 0.115 $\times$ 9 $\rightarrow$ 1.035 kg H_2O

Nitrogen coming in from air = 10.678 kg

Total gases 3.17 + 1.035 + 10.678 = 13.883 kg

Moles of combustion gases

CO_2 = 3.17/44 = 0.0720
H_2O = 1.035/18 = 0.0575
N_2 = 10.68/28 = 0.3813
Total moles = 0.5108

As 1 k mole occupies 22.4 m^3 (STP),

thus, volume of combustion gases
= 11.442 m^3/kg oil

As 1 kg oil gives 44 MJ

combustion gases per MJ

$$= \frac{1142}{44} = 0.260 \text{ m}^3/\text{MJ}$$

Volume of individual gases can be obtained from moles of gases on multiplying by 22.4

CO_2 = 0.0720 $\times$ 22.4 = 1.6128 m^3

H_2O = 0.0575 $\times$ 22.4 = 1.288 m^3

N_2 = 0.3813 $\times$ 22.4 = 8.541 m^3

The percentage composition of combustion gases by volume

$$\left.\begin{array}{l} CO_2 \; - 14.11\% \\ H_2O - 11.25\% \\ N_2 \; - 74.64\% \end{array}\right\}$$

Assuming that all the heat produced by combustion is taken up by the gases, and the specific heat of composite flue gas as 1.68 kJ/m³°C, then

$$Q^h = V_{cp} \times C_{cp} \times t_c$$

where V_{cp} = Volume of combustion products (m³/kg)
C_{cp} = Specific heat of combustion gases (at constant pressure) (kJ/m³°C)
t_c = Theoretical combustion temperature (°C)

Substituting

$$44 \times 10^3 = 11.4 \times 1.68 \times t_c$$

$$t_c = 2289°C$$

Air required for combustion is 10.75 m³/kg oil.

$N_2 : O_2$ volume ratio for air is 3.76 : 1

For 20% excess air the quantity is

$10.75 \times 0.20 = 2.150$ m³ which will contain
N_2 – 1.698 m³
O_2 – 0.452 m³

Combustion gases at theoretical air are

CO_2 – 1.6128 m³
H_2O – 1.288 m³
N_2 – 8.541 m³

Composition of combustion products will be

$$\left.\begin{array}{lll} CO_2 & - & 1.6128 \text{ m}^3 \\ H_2O & - & 1.288 \text{ m}^3 \\ N_2 & - & 8.541 + 1.698 = 10.239 \text{ m}^3 \\ O_2 & - & 0.452 \text{ m}^3 \\ \text{Total} & - & 13.592 \text{ m}^3 \end{array}\right\}$$

On percentage basis

$$\left.\begin{array}{lll} CO_2 & - & 11.84\% \\ H_2O & - & 9.47\% \\ N_2 & - & 75.73\% \\ O_2 & - & 3.32\% \end{array}\right\}$$

Assuming same specific heat, i.e., 1.68 kJ/m^3 °C the theoretical combustion temperature for 20% excess air is

$$44 \times 10^3 = 13.592 \times 1.68 \times t_c$$

$$t_c = 1927°\text{C}$$

Chapter 6

Fuel Burning Devices

CONTENTS

6.1 INTRODUCTION

Application of fuels for any practical heating process requires their continuous, sustained, and controlled burning. This involves bringing the required quantity of fuel and air to the combustion site and removing the combustion products.

Burners are devices used for this purpose. Combustion of gaseous and liquid fuels involve different techniques.

In this chapter we will discuss the dynamics of combustion processes and the main types of burners commercially available.

Once the fuel to be used is finalized and the quantity to be burned is known, it is necessary to choose a proper burner. Manufacturers of burners supply information about the range of burners available. Information in this chapter will be useful in making the choice.

6.2 COMBUSTION OF LIQUID FUELS

From a combustion point of view, liquid fuels can be classified into following two categories:

1. Fuels having low viscosity. These are also called distillate fuels. They vaporize very easily at comparatively low temperatures. Gasoline, petrol, diesel, and kerosene belong to this category. Their viscosity at 20°C is ~ 0.7–1.5 cSt. On heating, they vaporize and behave like gases that mix easily with air and burn. A common example of this type of combustion can be seen in a kerosene stove.

2. Fuels like light, medium, and heavy furnace oils have viscosity in the range 4–1200 cST at 20°C. These oils are generally called residual oils. They contain a multitude of heavy molecules of different constituent hydrocarbons, which do not burn easily. As the oil grade gets heavier the viscosity makes their flow in burner tubes and fuel pipes increasingly difficult.

The combustion of these oils is a surface phenomenon. Imagine a drop of combustible liquid heated to some temperature and surrounded by air. Combustion starts at the surface. Combustion gases move outward and air (oxygen) moves inward. The combustion slowly progresses toward the center. Heat evolved increases the temperature, and more and more molecules distillate.

It is thus necessary to increase the surface area and for some oils, to increase the temperature. The latter decreases viscosity thus making their flow less difficult.

The surface area is increased by converting or breaking the drop to minute globules. The process of breaking a drop or stream into globules is called atomization.

Atomization is shown schematically in Figure 6.1. The oil drop at the end of the burner tube meets a tangential, fast stream of the atomizing medium, which for light and medium capacity burners is air. For heavy oils and large output, steam is used. The drop is sheared by the atomizing medium and converted into minute globules. The shearing process is enhanced by low viscosity. In mechanical atomization, oil is introduced on the inner surface of a hollow cone revolving at high RPM. The oil globule is broken by centrifugal force.

The globules are surrounded by the atomizing air (primary air) and surrounding (secondary) air. This mixture of air and oil resembles an emulsion or mist. It has a forward moving velocity which depends on the atomization and secondary air pressure.

As the globule travels forward it catches fire and burns, giving a flame. First the lighter fractions burn, then the heavier, and finally the coke or carbon.

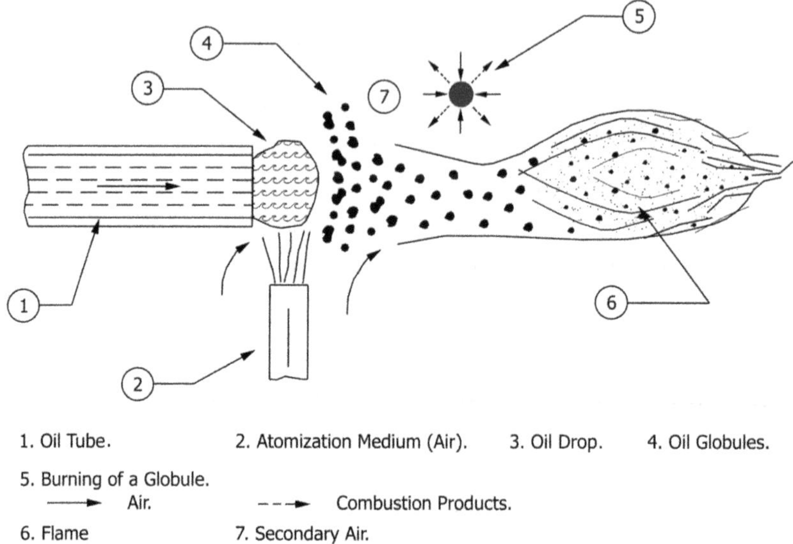

Figure 6.1 Oil atomization and combustion of liquid fuels.

The atomization and disappearance of the first drop from the tube makes the next drop appear and the process continues.

The pressure behind oil and the low pressure created by the removal of air in the vicinity, help to draw oil continuously.

For the globule-air to burn easily without the formation of smoke or soot it is necessary to ensure that each globule is surrounded by an adequate quantity of air and that the mixture is homogeneous. This is achieved by introducing excess air and making the mist turbulent.

A simple oil burner based on the principles discussed above is shown in Figure 6.2.

Oil with sufficiently low viscosity flows through the central tube toward the nozzle. Air is supplied through a coaxial outer tube. The oil and air streams meet at the nozzle opening and an atomized stream exits. Both the oil and air streams are controlled (volume, velocity) by proper valves in their path.

Commercial burners are variations of the basic design. Many different designs are available from different manufacturers. It is not possible to discuss all these designs in

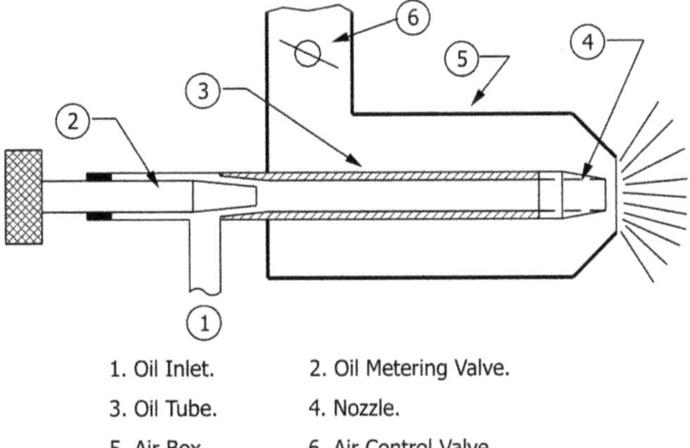

1. Oil Inlet. 2. Oil Metering Valve.
3. Oil Tube. 4. Nozzle.
5. Air Box. 6. Air Control Valve.

Figure 6.2 Basic oil burner.

this book. Only a few representative constructions are discussed in later sections.

An important characteristic of all burners is their turn down ratio (TDR). This is defined as

$$\text{TDR} = \frac{\text{Maximum fuel input rate for satisfactory burning}}{\text{Minimum fuel rate for satisfactory burning}}$$

Burners with fixed nozzles have air passages of fixed area which limits the flow rate. The orifice area is designed to supply adequate air for maximum rated oil flow.

When the oil flow is reduced (turned down) it is necessary to simultaneously reduce the air supply in the required ratio. This can be achieved by controlling the setting of the air valve. Control of air input through the valve is difficult.

If the air pressure is too low, atomization is inadequate. If it is too high for the reduced flow a lot of excess air is mixed with the oil. Both conditions reduce the fuel efficiency and hence the word “satisfactory” in the above definition of TDR.

There are some designs in which the area of air orifice is changed in proportion to the oil flow. This is achieved automatically. The velocity, and therefore the quantity, of air through a nozzle of fixed area is proportional to the square root of the air pressure. Suppose that a certain maximum quantity of oil

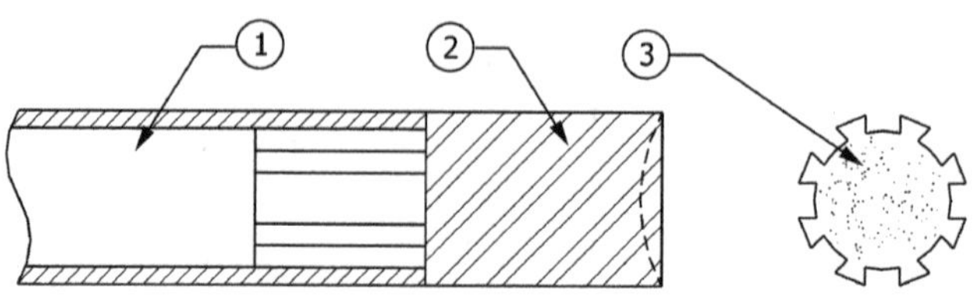

1. Oil Tube. 2. Swirler Tip with Grooves. 3. Hollow Tip Front.

Figure 6.3 Swirl on oil tube.

is supplied at pressure P. If the oil flow is reduced to half, the air flow required is $\sqrt{P}$.

Burners with air nozzle area adjustment coupled with oil valve are called self proportioning burners. Some burner manufacturers claim a TDR of 6.1 or better with fixed area nozzles. In these designs minimum air is supplied through the oil tube so that atomization is adequate. Remaining air for higher oil rates is supplied through the outer tube.

Burners of all types carry a swirler at the tip of the oil tube. This is a small cylindrical fitting with spiral or helical grooves on the outer surface. The oil flows through these grooves and acquires a swirl, i.e., turbulence combined with spiral motion to the spray. This improves the atomizing efficiency and mixing of air and oil globules. Further improvements are obtained by shaping the front face of the swirler, making it hollow and adding tangential grooves (see Figure 6.3). Swirl can also be given by adding spiral grooves to the inside of the air nozzle. Most manufacturers have their proprietary swirl designs.

Some burners have a small annular space between the swirl front end and the nozzle opening. This is called the swirl chamber. The swirled out globules get additional turbulence in the chamber.

6.3 CLASSIFICATION OF OIL BURNERS

Oil burners are classified according to the atomizing medium and/or the pressure (usually air) used for atomization. A classification scheme is given below.

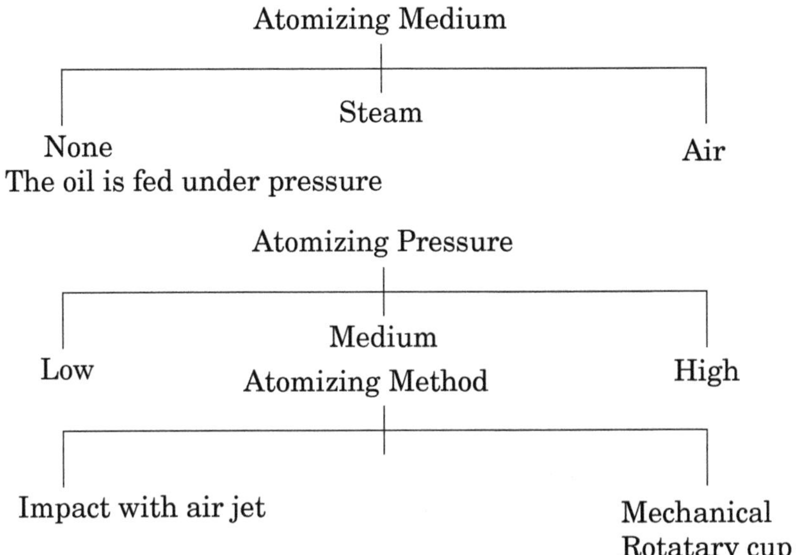

6.3.1 High Pressure Burners

In these burners the oil is under high pressure. It is necessary to preheat the oil to obtain low viscosity for proper flow through the burner. The pressure is developed by a positive displacement pump (e.g., gear pump, rotary vane pump).

Distillate oils easily atomize when sprayed under slight pressure and are used in small burners of this type (e.g., kitchen stoves).

Pressure jet burners for furnace oils are not suitable for industrial furnaces as they do not offer control over atomization and the flame shape.

Air for combustion is obtained from the surrounding atmosphere.

6.3.2 Low Pressure Burners

The air pressure in these burners is 15–110 kPa. They generally use a single stage blower to develop the required pressure and quantity of air.

The TDR normally available is about 2:1. All the air for combustion is supplied through the burner. This limits the

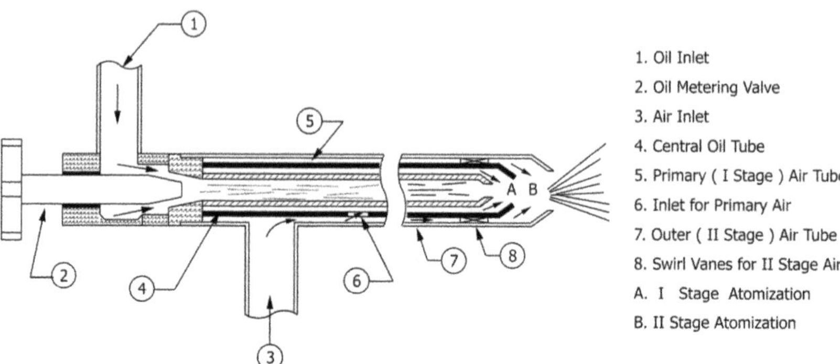

Figure 6.4 Typical low pressure oil burner with two stage atomization.

preheat to about 300°C. Mixing of globules and air is very good, assuring complete combustion with adequate excess air.

Figure 6.4 shows a typical low pressure burner. Part of the combustion air is supplied to the oil before it reaches the nozzle. Remaining air is supplied tangentially at the tip. This results in a conical atomized spray. The atomization takes place in two stages which results in a better globule-air mixing. There can be a separate control on the secondary air giving a turn down ratio of 5:1 or better.

Some designs based on the above principle provide only primary air (about 4–10 % of the total air) through the burner. The remainder of the air is introduced through a window around the burner. The window has a shutter to adjust the air flow.

Alternatively, the remaining air is preheated by exhaust gases to 300–400°C and introduced in the furnace at a slight positive pressure by using a single stage blower.

Oil preheating by recirculating combustion gases in the burner is avoided to prevent overheating. The preheating temperature cannot then be controlled, which results in dissociation and coking of oil, and clogging of the burner passages.

6.4 BURNERS FOR DISTILLATE FUELS

Many small appliances such as baking ovens use kerosene or diesel as fuels. These fuels are costly but they burn clean, i.e.,

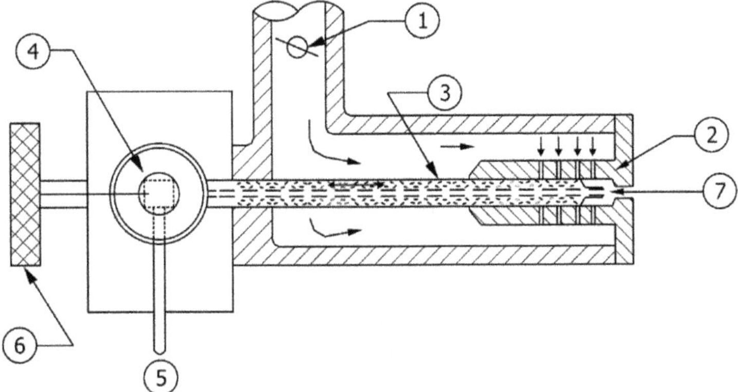

1. Burner body and air passage with damper
2. Outer nozzle with air holes
3. Central rod with oil passage and inner nozzle
4. Knob for lateral movement of central rod
5. Oil inlet.
6. Oil flow control
7. Mixing and Swirl chamber

Figure 6.5 Burner for distillate fuels.

do not produce or, smoke, or soot. As the viscosity of these fuels is very low, they gasify easily.

Figure 6.5 shows a typical burner for these fuels. Oil is admitted through the central passage. The central rod having the oil passage can be moved laterally. All the air (including excess air) is admitted through the burner. The air meets the oil through a number of holes in the outer nozzle. By moving the central rod the oil holes are successively opened or closed, thus adjusting the quantity of air. The air pressure is adequate at all settings of the air holes.

Oil flow is regulated by a metering valve. In some burners movements of the oil valve and central rod are linked so that the required air for the oil flow setting is automatically adjusted.

These are small burners capable of burning 1.7–3.6 l/h. Air pressure required is 2.5–12.5 kPa and oil pressure 7–8 kPa.

They produce about 3–3.3 kW and could be set to theoretical, excess, or deficient air supply and give a turn down ratio 6:1.

6.5 PREHEATING OF OILS

With the exception of light oils or gas oils (e.g., kerosene) all oils require preheating before they enter the burner. Preheating reduces their viscosity, resulting into a free and non-clogging flow through burner passages and nozzles. It also helps in atomization by reducing the globule size to 0.5–0.05 mm or smaller.

Changes in viscosity of furnace oils were discussed in Section 5.2 and Figure 5.1.

In general, preheating temperature is 80–110°C depending on the grade of oil. Overheating decomposes the oil with the evaporation of low boiling point components.

Oil-utilizing furnaces have a preheating arrangement supplementing the burner system. A typical system is shown in Figure 6.6.

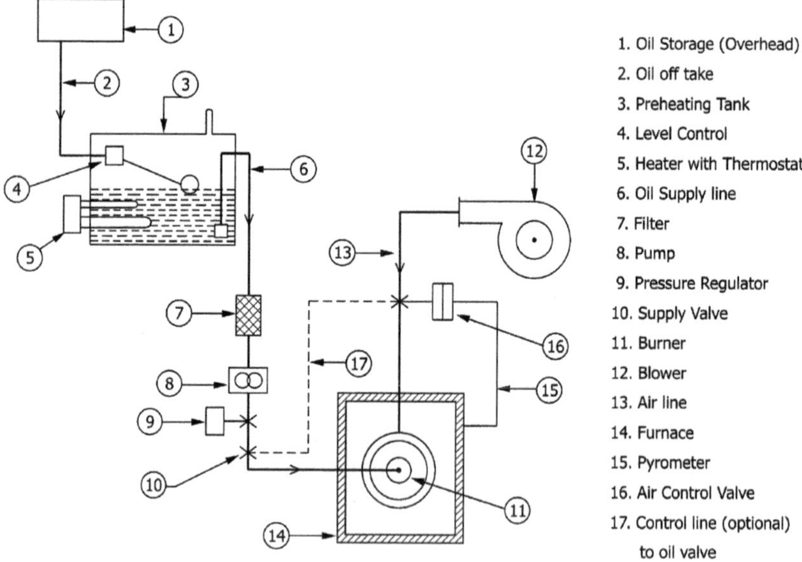

Figure 6.6 Typical layout of oil and air supply system for burners.

The oil is stored in an overhead tank from which it is drawn and taken to the preheater. Oil level in the preheater is controlled by a simple control device. A submerged electric heater heats the oil and there is a thermostat for temperature control. Heated oil is withdrawn by a pipe line, it is then filtered, and pumped. For low pressure a gear or vane pump is used. Before going to the burner the oil pressure and flow rate is adjusted.

Air is supplied by a blower and the flow rate is controlled by a damper. The furnace temperature is monitered by a pyrometer. Automatic control is obtained by using a motorized damper in the air passage and a servo valve in the oil line. The two valves may be linked to obtain simultaneous movement. The control devices are optional.

Figure 6.6 shows a generalized system and components may be added or deleted as per function demands.

The burner supplier gives information about the preheater and air supply system required. The above discussion is given with a view to make the furnace designer aware that he has to specify this system along with the burners.

6.6 KINETICS OF COMBUSTION OF GASES

In the previous chapter we reviewed the important gaseous fuels and related thermal calculations. In this section we will discuss the kinetics of the combustion process and the devices (burners) used to bring about an efficient combustion.

Fuel gases consist of hydrocarbon molecules such as CH_4, C_3H_8, C_4H_{10}, etc., and small amounts of hydrogen H_2. Besides these combustible components, commercially available gases may contain N_2, H_2O, H_2S, etc., in small amounts.

When a gas molecule burns the following conditions must be satisfied to obtain stable burning:

The gas molecules will require the immediate availability of the required amount of oxygen which is supplied by the surrounding air. Hence, the gas and the air must be thoroughly mixed. If we take a glance (see Table 5.1) at the densities of the fuel gases and air it will be seen that propane and butane are heavier than air while natural gas, methane, and biogas are lighter. Hence, the mixture of gas and air will have to be achieved mechanically or aerodynamically. Proximity of quiescent air gas masses (even in very slow moving streams) will only bring about molecular diffusion. This is a slow process assisted only by concentration difference and temperature.

To initiate combustion the air-gas mixture must have a certain temperature called the "ignition point" which depends on the air-gas ratio.

Thus two kinds of flames are possible. A slow burning diffusion flame and a fast burning turbulent flame.

A good and familiar example of diffusion flame is the laboratory or Bunsen burner shown in Figure 6.7(A). The gas is supplied at slight pressure at the base of mixer tube. It inspirates atmospheric air through the hole at the base of the mixer. Only part of the air is supplied through this hole. Mixing takes place in the tube. The mixture is ignited at the top. Remaining air for combustion is drawn from the surrounding air. Thus the combustion air is divided into two parts—primary air through mixer

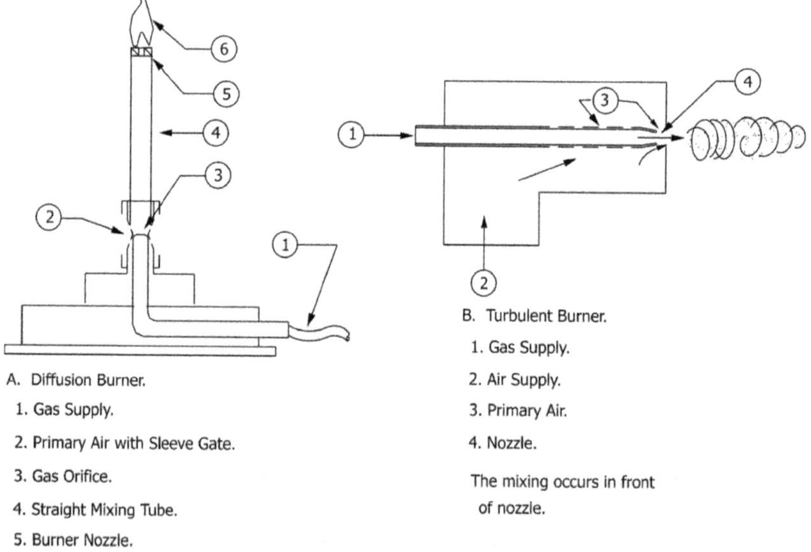

Figure 6.7 Basic gas burner.

and secondary air from the surroundings. A quiet slow burning flame (usually luminous) is produced. Diffusion flames are used in some applications requiring slow heating.

Almost all industrial burners use a turbulent flame. The air supplied may be in two components as discussed earlier or in a single source as shown in Figure 6.7(B). Turbulence is achieved by a variety of ingeneous ways which gives rise to a large number of commercial designs. To obtain a stable flame the air-gas ratio and velocities require control. To assure thorough combustion and control on flame length and shape, usually air is supplied in excess (50–500%). Turbulent flames are flickering or weaving and produce a noise when very large and strong.

If the air supplied to a flame is less than stoichiometric and if mixing is not good, a yellow flame is produced. It may also produce smoke and soot. Such flames may clog the burner passages. If mixing and air-gas ratio is correct the flame has a slight blue color or it is colorless. Excess air will produce a blue flame perhaps with yellow fringes.

Flames with deficient air will produce a reducing atmosphere containing unburned gases. In fact, such flames are used for production of protective atmospheres.

Flames with excessive air will contain unused oxygen in combustion gas but will give a flame of adjustable length and shape, though at a reduced flame temperature.

In subsequent sections we will review the properties of gases relevant to combustion and classification of burners.

There are a vast number of burner designs commercially offered. Only a few important ones are discussed.

6.7 BURNING PROPERTIES OF GASES

When a quiescent mixture of gas and air is ignited, the burning process or the combustion front progresses at a definite velocity known as ignition rate or the flame propagation velocity. This velocity depends on air-gas ratio and preheating temperature. Ignition rate of some gases is shown in Figure 6.8. It can be seen that there is a maximum velocity that increases with temperature. The maximum velocity occurs at excess air and not at the stoichiometric proportion.

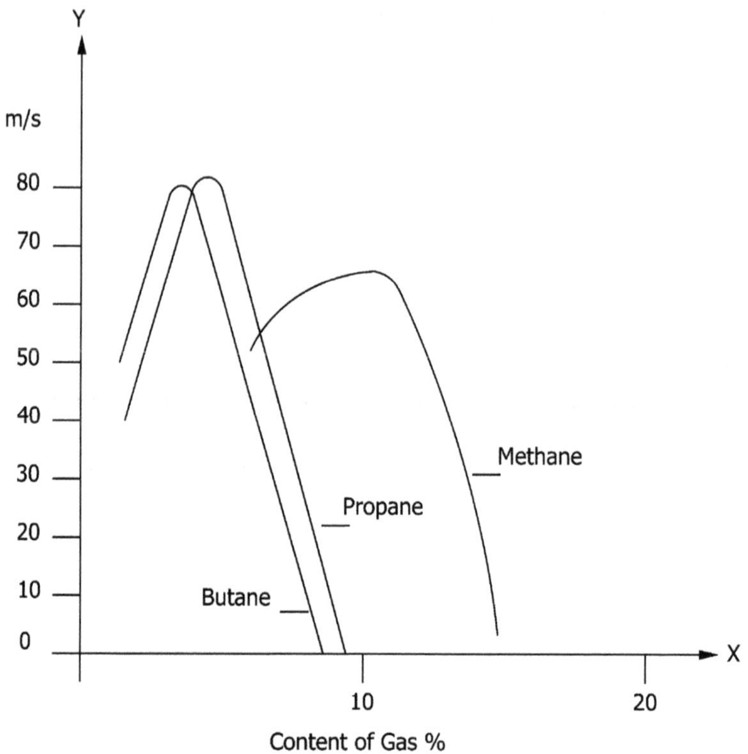

Figure 6.8 Ignition velocity of some gas-air mixtures.

If the air-gas mixture is turbulent the ignition rate increases with increasing turbulence (i.e., increasing N_{RC}). Under some combination of mixture proportion and N_{RC}, the propagation becomes unstable and there is a possibility of flame termination.

All fuel-air mixtures have a certain temperature at which the mixture catches fire. This temperature is known as the ignition point. At lower temperatures the ignition is not stable and the flame extinguishes. The ignition temperature depends on air-gas ratio and the homogeneity of the mixture.

When designing a burner and its subsequent operation, it is necessary to have a correct and homogeneous mixture and the right velocity to establish a continuous and stable flame.

6.8 CLASSIFICATION OF GAS BURNERS

In discussing the combustion of gases we have seen that for efficient combustion the mixing of gas and combustion air is important.

All gas burners are therefore classified primarily on the air supply and mixing method used. The classifications are as follows:

1. Neat gas burners — In these burners, only the gas is supplied through the burners and all the air is picked up by the gas stream (see Figure 6.9 (A))
2. Natural draft burners — Here the gas is supplied through an orifice at a slight or zero pressure. The gas stream enters in a mixing tube. This tube has one end open to the atmosphere and a nozzle at the upstream end (see Figure 6.9 (B)). The gas stream entrains or sucks some of the air (primary air) along with it. The remaining air (secondary air) is picked up from the furnace atmosphere as the combustion progresses.
3. High pressured draft burners — In this version of draft burners the gas is supplied at high pressure (7–210 kPa) through the orifice. The mixer is a venturi. The increased speed of the gas stream entrains air along with it. All the air required is supplied with the gas. Air quantity is controlled by an adjustable shutter (see Figure 6.9(C)) at the back end of the venturi. Thus, any desired gas/air mixture is obtained. The combustion takes place in a nozzle at the upstream end of the venturi.
4. Air blast furnace — The design of these furnaces is similar to draft burners, the only difference being that the air is supplied at high pressure (75–150 kPa) through the orifice into the venturi mixer (see Figure 6.9(D)). The gas is supplied at atmospheric pressure, near the throat. The air stream entrains the gas along with it. The quantity of gas is controlled by a needle valve near the throat. In some designs the throat cross section area is varied.
5. Nozzle mixing burners — All the air required for combustion and the gas is supplied to the nozzle through

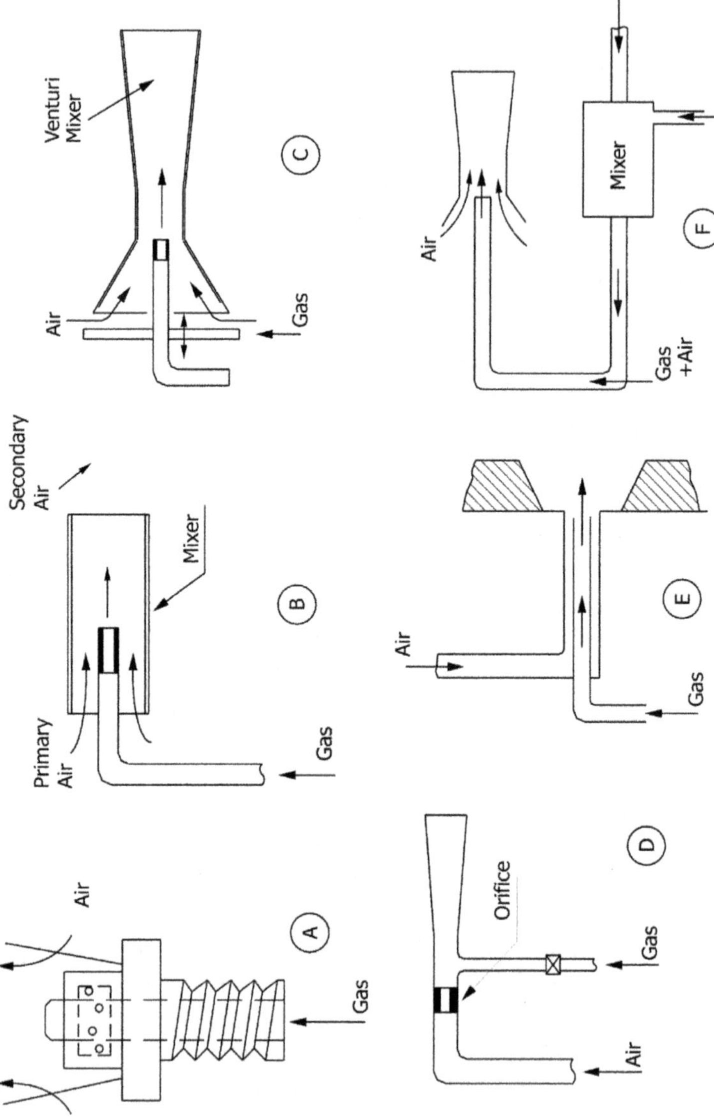

Figure 6.9 Classification of gas burners.

separate passages (see Figure 6.9(E)). The mixing takes place at the nozzle. Both air and gas are supplied under pressure. The air and gas quantities are controlled by valves in the supply lines. These valves are sometimes linked to obtain a good control during turn down.

6. Premix burner—Here the air and gas are thoroughly mixed before they enter the burners (see Figure 6.9(F)). The mixing is usually carried out in a centrifugal blower which supplies the air. The burners themselves are natural draft burners.

There are many designs and configurations based on the above classification. It is not possible to discuss all such designs in this book. Only some typical designs that pertain to furnaces and ovens are described in detail.

6.9 FLAME STABILIZATION, IGNITION, AND DETECTION

The stability of the gas flame depends on the amount and method of air supply. The air may be supplied singly where all the air is supplied along with the gas, as in nozzle mix burners. The air may be supplied in two parts. Some air is mixed and supplied with the gas. This is the primary air. The remainder, i.e., the secondary air is generally drawn from the surrounding air.

It is necessary that the emerging gas-air flow have a velocity that balances the burning velocity of the flame. A higher flow rate will lift off the flame. A lower flow rate will result in a backward movement of the flame, and it can be sucked in the mixer with disastrous results.

The thoroughness of mixing will also affect the stability. Poor mixing and a slow ignition velocity will lengthen the flame. Good mixing and high velocity will shorten the flame. From this point of view, a large amount of primary air or complete premixing with high velocity will produce a short flame.

With these factors in mind the burner designer incorporates several methods to ensure stability. Burner design is a specialized field but when choosing a burner it is necessary to check what the manufacturer is offering.

To prevent lift off or extinction every burner has a small auxiliary burner called the pilot burner. When lighting the main burner, the pilot is lit first. This provides the necessary ignition temperature to the air-gas mixture from the main burner. The location of the pilot in relation to the main flame is very important. Pilot flame should be very near the main flame. Pilot burners are of the neat gas type and are supplied through a separate gas line. They burn all the time and supply a constant ignition source which prevents flame extinction. Small burners (e.g., kitchen burners) do not require a pilot.

Pilot burners are ignited by a high voltage electric spark through a spark plug located near the burner tip. The electric circuit is not a part of the burner. It is situated in a separate enclosure.

To ensure that the pilot flame is lit, a flame detector or flame rod is situated directly in front of the pilot flame (see Figure 6.10). There are two types of detectors available. The ultraviolet (UV) detector senses ultraviolet radiation emitted by the pilot flame. The radiation is detected by a sensor that produces an electric signal. This signal is fed to the flame indicator and/or alarm.

The other type of detector is called a "flame rectification device." The flame rectifies an AC signal passing through it.

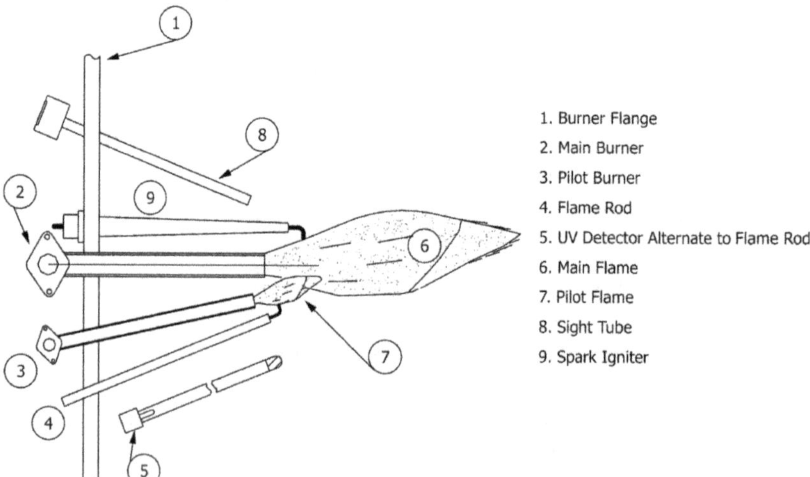

Figure 6.10 Flame ignition, detection, and stabilization.

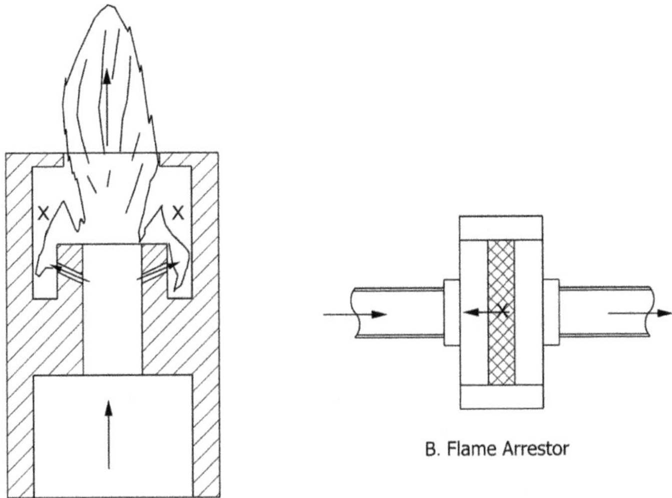

Figure 6.11 Auxiliary devices.

An electrode placed in front of the pilot flame detects the rectification of the burning flame and produces a signal.

Some burners have flame retention devices built in. These devices provide a small auxiliary pilot flame at the base of the main flame. One such design is shown in Figure 6.11(A). Other practical designs are discussed in a separate section.

The backward progress of the flame into the mixer is prevented by providing adequate minimum velocity to the mixer. To prevent the flame from being sucked into the gas tube, a flame arrester is provided in the gas supply pipe. This is a fine mesh or sintered metal screen providing small size multiple outlets for the gas (see Figure 6.11(B)).

6.10 ATMOSPHERIC GAS BURNERS

The majority of small appliances use "atmospheric burners" which are essentially natural draft burners, with the air drawn from the atmosphere.

These burners have two main parts. The first part is a mixer which may be a straight tube type but more commonly it is a venturi. The gas pressure is very low in straight tube mixers (0.25–5 kPa). In venturi mixers, the gas is supplied at 0.075–2.75 kPa for low-pressure venturis and 7.0–200 kPa for high-pressure venturis. The gas is supplied through an orifice of appropriate diameter. The air is inspired or entrained by the gas. The air inlet is adjustable to control the gas-air ratio. The gas orifice can also be changed to suit the gas. The gas-air mixture passes through the long enlarging passage and achieves a homogeneous mixture. A typical venturi mixer is shown in Figure 6.12.

In the application of the venturi for small appliance burners the venturi draws in only the primary air (~ 50%). The remaining air is drawn in at the flame from surrounding air. Hence, these burners cannot be used inside a furnace where the presence of air (or oxygen) is not desirable.

In larger venturis where the gas is supplied at a higher pressure, all the air for combustion is drawn in the venturi and the burner can be used inside a furnace.

The second part of the atmospheric burner is the nozzle. The exit end of the venturi is connected to the burner nozzle

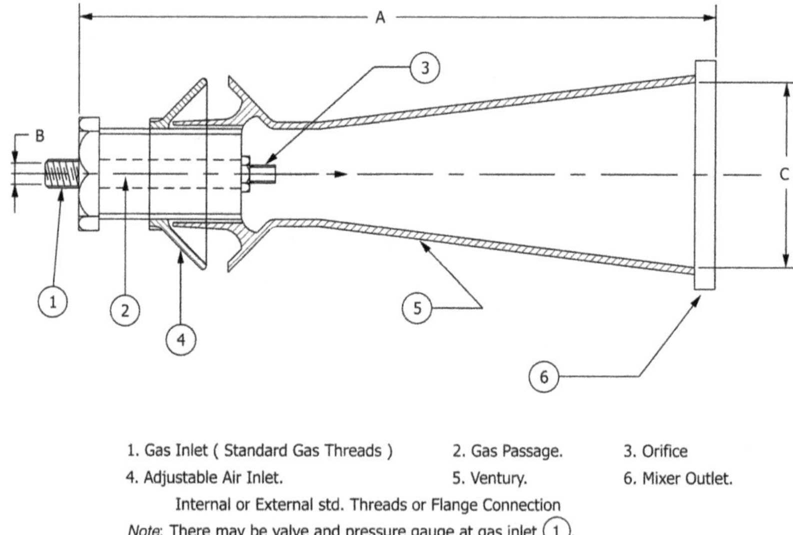

Figure 6.12 Venturi mixer for high and low pressure atmospheric burners.

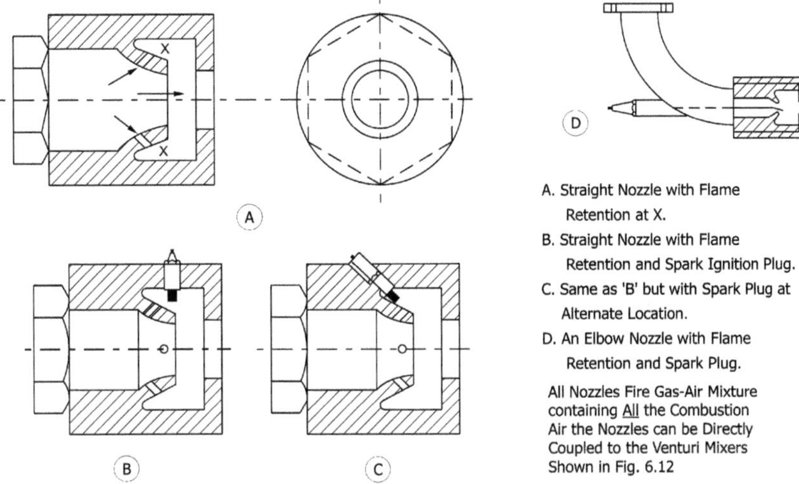

X
X
A
D
A. Straight Nozzle with Flame Retention at X.
B. Straight Nozzle with Flame Retention and Spark Ignition Plug.
C. Same as 'B' but with Spark Plug at Alternate Location.
D. An Elbow Nozzle with Flame Retention and Spark Plug.
All Nozzles Fire Gas-Air Mixture containing All the Combustion Air the Nozzles can be Directly Coupled to the Venturi Mixers Shown in Fig. 6.12
B
C

Figure 6.13 Typical burner nozzles (heads) for firing gas-air mixtures.

or head. Figure 6.13 shows some typical heads used for low pressure venturi mixers with primary air only. Figure 6.14(A) shows a rod burner consisting of a closed end tube with drilled holes for a gas outlet, Figure 6.14(B) shows a similar burner with raised nipples. There can be multiple row gas outlets as shown in Figure 6.14(C). A ribbon burner head is shown in Figure 6.14(D). This is a flat burner where the head is made of rows of two metal ribbons folded in opposite directions. This creates a succession of small and large gas openings. The small openings serve as pilot or retention burners and the larger ones as main burners. A ring burner head is shown in Figure 6.14(E). There can be one or many concentric rings all supplied by individual venturis. A kitchen-type burner head is shown in Figure 6.14(F).

Because of the low pressure used, there is a possibility of the flame being put off or blown off. To prevent this, retention rings or bars are provided as shown in Figure 6.14(G) and Figure 6.14(H). A small flame produced by small auxiliary orifices constantly burns below the retention ring and serves as a pilot flame.

Manufacturers offer many designs of these burners.

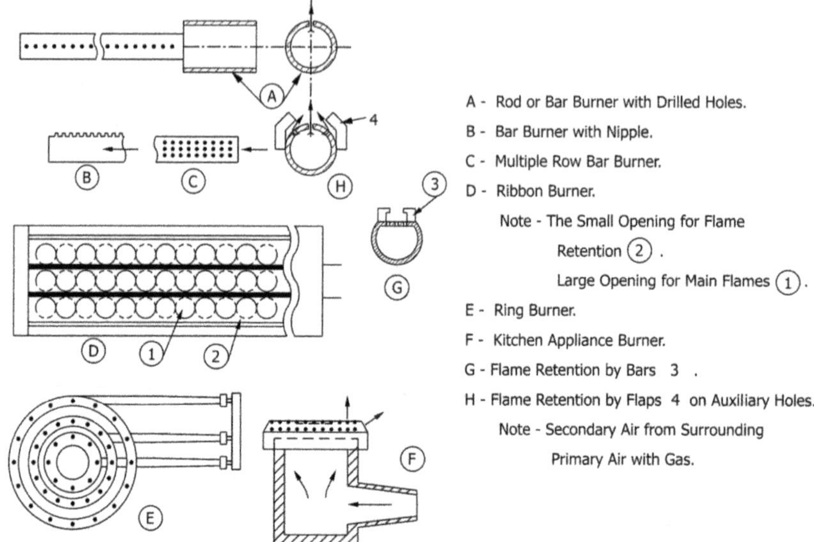

Figure 6.14 Burner nozzles or heads for low pressure atmospheric air-gas mixers.

High pressure gas venturis mix all the air with gas. Typical burner heads for these are shown in Figure 6.13. Note the flame retention device inside the burner opening. The head can be straight as in Figure 6.13(A), or an elbow, as in Figure 6.13(D). An igniter spark plug is fitted near the retention groove as shown in Figure 6.13(B), Figure 6.13(C), and Figure 6.13(D).

Venturi mixer gas inlets have standard gas threads (depending on national standards). Additionally, there may be a gas valve and pressure gauge at inlet. The outlet has male and female pipe threads, or a suitable flange. A large venturi can be used to supply multiple burner heads.

6.11 NOZZLE MIXING GAS BURNERS

Besides atmospheric burners discussed previously, nozzle mixing burners are used very widely in different types of furnaces.

A typical nozzle mixing burner is shown in Figure 6.15. There are many variations of the basic design offered by

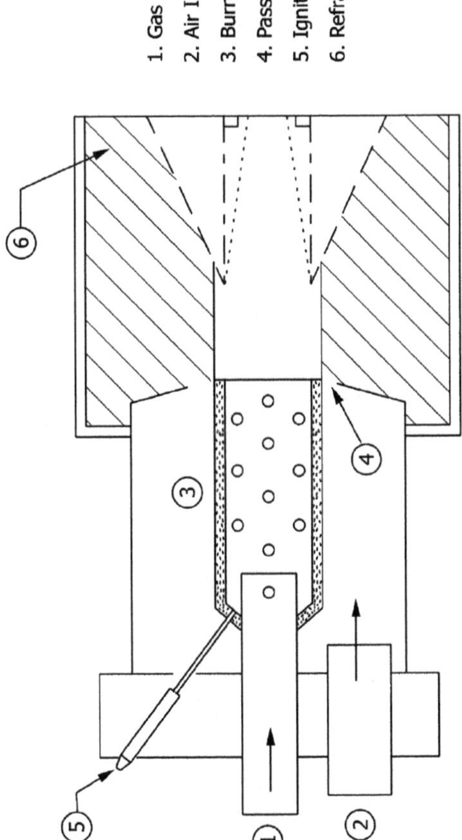

Figure 6.15 Nozzle mixing gas burner.

different manufacturers and all the features shown may not be in every burner. Note that all the combustion air is supplied through the holes in the nozzle or burner tube. Excess air, if provided, is supplied through the annular passage around the nozzle mouth. Excess air offers additional control over the flame length and shape but lowers thermal efficiency. However, it adds turbulence to gas. Some burners have swirl vanes in the air passage.

The quarl is generally sealed in the burner housing. It is made of refractory and can have different shapes of passages as shown by dotted and dashed lines. Some manufacturers offer thick metal quarls or ceramic flame tubes and quarls. These burners use cold atmospheric air but provision can be made for admitting preheated air.

Nozzle mixing provides very good control of the gas-to-air ratio. This helps to create the required atmosphere in the furnace. The air is supplied at 0.075–8.5 kPa (gauge). The gas pressure is about 40–50% of the air pressure.

6.12 RADIANT TUBES

There are many instances where it is not desirable to expose the furnace charge directly to the flame or combustion gas. Heat treatment furnaces and paint drying ovens are some of the examples.

Gas-fired radiant tubes are used to overcome this problem. A suitably designed gas burner fires inside a tube fixed inside the furnace (see Figure 6.16). Depending on the size of the furnace there may be more than one tube. The hot combustion gases heat the tube as they move toward the exit or the flue end. The hot tubes radiate heat to the furnace interior and the charge.

The burner used is a nozzle mixing one. It has a ceramic (silicon carbide) mixing tube or a refractory quarl. Some of the hot combustion gases travel back and heat the flame tube. This recirculation keeps the ceramic tube hot and helps in bringing about continuous mixing and burning at the nozzle. In some designs, part of the combustion gases are reintroduced inside

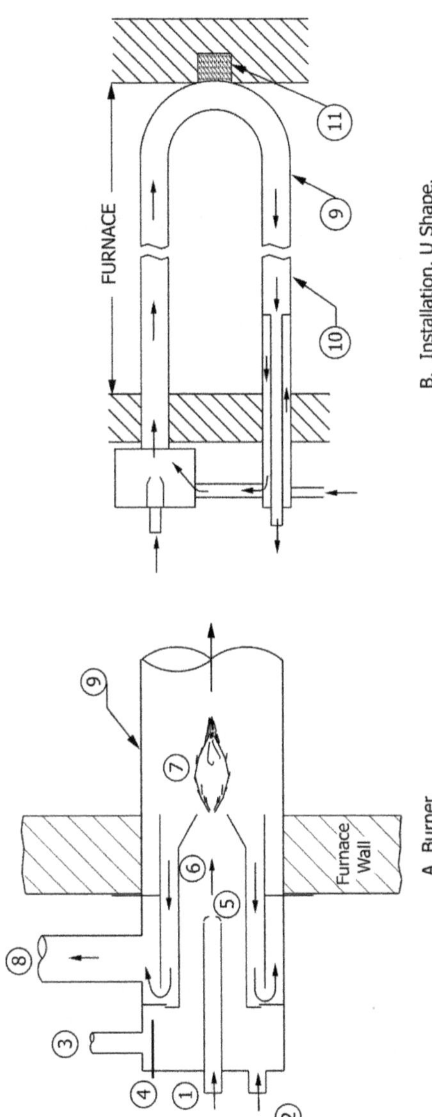

1. Gas Inlet.
2. Air Inlet.
3. Air Inlet (Alternate).
4. Flame Length Control.
5. Gas Tube & Orifice with Primary Air Inlet.
6. Ceramic Flame Tube.
7. Combustion Air.
8. Flue.
9. Radiant Tube.
10. Air Pre-heater in Flue.
11. Support.

Figure 6.16 Radiant tubes.

the flame tube. Flame length is adjusted by adjusting a shutter in the air inlet.

The outer (radiating) tubes are not supplied by the burner manufacturer. These tubes are made of heat resistant alloys such as Ni-Cr, Fe-Cr-Al, etc., alloys (see Chapter 8). The diameter and length of the radiating tubes are calculated from the expected temperature and the surface required to transfer the required heat to the charge. The tubes have a single bend U-shape, or a double bend S-shaped. The tubes are required to be properly supported to prevent sagging. They are fixed horizontally or vertically on the walls. The maximum useful temperature of the tube limits the furnace temperature. Presently available tubes can be operated at 1100–1300°C maximum. The burners for radiating tubes are rated from 7.5 to 150 kW.

6.12.1 Immersion Tubes

These tubes have construction similar to radiant tubes, i.e., the burner fires placed inside a close-ended ceramic tube. The combustion gases are exhausted from the burner end of the immersion tube as shown in Figure 6.17. Note the gas recirculation tube and the recuperation of heat from combustion gases. The tube is made from aluminous refractories or silicon carbide. These tubes are used for aluminum melting, zinc melting, galvanizing baths, etc.

6.13 DUAL FUEL BURNERS

As the name suggests, these burners can operate on both oil and fuel gas. Such a change of fuel may be dictated by availability and economics of gas and oil. Additionally, the heating process require of uninterrupted heating making, fuel changeover necessary when one is not available.

A typical dual fuel burner is shown in Figure 6.18. The oil is supplied through the central tube with an atomizing nozzle. Gas is supplied through a number of nozzles placed around the oil tube. Air is supplied around the oil and gas nozzles as shown. To obtain good atomization and fuel-air mixing the air stream

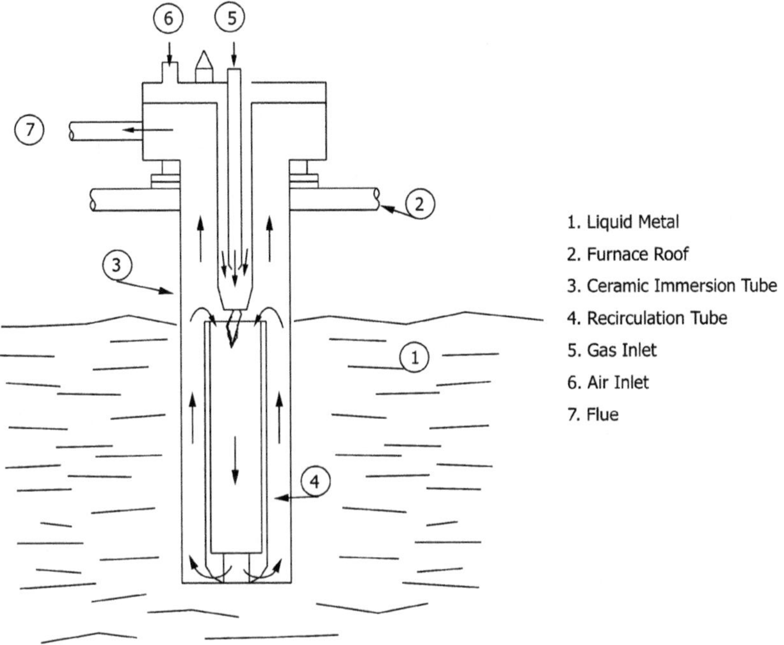

Figure 6.17 Single-ended recuperative burner.

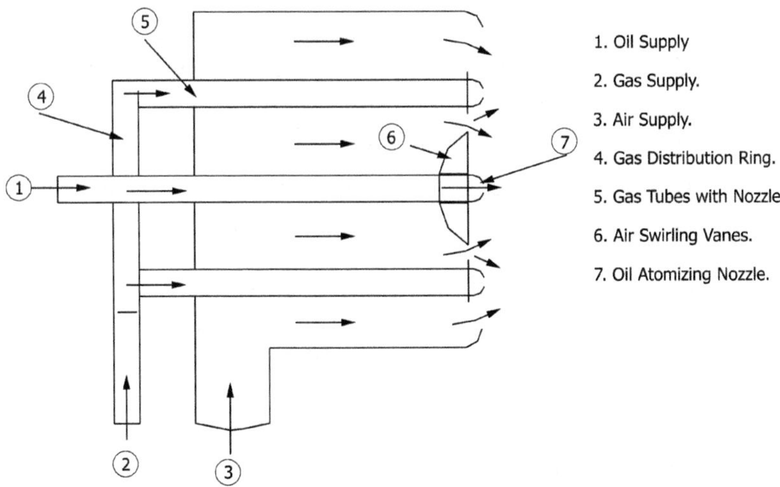

Figure 6.18 Dual fuel burner.

is given a swirl or rotational movement by the swirl vanes. There are many variations available commercially.

As burning characteristics of oil and gas are different, these burners do not operate equally efficiently on fuels. Their performance can be improved by using the proper quarl.

These burners are not likely to be useful for heating appliances and furnaces considered in this book and hence, they are not discussed further.

6.14 PACKAGED BURNERS

In previous sections we have reviewed various types of gas and oil burners. Heating appliances using these fuels require a number of auxiliary items for the efficient and safe operation of burners.

In the last two decades, the whole burner system has developed in the form of "packaged burners." These are compact assemblies which contain the burner and almost every ancillary piece of control equipment, and other auxiliary items such as air blowers, pumps, igniters, etc. Their construction is such that they can be bolted to the furnace and only external fuel and electricity supply are required to be connected to start the operation. Nowadays they also come with computer logic and menu operation facilities so that they can be programmed for any automatic operational sequences including system checkup, starting, and shut down.

Packaged burners contain the following components:

1. Burner—For gases, the burner is the nozzle mixing type. For light oils, low pressure atomizing burners with swirl are also used.
2. Combination air fan or blower which is directly coupled to the burner, i.e., there is generally no air line.
3. High voltage igniter with required circuit, and the flame detector or flame rod.
4. Pilot burner with separate fuel supply.
5. Safety (shutoff) valves.
6. Pressure governor (one each for the main line and the pilot line).

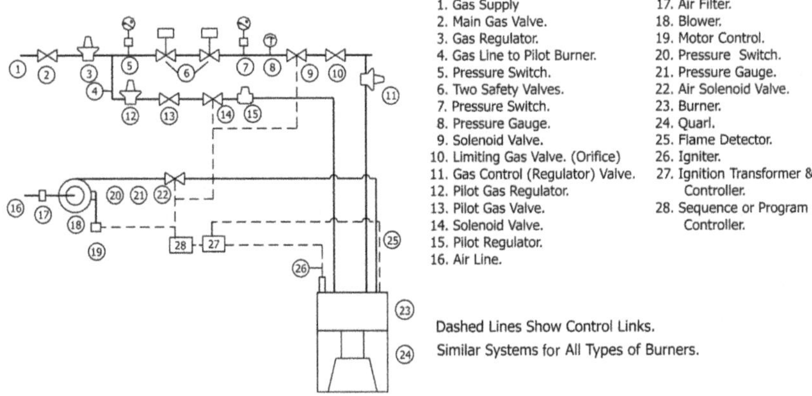

Figure 6.19 Supply and control system for a typical packaged gas burner.

7. Air purging valve.
8. Solenoid valves for automatic control.
9. All other control and regulating valves required by the law for safe operation.
10. Programming and control panels.
11. Light oils require a pump (gear or vane pump), which is incorporated in oil-based systems.
12. Heavy oils require preheating to lower the viscosity. Preheating systems are supplied separately in a package matched to the burner system.

Packaged burners are extremely convenient for the furnaces and appliances discussed in this book. Usually they are built around a single burner. If more than one burner is required, then many packages will have to be used. It is possible to link or synchronize their operation by programming.

A typical package burner system is shown schematically in Figure 6.19.

6.15 COMBUSTION OF SOLID WASTE AND GARBAGE

In the last chapter, we discussed the nature of solid wastes, their generation and incineration necessity, and associated problems.

From a combustion process point of view, solid wastes will broadly behave as solid fuels like coal, the only difference being the extreme heterogeneity of waste materials. Solid fuels are not considered in this book, hence, only a review of solid combustion process is presented here.

Solid fuels require a support on which they are burned. This support is called a grate. It may be a stationary grill or a moving endless belt made of metal Cast Iron (C.I.) slats. Some designs have a reciprocating movement to stoke the charge.

There are three stages of combustion as shown in Figure 6.20(A), Figure 6.20(B), and Figure 6.20(C). In practice, the three stages are not distinct but overlapping depending on design.

In the first stage the charge is heated by auxiliary burners to 500–600°C. The burners or heaters usually operate on 2–400% excess air. Additional air is supplied from the bottom through the grate. The charge loses moisture and some highly volatile components. Thus, stage one is essentially a drying operation.

The charge catches fire in the second stage. Combustion evolves the usual gases such as CO_2, H_2O, N_2, and vapors and

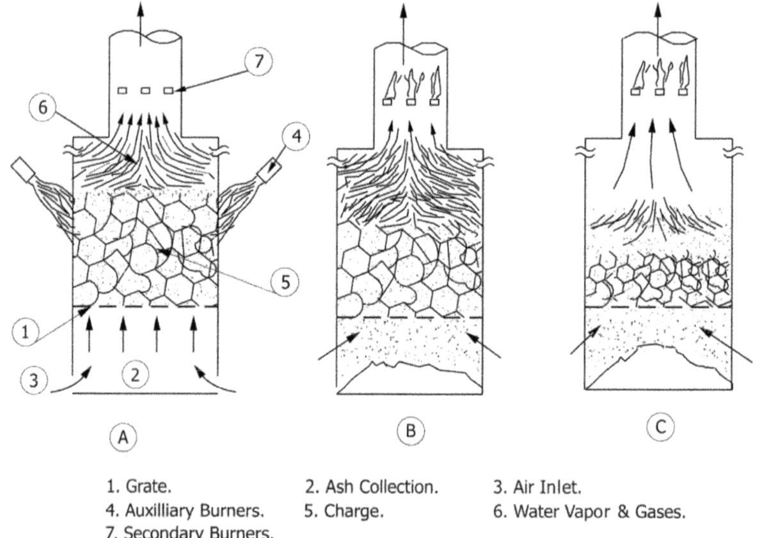

Figure 6.20 Stages in combustion of waste and garbage.

organic gases. The latter gases are highly toxic and odorous. The temperature attained is 800–1200°C depending on waste components. Frequently, auxiliary heating is also necessary to bring about continuous combustion. Ash and noncombustibles coagulate, travel downward, and fall through the grate. They are collected in an ash pit. This stage also requires excess air. The second stage is, thus, true incineration. However, the combustion is not complete as many constituents are pyrolyzed or gasified and pass into the flue.

In the third stage, the combustion is near completion. The flue gases are heated to 1000–1200°C by secondary burners where the pyrolyzed chemicals and gases are completely burned. This removes their odors and toxicity and the gases can be safely released in the atmosphere. The post combustion is also carried out in the second stage.

It can be seen that for efficient and quick incineration it is necessary that the charge has a good permeability for circulation of air and evolution of gases. At the collection stage, the waste is compacted for economical transport. The compaction will reduce permeability. Similarly, some arrangement for shaking or stirring the charge will help combustion.

In large incinerators, the waste heat is used for water heating or power generation. The ash is a good manure.

We will study typical designs of incinerators later.

6.16 BURNER AUXILLIARIES

6.16.1 Burner Blocks

All burners (gas or oil) discharge the fuel/air mixture into a refractory block known as a "burner block" or "quarl." The burner with the quarl is fixed to the furnace wall by bolts. The mixture burns in the quarl passage and a stream of hot combustion gases or part of the flame is projected into the furnace.

Typical quarl shapes are shown in Figure 6.21. It can be seen that the shapes of the quarl passages have different hydrodynamic shapes so that they change the velocity and the shapes of combustion gases or flames. An expanding shape (Figure 6.21(A)) will reduce the velocity and increase

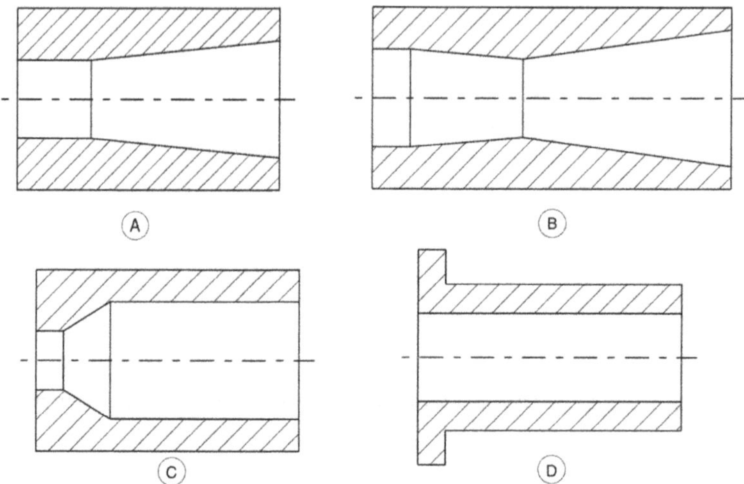

Figure 6.21 Typical burner blocks, quarls.

the pressure. A venturi-shaped burner block (Figure 6.21(B)) will first increase the velocity up to the throat and then slowly increase it toward the exit. An expansion in the entry followed by a straight passage to exit (Figure 6.21(C)) will project a flame with constant velocity. Some burner blocks have a straight passage throughout (Figure 6.21(D)). This will guide the flame without changing the velocity. There are many such shapes.

The quarl to be used with a particular burner for a given application is specified and frequently supplied by the burner manufacturer. Replacement blocks can be purchased or cast by the user by using a suitable castable refractory.

Quarls are made of dense refractories such as high duty fire clay. They hold a lot of heat, which is helpful in igniting the burner. Sometimes there is a special block fixed in or near the front of the quarl for this purpose.

Direct contact between the wall and the flame is also avoided by the quarl. Methods of fixing the burner and quarl to the wall are discussed in another chapter.

In a burner the ignition velocity has to be matched with the air-fuel discharge velocity. This can be achieved by using a proper quarl.

6.16.2 Ignition Devices

The air-fuel mixture discharged from the burner requires ignition, at least in the beginning. This can be carried out manually for small burners, either by using a matchstick or a small torch.

Industrial burners use an electric ignition device, which is a permanent fixture in the burner or occasionally the quarl. These are sparking devices operating electrically or electronically. The spark is produced in a spark plug situated in the flame path. A high voltage transformer and pulse transformer is used to produce a spark. The spark gap is 5–10 mm wide.

Many burners use a pilot flame. This is a small flame which is continuously burning irrespective of whether the main flame (burner) is on or off. Some burners use a pilot only to light the main flame and then shut it off. The fuel consumption of pilot is very small. The pilot flame uses gas and is placed in front of the burner to ensure easy ignition of the main flame. There is usually a view port to observe the ignition process. All burners have their ignition devices or an adjacent port into which recommended device can be fixed.

Well-atomized light fuels ignite easily. Heavy oil may require repeated attempts until the flame is well established.

6.16.3 Flame Protection Devices

These devices sense the pilot and main flames and ensure a correct ignition and extinction sequence. They are also called "flame rods." Industrial furnaces use a rod based on the electrical rectification property of a flame. Other devices operate on thermoelectric power, ultraviolet emission, or photoelectric transmission of flame.

Flame rods have to be correctly located in relation to the flame. They operate fuel flow valves to control the fuel flow in case of a malfunction. As with the ignition devices, the flame rod is usually an integral part of the burner.

Chapter 7

Refractories

CONTENTS

7.1 INTRODUCTION

"Refractory" means hard to melt. Construction of any heating apparatus or appliance requires material which withstands maximum temperature attained for a sufficient time. There are many other important properties that have to be taken into account such as electrical resistivity, porosity, and so on. As we will see in this and in the next chapter, both ceramic and metallic materials are used for this purpose.

Ceramics, which are traditionally called "refractories," are discussed in this chapter.

Refractories are inorganic compounds such as oxides, carbides, nitrides, and so on. These oxides form the bulk of commonly used materials as most of them are naturally available in large quantities.

It is necessary to understand the nature and properties of refractories to make a correct and economical choice. We will be discussing only those refractories which are of interest to us. For other refractories and details of properties, reference should be made to handbooks listed in the *Bibliography.*

7.2 CLASSIFICATION OF REFRACTORIES

Refractories are mainly inorganic oxides or their mixtures. These oxides, and hence the refractories, are classified into three classes — acidic, neutral, and basic. This classification is based on their resistance to the slags produced in metal extraction, glass making, cement, and other chemical manufacturing.

Acid refractories resist attack from slags predominant in silica (SiO_2). Basic refractories resist the attack of slags containing CaO and MgO. Neutral refractories are unaffected either by acid or basic slags. The degree of slag resistance varies among various oxides. Thus, silica is most acidic and calcium oxide is most basic. Based on these criteria the oxides forming refractories are arranged on a "scale" given below.

SiO_2, TiO_2, Al_2O_3, Cr_2O_3, Fe_2O_3, MgO, CaO

Acidic ← Neutral → Basic

Fire clays dominant in SiO_2 are acidic, but as their Al_2O_3 content increases they become increasingly neutral.

Pure Al_2O_3 is completely neutral. Refractories containing magnesite (MgO) and dolomite (CaO, MgO) are basic.

Other compounds used as refractories such as carbides, carbon, graphite, zirconia, etc., are considered neutral.

This book excludes all furnaces and reactors that produce slag. Hence, the above mentained method of classification is not discussed further. For general use in industrial furnaces fire clay refractories are mostly used in all forms and for parts of furnace structure (walls, roof hearths, doors, etc.). Other refractories that are occasionally used are pure alumina, pure silica, and silicon carbide among others. These refractories are discussed in more detail in the subsequent section.

7.2.1 Fire Clay Refractories

These mainly consist of silica (SiO_2) and alumina (Al_2O_3) in varying proportions. These two oxides form a continuous system of phases. These phases are

Tridymite—A silica (SiO_2) mineral stable between 870 and 1470°C.

Crystobalite—A silica mineral stable between 1470 and 1723°C.

Mullite—A silica-alumina mineral ($Al_6Si_2O_{13}$) having a melting point 1850°C and stable at all temperatures.

Fire clays also contain impurities and bonding agents in small amounts. These are classified according to their duty such as Super duty—Containing more than 45% Al_2O_3 and pyrometric cone equivalent (PCE) greater than 33. They have good resistance to thermal shock and good mechanical strength.

High duty—They have alumina content similar to super duty but a lower PCE (31–33).
Medium duty—They have a still lower PCE (29–31).
Low duty—They have a PCE of 15–28.

Thus, it can be seen that as the duty becomes less severe, the PCE value, i.e., the refractoriness, strength, and abrasion resistance, along with the cost, also lowers.

For general purpose medium-size furnaces, medium- or high-duty fire clay is adequate for temperatures up to 1550°C. For higher temperatures up to 1700°C, super-duty fire clay is the best.

Low-duty fire clays can be used up to 1400°C but have low strength and low thermal shock resistance. They are best used as backing layer for higher grades.

Almost all the above grades are also available as insulating bricks. These have similar thermal properties but due to a high porosity, they have low thermal conductivity and low strength.

Fire clay refractories containing more than 72% silica are known as "semisilica bricks." These can be used up to 1450°C. They have good dimensional stability.

7.2.2 High Alumina Refractories

These refractories contain 48–99% alumina and are classified according to alumina content. They consist of mullite (Al_6SiO_{13}) and corundum (natural Al_2O_3) along with bauxite and clays. Their impurity content is lowered to improve refractoriness.

High alumina bricks have a high density, PCE value higher than fire clays, good mechanical strength and creep resistance, and can be used up to 1850°C.

Their cost increases with increasing alumina content. These refractories find use in melting furnaces, incinerators, cement kilns and the like. In industrial heating where temperatures are as high as 1500–1800°C, high alumina refractories are the only choice.

They have good electrical resistance at high temperatures and are gas-tight (low porosity) and hence, are used for heating element supports, insulators and muffles for laboratories, and light duty high temperature industrial furnaces (e.g., semiconductor processing).

Other varieties of high alumina refractories are alumina-chrome and alumina-carbon refractories. Phosphate bonded alumina bricks are used for aluminium melting furnaces.

7.2.3 Silica Refractories

These types of refractories contain 94–98% silica and very low amounts of FeO and Al_2O_3. They have TiO_2 and CaO up to 1.5–2.5%.

Silica bricks have good resistance to acid slags and have good mechanical strength as well. If not processed properly, they show a permanent volume expansion up to 15% and have low thermal conductivity.

They are useful up to 1700°C and are mainly used for glass-melting furnaces.

Vitrified or fused silica is another variety containing >98% silica. They are produced from fused quartz. They have volume stability only up to about 1100°C. This type is resistant to most chemicals (except strong alkalies) and is widely used for laboratory ware, furnace tubes, etc. It can be polished to obtain good, glass-like transparency.

7.2.4 Carbon and Graphite Refractories

These contain 98% or more carbon in various crystal forms blended with pitch, resins, and petroleum derivatives.

Carbon is easily oxidized in atmospheres containing oxygen or oxidizing gases such as CO. Hence, these refractories can be used where there is no direct access to oxygen.

Such conditions exist in a vacuum or in inert atmosphere furnaces and also at places where carbon is covered by metal or slag (e.g., blast furnace bottom).

Carbon blocks prepared by blending and pressing can be easily machined to intricate shapes such as tubes and nozzles.

Currently carbon is also produced in fiber form which can be further processed to obtain tapes, cloths, blankets, and so on.

Carbon can be used (with above restrictions) for temperatures up to 2000°C. It has good electrical conductivity and high resistance to slags, chemicals, and gases. It has been used for electrical resistance heating elements.

Graphite bonded with clays and tar or pitch is used to manufacture foundry crucibles.

7.2.5 Silicon Carbide (SiC) and Carborundum

This is a major special refractory and is made by fusing silica and coke at 2000°C. It is available in various purity grades.

Silicon carbide is extremely inert and difficult to bond. It is bonded with clay or silicon nitride.

It is used in electrical heating elements (up to 1550°C) for furnace parts (up to 1800–2000°C). It has good spalling resistance and is very hard and strong. Compared to other refractories, it has very high thermal conductivity.

7.2.6 Zircon Refractories

These are zirconium silicates containing 67% ZrO_2 and 33% SiO_2 together with binders, etc. These are difficult to shape, have good resistance to acid slags, and good thermal shock resistance.

7.2.7 Zirconia Refractories

These are made from pure zirconium oxide (ZrO_2), which has a very high melting point (2680°C) but shows dimensional changes at high temperature. By certain stabilizing techniques, it is possible to control sudden dimensional changes and cracking. It is very costly and does not find wide use in general furnaces.

7.3 INSULATING REFRACTORIES AND MATERIALS

During its operation a furnace develops a high temperature on its inner side which is called the hot face. The outside, i.e., cold face temperature, is the atmospheric temperature. Hence, there is a considerable temperature difference between the hot and cold face under which heat flows out. This heat is totally wasted but unavoidable. However, this loss can be reduced by using a material with low thermal conductivity for furnace construction. Such low thermal conductivity materials are known as "insulating materials."

Many insulating materials have an upper useful temperature beyond which they cannot be used. They are also mechanically weak and cannot be loaded. This limits their direct use in furnaces. Materials of this class are cork, asbestos, glass fiber, and wood.

Fire clay materials (alumino-silicates) can be processed to give a low thermal conductivity. This is achieved by creating voids or pores in them. These pores reduce the density and make the bricks lightweight. Porous fire clay bricks are good insulators and are reasonably strong up to about 1000°C. Hence, they are more suitable than the materials discussed in the previous paragraph. Porous fire clays are not as strong as dense bricks and hence cannot be used for the hot face. These bricks are easily machinable.

Ceramic and mineral fiber materials in various forms constitute the third category of insulating materials. Their conductivity, density, and specific heat is low. Ceramic fibers can be used for both hot and cold faces; rock and mineral wool can be used at 700°C maximum. These materials have no mechanical strength. They are also very costly.

Hence, there is no one insulating material which will satisfy all requirements. We take advantage of the fact that the temperature difference between the hot and the cold face is in the form of a gradient. Hence, a wall or roof can be constructed

Table 7.1 Properties of Some Common Insulating Materials

Material	Density kg/m^3	Maximum Temperature °C	Thermal Conductivity W/m°C	Specific Heat kJ/kg°C
Light weight fireclay 65–80% SiO_2	450–800	950	0.15–0.25	0.835
Light weight fireclay 35–50% Al_2O_3	500–1000	1100–1250	0.45–0.55	0.837
Ceramic fiber blanket 40–50% Al_2O_3, SiO_2	64–130	900–1400	0.17–0.25	1.03
Ceramic fiber blanket 95% Al_2O_3	50–90	1600	0.09–0.21	1.07
Rockwool	150–200	750	0.033–0.037 @ 150–200 kg/m^3	0.418 –0.852
Asbestos	150–fiber 600–800 board	550	0.13–0.15	0.837
Glasswool	150–200	300–450	0.037–0.040	0.66
Glass fiber (mat)	12.0–80	350–450	0.035–0.040	0.66
Mica	2900	500–550	0.52–0.60	0.837

from a number of layers of different materials suitable for the temperature range of that layer. This is called a composite wall. The thickness of layers can be adjusted to give the most economic and effective loss reduction.

Various commonly used insulating materials and their important properties are given in Table 7.1. The design aspects are discussed elsewhere.

7.4 MANUFACTURE OF REFRACTORIES

7.4.1 Raw Materials

Besides a few exceptions, the raw materials for manufacturing refractories are natural minerals and clays. These are quartzite, quartz (SiO_2), kaolin ($AlSiO_4$), dolomite $CaMg(CO_3)_2$, calcium carbonate $CaCO_3$, magnesite $MgCO_3$, sillimanite $AlSiO_4$, alusite Al_2SiO_5, chrome ore (Fe, MgCrAl)O_4, Zircon sand $ZrSiO_4$, graphite, and many others.

The mined raw materials contain many impurities. Sea water is an important source of magnesium.

Pure alumina (Al_2O_3), silicon carbide (SiC), corundum (Al_2O_3), and mullite ($Al_6Si_2O_{13}$) are some of the raw materials which are prepared synthetically at great expense. Some refractories are made from carbon or contain carbon products, which are by-products of coke and oil production processes.

Besides the gross chemical compositions cited above, the actual mineral crystals that go into the formation of a raw material are also very important. For the same chemical composition, the crystal structures can be quite different and will show different properties. During the manufacturing (firing stage) and when used in a furnace, the original crystal structures may change. This will result in abrupt expansion or change in other properties such as mechanical strength. Hence, it is necessary to find out the actual crystal structure of the constituents.

Identification of the crystal structures and their bonding with other minerals constitute the separate science of minerology. There are vast numbers of mineral classes. The same is the case for different clays.

Broadly speaking, the manufacturing process of almost all refractories follow the same sequence. There are minor differences which are not discussed here.

All the mined raw materials are crushed, ground, and separated according to the lump or grain size. Either before or after the size reduction, it may be necessary to wash the mineral to reduce impurities, gangue minerals, and extraneous matter.

Next, a very important stage is to mix the ground constituents to obtain a proper blend of chemical composition, minerological content, and desired grain size distribution. It is the variation in blending that decides the future properties of the refractories. The mix is again ground and homogenized in mixers. It is at this stage that binders, such as clays, are also added to the mix. The required amount of water is also added at this stage.

There are many methods of manufacturing standard refractory shapes. The most common method is the forming of the shape in a mold under high pressure (~200 N/mm^2).

Complicated shapes which are difficult to press are shaped by slip casting. A slurry of raw material mix and water is poured in a plaster mold. Water is absorbed by the mold and a shape (semi-dry) is produced.

A few years ago the process of isostatic compaction was brought from the laboratory to the production floor. The raw material is filled in elastic rubber bags having the required shape. The bags are deaired and sealed. They are then charged into a pressure vessel and subjected to enormous hydrostatic pressure. The material is compressed from all sides. This results in a dense precision product. The process can also be carried out at high temperatures. Isostatically pressed shapes are far superior than the press formed ones. However, the process is costly and is used for special refractories only.

In the next stage, the green, lightly bonded shapes are carefully dried in large kilns. Drying produces steam which can damage the product by cracking or breaking. Removal of water also produces shrinkage, i.e., substantial change in dimensions. Hence, utmost care is necessary at this stage. The product is now strong enough for handling.

Final strength and shape is produced in the next stage of firing or sintering. Here the shapes are stacked on flat cars and passed through a tunnel kiln. The cars pass at a predetermined speed and are subjected to a definite cycle of preheating, firing (heating) at 1200–1800°C, and cooling. At this temperature the various constituent minerals react and produce a strong bond and some shrinkage. The actual thermal cycle used depends on the phases (crystal structures) of the constituents and their interrelationship (phase diagram). Many types of furnaces are used depending on the thermal cycle and the volume of production.

The shapes are then measured for dimensional accuracy. Their crystallographic structure is also checked along with the outer appearance. The shapes are then cleared for dispatch.

The shapes so produced are very hard and cannot be machined. Special refractory shapes are sometimes ground to produce a dimensional accuracy of 1.0 mm.

It is again pointed out that what is described above is only a broad outline of the manufacturing process. Individual refractories may require slightly varying procedures.

7.5 REFRACTORY SHAPES

There are many standard and nonstandard shapes of refractories. It is always economical to use the standard sizes (as far as possible) in furnace construction as they are readily available.

The basic shape available all over the world is the standard 9" brick which has the dimensions $9'' \times 4\frac{1}{2}'' \times 2\frac{1}{2}''$ (or $230 \times 115 \times 65$ mm) and has a volume of 101.25 cu. in. (~1660 mm^3).

Besides the standard brick, other standard shapes are small, straight, soap, key, jamb, tile, end skew, side skew, arch, split, circle brick, featheredge, neck, etc. Some of these shapes are shown in Figure 7.1. It is necessary to consult manufacturers' literature to check the standard shapes offered. The circle brick is a sector of a circle of given diameter and is available in many sizes. Similarly, the arch brick is used to build the furnace roof or arch of a given span and again a range of spans is offered.

Shapes are ordered by numbers or volume. Usually a 9" brick equivalent is used for this purpose. The standard 9" brick has a volume 101.25 in.^{3} The volumes of other shapes are determined in units of 101.25 in.^{3}

The nonstandard shapes available are special or modified standard shapes. Burner blocks (quarls) and corner pieces are some examples of this type. It is necessary to consult the manufacturer for these shapes. If he has already made the shape, the mold is ready and the cost will be less than a specially manufactured shape.

Along with the shapes, the manufacturer also supplies standard anchors or hooks with which the refractory walls are built.

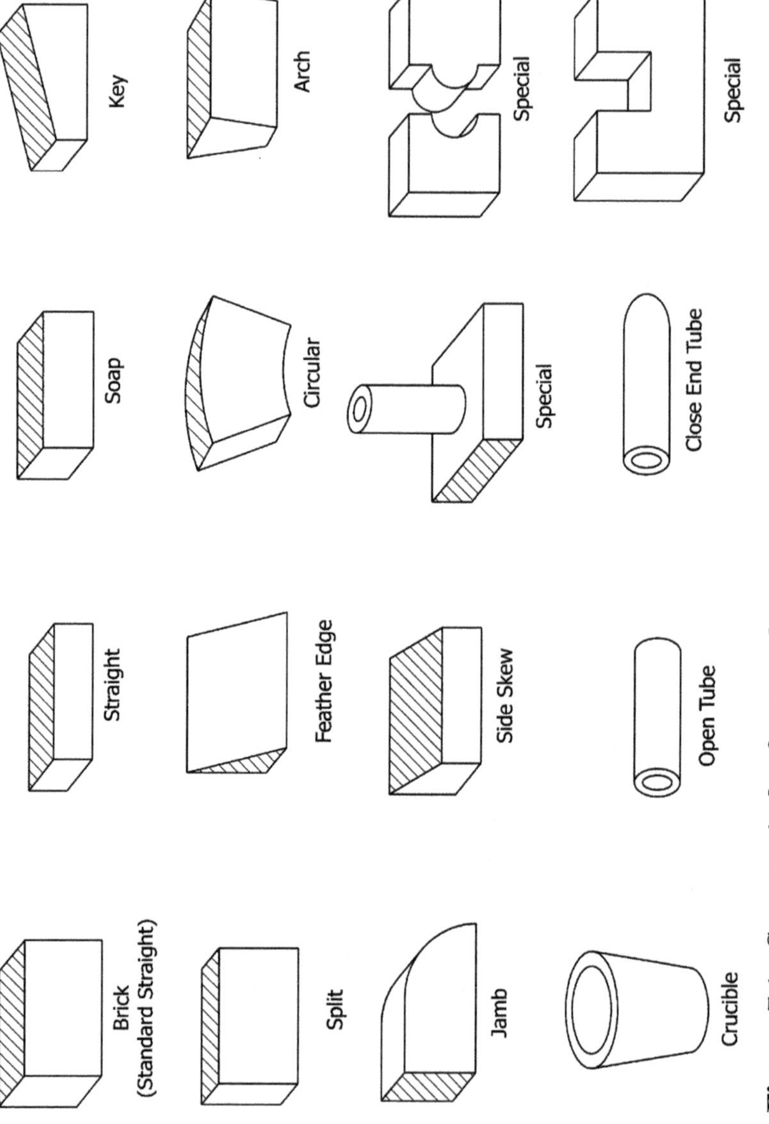

Figure 7.1 Some typical refractory shapes.

Almost all the refractory types (compositions) are available in standard shapes.

Many refractories such as pure oxides (Al_2O_3, SiO_2, ZrO_2), and alumino silicates (mullite, kayanite) are produced in tube form. Both rectangular and circular tubes (up to 150 × 150 mm or 150 mm diameter) are made. These are used for laboratory furnaces, pyrometer sheaths, etc. Other shapes available are rods, tiles, small crucibles, etc. All these shapes are manufactured by pressing or extrusion.

7.6 UNSHAPED REFRACTORY PRODUCTS

These products contain the same raw materials plus minor additions of plasticizing, bonding, and other surface active chemicals and minerals. These materials are offered in various formulations such as gunning and ramming mix, castables, mortars, cements, monolithics, and mastics. They are used for lining repairs, brick laying, hearth construction and so on. Almost all types of refractories are available in these forms. They can be used for hot or cold repairs. It is necessary to follow the manufacturers' instructions for their successful use. Castable mixes can be used to cast an uncommon shape at the site. Some formulations are heat-setting and some require careful curing or drying. These formulations are mainly used in process furnaces such as steel convertors, and cement kilns.

For the furnaces discussed in this book, unshaped products are mainly useful for hearth-making and crack-sealing.

An interesting unshaped refractory form is paints or washes. These are based on zircon ($ZrSiO_4$) and alumina or fire clay. They can be applied on refractories or metal surfaces. They improve the heat transfer characteristic of the lining by improving emissivity at high temperatures (up to or more than 1500°C). The lining life also increases on the use of paints and washes. They are applied by spraying or brushing.

7.7 REFRACTORY FIBERS

We have seen in previous sections that refractories are available in shaped and unshaped forms.

A third type of refractory form is fibers or filaments. From these we can get both shaped and unshaped products. The fiber form of refractories is a result of developments over the last twenty-five years.

The general outline of fiber manufacture is as follows:

A synthetic or mixed composition of the refractory is melted at a high temperature. The melt is then poured in the form of a stream. Air is blown into the stream so that it breaks the stream into fibers which solidify. By carefully adjusting the melting temperature, the speed of stream, and the velocity of air, the fiber length and diameter are controlled.

The fibers so produced have lengths of about 200 mm and diameters 2–4 μm.

By using textile manufacturing techniques, the fibers are converted into cloth and blankets of various thicknesses and lengths. By the addition of resins and other bonding materials the fibers are converted to rigid boards and paper. Spinning the fibers produces ropes and felts.

Along with these shapes the fibers are also formulated into unshaped forms such as mortars, moldable mix, wet wraps, and cements.

Unprocessed loose fiber is also offerred as a packing material for insulation. Presently, the following materials are available in fiber and fiber-made forms:

All fire clay compositions, alumino silicates such as mullite and kyanite, corundum, pure silica, and zircon.

Besides regular refractories, fiber materials made from glass, rock fiber, and slag are also available. This list is continuously growing due to new formulations and processes. The fiber form offers many distinct advantages over regular pressed solid forms. A few main points to be noted are:

1. For the same chemical composition the fibers have extremely low thermal conductivities. This is due to low density inherent to the porosity and voids.

2. Low density also gives low specific heat, hence, heat storage is also very low.
3. Very good thermal shock resistance as there is no expansion.
4. Good chemical, steam, and water resistance.
5. Can be easily cut into required shapes and quickly installed. No curing required.
6. Intricate shapes can be produced by molding or vacuum forming.
7. They can also be used for hot face.
8. They are resilient and make good acoustic barriers.
9. Due to their light weight, fiber boards make for much stronger and lighter roofs than parches.

Compared to these advantages a major disadvantage is that the mechanical strength, hardness, and abrasion resistance is poor. Hence, they can only be used in places having no mechanical load such as wall faces and roofs. If there is both mechanical load and elevated temperature, the structure is built with appropriate solid forms (bricks) and faced with fiber forms. Furnace hearths usually take the load from charge. Hence, fibers are not suitable for making hearths.

A short summary of the important properties of fiber products is given in Table 7.2. Detailed information about the properties and applications is best obtained from the manufacturers as there are some differences among various proprietary products.

7.8 PROPERTIES OF REFRACTORIES

The performance of a refractory in service depends on many factors. These factors are related to certain measurable properties. Similarly, the refractory construction is carried out at room temperature and is used for service at a higher temperature. Hence, the properties at both room and high temperature are required to be known and matched. These properties also serve as a guide for quality control in manufacturing.

Table 7.2 Properties of Ceramic Fibers and Rock Wool

Type	Ceramic fiber			Rock wool		
Form	Bulk fiber	Blanket	Board	Bulk fiber	Blanket	Board
Density kg/m^3	80–100 delivered 50–250 installed	64–128 Standard 96–128 High duty	250–300	150	150	50–200
Maximum Temperature °C	1260–1400	1260–1400	1260–1600	700–750	700–750	700–750
Thermal Conductivity W/m °C	0.140 0.133	0.126 0.120	0.079	0.03–0.05	0.04–0.06	0.04–0.1
Application	Expansion Joints, Sealing Molding	Lining Insulation	Hot, face lining Insulation	Insulation Acoustic and Thermal	Insulation Acoustic and Thermal	Insulation Acoustic and Thermal

7.8.1 Room Temperature Properties

1. *Chemical composition*—The major constituents and their contents in a refractory decide its class, i.e., acidic, basic, high alumina fire clay, etc. Minor constituents such as alkali oxides, carbonates, iron oxide, titanium oxide, etc., decide the chemical performance. The chemical analysis also indicates whether the raw materials chosen are correct and the mixture is homogenous.
2. *Minerological constituents*—These are determined by microscopic and X-ray techniques and provide information about the constituent minerals, their crystal structure, and phases. These decide the performance in firing as well as in service at high temperature.
3. *Bulk density and porosity*—Bulk density and porosity provide information about the heat storage capacity and porosity indicates the correctness of pressing and drying. Density is quoted in kg/m^3 and porosity is expressed in percentage. Porosity also indicates the insulation capacity (thermal conductivity). Highly porous refractories are good thermal insulators.
4. *Abrasion resistance*—Refractories are subject to rough handling during manufacture and also to abrasion with process materials (furnace charge, slags) and dust-laden gases at high temperature. Abrasion resistance determines the resistance to such erosion. It is determined by abrading the refractory on a standard surface (e.g., silicon carbide) when loaded under a standard weight.
5. *Cold crushing strength*—A refractory sample is loaded at a standard increasing rate and the load at the breaking point is noted. This breaking load divided by the area under load gives the cold crushing strength. This test gives a good indication about load bearing capacity, handling, and shipping capacity and also provides information about the strength of the bond obtained after firing.

7.8.2 High Temperature Properties

1. *Refractoriness* — This is the maximum temperature which a refractory withstands before undergoing permanent change of shape. Refractories do not have well-defined melting temperatures. They start softening rather than melting in a certain temperature range. Refractoriness is determined by a pyrometric cone test. A number of standard pyrometric cones having different softening points and a similar cone of the test refractory are heated in a furnace under standard conditions. At the softening temperature the test cone and a matching standard cone buckle or collapse. The number of this cone (ASTM standard) is reported as the refractoriness of the test piece, called the "pyrometric cone equivalent" (PCE) number. In practice, the actual maximum temperature of service is about 100–200°C less than that indicated by the PCE number.
2. *Dimensional stability*—All refractories show a permanent change in dimensions at the firing stage of manufacture. These changes are very slow at the end, and hence, remain incomplete. When fired in a furnace in service, the lining is heated to a high temperature for a long time. This tends to complete the remaining dimensional change. These changes, though small and slow, have to be considered in lining design. The test report indicates changes in linear dimensions on heating for a standard time and required test temperature. The dimensional change may be expansion or shrinkage depending upon the type and temperature.

   Apart from the permanent dimensional changes, the refractories undergo the usual linear expansion and contraction on heating and cooling. These dimensional changes are very important in lining design. They are required to be compensated for by designing proper expansion joints. Sudden changes in temperature will produce thermal shocks. If not taken care of by expansion joints and cushions, the lining will be subjected to high internal stress leading to cracking or collapsing.

Most of the refractories in fired condition show a fairly linear relation for expansion (contraction) and temperature. Silica bricks are an exception to this. They show a high expansion rate of up to about 800°C and then subside to a very slow linear rate.

Unfired refractories such as cements, gunning, ramming mixes and castables show drastic dimensional change due to both permanent and reversible expansion.

3. *Thermal expansion and creep under load* — All refractories are subjected to some load at high temperatures. For the furnaces considered in this book, the bottom will carry the load of the charge as well as self weight. The walls will be loaded mostly with self weight and will be under compression. If the roof is made of unsuspended brick arch, it will be quite heavy. Thus different parts of the furnace lining will be under different loading conditions. They will show a permanent change of dimension with time, i.e., creep. This permanent change (usually subsistence) is determined by loading a sample at standard (75×10^4 kg/m^2) load at increasing temperature and time. The phases present, density, porosity, and bonding between particles determine the creep properties.
4. *Thermal conductivity* — Due to the difference between the inside and outside temperatures, all refractories conduct considerable heat to their surroundings. This heat is unavoidable wastage. The conductivity is reported in W/m °C. It changes (usually increases) with increasing temperature and the values of the temperature coefficient of conductivity are also available.

   Insulating refractories are highly porous and have low thermal conductivity. The thickness of the lining, usually consisting of an inner layer of dense refractory and outer layer(s) of insulating refractory, is determined by the permissible heat loss and conductivity.

   For some applications such as recuperators and regenerators, a high thermal conductivity is desirable.

5. *Thermal spalling*—Some refractories, especially those undergoing thermal cyclic load and gas or slag attack, lose their exposed surface layers. This is known as spalling. Spalling can be reduced by proper choice and lining design.
6. *Specific heat*—A lining absorbs and retains a considerable amount of heat in service. The retention of heat depends on the specific heat and density, i.e., thermal mass. Thus, for quick heating and cooling we require a lining with low specific heat and density. Insulating refractories have a low thermal mass due to the large number of air pores that give low density.

There are many other properties of refractories that are occasionally useful in special designs.

Selected properties of some refractories relevant to the furnaces and considered in this book are given in Table 7.3, Figure 7.2, and Figure 7.3.

7.9 SELECTION OF REFRACTORIES

We have reviewed the main types of refractories and their important design and service properties in previous sections.

The choice of refractories chosen from the available materials basically requires matching the properties with the requirements determined at the design stage. Each design will have its peculiar demands and will need an individual choice. Most of the time there is no single refractory that is suitable. We then have to choose different refractories for different parts or make the best compromised selection. The following guidelines will be helpful.

7.9.1 Thermal Requirements

These are established by following data:

1. Type of furnace—such as horizontal muffle, car bottom or belt oven
2. Temperature—maximum temperature, rate of heating and cooling, temperature at different zones

Table 7.3 Properties of General Purpose Refractories

Class	Main constituents	Density kg/m^3	Refractoriness °C	Softening temperature under load °C	Cold compensation strength MN/m^2	Specific heat kJ/kg °C
Silica	96–97% SiO_2 Al_2O_3 < 0.5–1.0%	1800–1900	1700–1750	1600–1700	30–35	0.73–1.2
Semi Silica	88–90% SiO_2 8.0–9.0% Al_2O_3	1700	1710	1600–1700	30–35	0.8–1.4
Fire Clay	60–62% SiO_2 30–32% Al_2O_3	1900	1700	1450–1550	30–35	0.8–1.2
High duty	36–38% SiO_2 54–56% Al_2O_3	2400	1780	1550–1640	25–32	0.9–1.2
Mullite	20–22% SiO_2 70–73% Al_2O_3	2200	1850	1600–1700	65–70	0.65–0.75
High Alumina	> 85% Al_2O_3	2600–2900	1800–2000	1570–1920	65–73	0.8–1.2
Magnesite	85–90% MgO	2700–2800	> 1750	1450–1550	30–32	0.9–1.2
Chrome Magnesite	63–75% MgO 6–10% Cr_2O_3	3000	> 1700	1350–1500	30–50	0.8–1.2
Zircon	30–35% SiO_2 55–66% Zr_2O_3	3300	2580	1820–1870	98–196	0.6–0.8
Zirconia	80–98% ZrO_2 3–4% CaO/MgO	4000–4500	2480	2200	14–40	0.5–0.7
Silicon Carbide	93–50% SiC 4–30% SiO_2	2600	2000–2200	2000	40–75	0.8–1.1
Carbon Graphite	> 98% C	1350–1600	2000	2000	—	0.8–2.0

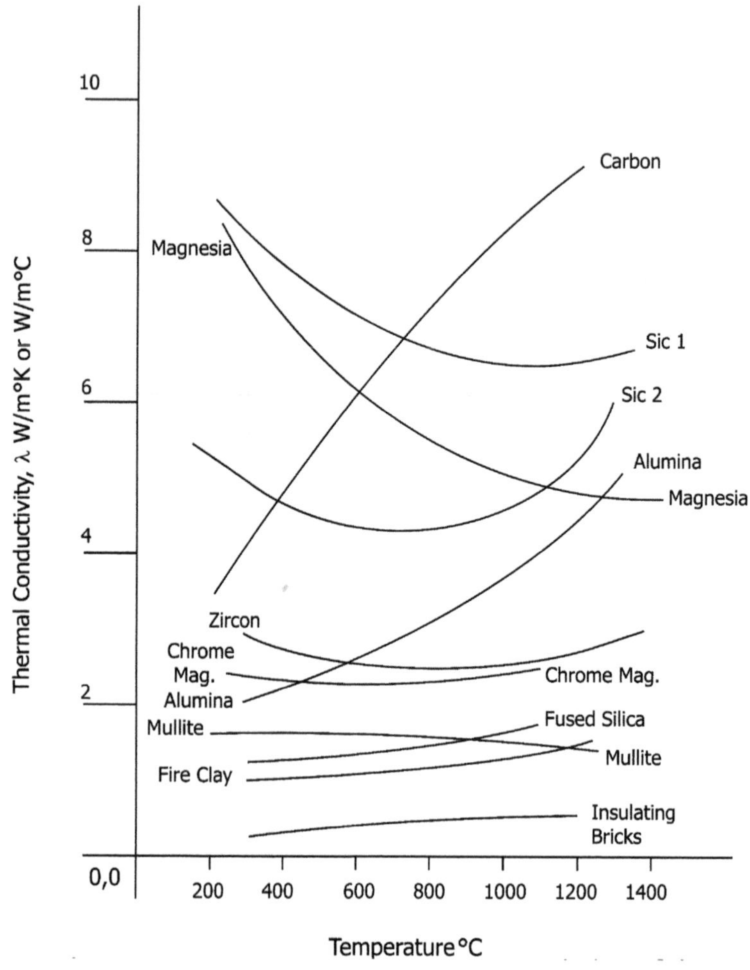

Figure 7.2 Thermal conductivity of selected refractories.

3. Size and shape of furnace—heating area height, cylindrical, or rectangular
4. Temperature gradient across the walls and general permitted heat losses to atmosphere
5. Operation cycle—continuous, or batch type
6. Heat stored in the walls and structure (this is very important for batch type, cyclic operation)

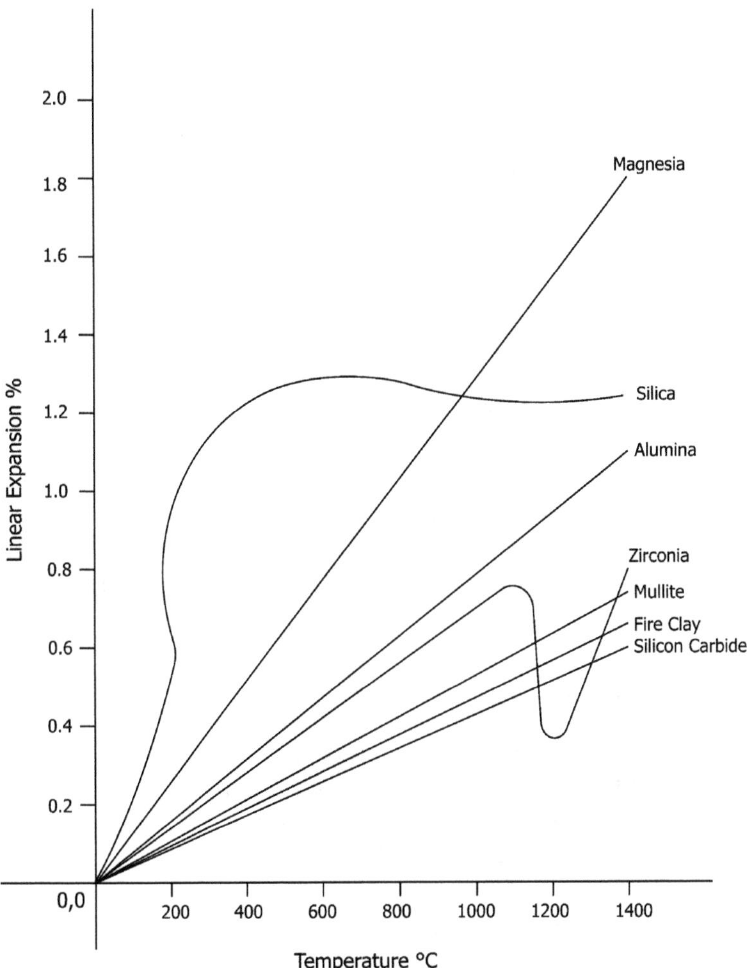

Figure 7.3 Expansion of common refractories temperature.

7. Total heat input, type of fuel, impact of flame on the walls, placement of electrical heaters
8. Load exerted by charge — single charge, small charge in containers, large heavy charge distribution of charge weight on hearth

Properties of refractories related to above requirements are:

1. Refractoriness and refractoriness under load
2. Dimensional stability—expansions and volume changes
3. Thermal conductivity, specific heat, and density
4. Creep and thermal shock resistance

7.9.2 Mechanical and Chemical Requirements

These are indicated by following design data:

1. Type of charge, expected and permissible oxidation, scale produced, type and quantity
2. Combustion products—gases, temperature and composition, ash and its composition
3. Methods of charging and discharging, mode of charging through door, roof, and bottom
4. Furnace construction and load on the walls, roof, hearth, at atmospheric and service temperatures, thermal cycles
5. Flue dust, rate of flow of flue gases
6. Heating, direct or indirect

Related properties are

1. Chemical composition and physical shape of charge
2. Cold and hot compressive strength
3. Surface texture of refractory — smooth, rough, medium
4. Size tolerances offered
5. Gas permeability
6. Resistance to slags, hot ash, dust

For most of the furnaces discussed in this book, fire clay refractories (of suitable grade) and ceramic fiber shapes are quite suitable for the design of hot face walls, roofs, and hearths. These are complemented with matching unshaped products such as cements, castables, coatings, etc. There will be a few exceptions which will be pointed out at that stage.

Insulating materials that are commonly used are ceramic, slag, rock, wool or loose fiber, asbestos, glass fiber, and insulating fire clay bricks.

Chapter 8

Metals and Alloys for High Temperature Applications

CONTENTS

8.1 INTRODUCTION

We have reviewed ceramics as construction materials for high temperature application in Chapter 7.

There are several instances where ceramics are not suitable in spite of their refractoriness. Principal disadvantages of ceramics are their low strength in tension, lack of elongation or ductility, brittleness or low shock resistance, and no formability.

Several furnace parts such as belts, conveyers, baskets, containers, boxes, gas tight muffles, etc. require strength, ductility, and easy formability. These properties are available only in metals and alloys. Hence, for such applications, the use of metals is the only choice.

Metals have several disadvantages as compared to ceramics. They react easily with air or process gases leading to oxidation or corrosion. Their strength decreases with temperature. These and other related properties restrict the use of metals to temperatures of about 1250°C.

In this chapter we will discuss the properties of common and special alloys at high temperatures. An understanding of the behavior of metals of such temperatures will help us choose the best and most economical alloy from a large number that are commercially available.

We will include in our discussion only such metals that are required for furnaces and appliances. Alloys specially developed for boilers, turbines, and jet engines are beyond scope.

8.2 MECHANICAL PROPERTIES OF METALS AT HIGH TEMPERATURE

The engineering stress-strain curve is the usual design guide for stress analysis of metals at or near room temperatures (20–100°C). The stress-strain test is a "short time" test conducted at a constant (room) temperature.

The curve is divided into two parts — elastic and plastic deformations. Information obtained is elongation (ϵ), yield stress (σ_y), ultimate tensile strength (UTS), and fracture stress σ_f. The design is based on these values.

If such a short time test is performed at higher temperature, it is observed that although the general shape of the σ, ϵ curve is essentially the same, the elastic deformation decreases and plastic deformation increases with increase in test temperature. At some temperature (near melting point) the elastic part vanishes and a large amount of plastic deformation appears. This shows that metals develop a plastic flow at high temperatures.

The design criteria such as the yield stress or UTS at room temperature are of no help at high temperatures. Even those values obtained from such a tensile test at design temperature are of little use as a new variable "time" appears on the scene.

Metals show a continuous deformation with time under constant stress and at constant temperature. This phenomenon is called "creep." Thus, if a rod specimen is located in a furnace at temperature T (Figure 8.1) and loaded at a stress σ, a continuous elongation (strain) is observed with time. The following points about creep should be noted:

1. Creep occurs in all metals and alloys.
2. At low temperatures (depending on the melting point) the elongation is negligible.
3. At a sufficiently high temperature creep becomes noticeable and becomes an important design factor.
4. Creep is not to be confused with linear expansion with temperature.
5. The stress at which creep becomes important is much less than the yield stress at that temperature as observed in the short time tensile test.
6. If load is continued for a long time, the specimen ruptures or fractures at that stress (which is much less than the fracture stress σ_f at room temperature).

A creep test is conducted in a manner shown in Figure 8.1. The results are presented in the form of elongation (or percent elongation) against time. This is the creep curve (Figure 8.1(A)).

At the beginning, a small strain appears instantaneously as elastic point 1 and plastic point 2 deformation. The strain rate then decreases up to point 3 and the region I is the primary area of creep. At point 3, the deformation or elongation

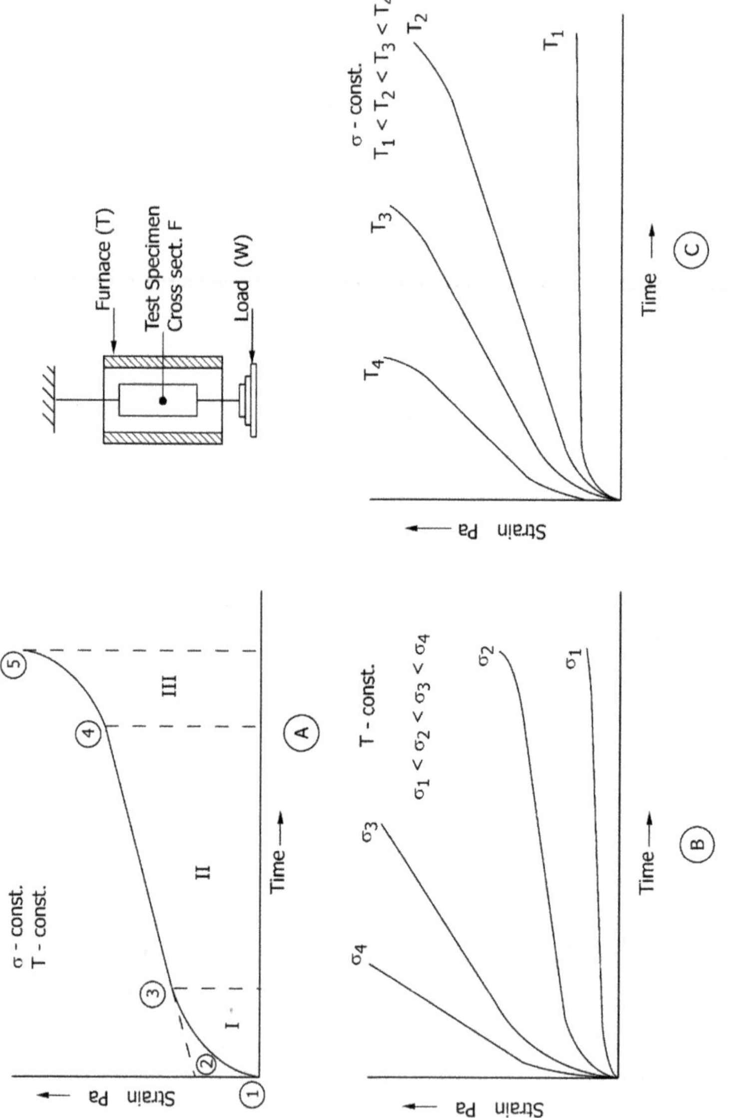

A. σ and T Constant B. T Constant, σ varying C. σ Constant, T varying

Figure 8.1 Typical creep test results.

rate becomes steady and is called "secondary creep," which continues up to point 4. This region II is the most important design parameter. In the next stage, the strain rate again increases, resulting in rapid flow, region III, and fracture at point 5. This is the region to be avoided in design.

As temperature and stress affect the creep behavior, the results obtained by varying the stress (at constant temperature) and the temperature (at constant stress) are shown in Figure 8.1(B) and Figure 8.1(C). As the test temperature and stress increase, the steady creep rate increases and the extent of secondary creep decreases, resulting in a sudden and early fracture.

Creep testing is quite time consuming and costly if carried out up to fracture. Some metals may take weeks or even years to fail. Creep behavior (like fatigue) is statistical. A large number of specimens is required to obtain reliable results.

The region of steady state creep is the region of interest as the time interval between points 3 and 4 in Figure 8.1(A) decides the useful "life" of the component, the applicable stress, and the elongation to be accommodated in design. For practical purposes, creep values are quoted for 0.1 or 1.0% elongation in designated time, such as 1,000 or 10,000 h and the applicable stress. The steady state creep can be mathematically included in the design equation as it is a line with constant slope.

Another way of considering creep behavior is to note the time it takes to rupture at different stress levels with constant temperature. This is called the "stress-rupture" test. A typical stress-rupture plot is shown in Figure 8.2(A). The plot is a downward sloping line with one or two segments which arise due to different fracture or deformation mechanisms. This test gives an indication of the time for which the specimen will last at a given stress level and temperature before it fractures. Similar curves obtained at different test temperatures are shown in Figure 8.2(B).

A typical design plot usually supplied by the material manufacturer combines percent elongation and rupture stresses as shown in Figure 8.2(C).

Both creep and stress-rupture tests show that for a lengthy high temperature service, we have to design for a certain guaranteed time before the component fails by excessive

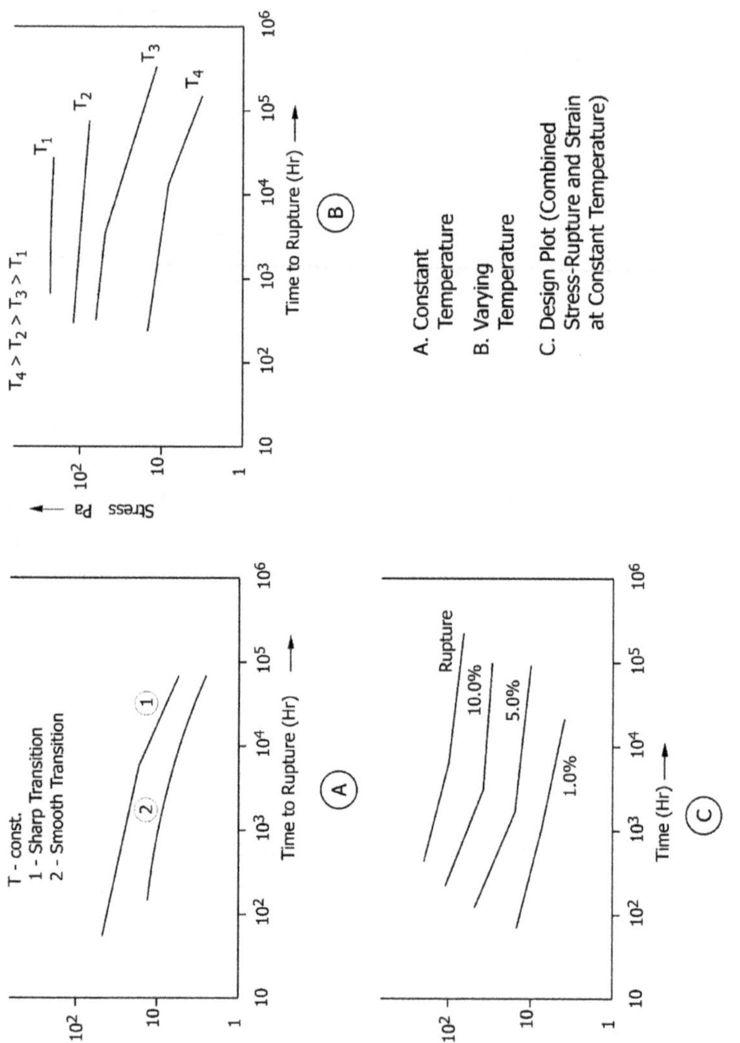

Figure 8.2 Typical stress-rupture test plots.

elongation or by fracture. After this "life span" it is necessary to replace it.

For a short service life we can use the short-time tensile test as design data. Creep or stress-rupture information is usually supplied by the manufacturers, along with advice gained from experience. Typical short-time tensile test data (UTS) at high temperatures for commonly used materials are shown in Figure 8.3. The data are presented for comparison only. Actual design figures will depend on composition and

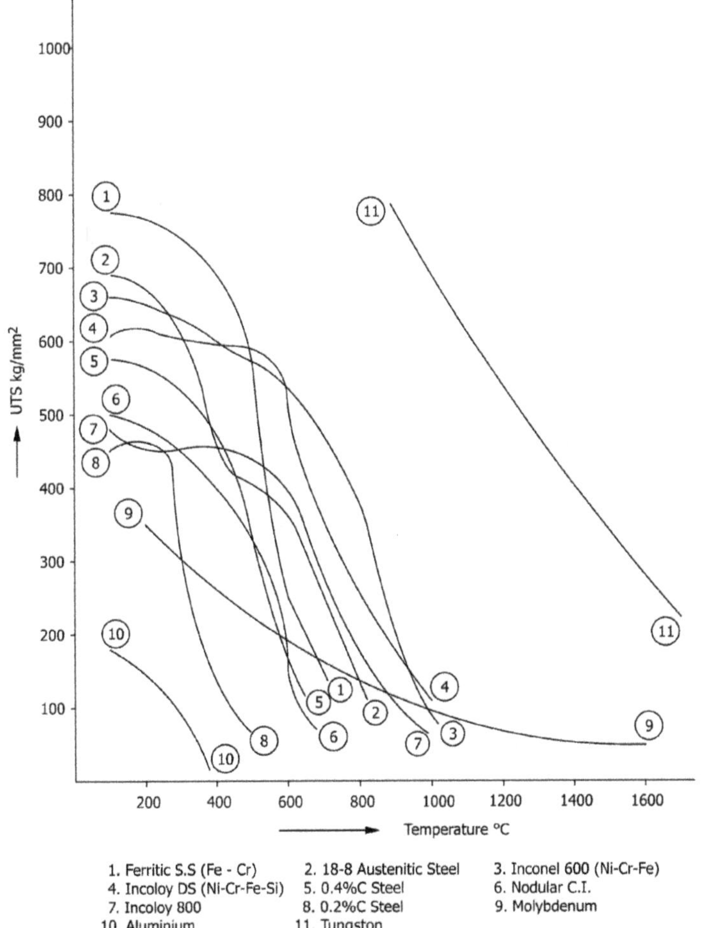

Figure 8.3 Short-time tensile strength (UTS) at high temperature for selected material.

pretreatment and should be obtained from handbooks (listed in the Bibliography) or manufacturers.

8.3 OXIDATION AND CORROSION

All metals and alloys react with oxygen (in air) and form an oxide layer on its surface. The thickness and morphology of the oxide layer is different for each metal and environment combination. Thus, some metals such as silver, gold, and platinum show very little oxidation (perhaps a monomolecular layer) and hence, are classified as noble metals. Alkali metals such as sodium, potassium, and lithium oxidize very easily and thoroughly and hence, are classified as base metals.

In between these two extreme categories are all metals and alloys used in engineering practice. These include metals of our interest such as steels, cast irons, chromium-nickel-iron alloys, stainless steel, aluminum, refractory metals (tungsten, molybdenum), and so on.

In corrosion parlance, oxidation of metals in air or similar reaction of metals with other gases (eg., CO_2, H_2, CO) are termed as "dry corrosion." A typical oxidation reaction is shown in Figure 8.4(A) and Figure 8.4(C).

Oxygen from the air is absorbed on the bare metal surface and is converted into metal oxide due to the electronic reactions with surface metal ions. This gives rise to a very thin surface film of oxide. The growth of the film occurs due to migration or diffusion of metal ions and electrons outward from the metal surface, and similar movement of oxygen (ions) in the opposite direction. These movements are assisted by the cracks and other irregularities in the film.

The overall reaction being

$$\mathrm{Me} \rightarrow \mathrm{Me}^{2+}\ 2\mathrm{e}^{-}$$

$$\mathrm{O}_2 \rightarrow 2\mathrm{O}$$

$$\mathrm{O} + 2\mathrm{e}^{-} \rightarrow \mathrm{O}^{2-}$$

$$\mathrm{Me}^{2+} + \mathrm{O}_2 \rightarrow \mathrm{MeO}$$

where Me is the metal and MeO is its oxide.

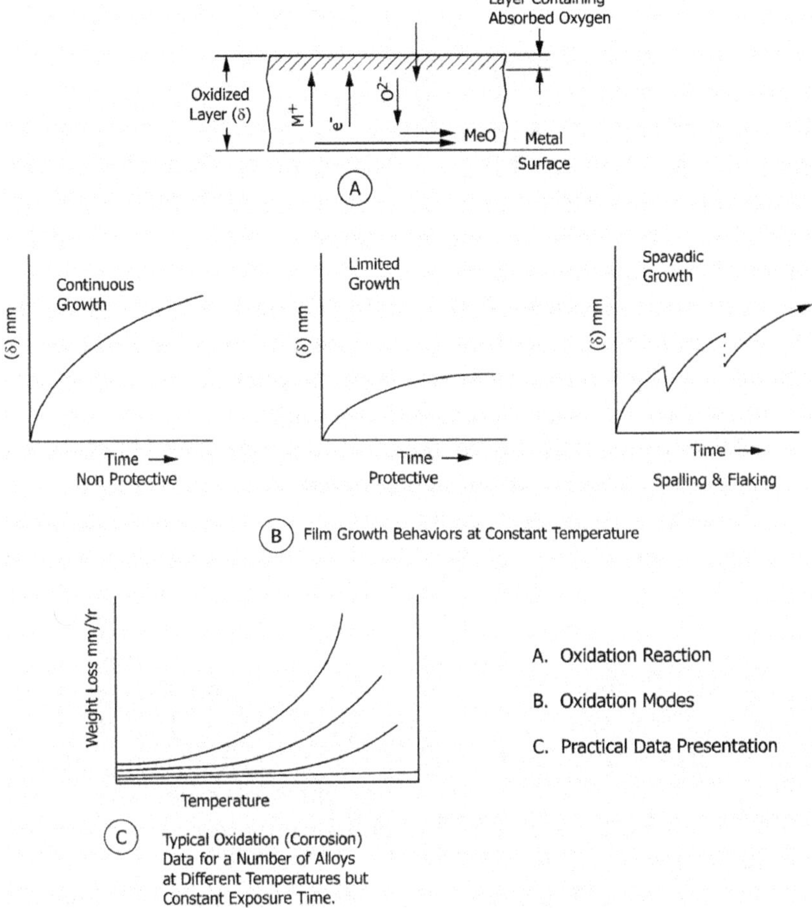

Figure 8.4 Oxidation of metals at high temperature in air.

The heat required for these reactions is available from the absorption process and the environment.

Consequently, oxidation reactions become more vigorous at higher temperatures. In the heating process, the use of metal poses the problem of extensive and rapid oxidation.

The rate of oxidation with temperature and the growth of oxide film show different patterns depending on metals, their crystal structure, and types of structural defects. The tenacity of oxide film is governed by its crystal structure and compatibility with underlying metal surface structure.

Metals such as aluminum, beryllium, chromium, and zinc form a tightly adherent, thin impervious layer of oxide. The oxide layer prevents oxygen from penetrating. This stops corrosion (oxidation) after the formation of the initial layer, which is about 10^{-6} m thick. Metals or alloys containing these metals are ideal for high temperature service. Out of the above-mentioned four metals, aluminum (658.7°C) and zinc (420°C) are basically low-melting-point metals and not directly useful for high temperature service.

Aluminum oxide (Al_2O_3, 2200°C) and chromium oxide (Cr_2O_3, 2300°C) films offer good oxidation resistance. Hence, almost all-high temperature alloys contain either chromium or aluminum or both as alloying elements.

Aluminum itself can be used in the construction of elevated or warm temperature application such as ovens, dryers, etc.

Ordinary unalloyed carbon steels are iron-based materials; the oxidation of iron at high temperatures shows a complex behavior. Iron forms three oxides. FeO near the metal surface, Fe_2O_3 on the outer face, and Fe_3O_4 in between. Thus, the layer of oxides is not tightly adherent and impervious. These steels show continuous oxidation, and therefore, in spite of the relatively high melting point of iron (1530°C), they can be used only up to 600°C for a limited time. Minor (0.2–1.5%) additions of chromium, molybdenum, and vanadium give a better strength but continuous maximum service temperature is still about 600°C.

Major groups of alloys that resist oxidation are nickel-chromium and nickel-chromium-iron alloys with or without minor additions of aluminum, titanium, vanadium, molybdenum, etc. Depending on the composition, they can be used up to 1150–1200°C without oxidation for a very long time. Their oxidation resistance arises from the chromium oxide layer that is developed on the surface.

Refractory metals and their alloys such as tungsten, molybdenum, and tantalum have very high melting points but their oxidation resistantance is very poor. They can be used only in a reducing atmosphere (e.g., H_2) or vacuum.

Graphite can be used in temperatures in excess of 2200°C but only in vacuum or reducing atmospheres.

Until now we have discussed the modes of oxidation and ways to improve oxidation resistance. Another problem that accompanies oxidation is the mechanical behavior of the oxide films. If the film is very thin, like that of Al_2O_3, and tightly adherent, there are no problems unless the film is mechanically abraded or removed.

Oxide films on many metals (e.g., steels) are not tightly adherent. They also contain many defects such as micro cracks through which oxygen can penetrate and cause further oxidation. If the oxides are complex they show differential expansion. The stress on the part, especially if it is cyclic, also leads to opening or widening of the cracks. All these factors cause breakdown of the film in the form of scales. This is known as "spalling." This exposes a new surface for oxidation. The spalling and fresh oxidation continues and leads to the failure of the part (Figure 8.4(B)).

Oxidation (corrosion) rates are measured by exposing a small piece (usually a sheet) to air at a desired temperature. The change in weight per unit area ($g/m^2 \cdot h$) or the thickness of the oxide film (mm/h) is measured and plotted against temperature. Corrosion rates for many metals and alloys are available in reference handbooks (listed in the *Bibliography*). These data, though useful as a guide for estimation, cannot be directly used in design. This is because the actual combination of stress and environment in practice is quite different than that used in the laboratory. It is safer to use data supplied by the manufacturer or conduct your own (short-time) tests.

Oxidation resistance of some commonly used metals and some alloys based on them is given with the maximum useful temperature in Table 8.1.

8.3.1 Corrosion by Other Gases

There are some instances when metals come in contact with gases other than air at high temperatures. Carburizing and

TABLE 8.1 Oxidation Resistance of Some Commonly Used Metals

No.	Metal	Nature of Oxide	Oxidation (Heat) Resistance in Air
1	Mg	Loose oxide	Very poor (150–200°C)
2	Ta, Mo, W, Nb, Zr, Ti	Thick, loose oxide spalling	Poor (500°C)
3	Cu, Fe, Ni, Co	Thick with defects spalling	Fair (400–500°C)
4	Al, Sn, Cr, Be, Mn	Dense, adherant oxide film	Good (400°C)
5	Ag, Au, Pt	No oxidation	Excellent but uneconomical. Only special or laboratory use (>1000°C)

nitriding operations are carried out in alloy muffles and containers. The atmosphere inside the muffle consists of a mixture of gases containing CO, CO_2, H_2O, H_2 N_2, and trace hydrocarbons like CH_4.

In such an atmosphere ordinary muffle alloys absorb carbon, which leads to the formation of carbides, especially chromium carbide (Cr_3C_2). The carbides so formed migrate toward grain boundaries and weaken the metal. Similar reactions take place in ammonia (NH_3) leading to the formation of nitride gases containing sulfur in the form of SO_2 and H_2S which will react with metals, forming sulphide scale. Sulphide scale formation leads to flaking and spalling.

Nickel does not form carbide and nitride while iron and chromium easily react with both, forming stable carbide and nitride.

Hence, alloys for resisting carbon and nitrogen penetration contain large amounts of nickel and little or no iron.

Nickel is prone to sulphide formation. Alloys to resist sulfur attack contain high chromium and 1.0–2.0% aluminum.

The above discussion shows that there is no one alloy which will resist the attack of all harmful gases and a proper choice must be made to get satisfactory service life. Note that in many applications, the part (muffle) has to resist hot gases from flame, oxygen attack from the outside, and process gas from the inside. The metal choice then becomes quite critical.

8.4 MELTING POINT AND PHYSICAL STABILITY

All metals have a melting temperature. Alloys have a melting temperature range. Melting restricts the maximum temperature to which a metal can be exposed in any form or environment.

Later, we will see that the strength of metals decreases with increasing temperature. Near its melting point, the metal shows practically no strength and starts to exhibit a viscous flow. Hence, the actual temperature to which a metal can be used as an engineering component is much less than the melting temperature. There are other complicating factors such as stress and environment, which restrict the maximum useful service temperature. It is therefore not possible to quote any rule. It is observed though, that many metals and alloys can be safely used up to about 0.3–0.4 × T_m where T_m is the melting temperature (°C). Alloying and other strengthening techniques have extended the useful temperature much above the iron-chromium-aluminum alloy (Kanthal), which has a melting temperature (T_m) 1500°C and maximum service temperature (T_s) 1300–1400°C.

Approximate melting temperatures of some metals and alloys are given in Table 8.2.

Table 8.2 Melting Temperature of Metals and Alloys

Temperature Range °C	Metal Melting Temperature (°C)
3300–4500	Graphite (Sublimes ~ 4200), Tungsten (3370)
2750–300	Tantalum (2870), Rhenium (3180)
2200–2750	Columbium (2415), Molybdenum (2620), Boron (2300)
1650–2200	Platinum (1750), Rhodium (1865), Titanium (1795), Chromium (1650), Zirconium (1850), Vanadium (1710)
1100–1650	Manganese (1260), Cobalt (1845), Nickel (1450), Iron (1540), Nickel alloys (1460), Stainless Steel (Austenitic 1450), Berylium (1285), Steels (1450)
700–1100	Copper (1085), Brass, Bronze (1040), Silver (960), Other Bronzes (1000), Aluminum Bronze (1085)
200–700	Aluminum and alloys (650), Magnesium alloys (600), Zinc (420)

The table shows that there is a very little choice of metals for use above 1600°C. Only refractory metals and graphite are available but they all require a special atmosphere or vacuum. There are no metals available above about 2200°C. The choice for temperatures up to 1400°C is very wide, especially in the range of 600–1000°C. All commonly used metals are in this range.

8.5 LINEAR EXPANSION

All materials show linear expansion when heated. This arises due to increasing lattice or molecular vibration. This expansion or dilation is reversible, i.e., increased dimension is recovered when original temperature is restored.

Typical linear expansion coefficients for metals and alloys of interest to us are given in Table 8.3. Note that the values given are for comparison only. Actual values will depend on composition and pretreatment. The coefficient has no units as it is measured in the same units, e.g., (m/m length $\times 10^{-6}$).

TABLE 8.3 Mean Coefficient of Linear Expansion for Selected Metals

Metal/Alloy	Temperature range°C	Coefficient of linear expansion $\times 10^{-6}$/°C
Inconel 600	20–1000	16.7
Inconel 601	20–1000	17.8
Incoloy DS	20–1000	18.7
Aluminum	20–600	28.7
Graphite	20–600	3.5
Molybdenum	20–2100	7.2
Platinum	20–1000	10.2
Tantalum	20–2200	7.8
Tungusten	20–2400	5.8
Carbon and low alloy Steel	20–600	13.5
Austenitic S.S.	20–600	15.5
Ferritic S.S	20–750	11.2
Cast Iron Spheroidal Graphite and Alloy	20–400	12–18

In the design of furnace parts or components, this expansion, though small, must be taken into account and accommodated for. If restricted, considerable internal stresses are set up and give rise to bending, buckling, and distortion. Many high temperature applications involve fabrication by welding. The location of the weld joint and room for expansion are very important for long life.

Linear expansion should not be confused with creep. The latter arises due to plastic flow under stress and is irrecoverable.

8.6 CAST IRONS

There are many types of cast irons. They are generally classified according to the shape of the graphite and the structure of the matrix. There is also a class of alloy cast irons which contains nickel, chromium, silicon, or molybdenum.

Cast irons are cheap compared to special heat resistant alloys but their use in heating applications is limited due to low tensile strength and difficulties in fabrication. In a sense, they are brittle in tension but can take large compressive loads (150–300 N/mm^2) depending on type. Fabrication is possible only by casting, machining, and welding. Hence, they are suitable only for heavy sections under compression.

Ordinary flake graphite cast irons can be used up to about 400°C, beyond which they show extensive dimensional growth and flaking. "Growth" of cast irons is different from thermal expansion and creep.

Nodular cast irons contain graphite in nodular form and have a pearlitic or ferritic matrix. They have limited tensile strength and are called "ductile irons." These irons are suitable for service up to about 500°C and show little growth and flaking.

Alloy cast irons (nodular) have been used up to 800°C. They contain nickel, silicon, and molybdenum. Chromium cast irons (C.I.) have good oxidation resistance. The properties of selected C.I.s are given in Table 8.4.

Table 8.4 Cast Irons for High Temperature Service

For mechanical and other properties refer to handbooks.
(Source - Metals Handbook, Vol. 1, ASM 8th Ed., 1961)

No	Type	Composition						Maximum service Temperature°C
		C	Si	Mn	Ni	Cr	Cu/Mo	
1	Ordinary flake graphite	3.4–3.8	1.8–2.8	0.5–0.8	—	—	—	450–500
2	Ordinary nodular graphite	3.2–4.1	1.8–2.8	0.5 0.8	—	—	—	500
3	Medium silicon flake graphite	1.6–2.5	4.0–6.0	0.4 0.8	—	—	—	900
4	High chromium flake graphite	1.8–3.0	0.5–2.5	0.3–1.5	5.0	15–35		1100
5	High nickel flake graphite	1.8–3.0	1.0–2.75	0.4–1.5	14.0–30.0	1.75–5.5	7.0/1.0	816
6	Ni-Cr-S: flake graphite	1.8–2.6	5.0–6.0	0.4–1.0	13–32	1.8–5.5	10/1.0	950
7	Aluminum flake graphite	1.3–1.7	1.3–6.0			18–25 Al		1100
8	High Si Nodular	2.8–3.8	2.5–6.0	0.2–0.6	1.5	—	—	900
9	20% Ni Nodular	2.9	1.75–3.2	0.8–1.5	18–22	1.75 2.50	—	700
10	30% Ni Nodular	3.0	2.0 3.0	1.8 2.4	21–24	0.50	—	600

8.7 STEELS AT HIGH TEMPERATURE

Plain carbon steels containing 0.2–0.4% carbon and small amounts of silicon and manganese are ferritic or pearlitic. They are cheap, weldable, and available in all forms. They oxidize above 400–500°C and are suitable for low or medium (elevated) temperature parts such as structural frames, supports, panels, etc. Painting improves corrosion resistance.

Low alloy steels containing 0.15% carbon and 1.2% molybdenum or chromium show slightly better properties and can be used for parts requiring intermittent service at temperatures up to 500–600°C.

Chromium steels containing up to 20–24% Cr have good oxidation resistance but will show spalling under thermal cycling. They can be used up to 800°C.

Austenitic stainless steels are resistant to corrosion by oils, fumes, moisture, and many gases up to 450°C. They retain their surface shine and cleanliness up to 450°C after which the surface may tarnish or blacken, but corrosion resistance is good up to 650–800°C. High alloy steels contain large amounts of chromium and nickel along with small amounts of Mo, Cb, Ti, Al, etc. They can be used up to 1150°C.

Table 8.5 shows the properties of selected heat resistant steels. Reference should be made to handbooks as the number of commercially available heat resistant steel is very large.

8.8 SELECTION OF METALS FOR HIGH TEMPERATURE APPLICATION

We have seen in previous sections that at high temperatures, metals will fail by excessive stress or by reaction with their environment. Hence, these two cases become the deciding factors in the choice of metals.

The following general remarks will help to make a preliminary choice. Note that there can be exceptions due to some combinations of temperature, stress, and environmental attack.

Table 8.5 Constitution and High Temperature Properties of Selected Plain Carbon and Alloy Steels

Type	Main Constituents %				Tensile Strength N/mm^2 (MPa),°C				Stress for 1.0% elongation in 10,000 h MPa/@ °C		Stress for rupture in 10,000 h MPa/@ °C		Oxidation temperature °C and Applications
	C	Cr	Ni	Mo/V	25	200	400	600					
Plain carbon	0.2	—	—	—	450	463	371		136/400	39/500	192/400	54/500	Oxidation > 400°C Furnace structure
Plain carbon	0.4	—	—	—	602	630	460		do	do	do	do	Oxidation > 400°C Furnace structure
Low alloy	0.15	1.2–2.0	—	0.6–1.0	510	510	500	293	216/450	85/530	304/450	105/530	Oxidation > 550°C
Cr steel	0.06	20	—	—	541	400	432	216	—	—	—	—	Oxidation > 600°C Refractory supports
Austenitic S.S	0.1	18	8–12	1–2 Mo/Ti	630	—	432	378	85/600	39/700	116/600	54/700	Oxidation at 850°C Resistant to organic corrosion
High alloy	0.1	25	12		656	—	556	432	69/650	48/700	93/650	62/700	Oxidation at 1000°C. Muffles, belts, pots and baskets.

(All Steels in annealed or normalized condition.)

1. Aluminum is the cheapest choice if the temperature does not exceed 300°C. It is quite resistant to oxidation and attack by CO_2 and moisture, and finds application in ovens, dryers etc.
2. Ordinary (plain carbon) steels have adequate strength up to 400–500°C, but do not possess a high oxidation resistance. They are cheap and easy to fabricate in a variety of forms such as rods, strips, and wires, and structurals.

   Ordinary or nodular cast irons are sometimes better than steels in oxidation. They have two limitations. They cannot be formed; and the cast section is thicker and heavier than steels. Oxidation and dimensional instability is the second limitation. Inability to take tensile loads limit their use to compressive stress.
3. Low alloy steels containing 1–3% chromium and small amounts of molybdenum and vanadium can be used up to 600–700°C. They can be easily fabricated and have better oxidation resistance than plain carbon steels or cast irons. They should be used in normalized or annealed conditions.
4. Low corrosion resistance of steel at high temperature can be improved by plating or impregnation. Chromium plating is frequently used. It helps to reduce oxidation and improves reflectivity. Impregnation of the surface with an oxide layer improves resistance to oxidation and attack by molten salts and carburizing chemicals.
5. In some applications, steel parts can be cooled by water or air circulation. This keeps their temperature at an acceptable level.
6. Metallurgically, the metal to be used should have a single phase (e.g., ferrite and austenite) and a large grain size. Generally, alloys hardened by heat treatment are not suitable because a prolonged high temperature breaks the hardening phase such as with martensite. The exception is dispersion-strengthened alloys.
7. For design purposes, short time tensile test data (e.g., Figure 8.3) and combined creep-rupture data (e.g.,

Figure 8.2(C)) are useful. Such data are available in handbooks or manufacturers' publications. Before making use of such data it is necessary to make an estimate of the desired "life" of the part and expected stresses.

8. In design and fabrication keep in mind the linear expansion. Some useful hints are:

    - Avoid sharp corners or edges.
    - There should be no welds at corners or edges.
    - Corrugated sheets are better than flat ones.
    - Use the same section throughout.
    - Allow free expansion at the ends to prevent bending, buckling, distortion, or sagging.
    - Baskets, fixtures, trays, etc. used in heat treatment undergo repeated heating and cooling under a stressed condition. The effects of expansion and thermal fatigue are more severe in such cases.

9. Next in cost to plain and low alloy steels are the various stainless steels (S.S.). There are two main types of S.S. that are suitable for our purpose.

    Ferritic S.S.—These contain 8–25% chromium and about 0.15% carbon. The chromium forms an oxide layer and protects from oxidation and furnace gases. Ferritic steels are used up to 600°C. The chromium oxide layer breaks off after repeated heating and cooling.

    Austenitic S.S.—These contain 18–25% chromium and 8–20% nickel with about 0.08–0.1% carbon. These steels can be used up to 700°C without appreciable oxidation. However, they are not suitable for sulfur bearing gases.

    Austenitic stainless steels have excellent resistance to oxidation up to about 400°C and are not affected by gases and substances associated with food processing. Hence, they find major applications in ovens for baking, etc.

10. Nickel and nickel chromium alloys dominate the temperature range from 600 to 1100°C and are the costliest. They offer several advantages if properly fabricated and used and thereby prove economical. There are many alloys based on Ni and Cr. They contain small amounts of aluminum, titanium, molybdenum, and balance iron. Most of these alloys are proprietary and are recommended for individual situations. The alloys are available in all forms such as rod, plate, sheet, wire, and tube. They can be easily welded. They have austenitic, single-phase structure. Some of the alloys can be hardened by dispersion or solution strengthening. Some alloys offer excellent resistance to process gases and molten salts.

    The properties of some nickel alloys frequently used in high temperature furnaces are given in Table 8.6. There are many other heat-resistant alloys used in machinery such as boilers, turbines, jet and automotive engines, etc. These are not considered here.
11. Beyond about 1200°C, the only metals available are refractory metals — tungsten, molybdenum, and tantalum. They are very costly compared to nickel alloys. Their availability is limited to simple shapes such as sheets, rods, and wires. They oxidize easily beyond about 600°C and have to be protected by special gases, or inert gases, or vacuum. At temperatures beyond 1200°C they have low strength. The maximum useful temperature when properly protected are tungsten–2560°C, molybdenum–1900°C, and tantalum–2400°C. Their main use is in radiation shields, resistors, furnace tubes, boats, crucibles, etc.

    For processing high purity metals as in semiconductor manufacture, platinum or platinum alloys are used up to 1600–1700°C. They are extremely costly but can be used in air and are easy to fabricate.
12. Metals have high specific heat and conductivity. Heat absorbed or conducted away represents thermal loss. Hence, the amount of metal fittings and furniture in the heating zone should be the minimum required.

TABLE 8.6 Selected Nickel Base Alloys, Properties, and Applications

Alloy	Composition						Melting Point °C	Tensile Strength N/mm^2			Elongation %			Applications
	Ni	Cr	Fe	C	Si	Al		20°C	700°C	1000°C	20°C	700°C	1000°C	
Inconel 600	72	14–17	8–10	0.15	0.5	—	1370–1425	620	365	75	50	51	61	Good oxidation resistance up to 1150°C. Heat treatment furnaces, salt baths, rollers, belts.
Inconel 601	58–63	21–24	B	0.1	0.5	1.0–	1300–1370	750	360	50	52	53	—	Oxidation resistance up to 1250°C including sulfur gases. Boxes, baskets, muffles.
Incoloy 800 H	30–35	19–23	B	0.05–0.1	—	0.15 0.6	1355–1385	520	300	—	55	70	—	Resistant to oxidation, sulfur, and carburizing. Fixtures, sheaths, conveyor belts.
Incoloy D S	34–41	17–19	B	0.08 0.15	1.9 2.6	—	1330–1400	620	335	65	38	49	75	Muffles, furnace parts, rollers, chains, fans, fixtures.
Nimonic 75	B	18–21	5	0.08 0.15	1.0	0.2 Ti	1340–1380	850	420	80	30	57	50	Muffles, furnace parts, rollers, chains, fans, fixtures.

(Courtesy—*Henry Wiggin & Co. Ltd. U.K*)

Summing up, the choice of metals and alloys should be made by considering the temperature, type of service (continuous or intermittent), expected life, oxidation, and possible corrosion, the most important criteria being the cost. In many cases the choice is a compromise.

Metals and alloys used for resistance heating elements are discussed in chapter 9.

Chapter 9

Electric Resistance Heating

CONTENTS

9.1 INTRODUCTION

When an electric current flows through a material (preferably a good conductor) some energy is dissipated in the form of heat due to the resistance offered by the material. This phenomena offers a good, clean, and easily controllable source of heat, and is used on a very large scale in industrial and domestic heating. The object or work can be heated "indirectly" by exposing it to an electrically heated radiator or heater, or element. Alternatively, the object itself can be used as a resistance or heater in which heat is generated internally. In this chapter we will discuss both modes of electrical resistance heating.

9.2 INDIRECT ELECTRICAL HEATING

9.2.1 Principles of Indirect Electric Heating

This is the most common mode of electrical heating. It is used in small and large furnaces, ovens, water and space heaters, soldering guns, cartridge heaters for plastic dies, home appliances, and countless other heating processes in various fields.

Indirect heating is based on heat produced in a conductor carrying a current. The heater is away from the work and heat transfer takes place by conduction or radiation (see Figure 9.1). If the resistance of the conductor is R (ohms) and a current I (amp) flows through it, for a time t (sec), the heat produced (H) is

$$H = I^2Rt \quad \text{J} \tag{9.1}$$

and the power dissipated P (watts) is

$$P = I^2R \quad \text{W} \tag{9.2}$$

By using Ohms law ($I = V/R$) other expressions for power are

$$P = I \times V = V^2/R \tag{9.3}$$

where V is the voltage across the conductor.

The conductor is called the "heater" or "heating element," or "element."

9.2.2 Material for Heaters

The heaters are specially developed materials usually available in wire or strip form and can be given the desired shape by bending. These are metallic heaters.

Other types of heaters are made from nonmetallic materials such as silicide or carbide in ready-made forms such as rods or hairpins.

Following are some of the desired properties of the materials to be used as heaters:

1. High melting temperature
2. Stability in the atmosphere
3. Constant temperature coefficient ($\propto$) of resistivity (θ)
4. Forming possibility or ready-made forms available
5. Resistance to thermal and mechanical shocks

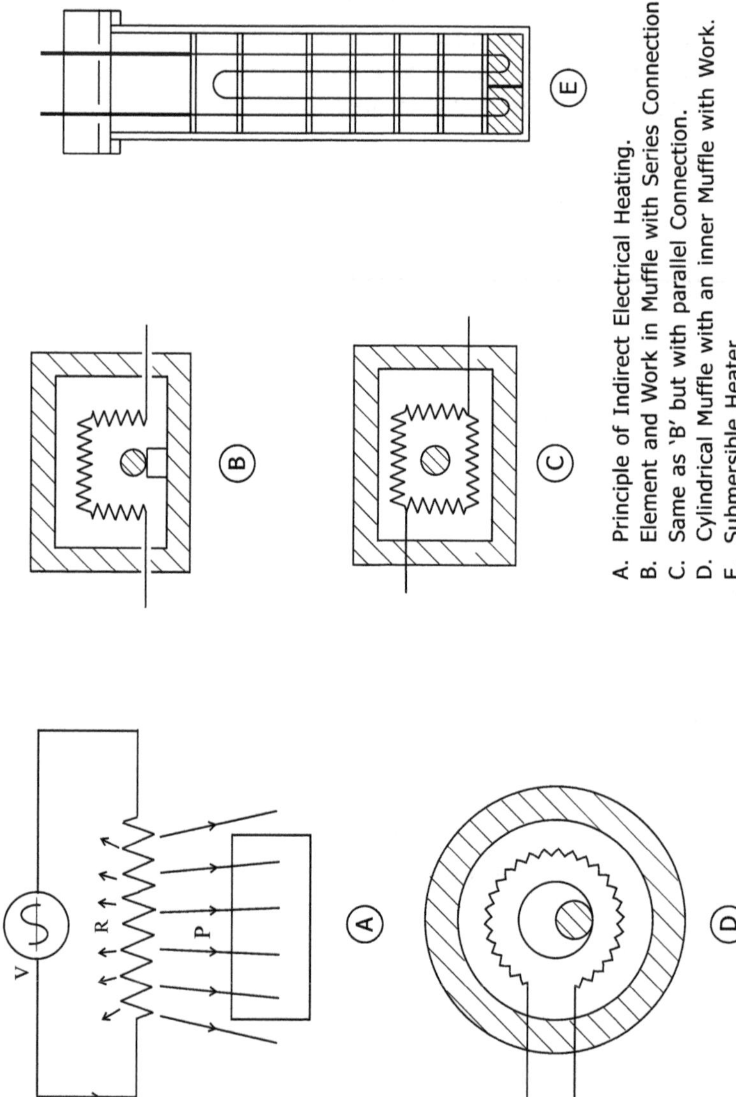

Figure 9.1 Indirect electrical heating and some typical enclosures (muffle).

Resistance materials are available up to a maximum useful temperature of 2000°C. However, there is no one material for the whole range and all types of atmospheres. Each material has its own maximum useful temperature in a given atmosphere. For example, a metallic material which can be used up to 1500°C in air (oxidizing) can be used only up to 1150°C maximum in a reducing atmosphere.

Available materials can be classified as follows:

1. *Metallic alloys*—These are Fe-Cr-Al alloys or Ni-Cr-Fe alloys and can be used in the range 1000–1400°C depending on the grade. Metallic alloys are the cheapest heating materials and are used in the majority of industrial and domestic heating applications.
2. *Refractory metals*—Tungsten, molybdenum, and tentalum can be used from 1500 to 2000°C.
3. *Noble (precious) metals and alloys*—These are platinum or platinum-rhodium alloys useful in the range 1200–1800°C.
4. *Nonmetallic elements*—These are readymade elements made from silicon carbide or molybdenum disilicide and are useful in the 1200–1750°C range.

Important properties of these materials are given in Table 9.1, Table 9.2, Table 9.3, and Table 9.4.

Because of availability and cost considerations refractory and precious metals are used only for special application. For even higher temperatures (>2200°C), graphite is useful if protected by a suitable atmosphere (reducing).

9.2.3 Special Insulating Materials in the Construction of a Heater

Heating elements require a support which must be nonconducting and capable of withstanding the maximum temperature without any reactions with the element. These are usually mullite (Al_2O_3-SiO_2) or pure alumina (Al_2O_3 > 99%) shapes. Some of these are shown in Figure 9.2. Presently, commonly available refractory supports can be used up to 1800°C. For higher temperatures, the element is free-hanging.

TABLE 9.1 Selected Properties of Electrical Resistance Alloys

Composition %	Commercial name	Max. and continuous service temp. °C (Air)	Forms available	Forming	Electrical resistivity (20°) Ω.cm × 10^6	Atmosphere and useful temp.°C
22 Cr, 5.8 Aℓ Fe - Rem	Kanthal A1 Kanthal AF	1400	Rod, Wire, Strip	Bending Cutting Welding	145	Moist air, Nitrogen, Hydrogen, Exogas - 1100–1200 Reducing- 1050–1200. Vacuum-1100
22 Cr, 5.3 Aℓ Fe - Rem	Kanthal A	1400 (1300)	do	do	139	do
22 Cr, 4.8 Aℓ Fe - Rem	Kanthal D	1300 (1200)	do	do	135	Moist air, Nitrogen, Hydrogen-1100–1200 Vacuum-1050
20 Cr 80 Ni	Nikrothal 80+, Brightroy Nichrome	1200	do	do	109	Hydrogen 1250, Neutral, Reducing 900–1100
15 Cr, 25 Fe, 60 Ni	Nikrothal 60+, Brightroy B Nichrome	1250	do	do	111	Moist air 1100, Nitrogen, Neutral Hydrogen 1200
20 Cr, 45 Fe, 35 Ni	Nikrothal 40+, Brightroy F Nichrome	1100	do	do	104	Moist air 1050 Neutral, Exogas- 1050 Hydrogen-115

TABLE 9.2 Properties of Refractory Metals for Heaters

Metal	Density g/cm^3	Melting Point °C	Max. Useful Temp.°C	Resistivity $\times 10^6$ ohm.cm (Temp. coef. $\alpha \times 10^{-3}$)	Refractory stability Temp.°C	Surface load W/cm^2 T > 1200°	Forming	Forms	Atmosphere
Molyb-denum	10.20	2630	2000, 2500	5.15 (4.35×10^{-3})	Al_2O_3 – 1900 MgO – 1800 ZrO_2 – 1900	20–40	Very good	Wire Rod Sheet	Hydrogen, Cracked NH_3 Inert, Vacuum
Tungsten	19.30	3380	3000	5.2 (4.8×10^{-3})	Al_2O_3 – 1900 MgO – 1800 ZrO_2 – 1600	20–40	Good	do	Dry H_2 Vacuum – 2400°C
Tungsten	19.30	3380	3000	5.2 (4.8×10^{-3})	Al_2O_3 – 1900 MgO – 1800 ZrO_2 – 1600	20–40	Very good	do	Inert or Vacuum

Table 9.3 Properties of Precious Metals and Alloys Used for Heating

Metal/Alloy	Density g/cm^3	Melting Point °C	Max. Useful Temp.°C	Resistivity $\times 10^6$ ohm.cm (Temp. coef. $\alpha \times 10^{-3}$)	Forms	Refractory stability °C	Surface load W/cm^2 T >1200°	Atmosphere
Pure Platinum	21.45	1768	1400 (1550 short time)	10.65 (3.9)	Wire, Mesh, Foil, Strip	Al_2O_3 – 1800 MgO – 2400 (Oxidizing) ZrO_2 – 2300 ThO_2 – 2600	3–5	Oxidizing only, Air. Avoid contact with SiO_2 (>900°C) reducing atm.
Platinum 10% Rh.	20.0	1825	1550, 1600	18.5	Wire	do	do	do
Platinum 40% Rh.	16.6	1900	1750	16.8	Wire	do	do	do

TABLE 9.4 Selected Properties of Nonmetallic Resistance Materials

No.	Composition	Commercial Name	Max. Service Temp.°C	Forms Available	Forming	Atmosphere and useful temperature °C
1	Molybdenum disilicide $MoSi_2$, SiO_2	Kanthal ST Super N Sup 33	1700 (1450) 1700 (1375) 1800 (1600)	Hair Pin Single Shank Multiple Shank	NO	Air – 1700 Inert – 1600 Reducing – 1400 Vacuum – Not Suitable
2	Silicon Carbide SiC	Crucilite Globar Crystalon	1540 (1370)	Rods-Double Helical, Hollow	NO	Air – 1500 Hydrogen and Reducing – 1200 Inert – 1500 Vacuum – Not Suitable
3	Carbon Graphite	—	2480 (2200) (Inert Atm. only)	Rods Plates Fiber tapes	NO Only machining	Only Reducing or Inert atmosphere (2200)

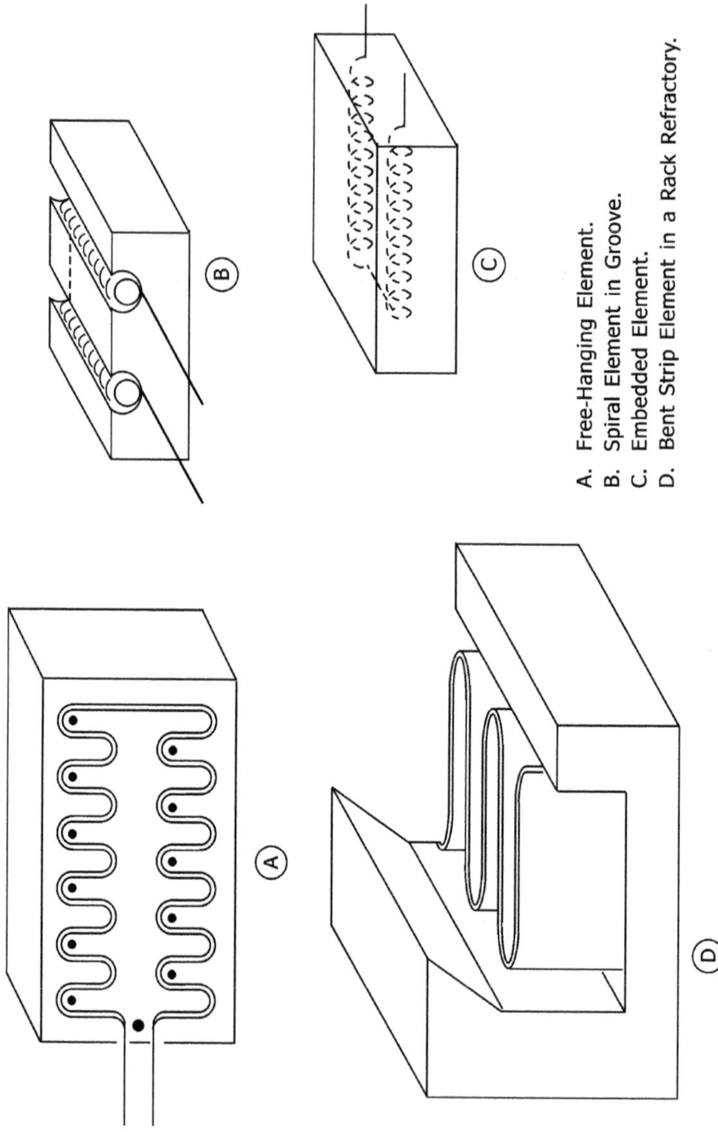

Figure 9.2 Some types of heating electrical supports.

The refractory supports also serve (at least partly) as thermal insulaters. Free-hanging elements are thermally insulated by using radiation shields (see Chapter 3).

Domestic and other low temperature appliances make extensive use of mica, asbestos, and their synthetic bonded forms.

The heating arrangement principally shown in Figure 9.1(A) shows that only a part of the heat from the element reaches the work. This results in considerable loss. It is therefore necessary to enclose the heater-work system in an enclosure or a muffle which is insulated from atmosphere. Typical enclosures are shown schematically in Figure 9.1(B), Figure 9.1(C), and Figure 9.1(D). Furnaces are also classified according to the shape of the enclosure (e.g., horizontal muffle type, vertical cylindrical with open top, etc.).

Mica—This is a mineral with a crystalline structure. There are two main varieties. Muscovite is K_2O, $3Al_2O_3$, 6SiO2, $2H_2O$; and phlogopite K_2O, 6MgO, Al_2O_3, $6SiO_2$, $2H_2O$. The water content is part of the crystal structure. Mica occurs as a stack of thin sheets (mica books) in other minerals such as quartz, feldspar, pegmatite, etc.

After removal of extraneous materials, the stacks are split into irregular sheets of various thicknesses. Pure mica is clear and transparent. Impurities such as FeO appear as black streaks. Large amounts of impurities make mica opaque or milky.

Mica is relatively flexible. It can be bent or cut with shears, or die-punched. Depending on purity, mica can be used in the 600–1000°C range. Its electrical resistivity is 10^{13}–10^{15} ohm.cm and thermal conductivity 0.47–0.58 W/m °C.

Besides cut sheets, mica is produced in various forms such as paper, tapes, washers, tubes, etc. For this, mica powder or scrap is mixed with a suitable resin, formed, and heat-cured. These forms are good electrical insulators but do not withstand temperatures greater than 300–400°C.

Mica sheet is extensively used as a supporter for resistance heaters in domestic and other appliances such as toasters, irons, and ovens. Mica is also used as a dielectric for low and high frequency capacitors.

Glass-bonded and synthetic micas are also available. Synthetic mica has better electrical and thermal properties and is available in sheet form. Glass-bonded mica is available in sheet and rod form and is machinable.

Asbestos—This is a mineral with fibrous forms having a chemical formula $3MgO, 2SiO_2 \cdot 2H_2O$. The fibers are 1.0–50 mm long. The water is in a combined form. After washing, the fibers are converted into various forms such as hard or flexible sheets, tapes, paper ropes, cements, etc. These forms can be readily cut into suitable shapes.

The electrical resistivity of asbestos is 10^8 ohm.cm (much less than mica) and its thermal conductivity is 0.12–0.18 W/m °C. Asbestos can be used as a thermal insulator in many appliances such as hot plates, irons, etc. It can be used up to 500°C, above which it loses its water content which drastically lowers the mechanical strength. Its flexibility makes it very useful as a gasket for sealing. Asbestos can be mixed with resins to obtain many forms and molding compounds.

Now, both mica and asbestos shapes bonded with glass are available. They are easily machined and can withstand temperatures up to 600–800°C. They are extensively used as element supports in low temperature ovens and appliances. Epoxy and other resins are also used for bonding.

9.3 CONSTRUCTION AND PLACEMENT OF HEATERS

For wire and strip elements, the size and length for dissipating the required power is calculated as discussed earlier. Usually this length is quite large and poses an accommodation problem within the space available inside the muffle.

A rectangular muffle has six sides. There are some designs in which all six sides carry the elements, and accommodation is easy as a large area is available. In common designs, the front side, which has a door in it, is avoided for operator safety and for avoiding moving contacts. The roof is also avoided as there is a possibility of element breakage resulting in a live piece hanging from the top or short-circuiting with the work.

The bottom has to carry the weight of the work, and if the element is placed there, it has to be hidden or insulated from the work by a bottom plate, resulting in slow heat transfer. The back wall is avoided to minimize the risk of the work striking the back wall and damaging the element. Continuous process muffles require both front and back to be open for thorough movement.

Hence, in a majority of furnaces the heating elements are located in vertical walls, thus limiting the area available for placement. If this area is not sufficient, the bottom and back wall are considered, and then, if necessary, the hearth or bottom is considered.

In cylindrical enclosures, the element is always located around the circumference, with the work placed at the center.

For appliances such as electric irons, hot plates, toasters, soldering guns, and immersion heaters, the area or space available for the element is always fixed.

To ease the accommodation problem the wire is coiled so that the effective or accommodation length is shortened. If a strip is to be used, it is bent zigzag to achieve the same effect. Strip and wire manufacturers supply information about the bending radius, coil diameter, etc., for a given wire diameter or strip thickness.

There are many ways of locating the prepared coil (or strip) on the muffle walls. Some of these are shown in Figure 9.2(A), Figure 9.2(B), Figure 9.2(C), and Figure 9.2(D). The coil is located in specially grooved refractory shapes or is embedded within them. Coiling and placement or embedment poses heat transfer problems.

The best way to obtain good heat transfer is to freely hang the wire or strip on the wall. Refractory metal or ceramic hooks can be used for hanging. If fiber board is used for walls, the element can be stapled to the wall. The length of the wire (strip) may not permit free-hanging.

Ready-made metal–sheathed embedded elements are available for domestic applications. These are made by co-drawing a coiled element surrounded by an alloy (usually inconel) tube filled with insulation powder (MgO, Al_2O_3). Typical shapes are shown in Figure 9.3. They give a surface temperature of about

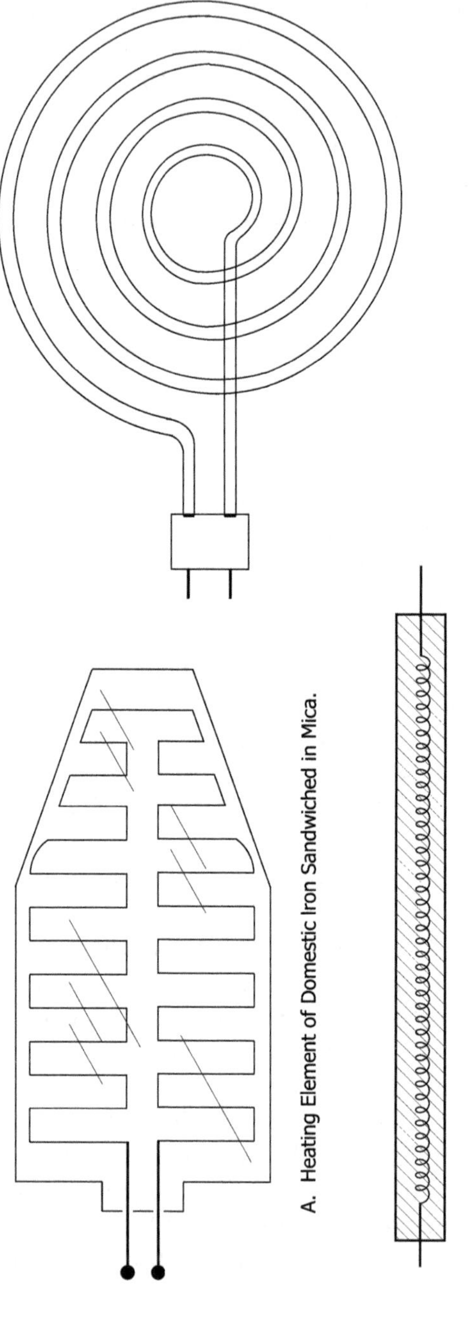

A. Heating Element of Domestic Iron Sandwiched in Mica.
B. Sealed, Embedded Rod Element.
C. Sealed Element for Domestic Kitchen Heater.

Figure 9.3 Some ready-made heating elements for domestic applications.

600°C–900°C. As the heating element is fully supported and enclosed in insulation, sheathed elements can be designed to a much higher surface loading (2–10 w/cm^2). They are available in 750, 1000, 1500, and 2000 W power, in many shapes such as flat, spiral, hair pin, simple cylindrical, etc. They have a long life and are safe.

To make maximum use of the available accommodation area, heating elements are often designed for series or parallel combinations. By switching the elements to a desired combination, some control can be exerted on the delivered power. This enables the appliance to be operated on "high" or "low" heat modes. One advantage of parallel combination is that the device keeps working (though at a lower power) even after one element fails. For large furnaces operating on three phases, supply can be designed to operate either in "star" or "delta" modes (Figure 9.4).

Many tubular or rectangular long muffle furnaces require a definite thermal gradient or a constant temperature inside. These furnaces are used for crystal growing or semiconductor processing and have a maximum temperature about 1250°C. They use a special atmosphere in the process zone to which the heating elements cannot be exposed. The construction uses a process muffle (tube) inside an outer muffle having the heating element. Such construction gives rise to considerable heat loss from the ends.

To compensate for the end loss, the spacing between the elements at the end is kept less than that in the middle.

To obtain a constant temperature or a gradient in the middle region, the heating element is "tapped" at regular sections or a number of separate elements are used in place of one continuous element. A number of thermocouples (see Figure 9.4(B) and Figure 9.4(C)).

Monitor the temperature at different spots along the length. The elements (if "tapped") are shunted so that they are switched on or off for control. Both of the arrangements require a multichannel controller and a power regulator.

A number of solved design problems follow to illustrate various heating element designs.

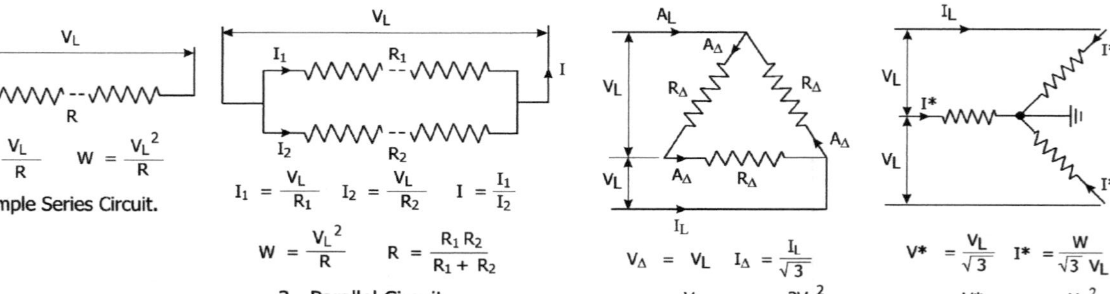

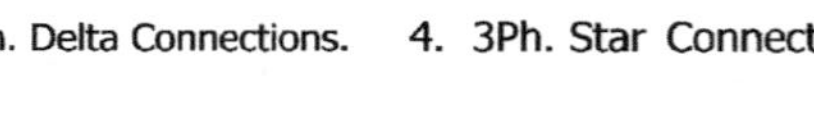

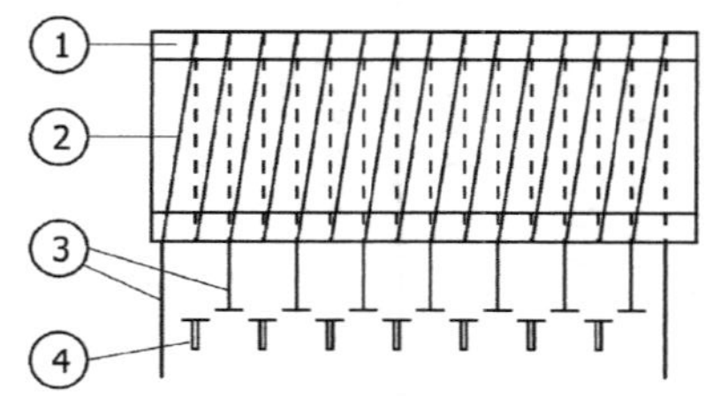

Figure 9.4 Some heater circuits and special windings.

9.4 DESIGN OF METALLIC ELEMENTS

Consider a wire heating element as shown in Figure 9.5(C).

Let d = Wire diameter (cm)
ℓ = Wire length (cm)
ρ = Resistivity (ohm.cm)
V = Voltage across wire (V)
I = Current in wire (A)
R = Wire resistance (ohm)

The power $P = I^2R = V^2/R\text{W}$

where $$R = \rho\frac{\ell}{\pi d^2/4} \text{ ohm} \tag{9.4}$$

Wire surface area $A = \pi \cdot d \cdot \ell \text{ cm}^2$

Power radiated per unit area

$$W = \frac{P}{A} \text{ W/cm}^2$$

A similar expression can be developed for a strip heater.

The thermal mass of wire (specific heat × w × temp diff.) is very small and practically all the heat generated will be radiated or given out.

It is now necessary to analyze the distribution of this power to the surroundings.

Some heat will directly strike and heat the work. In an efficient process this component should be maximum.

A large component will strike the lining. Some of this will be reflected by it to the work. This is the second useful component and will depend on lining geometry and surface properties.

The lining will absorb some heat, which will raise the lining temperature and more heat will be lost through the lining and insulation to the surroundings. This loss is unavoidable and will depend on the thermal properties and design of the lining.

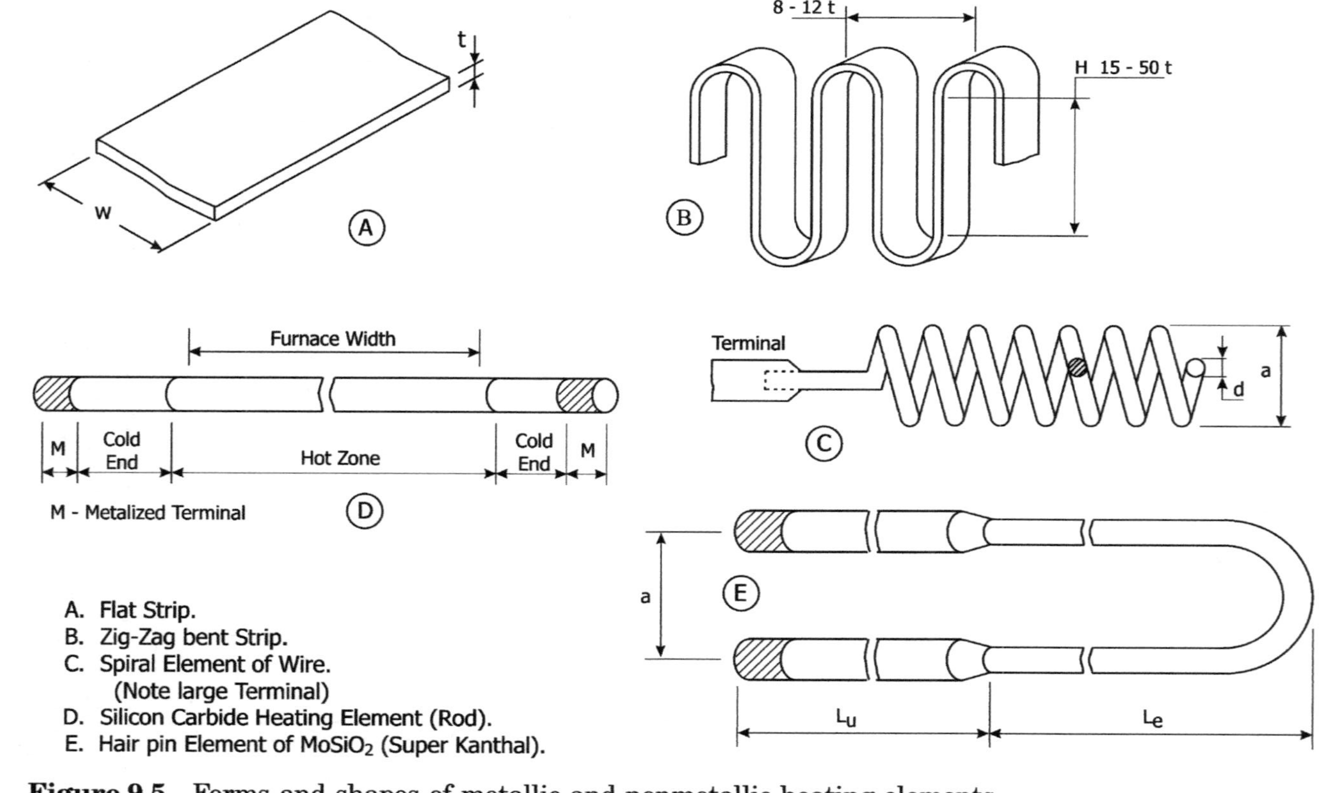

Figure 9.5 Forms and shapes of metallic and nonmetallic heating elements.

The shape of the heating element and the method of support will also affect the heat transfer process. If the heating element is freely hanging or open (e.g., Figure 9.2(A)), heat will be radiated out in all directions. If it is supported in a groove only a small part of the heat will be radiated through the groove opening. The remaining part will be conducted to the refractory. If the element is embedded (Figure 9.2(B) and Figure 9.2(C)), all the heat will be conducted to the refractory.

In the majority of applications the element is in the form of a coil or zigzag strip. These elements will radiate and circulate some heat on themselves.

The above discussion shows that the heat transfer from the element to the work is quite complicated. The same applies to the object. It rarely has a simple shape and uniform surface. The emissivity and reflectivity is uncertain; so is the state of the surface. It is, therefore, impossible to exactly calculate the heat reaching and absorbed by the work.

In an ideal case (see Chapter 3) the power radiated by a unit area of heater to the object is

$$W = \frac{\sigma_o}{\frac{1}{\epsilon_h} + \frac{1}{E_o} - 1}\left(T_h^{\,4} - T_o^{\,4}\right) \quad \text{w/m}^2 \tag{9.5}$$

where ϵ_h and E_o are the emissivities and T_h and T_o are the absolute temperatures of heater and object, respectively.

The assumptions involved are:

1. The lining has no specific heat and has zero conductivity, i.e., there are no losses.
2. The heater surrounds the object from all sides.
3. The surface area of the heater and the work are equal.

In any practical situation, none of the above assumptions are valid. The Equation (9.5) indicates the maximum power transfer that is possible.

The power required to heat the work to the desired temperature in a given time and in a given enclosure is primarily decided by heat transfer calculations. These are discussed in Chapter 3 and Chapter 4. Once the power is known, it is necessary to decide the heater size (e.g., wire diameter and length).

The heater has to withstand the surface loading, i.e., the power that can be dissipated through its surface (W/cm^2) safely (without damage) and for the designed lifetime.

The heating element has many constraints affecting its power load capacity.

1. The melting temperature limits the maximum working temperature. The mechanical strength and rigidity are very low near the melting point and a practical maximum temperature is 100–200°C less than the melting point.
2. Freely suspended heaters have to support their own weight. Coiled elements in grooves expand and are stressed by the groove walls. The heater has to withstand complex stress at high temperature and creep is also to be considered.
3. Heaters exposed to furnaces are affected by the atmosphere and corrode under gases such as oxygen, reducing gases, moisture, and special protective atmospheres. The corrosion reactions are enhanced by high temperatures.
4. Contact of the heating element with refractories produces slagging and oxidation. This can be considered as contact corrosion. It reduces the cross-section and increases local resistance, giving rise to hotspots and local melting.
5. Impurities and micro-cracks produce local stresses and hotspots, leading to sudden breakage.

In general, the heating elements are constrained by their own properties, method and material of support, and the environment.

One of the major requirements of heaters is an adequate service life (if operated properly). The conditions discussed here make it impossible to correctly predict the life span. Commercial literature gives expected life (in hours) at various temperatures.

The preceding discussion shows that the working life can be increased by operating the element at a power density less than that given by Equation (9.5). As it is impossible to correctly predict both the heat transfer and the physical state of the element, it is not possible to give exact figures of applicable power density (W/cm^2). Manufacturers supply recommended values at various temperatures, shapes, and support. These values are obtained from long experience and are recommended or suggested values. The designer has to make a choice based on his own experience.

Surface loading values recommended by manufacturers are given in Figure 9.6. Roughly, the maximum value obtained from the equation should be multiplied by a factor k.

k = 0.15–0.3 for elements in grooves

k = 0.6–0.8 for free-hanging coils or strips

9.4.1 Determination of Wire or Strip Size

Theoretical method

When the permissible surface loading q is decided then for round wire

$$q = \frac{watts}{cm^2} = \frac{W}{\pi d \ell} \tag{9.6 A}$$

$$\ell = \frac{W}{q \pi d} \tag{9.6 B}$$

Combining Equation (9.6 A) with Equation (9.6 B) we get

$$d = \sqrt[3]{\left[\frac{P}{V}\right]^2 \times \frac{4\rho}{\pi^2 q}} \quad \text{cm} \tag{9.7}$$

$$= 0.74\sqrt[3]{\left[\frac{P}{V}\right]^2 \times \frac{\rho}{q}} \tag{9.8}$$

P = Power (W)
ρ = Specific resistivity (ohm.cm)
V = Voltage applied
q = Permissible surface load (W/cm^2)

A similar expression can be developed for strip a having a width b and thickness t cm.

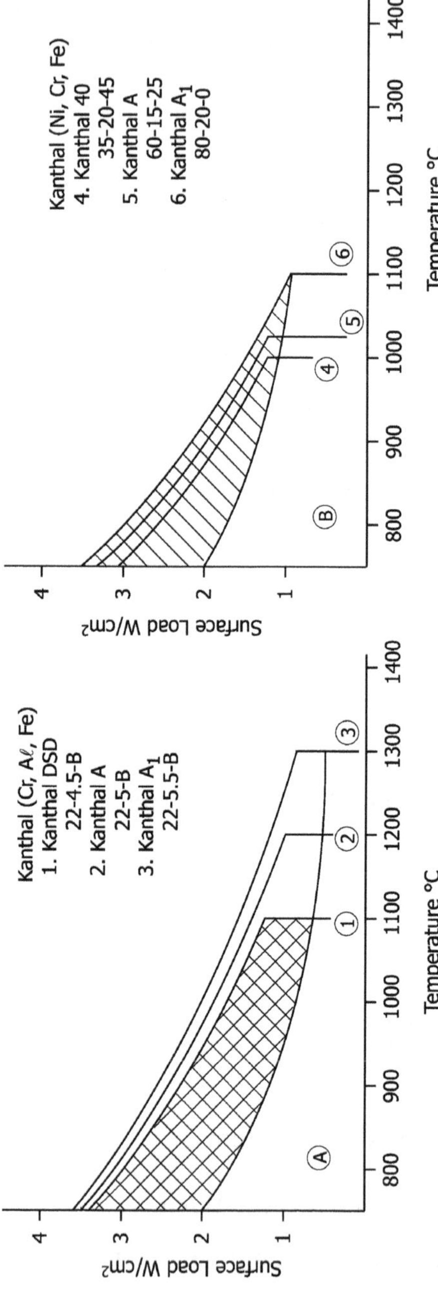

Figure 9.6 Permissible surface load for Fe-Al-Cr and Ni-Cr-Fe resistance heating alloys. (*Courtesy:* Kanthal AB Sweden.)

so that

$$\left.\begin{aligned} t &= b/t \\ t &= k_t \sqrt[3]{\left[\frac{P}{V}\right]^2 \times \frac{\rho}{q}} \\ b &= k_b \sqrt[3]{\left[\frac{P}{V}\right]^2 \times \frac{\rho}{q}} \end{aligned}\right\} \qquad (9.9)$$

The factors k_b and k_t depend on the strip width to thickness ratio k defined above.

Commonly used values of k_b and k_t are given below:

$k =$	5	8	10	12	15
$k_t =$	$\frac{1}{963}$	$\frac{1}{1290}$	$\frac{1}{1490}$	$\frac{1}{1670}$	$\frac{1}{1930}$
$k_b =$	$\frac{1}{193}$	$\frac{1}{162}$	$\frac{1}{149}$	$\frac{1}{140}$	$\frac{1}{129}$

From the diameter or strip width and thickness ratio determined from the above equations, we have to choose the nearest size commercially available.

Manufacturers of heating alloys supply the information required for design in the form of tables for each alloy, the wire diameters, strip thicknesses, and width available. Wire diameters are quoted in millimeters or standard wire gauge (SWG) or other appropriate wire gauges. The information in the tables typically contain, wire diameter, resistance (ohm/m at 20°C), surface area per ohm (cm^2/ohm at 20°C), weight (g/m), surface area (cm^2/m), cross-sectional area (mm^2), etc. This information makes the design calculations easy. They are given as shown in Table 9.5 and Table 9.6. For details, consult manufacturers' handbooks (e.g., Kanthal Handbook). Similar tables are available for strips and ribbons.

Sample tables and some information about these factors are given further.

TABLE 9.5 Design Data for Nikrothal (80Ni-20Cr) Heating Wires (*Courtesy*: Kanthal AB)

Wire mm	6.5–0.02 mm Ø Resistivity Ω mm² m⁻¹ 1.09 Density, gcm⁻³ 8.30	$cm^2/\Omega \frac{I^2 C_t}{P}$	I = Current C_t = Temperature factor P = Surface load W/cm²

To obtain resistivity at working temperature multiply by the factor C_t in the following table.

°C	20	100	200	300	400	500	600	700	800	900	1000	1100	1200	1300	1400
C_t	1.00	1.01	1.02	1.03	1.04	1.05	1.04	1.04	1.04	1.04	1.05	1.06	1.07		

Diameter mm	Resistance Ω/m 20°C	cm²/Ω 20°C	Weight g/m	Surface area cm²/m	Cross-sect. area mm²	Diameter mm
6.50	0.0328	6220	275	204.0	33.2	6.5
6.00	0.0386	4890	235	188.0	28.3	6.0
5.50	0.0459	3770	197	173.0	23.8	5.5
5.00	0.0555	3770	163	157.0	19.6	5.0
4.75	0.0615	2430	147	149.0	17.7	4.75
4.50	0.0685	2060	132	141.0	15.9	4.5
4.25	0.0768	1740	118	134.0	14.2	4.25
4.00	0.0867	1450	104	126.0	12.6	4.0
3.75	0.0987	1190	91.7	118.0	11.0	3.75
3.50	0.113	971	79.9	110.0	9.62	3.5
3.25	0.131	777	68.9	102.0	8.30	3.25
3.00	0.154	611	58.7	94.2	7.07	3.0
2.80	0.177	497	51.1	88.0	6.16	2.8
2.50	0.222	354	40.7	78.5	4.91	2.5
2.25	0.274	258	33.0	70.7	3.98	2.25
2.00	0.347	181	26.1	62.8	3.14	2.0
1.80	0.428	132	21.1	56.5	2.54	1.8
1.70	0.480	111	18.8	53.4	2.27	1.7
1.60	0.542	92.7	16.7	50.3	2.01	1.6
1.50	0.617	76.4	14.7	47.1	1.77	1.5
1.40	0.708	62.4	12.8	44.0	1.54	1.4
1.30	0.821	49.7	11.0	40.8	1.33	1.3
1.20	0.964	39.1	9.39	37.7	1.13	1.2
1.10	1.15	30.1	7.89	34.6	0.95	1.1
1.00	1.39	22.6	6.52	31.4	0.785	1.0
0.95	1.54	19.4	5.88	29.8	0.709	0.95
0.90	1.71	16.5	5.28	28.3	0.636	0.90
0.85	1.92	13.9	4.71	26.7	0.567	0.85
0.80	2.17	11.6	4.17	25.1	0.503	0.80
0.75	2.47	9.55	3.67	23.6	0.442	0.75
0.70	2.83	7.76	3.19	22.0	0.385	0.70
0.65	3.28	6.22	2.75	20.4	0.332	0.65

TABLE 9.5 Design Data for Nikrothal (80Ni-20Cr) Heating Wires (*Courtesy*: Kanthal AB) (Continued)

Diameter mm	Resistance Ω/m 20°C	cm²/Ω 20°C	Weight g/m	Surface area cm²/m	Cross-sect. area mm²	Diameter mm
0.60	3.86	4.89	2.35	18.8	0.283	0.60
0.55	4.59	3.77	1.97	17.3	0.238	0.60
0.50	5.55	2.83	1.63	15.7	0.196	0.50
0.475	6.15	2.43	1.47	14.9	0.177	0.475
0.45	6.85	2.06	1.32	14.1	0.159	0.45
0.425	7.68	1.74	1.18	13.4	0.142	0.425
0.40	8.67	1.45	1.04	12.6	0.126	0.40
0.375	9.87	1.19	0.917	11.8	0.110	0.375
0.35	11.3	0.971	0.799	11.0	0.0962	0.35
0.325	13.1	0.777	0.689	10.2	0.0830	0.325
0.30	15.4	0.611	0.587	9.42	0.0707	0.30
0.28	17.7	0.497	0.511	8.80	0.0616	0.28
0.26	20.5	0.398	0.441	8.17	0.0531	0.26

TABLE 9.6 Design Data for Kanthal A, AF (Cr-Al-Fe) Alloy Wires (*Courtesy*: Kanthal AB)

Wire mm

A : 12.0–0.05 mm Ø
AF : 12.0–0.1 mm Ø

$$\text{cm}^2/\Omega \; \frac{I^2 C_t}{P}$$

I = Current
C_t = Temperature factor
P = Surface load W/cm²

Resistivity Ω mm² m⁻¹ 1.09 — Density, gcm⁻³ 7.15

To obtain resistivity at working temperature multiply by the factor C_t, in the following table.

°C	20	100	200	300	400	500	600	700	800	900	1000	1100	1200	1300	1400
C_t	1.00	1.00	1.01	1.01	1.02	1.03	1.04	1.04	1.05	1.05	1.06	1.06	1.06	1.07	1.07

Diameter mm	Resistance Ω/m 20°C	cm²/Ω 20°C	Weight g/m	Surface area cm²/m	Cross-sect. area mm²	Diameter mm
12.0	0.0123	30700	809	377	113	12.0
10.0	0.0177	17800	562	314	78.5	10.0
9.5	0.0196	15200	507	298	70.9	9.5
8.0	0.0277	9090	359	251	50.3	8.0
7.5	0.0315	7490	316	236	44.2	7.5
7.0	0.0361	6090	275	220	38.5	7.0
6.5	0.0419	4870	237	204	33.2	6.5
6.0	0.0492	3830	202	188	28.3	6.0
5.5	0.0585	2950	170	173	23.8	5.5

(*continued*)

Table 9.6 Design Data for Kanthal A, AF (Cr-Al-Fe) Alloy Wires (*Courtesy*: Kanthal AB) (Continued)

Diameter mm	Resistance Ω/m 20°C	cm²/Ω 20°C	Weight g/m	Surface area cm²/m	Cross-sect. area mm²	Diameter mm
5.0	0.0708	2220	140	157	19.6	5.0
4.75	0.0784	1900	127	149	17.7	4.75
4.5	0.0874	1620	114	141	15.9	4.5
4.25	0.0980	1360	101	134	14.2	4.25
4.0	0.111	1140	89.8	126	12.6	4.0
3.75	0.126	936	79.0	118	11.0	3.75
3.5	0.144	761	68.8	110	9.62	3.5
3.25	0.168	609	59.3	102	8.30	3.25
3.0	0.197	479	50.5	94.2	7.07	3.0
2.8	0.226	390	44.0	88.0	6.16	2.8
2.5	0.283	277	35.1	78.5	4.91	2.5
2.25	0.350	202	28.4	70.7	3.98	2.25
2.0	0.442	142	22.5	62.8	3.14	2.0
1.9	0.490	122	20.3	59.7	2.84	1.9
1.8	0.546	104	18.2	56.5	2.54	1.8
1.7	0.612	87.2	16.2	53.4	2.27	1.7
1.6	0.691	72.7	14.4	50.3	2.01	1.6
1.5	0.787	59.9	12.6	47.1	1.77	1.5
1.4	0.903	48.7	11.0	44.0	1.54	1.4
1.3	1.05	39.0	9.49	40.8	1.33	1.3
1.2	1.23	30.7	8.09	37.7	1.13	1.2
1.1	1.46	23.6	6.79	34.6	0.950	1.1
1.0	1.77	17.8	5.62	31.4	0.785	1.0
0.95	1.96	15.2	5.07	29.8	0.709	0.95
0.9	2.18	12.9	4.55	28.3	0.636	0.9
0.85	2.45	10.9	4.06	26.7	0.567	0.85
0.8	2.77	9.09	3.59	25.1	0.503	0.8
0.75	3.15	7.49	3.16	23.6	0.442	0.75
0.7	3.61	6.09	2.75	22.0	0.385	0.7
0.65	4.19	4.87	2.37	20.4	0.332	0.65
0.55	5.85	2.95	1.70	17.3	0.238	0.55
0.5	7.08	2.22	1.40	15.7	0.196	0.5
0.475	7.84	1.90	1.27	14.9	0.177	0.475
0.45	8.74	1.62	1.14	14.1	0.159	0.45
0.425	9.80	1.36	1.01	13.4	0.142	0.425
0.4	11.1	1.14	0.898	12.6	0.126	0.4

9.5 NONMETALLIC HEATING ELEMENTS

Metallic alloy heating elements have a temperature limit of about 1400°C. They are mostly used up to about 1300°C for continuous service. The alloys lose their strength beyond this and have a relatively short life.

Currently commercially available nonmetallic elements are based on silicon carbide (SiC) or molybdenum silicide ($MoSi_2$) and are useful for service up to 1550°C and 1700–1800°C, respectively. Their important properties are given in Table 9.4. The maximum temperatures given here are subject to proper design and compatible atmospheres.

The heating elements differ from metallic elements in many respects. An important difference to be noted is that they have no formability. The manufacturer supplies the elements in a fixed form (e.g., rods, hair pins, etc.) and the furnace has to be designed around these "ready-made" forms (see Figure 9.5 (D), Figure 9.5 (E). In a given form there is a wide range of sizes to choose from.

9.5.1 Silicon Carbide Heating Elements

Silicon carbide has a negative temperature coefficient (α) up to about 788°C, above which it is positive (see Figure 9.7). Molybdenum silicide has a positive coefficient throughout. The resistivity is much lower than those of metallic alloys and in most cases, a step-down transformer with suitable tapes is required. The maximum surface loading is much higher (~20 w/cm^2) than that of metallic elements (see Figure 9.8).

As the operating temperature range of these elements is 1400–1500°C, the choice of refractories to be used is limited to high alumina types such as mullite, corundum, and pure alumina (99.8 +). These refractories are discussed in detail in Chapter 7. The specifications and shapes available can be obtained from the manufacturer who may supply special shapes from his standard inventory. Fiber materials of the same grade in the form of boards, wool, blankets, or vacuum-formed shapes are also extensively used in muffle construction.

Silicon carbide elements (e.g., rods) have two distinct zones. The central zone is the heating zone. Two adjoining lengths are

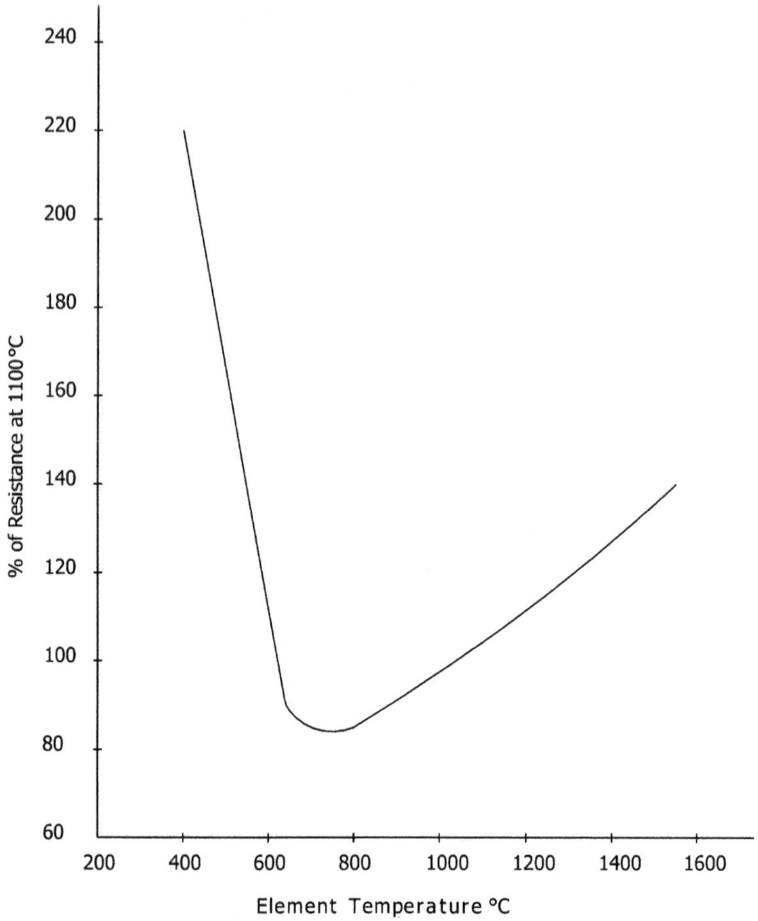

Figure 9.7 Temperature and resistivity of typical silicon carbide bar heating element. Heating element resistivity at 1100°C ~ 0.016 ohm.cm.

terminal zones. The resistivity in the heating zone is higher than that of the terminal zones. Thus, the resistivity of the whole rod is not uniform. It is necessary to consult manufaturers' data for that element. The two side zones remain cooler than the central part. They are used to support the rod on refractory walls and carry the power connections (see Figure 9.9(C)). The end parts are aluminum spread for good electrical contact.

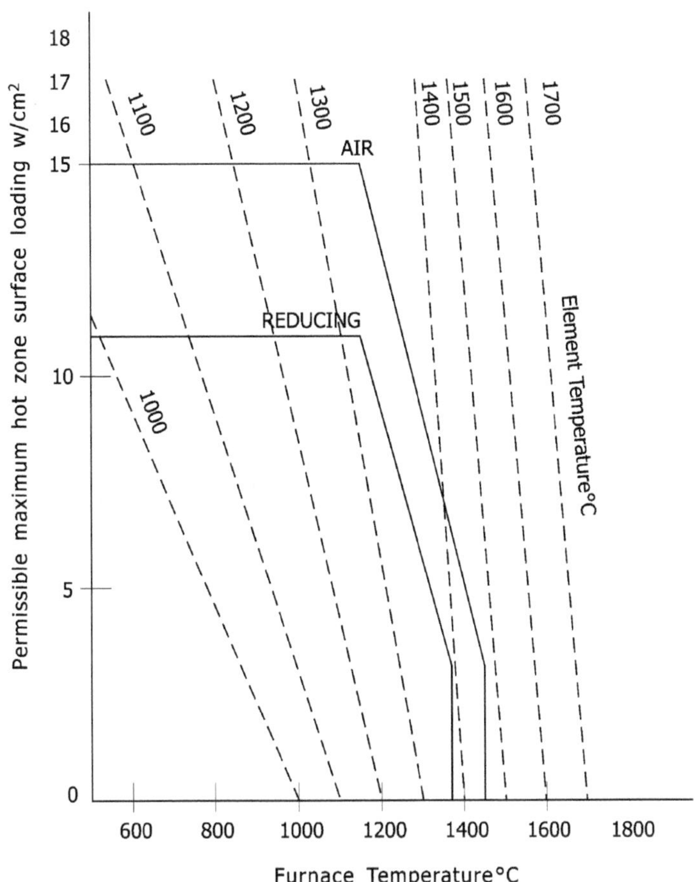

Figure 9.8 Surface load for silicon carbide elements.

Silicon carbide elements show a change in resistance (aging) with time. The resistance increases by 50% after 1000 h and by 70% in 3500 h at high (more than 1200°C) temperatures.

9.5.2 $MoSi_2$ Heating Elements

Molybdenum disilicide (Kanthal Super) elements are available in single or multiple bend hairpin shapes. They are very brittle when cold and thus require delicate handling. They are available in two commercial grades. Super ST and N grades can be used up to 1700°C maximum. Super 33 grade can be used up to 1800°C.

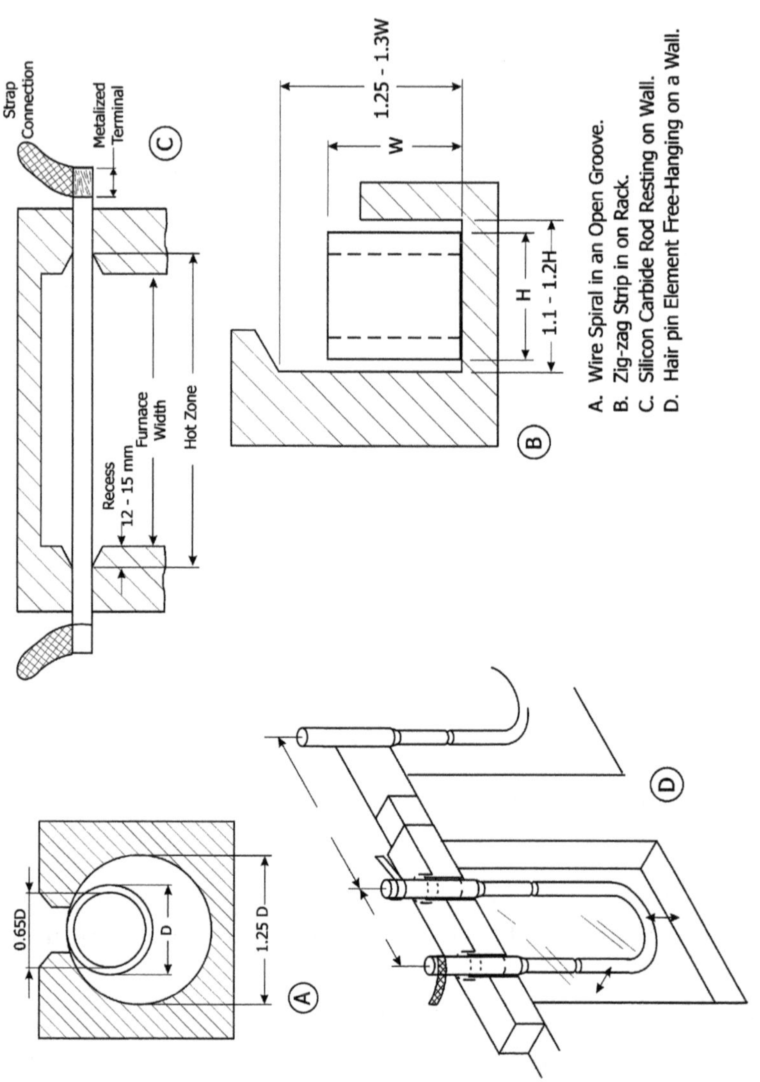

Figure 9.9 Heating element supports.

The shanks (end parts) have a diameters twice that of the heating (middle) zone. The connecting zone of the end is aluminized for better contact (see Figure 9.9(D)).

$MoSi_2$ has an almost linear, positive resistance temperature relation (Figure 9.11). A stepdown multiple tap transformer is usually necessary. These elements are used in a free-hanging fashion supported in the roof or on the walls (see Figure 9.9(D)). The manufacturer supplies information regarding the support design and placement restrictions such as distance from walls, and separation between adjacent elements.

Modern furnaces with these elements are constructed with fiber walls and roofs and the elements cannot be directly supported by them. The whole element assembly, in such cases, is supported by the outer metal cover of the furnace.

Typical data supplied by the manufaturer regarding sizes, surface loading, resistivity, etc. for nonmetallic elements are given in Table 9.8 and Figure 9.10 and Figure 9.11. For full details refer to manufacturers' handbooks.

9.6 DESIGN CALCULATIONS FOR NONMETALLIC ELEMENTS

As stated before, the design calculations for these elements consist of choosing the shape, size, and number of elements suitable for the size and service temperature of the required furnace.

Once these parameters are decided, the geometrical placement (e.g., distance between elements, etc.) is carried out according to the suggested data from the manufacturer (see Figure 9.9(B)). A wide collection of such data are available in the manufacturers' handbook. Sample pages of such data are given in Table 9.7 and Table 9.8.

The current and voltage are calculated on the basis of which suitable transformer is specified.

The change in resistance with temperature and time makes it necessary to use a multiple tap transformer. Taps lower than nominal design voltage are required to take care of the temperature effect. Taps with higher voltage are required to take care of aging.

Manufacturers of both elements, SiC and $MoSi_2$ supply the special hardware required for electrical connections for

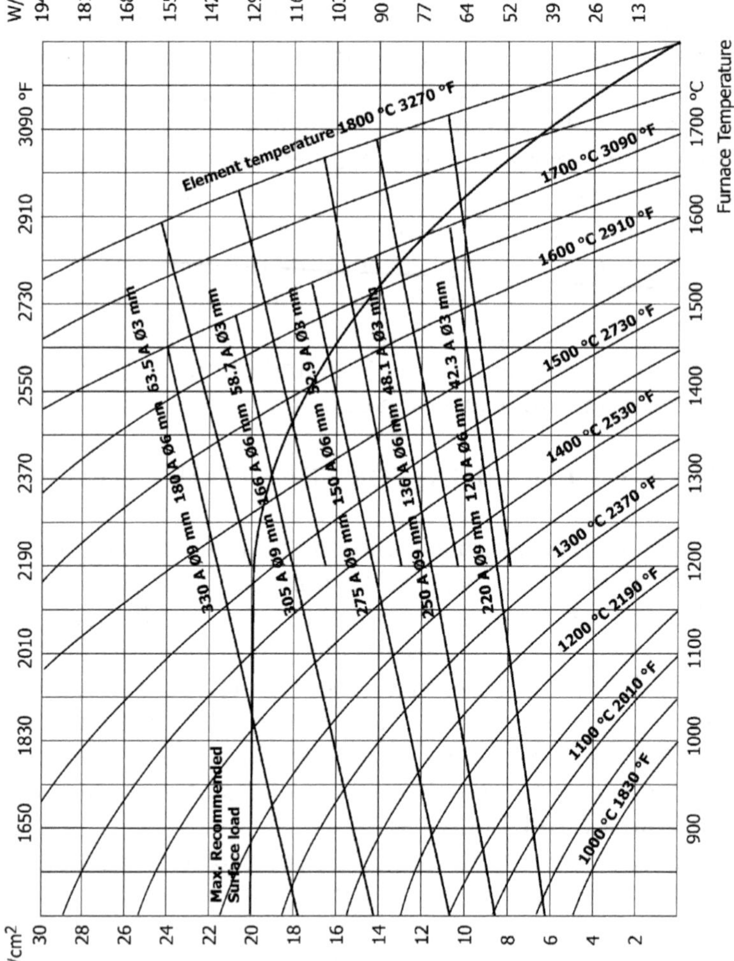

Figure 9.10 Temperature-loading diagram for Kanthal Super ST, N.

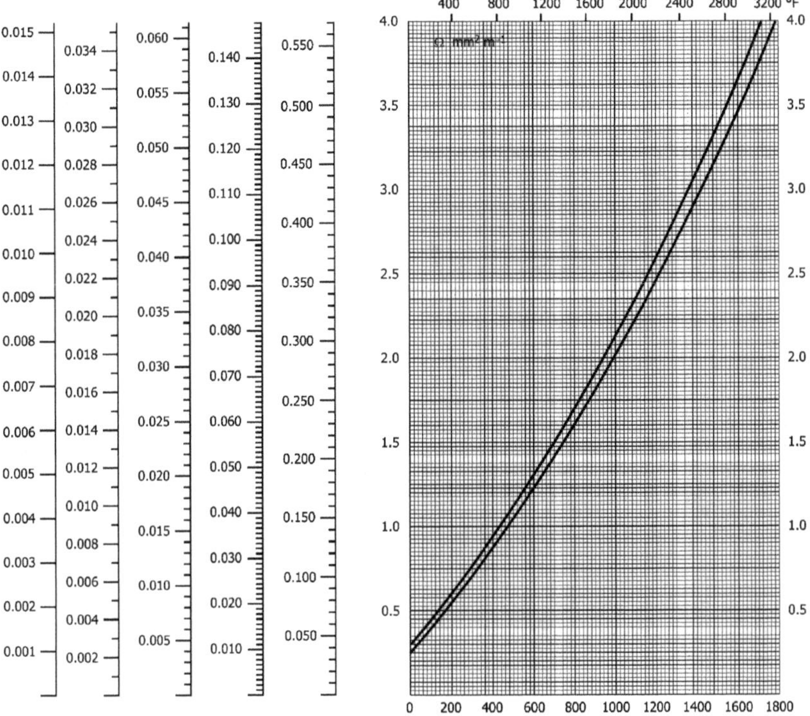

Figure 9.11 Resistivity for Kanthal Super ($MoSi_2$).

the elements. These are flexible metal straps and clamps. Special refractory shapes for supporting the elements are also available. Consult manufacturers' handbooks for details of connection practices and the special accessories available.

EXAMPLE 9.1

A resistance-heated furnace is to be designed with following requirements:

> Working space = $30 \times 40 \times 100$ cm
> Occasional maximum temperature = 1250°C
> Continuous working temperature = 1150°C
> Atmosphere = Air
> Estimated power = 24 kW
> Available voltage = 440 V–3ph
> Specific resistivity (average) = 150×10^{-6} ohm.cm

Table 9.7 Extract from Specification Data for Silicon Carbide Heaters (Norton)

Effective heating length (mm)	Width of furnace chamber (mm)	Diameter (mm)	Nominal radiating surface (mm^2)	Thickness of furnace wall (mm)	Overall length (mm)	Nominal resisitance (ohm)
—	—			76	483	4.55
		9.5	9,120	127	584	4.62
				165	600	4.87
				76	483	3.24
				127	584	3.40
—	—	11.1	10,639	165	600	3.43
				191	711	3.46
				229	787	3.50
				76	483	2.50
				127	584	2.54
305	279	12.7	12,161	165	600	2.57
				191	711	2.59
				229	787	2.62
				76	508	1.57
				127	610	1.60
—	—	15.8	15,201	165	686	1.62
				191	737	1.63
				229	813	1.65
				76	508	1.07
				127	610	1.08
—	—	19.0	18,239	165	686	1.10
				191	737	1.11
				229	813	1.12
				191	787	0.063
		25.4	24,323	229	864	0.064
				343	1,092	0.066

TABLE 9.8 Power Rating, Voltage and Dimensions of Kanthal Super ($MoSi_2$) Hairpin Elements (*Courtesy*: Kanthal AB.)

2-Shank Elements KANTHAL SUPER ST and N 9/18 mm
Max. element temperature 1700°C 3090°F
Heating zone, Le

Power, W
Resistance at 1500°C 2730°F, W
Nominal Voltage V

Terminal, Lu (in / mm)	4.9 125	6.3 160	7.1 180	7.9 200	8.9 225	9.8 250	11 280	12.4 315	14 355	15.8 400	17.7 450	19.7 500	22 560	24.8 630	26.4 670	28 710	31.5 800	35.4 900	39.4 1000	44.1 1120	49.2 1250
9.8 250	1400 0.019 5.1	1690 0.022 6.1	1850 0.024 6.7	2010 0.027 7.3	2220 0.029 8.1	2420 0.032 8.8	2660 0.035 9.7	2950 0.039 10.7	3280 0.043 11.9	3640 0.048 13.2	4050 0.054 14.7	4460 0.059 16.2	4940 0.065 18.0	5510 0.073 22.0	5840 0.077 21.2	6160 0.082 22.4	6900 0.091 25.1	7710 0.102 28.0			
11 280	1440 0.019 5.2	1720 0.023 6.3	1880 0.025 6.9	2050 0.027 7.4	2250 0.030 8.2	2460 0.032 8.9	2700 0.036 9.8	2980 0.039 10.8	3310 0.044 12.0	3680 0.049 13.4	4080 0.054 14.8	4490 0.059 16.3	4980 0.066 18.1	5550 0.073 20.2	5870 0.078 21.4	6200 0.082 22.5	6930 0.092 25.2	7740 0.102 28.2			
12.4 315	1480 0.020 5.4	1760 0.023 6.4	1920 0.025 7.0	2090 0.028 7.6	2290 0.030 8.3	2490 0.033 9.1	2740 0.036 10.0	3020 0.040 11.0	3350 0.044 12.2	3720 0.049 13.5	4120 0.055 15.0	4530 0.060 16.5	5020 0.066 18.2	5590 0.074 20.3	5910 0.078 21.5	6240 0.082 22.7	6970 0.092 25.3	7780 0.103 28.3	8600 0.114 31.3		
14 355	1520 0.020 5.5	1810 0.024 6.6	1970 0.026 6.7	2130 0.028 7.8	2340 0.031 8.5	2540 0.034 9.2	2780 0.037 10.1	3070 0.041 11.2	3390 0.045 12.3	3760 0.050 13.7	4170 0.055 15.2	4570 0.060 16.6	5060 0.067 18.4	5630 0.074 20.5	5960 0.079 21.7	6280 0.083 22.8	7020 0.093 25.5	7830 0.104 28.5	8640 0.114 31.4	9620 0.127 35.0	
15.8 400	1570 0.021 5.7	1860 0.025 6.8	2020 0.027 7.4	2180 0.029 7.9	2390 0.032 8.7	2590 0.034 9.4	2840 0.037 10.3	3120 0.041 11.3	3440 0.046 12.5	3810 0.050 13.9	4220 0.056 15.3	4620 0.061 16.8	5110 0.068 18.8	5680 0.075 20.7	6010 0.079 21.8	6330 0.084 23.0	7070 0.093 25.7	7880 0.104 28.7	8690 0.115 31.6	9670 0.128 35.2	10730 0.142 39.0
17.7 450			2080 0.027 7.6	2240 0.030 8.1	2440 0.030 8.9	2650 0.035 9.6	2890 0.038 10.5	3180 0.042 11.6	3500 0.046 12.5	3870 0.051 14.1	4280 0.057 15.5	4680 0.062 17.0	5170 0.068 18.8	5740 0.076 20.9	6060 0.080 22.1	6390 0.085 23.2	7120 0.094 25.9	7940 0.105 28.9	8750 0.116 31.8	9230 0.129 35.4	10780 0.142 39.0
19.7 500					2500 0.033 9.1	2700 0.036 9.8	2950 0.039 10.7	3230 0.043 11.8	3560 0.047 12.9	3920 0.052 14.3	4330 0.057 15.8	4740 0.063 17.2	5230 0.069 19.0	5800 0.077 21.1	6120 0.081 22.3	6450 0.085 23.4	7180 0.095 26.1	7990 0.106 29.1	8810 0.116 32.0	9780 0.129 35.6	10840 0.143 39.4
22 560					2570 0.034 9.3	2770 0.037 10.1	3020 0.040 11.0	3300 0.044 12.0	3630 0.048 13.2	3990 0.053 14.5	4400 0.058 16.0	4810 0.064 17.5	5300 0.070 19.3	5860 0.078 21.3	6190 0.082 22.5	6520 0.086 23.7	7250 0.096 26.4	8060 0.107 29.3	8880 0.117 32.3	9850 0.130 35.8	10910 0.144 39.7
24.8 630							3100 0.041 11.3	3380 0.045 12.3	3700 0.049 13.5	4070 0.054 14.8	4480 0.059 16.3	4490 0.065 17.8	5370 0.071 19.5	5940 0.079 21.6	6270 0.083 22.8	6600 0.087 24.0	7330 0.097 26.6	8140 0.108 29.6	8960 0.118 32.6	9930 0.131 36.1	10990 0.145 40.0
26.4 670							3140 0.042 11.4	3430 0.045 12.5	3750 0.050 13.6	4120 0.054 15.0	4520 0.060 16.5	4930 0.065 17.9	5420 0.072 19.7	5990 0.079 21.8	6320 0.084 23.0	6640 0.088 24.1	7370 0.097 26.8	8190 0.108 29.8	9000 0.119 32.7	9980 0.132 36.3	11030 0.146 40.1
28 710									3800 0.050 13.8	4160 0.055 15.1	4570 0.060 16.5	4980 0.066 18.1	5460 0.072 19.9	6040 0.080 21.9	6360 0.084 23.1	6690 0.088 24.3	7420 0.098 27.0	230 0.109 29.9	9050 0.120 32.9	10020 0.133 36.4	11080 0.147 40.3
31.5 800									3900 0.052 14.2	4260 0.056 15.5	4670 0.062 17.0	5080 0.067 18.5	5570 0.074 20.2	6140 0.081 22.3	6460 0.085 23.5	6790 0.090 24.7	7520 0.099 27.3	8830 0.110 30.3	9150 0.121 33.3	10120 0.134 36.8	11180 0.148 40.7
35.4 900										4380 0.058 15.9	4790 0.063 17.4	5190 0.069 18.9	5680 0.075 20.7	6250 0.083 22.7	6580 0.087 23.9	6900 0.091 25.1	7630 0.101 27.8	8450 0.112 30.7	9260 0.122 33.7	10240 0.135 37.2	11300 0.149 41.1

Data at 1500 °C 2730 *°F* element temperature corresponding to a furnace temperature of 1300 ° 2370 *°F*. (Freely radiating vertically mounted elements.)

Amperage : 275 A
Surface load : 14.5 W/cm^2
93.5 W/sq.m
Standard dimensions see page 24

Element Dimensions (mm)	a	c	d	f	g	$k^2_{90°}$	$k^2_{45°}$	m	n	r
9/18	50, 60	18	9	75	30^4	71	35	90^3	$135^{3)}$	45
6/12	40, $45^{1)}$, $50^{1)}$, 60	12	6	40	20	47	24	60	90	30
3/6	20, 25, 40	6	3	25	15	10	5		45	

9/18, 6/12, 3/6 mm bent at terminals (see Figure 8 (a), Figure 8 (b) (ST and N)5)

9/18, 6/12 mm bent at taper (see Figure 8 (c) (ST and N)5)

3/6 mm bent at taper zone (see Figure 9)

Now determine:

1. Material for heating resistors
2. Surface loading
3. Suggest method and material for construction
4. Wire or strip size
5. Element distribution, shape, and construction
6. Heating rate achievable

Solution

The furnace atmosphere is oxidizing. Working temperature 1150°C and occasional maximum temperature 1250°C suggest Fe-Cr-Al alloy (Kanthal A1 or Kanthal AF or other similar alloy) having a maximum permissible temperature of 1400°C and a probability of long service life (>1000 h).

We can use nonmetallic elements such as SiC or $MoSi_2$ (Kanthal Super). They have a much higher temperature capability and are costly. Higher temperatures (> 1250°C) are not specified.

The recommended surface load (q) in the range 1150–1250°C is 1–1.8 W/cm^2. For trial, a preliminary load of 1.0 W/cm^2 is assumed. The chosen alloy is available in both wire and strip form. For the first trial choose a round wire of diameter d cm.

The power requirement given is 24 kW with a 380 V 3 ph supply. Assume a star connection so that each phase has 24/3 = 8 kW power at 220 V.

Diameter of Wire

Voltage per phase (V) = 220 V

Power dissipated in each phase (P) = 8 kW

Surface loading q = 1.0 W/cm^2

Specific resistance = 150×10^{-6} ohm.cm

$$d = 0.74\sqrt[3]{\left[\frac{8000}{220}\right]^2 \times \frac{150 \times 10^{-6}}{1.0}}$$

$$= 0.43 \text{ cm } (4.3 \text{ mm})$$

Alternatively,

$$\text{Resistance} \quad R = \frac{V^2}{P}$$

$$= \frac{220^2}{8000} = 60.5 \text{ ohm}$$

$$\text{Surface area} = \frac{P}{q} = 8000 \text{ cm}^2$$

$$\frac{\text{ohm}}{\text{cm}^2} = \frac{8000}{6.05} = 1323 \text{ cm}^2/\text{ohm}$$

From Table 9.5, nearest wires are

Kanthal A1 – 1550 cm^2/ohm – 4.5 mm

Kanthal AF – 1363 cm^2/ohm – 4.25 mm

As the diameters determined by both methods are of the same order, we choose Kanthal A1, 4.5 mm diameter wire. This wire has following data:

0.0912 ohm/m at 20°C 113 g/m

Length required for 6.05 ohms is

$$\frac{6.05}{0.0912} = 66.33 \text{ m (663 cm)}$$

At 113 gm/m the weight of wire is

$$66.33 \times 113 \times 10^{-3} = 7.5 \text{ kg}$$

Add 10% for terminals, breakage etc.

Wt = 8.25 kg

EXAMPLE 9.2

A muffle furnace is to be designed for silicon carbide heaters.

Size of heating space (nominal)	28 $(W) \times 22\,(H) \times 120\,(D)$ cm
Maximum temperature	1400°C
Atmosphere	Air

Estimated power	15 kW
Power supply	380 V, 3 phase

Heating elements to be situated at the top across the width.

Now determine:

Number of heating elements required.
Size of elements.
Current and resistance of elements.
Transformer required.

Solution

The width of the chamber is 28 cm. The heating element chosen must have a hot (central) zone about 28 cm.

According to Figure 9.8 the permissible surface loading (P) at 1400°C and in air atmosphere is

$$P = 6 \text{ W/cm}^2$$

Element temperature ~ 1,475°C

Total heating surface required

$$= \frac{1{,}5000}{6} = 2{,}500 \text{ cm}^2$$

For equal distribution in three phases the number of elements must be divisible by 3.

For trial consider 18 elements so that power per element will be

$$= \frac{15{,}000}{18} = 833 \text{ W/element}$$

Hot surface for each element

$$= \frac{2{,}500}{18} = 138.89 \sim 140 \text{ cm}^2$$

Referring to Table 9.7 it can be seen that the nearest resistor has a heating length 305 mm for furnace width 279 mm.

The surface area required per element is 140 cm^2. The nearest area of the heating zone is 152 cm^2 for a 15.8 mm diameter rod.

We have no information about the thickness of the insulation. There are four options available. They are 76, 127, 166, 191, and 229 mm. Assume a wall thickness of about 200 mm. The nearest size is 191 mm.

The element chosen has following specifications:

Heating length	305 mm
Furnace width	279 mm
Element diameter	15.8 mm
Radiating surface	15201 mm^2
Resistance	1.63 ohm

Let us try a combination of six elements in parallel. The total resistance in each phase will be

$$\frac{1}{R} = \frac{6}{1.63}$$

$$\therefore\ R = \frac{1.63}{6} = 0.271 \text{ ohm}$$

$$\text{Current} = \sqrt{5000/0.271} = 136 \text{ A/phase}$$

$$\text{Voltage} = 136 \times 0.271 = 36.8 \text{ V/Phase}$$

$$\text{kVA} = 5 \text{ kW}$$

$$\text{Primary voltage} = 380/\sqrt{3} = 220 \text{ V}$$

The transformer required is 380 V, 3 ph, 30 amp-primary; and 20, 30, 40, 50, 80 V taps on secondary with current 200 A, kVA ~ 20–25

EXAMPLE 9.3

Design an indirect resistance furnace as specified below.

Working Space	60 × 45 × 90 cm
Continuous temperature	1550°C
Occasional maximum temperature	1600°C
Estimated power	52 kW
Electric supply	380 V, 3 ph

Now determine:

Heater material
No. of heaters and connections
Voltage and current
Transformer specification
Heater dimension
Approximate inner and outer dimensions

Solution

Working volume

$= 45\ (W) \times 60\ (H) \times 90\ (D)$
$2.5 \times 10^5\ cm^3 = 0.25\ m^3$

As this is the working volume, actual volume will be a little larger

$\simeq 0.27\ m^3$

Maximum continuous temperature = 1550°C

Occasional temperature = 1600°C

For these temperatures we select Kanthal Super ST grade which has a maximum element temperature 1700°C.

The power requirement is 52 kW. The muffle height is 600 mm. The heating portion *Le* of the element should be about 600/10 = 60 mm from bottom hearth, i.e.,

Le = 540 mm

Nearest to shank heater is 560 mm (*Le*) and has terminals 60 mm apart.*

To keep proper clearance with the bottom, the height will be increased to about 616 mm.

Assume that the distance between adjacent elements is about 100 mm (more exact calculations can be done from the handbook). Heaters are to be located on the two vertical sides, hence, the number of heaters should be divisible by 2.

The supply available is three-phase. To distribute the load equally, the number of heaters should be divisible by 3.

* Values can be found in the hand books listed in the *Bibliography*.

Hence, the number of heaters to be used must be 6 (3 × 2) or 12, or 18, and so on.

The required depth is 900 mm (each side) and one element will occupy 60 + 100 mm. If the number of heaters chosen is 12, there will be 6 heaters on each side. We will have to increase the depth by about 100 mm. Final depth is 900 + 100 = 1000 mm.

Note: *If suitable heaters are available we can design for six heaters (two per phase) and keep the depth at 900 mm.*

At 52 kW power each heater will be rated at 52/12 = 4.33 kW.
Assume an insulation thickness *W*/2

= 45/2 = 22.5 cm (225 mm).

The element shanks will have to be about 225 (in insulation) + 150 (terminals) which equals to 375 mm.

The catalog extract (See Fig 9.8) shows that the nearest elements are

1. Heating length 500 mm Shank 355 mm 4.57 kW
2. Heating length 560 mm Shank 355 mm 5.06 kW
3. Heating length 500 mm Shank 400 mm 4.60 kW

We will have to choose an element from the three sizes above.

The following element is chosen (refer to Table 9.8) is

Kanthal Super ST 9/18 mm
Heating zone $Le = 500$ mm
Tapered zone $g = 30$ mm
Shank $Lu = 400$ mm
Distance between shanks $a = 60$ mm
Diameter between elements $b = 100$ mm
Diameter of heating zone = 9 mm

Power per element $= \dfrac{52}{12} = 4.33$ kW.

Surface load:

Distance of bend = a = 60 mm
Heating length $2Le + (\pi/2 - 1)\,\alpha = 2 \times 500 + 0.57 \times 60$
$= 1034$ mm (103.4 cm)

Heated surface

$= 103.4 \times \pi \times 0.9 = 292\ \text{cm}^2$

Surface load, (assuming 10% power lost in heating terminals)

$$= \frac{0.9 \times 4.33}{292} = 13.3\ \text{W/cm}^2$$

The surface load diagram (Figure 9.10) shows that for a furnace temperature 1550°C and surface load 13.3 W/cm^2 current of 9 mm element is 250 A and the element temperature is 1650°C.

Electrical resistivity

Room temperature (20°C) = 25×10^{-6} ohm.cm

at 1650°C = 375×10^{-6} ohm.cm

at 900°C = 180×10^{-6} ohm.cm

Length of hot zone = 103.4 cm

Resistance of hot zone = $103.4 \times 375 \times 10^{-6}$

$= 0.0387 \sim 0.04$ ohm.

Assume average terminal (shank) zone temperature = 900°C

Terminal length = $2 \times Lu = 2 \times 400 = 800$ mm (80 cm)

Resistance = $80 \times 180 \times 10^{-6}$

$= 0.0144$ ohm.

Total Resistance = 0.04 + 0.0144 = 0.0544 ohm.

Voltage required = current × resistance

$= 250 \times 0.0544 = 13.6$ volts/element.

There are four elements in series in each phase

Phase voltage (secondary) = $4 \times 3 \times 13.6$

$= 94.2$ volts ~100 V

This is the highest voltage for starting and control; suggested taps are 40 and 60 V.

As a higher current will be drawn when starting (because of low resistance) the current rating of the transformer should be higher than 250 A. Let it be 350 amp.

The power rating of the transformer

$$= \sqrt{3} \times 350 \times 100 = 60 \text{ kVA}$$

The above calculations should be repeated for other element sizes and insulating materials to arrive at an optimum design. The outer dimensions and exact interior size can then be calculated.

EXAMPLE 9.4

Design a heating element for heating a metal specimen in a hot stage microscope.

Specimen size	25 mm diameter 15 mm thick
Maximum temperature	1500°C
Atmosphere	Air or Helium/Argon
Power estimate for hot chamber	800 W

Solution

The heating element required has to be very compact. Considering the high temperature and oxidizing environment, precious metals appear to be the only choice.

A glance at Table 9.3 shows that platinum, 10 or 40% rhodium alloy, is suitable.

Pt – 10% Rh has higher resistivity, hence we will try this alloy.

The heating element will be spiral wound, free coil, with pure Al_2O_3 pieces to separate turns.

Considering a specimen size (25 diameter × 15 ht), let the element size be 40 mm diameter and 50 mm ht.

The recommended surface loading is 2.5 W/cm^2. As the service expected is intermittent, 5 W/cm^2 is chosen as the permissible load.

$$\text{Surface area required} = \frac{800}{5} = 160 \text{ cm}^2$$

As the heater is essentially free for rigidity considerations, let the wire diameter be 1.0 mm and the coil pitch 3 mm.

$$\text{Wire cross-section area} = \frac{\pi \times 0.1^2}{4} = 7.8 \times 10^{-3}\ \text{cm}^2$$

The specific resistivity of Pt–40% Rh alloy is 18.5×10^{-6} ohm.cm. The resistance of 1 cm length of wire will be

$$= \rho\frac{\ell}{a} = \frac{18.5 \times 10^{-6} \times 1}{7.83 \times 10^{-3}} = 2.37 \times 10^{-3}\ \text{ohm}$$

The coil diameter is 40 mm, hence, the length of wire in one turn is $\pi \times 4.0 = 12.6$ cm.

The number of turns in 50 mm height at 3.0 mm turns is 50/3 = 16.7 turns.

Total length of wire in coil = 12.6 × 16.7 = 210 cm

Resistance $R = 210 \times 2.37 \times 10^{-3} = 0.497 \sim 0.5$ ohm

Assume a voltage 20 V, the current is

$I = 20/0.495 = 40$ A

Power dissipated $= I^2 R = 40^2 \times 0.5 = 800$ W

Current carrying capacity of Pt–40% Rh alloy at 1600°C is 2240 Amp/cm^2, hence, the calculated current is much lower.

Surface area for 0.1 cm diameter and 210 cm length $= \pi \times 0.1 \times 210 = 66$ cm^2

Surface load is = 800/66 = 12 W/cm^2

Volume of wire = area × length

$= 7.8 \times 10^{-3} \times 210 = 1.64$ cm^3

Weight of wire at 20.0 gm/cm^3

$= 20 \times 1.64 = 32$ gm

We can repeat the calculations for different power ratings and element weight reduction. It is important to design the hot chamber properly in order to reduce heat loss. The power supply terminals (usually copper rods) will have to be water-cooled.

9.7 DIRECT RESISTANCE (CONDUCTIVE) HEATING (DRH)

Among all the electrical heating methods, DRH is the simplest method. The object to be heated is connected to a suitable electric supply. It acts as a direct resistance and heats up by providing internal resistance. Though conceptually simple, the process has many practical problems. Due to these, DRH has limited use but is still an important heating process for quick and clean heating of conductive (metallic) objects.

9.7.1 Principle of DRH

The work, which is usually a billet with a uniform cross section, is held between two grips which serve as electric contacts. A variable voltage is applied to the grips as shown in Figure 9.12(A).

If the voltage is V and the resistance R (ohm), a current I (amp) flows through the billet and power P is evolved in heat form.

$$P = VI \tag{9.10}$$

or,

$$P = I^2R \tag{9.11}$$

The resistance R of a billet with uniform cross section a and length ℓ (cm) is

$$R = \rho \frac{\ell}{a} \text{ ohm} \tag{9.12}$$

where ρ is the resistivity (ohm × cm).

The power consumed appears as heat, which is used up in the following two components:

Heat of raising the temperature of the work from t_1 to t_2 which is given by

$$H_1 = w \cdot C(t_2 - t_1) \text{ J} \tag{9.13}$$

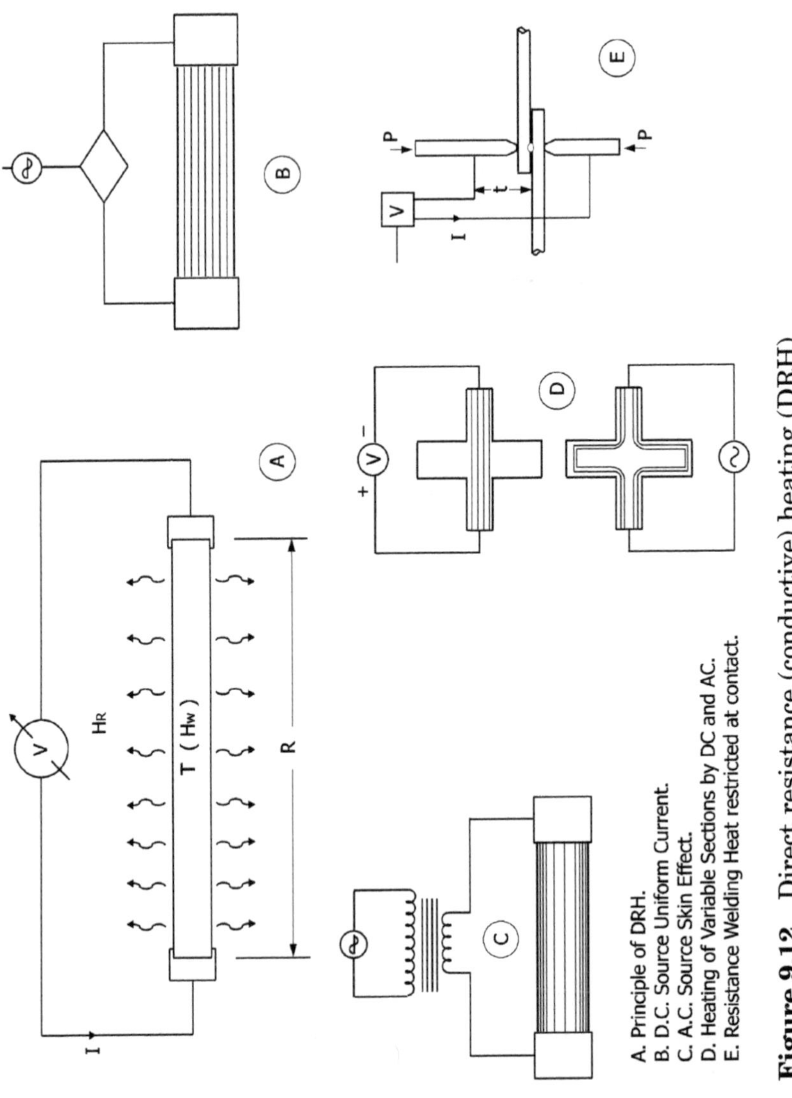

A. Principle of DRH.
B. D.C. Source Uniform Current.
C. A.C. Source Skin Effect.
D. Heating of Variable Sections by DC and AC.
E. Resistance Welding Heat restricted at contact.

Figure 9.12 Direct resistance (conductive) heating (DRH).

where C is the specific heat J/kg °C and w the weight (kg) of the work.

The second component of heat is the loss due to radiation, which is given by

$$H_2 = 5.67 \times \left[\left(\frac{T_2}{100} \right) - \left(\frac{T_1}{100} \right) \right] \times S \text{ w} \tag{9.14}$$

where T_1 and T_2 are absolute temperatures O.K. and S the radiating surface area cm^2.

If heating from t_1 to t_2 is achieved in time τ, the total heat will be

$$H = \frac{H_1}{\tau} + H_2 \text{ w} \tag{9.15}$$

From Equation (9.10) and Equation (9.12) above, the voltage or current required can be calculated.

9.7.2 Design for DRH

For DRH we have a choice of AC or DC power. AC is readily available. By using a tapped transformer or by inserting an auto transformer in the primary of the power transformer, any desired voltage can be obtained easily and economically.

However, in AC we come across skin effect and inductance. Details of skin effect are discussed in Chapter 10. It is sufficient to state here that the current in AC conductors tends to be localized on a thin surface layer, leaving the interior with practically no current. Hence, the heat is produced only in the surface layer (the skin). Thus heating is uneven. The skin depth (depth of penetration δ) depends on the frequency, permeability, and resistivity.

Nonferrous metals and steels (above 760°C) are nonmagnetic and have a relative permeability 1. Iron and most steels are magnetic up to 760°C (Curie temperature) and have a permeability between 200 and 7000. Hence, materials will have a varying skin and will show uneven and changing evolution of heat (see Figure 9.12 (B) and Figure 9.12 (C)).

AC will also produce inductance in the work and give rise to local current circulation and changing magnetic field. The resistivity of all metals increases with temperature.

Hence, the power generated by an AC supply is less than that in DC because of the skin effect and inductance. In both AC and DC heating, the effect of change in resistivity with temperature will have the same influence of resistance R.

If a rectified (AC) DC is used it will contain an AC ripple which will give a mixed heating (surface heat and through heat).

The ratio (r/δ) of radius of the work and the skin depth is also important. For (r/δ) $\leq$ 1.0 the skin effect is not noticeable and either AC or DC can be used. At (r/δ) > 1.5 the skin effect will have the influence of heat generation.

In most installations mains frequency (50 Hz) is used and for relatively thin jobs ($r/\delta \geq$ 1.5–2.0) AC or DC will give the same heat generation.

If the work has an uneven cross section (see Figure 9.12(D), Figure 9.12(E)), the heating will be uneven. In case of DC the heating (i.e., the current) will be concentrated in the central portion. (Recall the basic principle that current always takes the shortest path). If the object is heated by AC, the skin effect will make the current flow along the contour or skin and the central portion will not be directly heated. This is subject to the r/δ ratio discussed above.

The effect of AC and DC heating discussed above points out that DRH is suitable for long, thin, and even section work.

Generally DC is preferable to AC. It is again pointed out that DC will be costly as compared to a transformer-based AC source. Also depending on r/δ ratio either AC or DC can be used for many applications.

The grips used to hold the work and supply the current also pose a design problem. The contact surface between the grip and the work is uneven. It will have some high points (peaks) which will be the real contact surfaces. At high currents these contact points will have a high resistance and considerable heat will be developed there. The contacts will be heavy and may also be water cooled. Effectively, this makes for slightly cooler work ends.

Due to high temperature effects on strength, the contacts or grips may become sticky or distorted due to expansion, and will have a short life. They may develop inductance which will lower the efficiency. Grips for simple-shaped sections like square or round can be easily designed to hold at high temperatures (~1300°C). They may be improved by hydraulic, pneumatic, or mechanical pressure.

Grips are usually made from copper-based alloys such as copper-chromium, copper-beryllium, copper-tungsten, etc. Between two jobs the contacts are cooled and cleaned by compressed air blasts.

Difficulties in practical and long life contacts have limited the growth of DRH.

9.7.3 Advantages and Limitations of DRH

1. The process is (at least theoretically) very simple, though there are some practical problems peculiar to each job.
2. Heating is very quick (a few seconds).
3. DRH is useful for production jobs as it is quick and can be standardized for a given job by designing proper grips, power supply, etc.
4. It is more suitable for even and simple cross-section jobs.
5. Depending on r/δ ratio both AC and DC can be used, although DC is preferable.
6. The heating rate can be easily controlled by adjusting the current.

EXAMPLE 9.5

Steel billets of 35 mm diameter and 500 mm length are to be heated to 1200°C by direct resistance heating.

Heating time (sec)	30
Density (g/cm^3)	7.6
Mean thermal conductivity (λ.w/m °C)	35.7
Mean specific heat (kJ/kg °C)	0.576
Mean resistivity ($\sigma \cdot$ ohm $\cdot$ cm)	61×10^{-6}
Ambient temperature (°C)	30
Magnetic permeability (μ_r)	100
Permeability of air (μ_o)	$4\pi \times 10^{-7}$

Now determine:

1. Power required to heat the billet from 30 to 1200°C.
2. Power lost by radiation and total power required.
3. Current required from a 100 V DC or AC supply.
4. Effective resistance for DC and AC heating.

Solution

Volume of billet

$$= \frac{\pi \times 3.5^2 \times 50}{4} = 480 \text{ cm}^3$$

Surface area (curved surface)

$= \pi \times 3.5 \times 50$

$= 550 \text{ cm}^2$

$= 0.055 \text{ m}^2$

Weight

$= 480 \times 7.6$

$= 3648$ gm

$= 3.648$ kg

Resistance

$$\sigma \frac{\ell}{A} = \frac{61 \times 10^{-6} \times 50}{38.48}$$

$$= 80 \times 10^{-6} \ \Omega$$

Heat required to raise the temperature from 30 to 1200°C = mass × sp. heat × temp. difference

$= 3.648 \times 10^3 \times 0.576 \times (1200 - 30)$

$= 2.458 \times 106$ J

If heating is done in 30 sec

$$\text{Power} = \frac{2.458 \times 10^6}{30}$$

$$= 85 \text{ kW}$$

Heat radiated

Assume a mean surface temperature 800°C and an emissivity 0.8

$$Q = 5.67 \times 0.8 \times \left[\left(\frac{1200+273}{100}\right)^4 - \left(\frac{30+273}{100}\right)^4\right] \times 0.055$$

$$= 1.1785 \times 10^4 \text{ W}$$

$$\simeq 12 \text{ kW}$$

Total power

= Heating load + radiated loss

= 85 + 12

= 97 kW~100 kW

Note*: There may be additional losses in grips.*

If a 100 V DC power supply is available

$$\text{Current required} = \frac{100 \times 10^3}{100}$$
$$= 1000\ A$$

The radiation loss is very high. It can be reduced by keeping the billet in an enclosure or heating for a shorter duration.

Note that the calculated radiation loss is at 1200°C surface temperature which is attained at the end of the heating time.

If the power available is AC at 50 Hz, the effective resistance will change from R_{DC} to R_{AC}.

$R_{DC} = 80 \times 10^{-6}\ \Omega$ (as calculated earlier)

$\rho \cdot (30\text{–}760°\text{C}) = 57 \times 10^{-6}$ ohm.cm

$\rho \cdot (760\text{–}1200°\text{C}) = 120 \times 10^{-6}$ ohm.cm

$\mu_o = 4\pi \times 10^{-7} \quad \mu_r = 100$

We can get R_{AC} by using the formula

$$R_{AC} = \rho \frac{\ell}{A}$$

where a is the area of annulus through which most of the current flows. This will be circular ring of outer diameter 3.5 cm and inner diameter = 3.5 – 2δ, where δ is the depth of penetration.

For the temperature range 30–760°C

$$\delta = 5030\sqrt{\frac{\rho}{\mu r \cdot f}}$$

$$\delta = 5030\sqrt{\frac{57 \times 10^{-6}}{100 \times 50}}$$

$$= 0.53 \text{ cm or } 5.3 \text{ mm}$$

$$\text{Inner diameter} = 3.5 - 2(2 \times \delta)$$

$$= 3.5 - 2(2 \times 0.53)$$

$$= 1.38 \text{ cm}$$

$$\text{Area} = \frac{\pi(3.5^2 - 1.38^2)}{4}$$

$$= 8.125 \text{ cm}^2$$

$$R_{AC} = \frac{57 \times 10^{-6} \times 50}{8.125}$$

$$= 350 \times 10^{-6}$$

or $= 35 \times 10^{-5}$ ohm (30–760°C).

9.8 STORED ENERGY HEATING (SEH)

This is one of the methods of DRH in which the energy for heating is supplied from a source containing the energy in a stored form. This energy is conveyed to the work in one shot or "pulse." The duration of the pulse is a fraction of a second to a few seconds.

9.8.1 Principle of Stored Energy Heating

The device for storing the energy is a capacitor or a bank of capacitors. These are charged to a voltage V_c by a suitable DC source. After acquiring the required charge C (farads) the capacitor is isolated from the charging source and discharged through the work. The discharge produces a decreasing pulse of energy in the work and heats it within a very short time (see Figure 9.13(A), Figure 9.13(C)).

When a capacitor of capacity C (farads) is charged to a voltage C the stored energy is

$$W = \frac{1}{2}CV^2 \text{ J} \tag{9.16}$$

the charge required is

$$q = C \cdot V \text{ coulombs} \tag{9.17}$$

If the charged capacitor is discharged through a resistance R_D, the stored charge, voltage, and current show a logarithmic decay according to the following equation:

$$I = I_O(1 - e^{t/\lambda}) \tag{9.18}$$

where

I_O = Current at $t = 0$
I = Current at time t sec
λ = Time constant of the R_C circuit

$$= C \times R_D \text{ sec} \tag{9.19}$$

The charging cycle also obeys a similar equation for current, voltage, and charge at a time t after the beginning of charging at $t = 0$. If the resistance in the charging circuit is R_C

$$\lambda = C \times R_C \text{ sec} \tag{9.20}$$

In discharge the current flows in a direction opposite to that in charging.

The time constant λ is the time required to reduce the charge to 0.632 of its original value. For a given capacity C, the

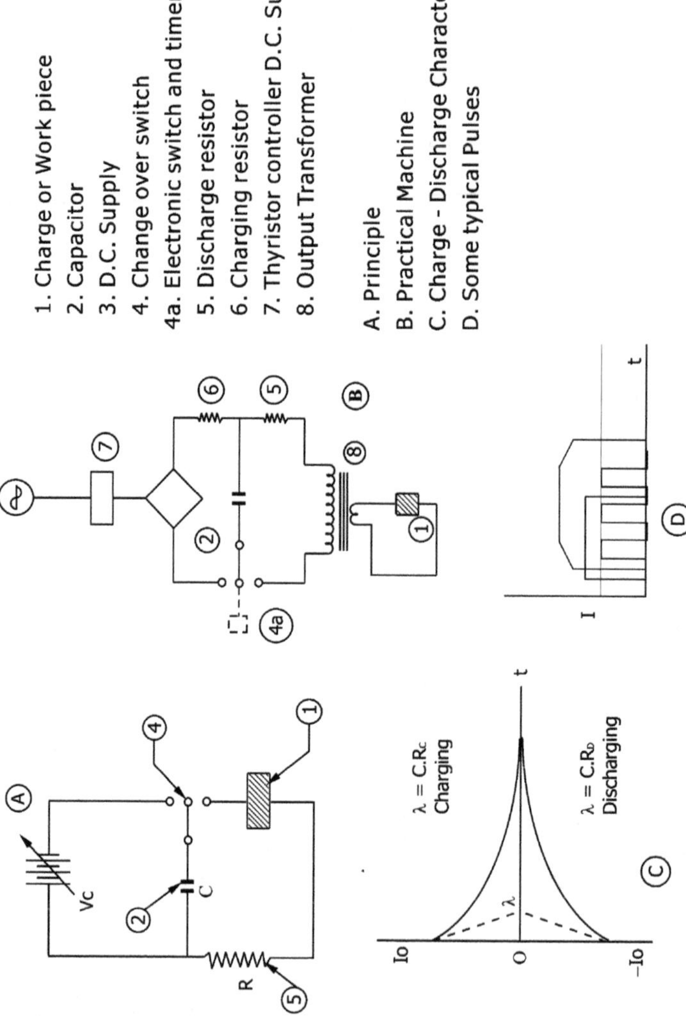
1. Charge or Work piece
2. Capacitor
3. D.C. Supply
4. Change over switch
4a. Electronic switch and timer
5. Discharge resistor
6. Charging resistor
7. Thyristor controller D.C. Supply
8. Output Transformer
A. Principle
B. Practical Machine
C. Charge - Discharge Characteristics
D. Some typical Pulses
Vc
C
R
λ = C.Rc
Charging
λ = C.RD
Discharging
Io
O
–Io
t
I

Figure 9.13 Stored energy heating.

time constant depends on the circuit resistance R. By introducing different resistances in charging and discharging circuits, the time constants can be adjusted to required values for charging and discharging. Note that the time constant repeats after every 0.632th step so that full charging or discharging requires (theoretically) infinite time. This is similar to the half life period of a radioactive material. The charge-discharge characteristics are shown in Figure 9.13(C).

9.8.2 Practical Heating Circuit

The simple stored energy heater shown in Figure 9.13(A) has many shortcomings which are overcome by incorporating many modifications and devices as shown in Figure 9.13(B).

The mains AC source is a three-phase 50 Hz supply which is rectified by a rectifier bridge based on thyristors or SCRs. This gives a variable and high DC voltage.

Condensers are the electrolytic type and usually, a bank of capacitors is provided so that the capacitance can be adjusted to obtain the required charge. The charging voltage is a few thousand volts. The choice of C and V_c makes it possible to obtain the desired energy storage. The chargeover from charging to discharging is obtained by an ignitron (vacuum tube) or a triggered thyristor. These devices can handle large currents and their switching time is very short and adjustable.

The output is usually through a transformer having a few (usually 1–3) secondary turns. The voltage applied to the job is low and the current is high (a few thousand amp).

Thus, in a practical machine, the charging voltage, time constant, and switching time is adjustable. Note that all the electronic details are omitted for simplicity. In conclusion, the energy pulse is “tailored” to the work requirements.

9.8.3 Some Peculiarities of SEH

There is a practical limit to the capacitance that can be provided. Usually this is a few thousand microfarads. This places a limit to the energy that can be stored in a practical machine. Thus the process is suitable for heating small jobs (a few hundred grams) or welding of relatively thin sheets, wires, etc.

The process is very quick (a fraction to a few seconds) and hence the heat loss oxidation is negligible. SEH is very good for heating a large number of small jobs.

The work is held under pressure to obtain a good contact. Depending on the space available any job can be heated, i.e., usually no special grips or fixtures are required.

An important difference between SEH and capacitive heating is that in the latter, the work becomes a capacitor and hence, only nonconducting jobs can be heated.

With the incorporation of electronic devices and circuits, very good control is obtained, and the energy or current pulse is shaped to requirements.

EXAMPLE 9.6

Brass slugs are to be heated to 600°C in about 1.0 sec by capacitive discharge.

Slug dimensions	10 mm diameter, 15.0 long
Sp. heat of brass (C)	0.385 J/gm K
Density (r)	8.72 g/cc
Ambient temperature	30°C
Assume a charging time	5 sec

Now determine:

1. Heat required
2. Condenser required
3. Voltage required
4. Resistance required in charging and discharging circuits

Solution

Slug size 10 mm diameter, 15 mm long

Material-brass

$$\text{Volume} = \frac{\pi \times 3.5^2 \times 50}{4} = 1.18 \text{ cc}$$

$$\text{Density} = 8.72 \text{ g/cc}$$

$$\text{Weight} = 8.72 \times 1.18$$

$$= 18.3 \text{ gm}$$

Specific heat $C = 0.385$ J/g °C
Initial temperature = 30°C
Final temperature = 600°C
Heat required = $18.3 \times 0.385(600 - 30)$
= 4.015 kJ

Thermal losses and contact losses assumed = 25%, i.e., $0.25 \times 4.015 = 1.0$ kJ

Total heat required

$$= 4.015 + 1.00$$

$$\simeq 5 \text{ kJ}$$

The capacitor bank has a 3000 μF capacity. If the charging voltage is V the energy stored is

$$= \frac{1}{2} CV^2 \text{ J}$$

This should be equal to the energy required for heating

$$= \frac{1}{2} \times 3000 \times 10^{-6} V^2 = 5 \times 10^3$$

$$V = \sqrt{\frac{2 \times 5 \times 10^3}{3000 \times 10^{-6}}}$$

$$= 1800 \text{ V}$$

The slug is to be heated in about 1.0 sec

Let the time constant $\lambda = C\ R_D = 0.1$ sec

$$3000 \times 10^{-6} \times R_D = 0.1$$

$$R_D = \frac{0.1}{3000 \times 10^{-6}}$$

$$= 30 \text{ ohm}$$

If the charging is to be done in 5.0 sec, then assume a time constant $\lambda = C \ R_C = 0.2$ sec

$$3000 \times 10^{-6} \times R_D = 0.1$$

$$R_D = \frac{0.2}{3000 \times 10^{-6}}$$

$$= 66 \text{ ohm}$$

If we use a step-down transformer with a few turns (1–3) on secondary, the heating current will be increased and voltage will be decreased.

9.9 SALT BATH FURNACES

9.9.1 Introduction

In this type of furnace, the heating medium is a bath of molten salt mixture. The object to be heated is immersed in the bath. It is surrounded on all sides by the liquid salt and the heating takes place mainly by conduction. In some furnaces the salt bath is stirred by electromagnetic forces and the conduction is supplemented by convection. Heating is very fast as the heat transfer coefficient is of the order of 10^3 w/m^2 °C. These furnaces are therefore direct conduction heaters. In this section we will discuss special features and design aspects of these furnaces.

Broadly, salt bath furnaces are of two types. In the indirect salt bath furnace the salt is heated from the outside, usually by electric resistors. A typical furnace of this type is shown in Figure 9.14. These furnaces are similar to other indirect heating furnaces and their design principles are the same as those discussed in Section 9.2. Hence, they are not discussed further.

In the other type, the salt bath is heated directly by passing a current through it, using immersed or submerged electrodes. These are “electrode-type” salt bath furnaces and are discussed here in detail.

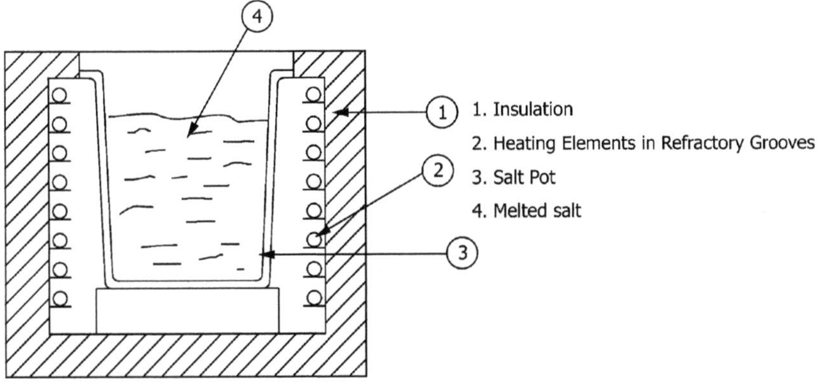

Figure 9.14 Indirectly-heated salt bath furnace.

9.9.2 Construction and Working of Electrode Furnaces

A typical immersed electrode furnace is shown in Figure 9.15. The salt bath container is made from welded carbon or alloy steel, or inconel plate (1–2 cm thick). The bath is usually rectangular. The sides and bottom are insulated by insulating bricks and refractory blankets and about a third of the top surface on one side is occupied by the electrodes. The remaining top oven is covered by a sliding door. The bottom of the bath is well supported by using dense insulating bricks.

The electrodes have a square or rectangular section and are made of MS or inconel. The inter electrode distance (ℓ) is 10–20 cm. They are supported on the top by using electrical and thermal insulating separators. The legs of the electrodes are immersed in the molten salt. The depth of immersion (h) is up to 2/3 of the bath depth leaving ample space near the bottom for the salt circulation. Other ends of the electrode (the outer ends) are carried and connected to the transformer and tap changing/switching gear. Large baths may have two pairs of electrodes. Furnaces having more than about 6 kW power use a three-phase input and a 3:1 phase transformer.

Salt baths have a temperature in the range 300–1300°C. An open top will radiate tremendous heat. To prevent this loss the furnace has a sliding insulated cover. Similarly, the surface

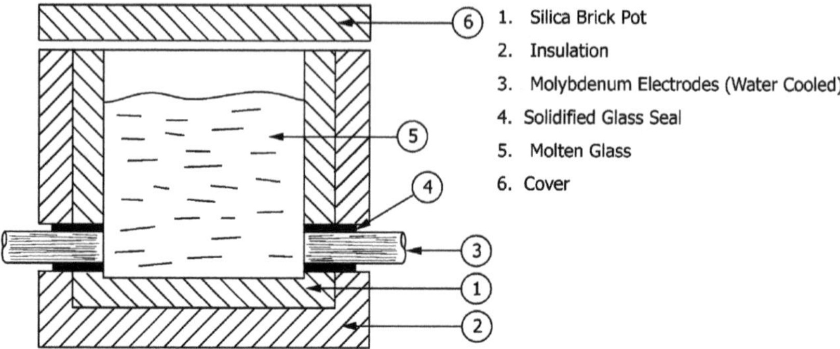

Figure 9.15 Submerged electrode glass melting furnace.

carrying the electrodes has a movable refractory cover. The top plate has rails to support the fixtures carrying the work.

The furnace is equipped with a suitably sheathed thermocouple to measure and control the bath temperature.

Voltages commonly used are 3–30 V AC. The current depends on bath size and power and may be in the range 2–50 kA. Only AC is used. Molten salts are ionic and will electrolytically decompose on DC. The bath size commonly available has a range 1–50 m^3. Larger baths are also made.

9.9.3 Bath Salts

The bath consists of a mixture of inorganic salts such as NaCl, $BaCl_2$, $NaNO_2$, $NaNO_3$. These are "neutral" salt baths and have a temperature range of applications. Some mixtures contain cyanides such as NaCN or KCN, and carbides, SiC. These salts are used for carburizing, nitriding, cyaniding, etc. Some baths contain sodium hydride which is used for descaling.

Salt mixtures may contain trace chemicals such as calcium and strontium chlorides, manganese-dioxide, sodium fluorides, etc. These chemicals act as accelerators, bath correctors, and defoaming agents.

Commercially available bath salts are of proprietary nature. Some salt mixtures used for various applications are given in Table 9.9 along with their temperature ranges.

Table 9.9 Some Commercial Salt Mixtures

Composition	Melting temperature °C	Useful temperature range °C	Applications
Common Salt NaCl 100%	800	830–1100	Heating for hardening
Barium chloride $BaCl_2$ 100%	962	1100–1350	——do——
NaCl 20–30, BaCl 70–80	700	760–900	—— do ——
NaCl 10–20, BaCl 80–90	760	800–1100	—— do ——
$NaNO_2$ 50, KNO_3 50	143	160–650	—— do ——
NaCl 15–20, $Na2CO_3$ 350 NaCN 40–45	550	600–800	Heating, liquid carburizing
NaCl 20–40, $Na2CO_3$ 350 BaCl2 0–40 NaCN 10–25	780–820	850–930	Liquid carburizing
NaCN 30–90, $Na2CO_3$ 2–5 NaCl 2–30	560–600	600–860	Liquid cynading
NaCl 10–20, KCl 2 5–40, $BaCl_2$ 40–50	500	550–700	Tempering
$NaNO_3$ 30–50, $NaNO_2$ 10–40 KNO_3 50–60	135	150–500	Tempering

The most important property of salts used in electrode furnaces is their resistivity. All salts are nonconducting in the solid state. They start conducting upon melting. The resistivity decreases with increasing temperature. It is also a function of component salts. Due to the proprietary nature of the salt mixtures, no standard resistivity figures can be quoted. Actual values should be obtained from the manufacturer.

Resistivity and temperature data for some commonly used pure salts and mixtures is shown in Figure 9.16. These data will be useful for design purposes.

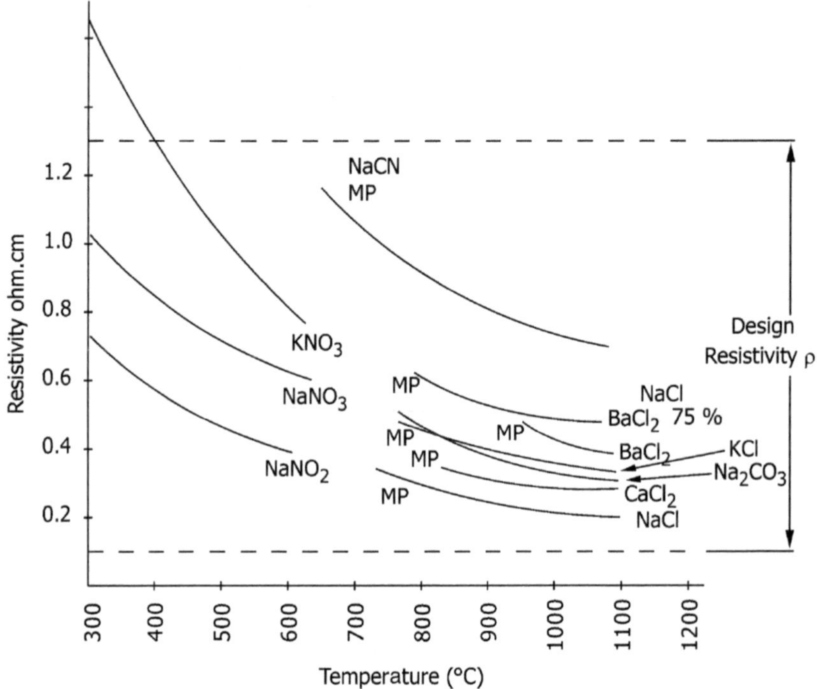

Figure 9.16 Resistivity changes with temperature for salts used for salts baths.

Other desired properties of salt baths are:

1. Stability in the application temperature range.
2. Nontoxicity. Neutral salts are generally nontoxic. However, cyanides are extremely poisonous and harmful even in small quantities. They should be used and disposed of with great caution.
3. Nonhygroscopic.
4. Noncorroding to the pot and the work.
5. Good fluidity, conductivity, and high specific heat.
6. No sludge forming.
7. Nonadhering to the work.

Commercial salt mixtures satisfy most of the requirements. Salt baths require replacement or replenishing. This is to be done by analyzing the bath.

Contact between moisture and molten salt can cause an explosion. Wherever possible the bath surface should be covered with graphite or other proprietary covering flux. Salt baths require frequent stirring.

9.9.4 Some Peculiarities of Salt Baths

Electrode salt bath furnaces are not self starting. Salts, when solid, are nonconducting. One way to start the bath is to melt the salt outside (in another furnace) and use it as a starter.

Another, more common way, is to charge the furnace with solid charge. The electrodes are then short circuited with a piece of metal. The intense heat creates a small molten pool between the electrodes. Electric conduction starts through the pool and melting progresses.

An outcome of the starting problem is the necessity of keeping the bath liquid when the furnace is not in use, i.e., at night. For this purpose the power is always on and kept on a minimum tap to ensure that the both remains liquid.

Due to heavy currents and the electric field around electrodes, the bath produces convection currents as shown in Figure 9.17. This circulation is beneficial as it makes heating faster and more uniform.

Articles to be treated are held in fixtures. The fixtures rest on the top plate. The operation is manually assisted by a chain pulley. The fixtures are made from steel or inconel wire. The work objects should be sufficiently separated from each other to assure salt circulation from all sides.

9.9.5 Applications of Salt Baths

As the temperature range of various bath mixtures is very wide, salt baths can be used for virtually all types of heat treatment. Another advantage is very quick heating from all sides. This makes it possible to use this method for treating large, small and medium size works having complex shapes. Following is a list of some of the common applications:

1. Annealing
2. Normalizing
3. Direct or interrupted quenching, hardening, martempering

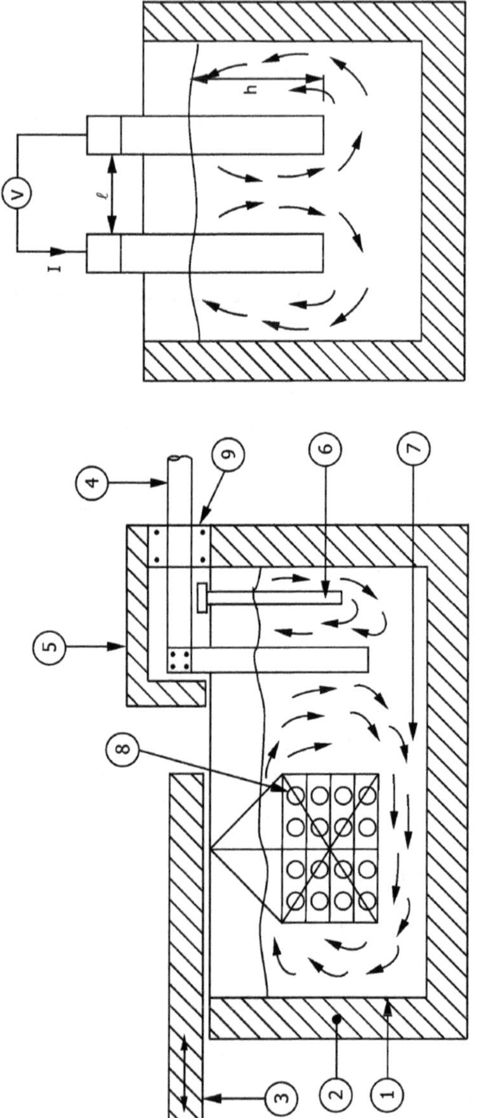

1. Welded Steel Pot (Bath).
2. Insulation.
3. Sliding Door.
4. Steel Electrodes.
5. Refractory Cover over Electrodes.
6. Thermocouple.
7. Circulating Salt Bath.
8. Fixture with Work.
9. Insulating support for Electrodes.

Figure 9.17 Electrode-type salt bath furnace.

4. Case carburizing, cyaniding

   Note that the case depth is limited to 25–50 microns even after a long time.

5. Tempering and colaring
6. Drawing
7. Solution treatment and precipitation hardening
8. Mold and die treatment
9. Descaling
10. Brazing

Both ferrous and nonferrous materials can be treated.

9.9.6 Other Bath Furnaces

Besides the salt baths there are other "bath" type furnaces used for special purposes.

Molten glass is a good electric conductor. There are many types of glasses and the resistivity varies over a wide range. Like salts, the resistivity decreases with temperature. At 1100°C it is 1.2 to 3 ohm.cm and at about 1500°C it is 0.4 to 1.2 ohm.cm.

Glass can be used as a resistor in place of salts. However, molten glass is very reactive and attacks the furnace lining. Different types of glasses exhibit different reactivity and the refractories have to be chosen carefully.

Electrode type furnaces are widely used for glass melting. The electrodes are submerged and located near the bottom. Molybdenum or graphite is used for electrodes. In glass making, the color is very important. Hence, there should be no reaction between the melt, the electrodes, and the lining. Platinum electrodes are used in making special (optical) glasses. The electrodes are water-cooled. A solidified portion of glass near the cooled end prevents leakage. For lining, usually 100% silica bricks are used. A typical furnace is shown in Figure 9.15.

We will consider detailed designs of salt bath furnaces in the next volume.

Chapter 10

High Frequency Heating

CONTENTS

High frequency electromagnetic radiation is extensively used for special heating purposes. In these methods of heating, electromagnetic radiation from 10^3 to 10^{10} Hz are used to induce heat in the object (work). There are three such methods viz. induction heating, di-electric heating, and microwave heating. We will discuss these three methods in detail in this chapter.

10.1 INDUCTION HEATING

10.1.1 Introduction

Induction heating is contactless electrical heating. It is a costly but extremely versatile process with innumerable applications. It can be used for partial and through heating, melting, and brazing of all metallic materials. In this section, we will review the basic principles, role of various related variables, and the equipment used. Large-scale melting applications are not included in the discussions. Preliminary design calculations are outlined, though the detailed design is best left to specialists.

10.1.2 Principles of Induction Heating

Induction heating is based on the principles of an electric transformer. The job or work-piece is the secondary while a surrounding copper coil is the primary. The two are linked or coupled by air. Thus, this is an air core transformer with a single turn secondary. High frequency (HF) current (1,000–100,000 Hz) is passed through the primary (coil) by connecting it to a suitable HF generator (see Figure 10.1).

A similar HF current is induced in the job, i.e., secondary. This current circulates and produces heat. These are induced eddy currents and circulate circumferentially as shown. Eddy currents, and therefore the heating, is concentrated in a thin outer layer or skin of the work.

The primary coil gets heated due to the I^2R losses in it. Some heat is also absorbed from job radiation. The coil is therefore made of copper tubing through which cooling water is circulated.

The job thus gets heated by the induced current and there is no contact between the primary and secondary.

Iron and other soft magnetic materials are used in ordinary transformers to act as susceptors, i.e., to concentrate the magnetic linking flux. They cannot be used in high frequency fields as they heat up excessively. Hence, air is used as the coupling medium. A few large scale induction heating processes (melting, heavy billet heating) that operate at mains or low frequencies (50–150 Hz), do use iron or alloy cores.

The concentration of induced currents in the surface layer is called the "skin effect." It is discussed in detail in subsequent sections. Skin effect plays an important role in the design and operation of the induction heating process.

10.1.3 Advantages and Disadvantages of Induction Heating

1. Like other electrical heating processes this is a "clean" process, i.e., there are no combustion gases and attendant problems.
2. The heat is produced directly in the job. Hence, no enclosure of furnace is required.

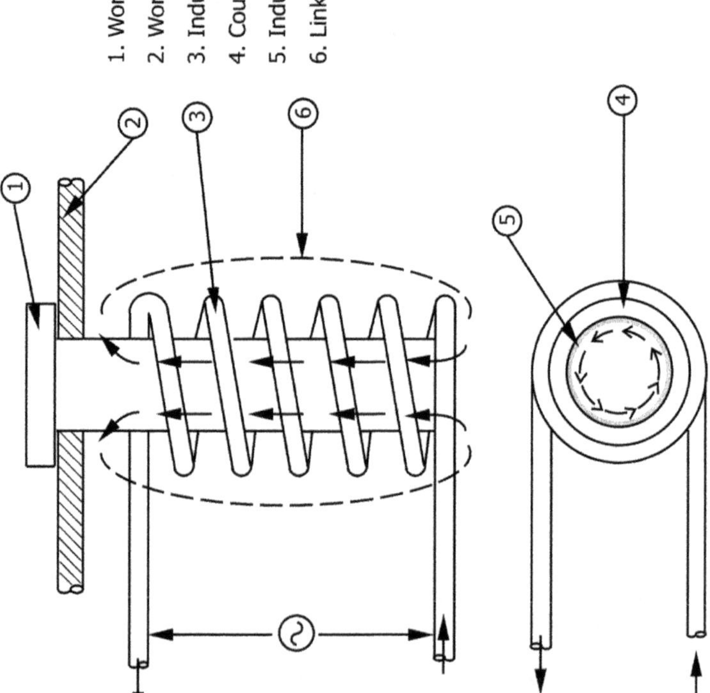

Figure 10.1 Induction heating.

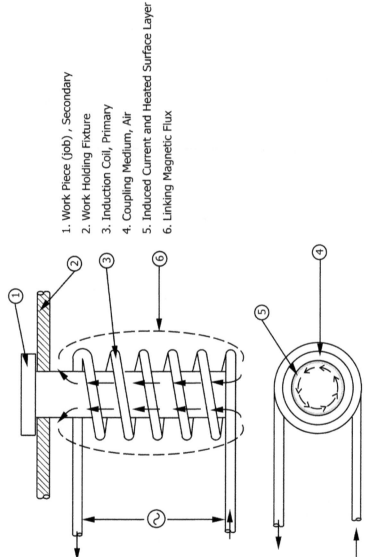

3. Induction heating is applicable only to conducting materials, i.e., metals and alloys. It can be applied for heating both magnetic (ferromagnetic) and nonmagnetic (paramagnetic) metals. The two materials behave differently as will be discussed later.
4. As there is no contact between the coil (the inductor) and the job, the two can be located at a (reasonable) distance from each other. Hence, the process can be used in a vacuum or protective environment. A very wide coil-job spacing lowers the efficiency.
5. The process uses very high power densities (~ 0.01 to 2 kW/cm^2), therefore the heating is very fast (~ a few seconds). This makes the process suitable for high production.
6. The coil is to be designed and shaped to suit a given job. Induction heating is therefore a special production process and is not economical as a universal heating process (as in a furnace).
7. Quick heating virtually eliminates oxidation of the work-piece. There is also no distortion and oxidation.
8. There is excellent control on the power input and time, hence, the heating temperature and depth can be accurately controlled. Fast heating and good control makes the process suitable for automation by using proper fixtures.
9. As the heating (at least in the beginning) is confined to the job surface, this is an ideal process for case hardening of steel. The "case depth" can be controlled by controlling the power and time; and choosing the proper frequency.
10. The heating effect is located on the surface opposite the coil. This makes it possible to restrict heating only at a desired location. The coil has to be correctly shaped to shadow the desired location. This feature makes the process suitable for brazing.
11. There is no temperature limit. With proper design and adequate power, any desired temperature can be attained.

12. Once set by adjusting power and other parameters, the process can be operated by relatively unskilled labor.
13. Induction currents produce vigorous stirring in a molten metal. This gives a homogeneous melt.
14. Compared to other processes, induction heating requires higher capital investment. The overall efficiency is 30–65% depending on the job, generator, and coil design.
15. The process can be applied to long articles also. In such situations the job (or rarely the coil) is moved or rotated and treated successively.
16. By proper combination of electric and magnetic fields it is possible to "suspend" a small metallic mass and melt it without a container. This is called "levitation."

10.1.4 Skin Effect

The induced electromotive force (e.m.f.) in the work-piece or job sets up eddy currents in it. These currents are not uniform all over the cross section but tend to concentrate near the surface facing the coil. Thus, the current is highest at the surface and diminishes or attenuates progressively toward the interior. The heating, i.e., the I^2R loss, is highest at the surface and diminishes toward the interior. This selective concentration of current at the surface layer is called the "skin effect."

The current diminishes exponentially and can be expressed as shown in Figure 10.2,

$$I_x = I_o e^{-x/\delta} \tag{10.1}$$

where

I_x = Current at a depth x from the surface
I_o = Current at surface ($x = 0$)
δ = "Depth of penetration"

Thus at a distance $x = \delta$ the current is

$$I_\delta = I_o e^{-\delta/\delta} = \frac{I_o}{e} \tag{10.2}$$

$$I_\delta = 0.368\ I_o$$

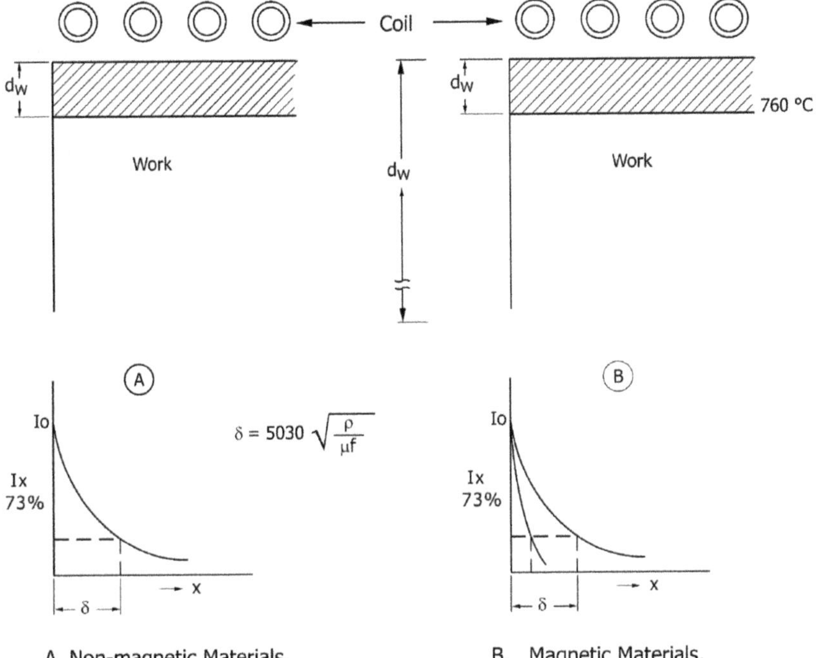

Figure 10.2 Skin effect and the depth of penetration.

The depth δ at which the current is 0.368 of surface current is defined as the "depth of penetration." It is given by

$$\left.\begin{aligned} \delta &= 5033\sqrt{\frac{\rho}{\mu_\gamma f}}\ \text{cm} \\ \text{or more conveniently} \qquad & \\ \delta &= 5000\sqrt{\frac{\rho}{\mu_\gamma f}}\ \text{cm} \end{aligned}\right\} \tag{10.3}$$

where

δ = Depth of penetration (cm)
ρ = Resistivity of object (ohm.cm)
μ_γ = Relative permeability of object
μ_γ = 1 for paramagnetic materials
μ_γ = 25–10,000 for ferromagnetic materials
f = Frequency cps or Hz

Figure 10.2 shows that about 86.5 % of the heat is created in a surface layer of thickness δ and more than 96% in a layer of thickness 3δ.

The above equation shows that the depth of penetration decreases with increasing frequency (if ρ and μ are constant). Qualitatively, this indicates that for deep or through heating, a low frequency is preferred while a high frequency should be used for surface or shallow heating. The statement is to be qualified by γ/δ ratio where γ is the diameter or half-width of the work.

Magnetic permeability is constant ($\cong$ 1.0) for non (para) magnetic materials. Hence, μ has no effect on the heating of materials such as copper, aluminum, or nonmagnetic steels.

For ferromagnetic materials (common carbon and low alloy steels) the permeability is high and decreases with temperature. At a temperature higher than Curie temperature (~ 760°C) the permeability vanishes. Thus the depth of penetration for ferromagnetic material will be very low initially and will increase as temperature increases. It will stabilize at $\tau = t$ curie.

Resistivity ρ and δ are directly proportional. Hence, copper or aluminum will have a larger δ than brass, steels, and other alloys.

The depth of penetration is thus a complicated function of ρ, μ, and f. For a given metal the only variable of choice is the frequency. The skin effect is the main difference of heat generation in resistance and induction heating. It plays a critical part is all aspects of design in induction heating.

EXAMPLE 10.1

Determine the depth of penetration for 0.4% carbon steel heated from 30–900°C at 1000 Hz. The resistivity and permeability at various temperatures is given below.

Temperature °C	Resistivity $\rho \times 10^{-6}$ ohm.cm	Permeability $\mu\gamma$
30	16.0	40
200	29.6	36
400	49.3	30
600	76.6	15
760	104.2	1.0
900	114.6	1.0

Solution

$$\delta = 5030\sqrt{\frac{\rho}{\mu_\gamma f}}\ \text{cm}$$

For 30°C

$$f = 1000\ \text{Hz},\ \rho = 16 \times 10^{-6}\ \text{ohm.cm},\ \mu_\gamma = 40$$

Substituting

$$\delta_{30} = 5030\sqrt{\frac{16 \times 10^{-6}}{40 \times 10^{3}}}$$

$$= 0.101\ \text{cm}$$

Similarly

$$\delta_{200} = 0.135\ \text{cm}$$

$$\delta_{400} = 0.174\ \text{cm}$$

$$\delta_{600} = 0.219\ \text{cm}$$

$$\delta_{760} = 1.62\ \text{cm}$$

$$\delta_{900} = 1.70\ \text{cm}$$

EXAMPLE 10.2

What will be the depth of penetration of a nonferrous alloy heated from 30–600°C at 10,000 Hz. The resistivity of the alloy is same as that of steel in the previous problem.

Solution

For the concerned alloy $\mu_\gamma = 1$

$$\delta = 5030\sqrt{\frac{\rho}{1\times 10^{4}}} = 50.30\times\sqrt{\rho}$$

For 30°C

$$\delta_{30} = 50.30 \times \sqrt{16 \times 10^{-6}}$$

$$= 0.2 \text{ cm}$$

Similarly

$$\delta_{200} = 0.276 \text{ cm}$$

$$\delta_{400} = 0.353 \text{ cm}$$

$$\delta_{600} = 0.440 \text{ cm}$$

It can be seen that the depth of penetration increases very slowly and can be treated as constant at some mean value.

In the last example, the depth of penetration, affected by changes in resistivity and permeability shows an abrupt change at the Curie point (760°C).

10.1.5 Ferrous and Nonferrous Heating

In induction heating, ferrous and nonferrous, i.e., ferromagnetic and paramagnetic materials behave differently.

The difference arises from their inherent (magnetic) properties. The depth of penetration δ, as discussed in the previous section, depends on the relative magnetic permeability.

$$\delta = 5030\sqrt{\frac{\rho}{\mu_{\gamma} f}} \tag{10.4}$$

At a constant frequency f the above equation can be written as

$$\delta = \frac{5030}{\sqrt{f}}\sqrt{\frac{\rho}{\mu}} \tag{10.5}$$

The resistivity ρ of all metals increases with temperature. In ferrous materials ρ is about 15×10^{-6} ohm.cm at 100°C and 117×10^{-6} at 1000°C. In nonferrous materials, (e.g., copper), it is 2.22×10^{-6} at 100°C and 8.2×10^{-6} at 950°C.

The relative permeability μ_γ of magnetic materials depends on the composition, magnetizing force H, and temperature. The practical representative values are 10^{-50} at 100°C and 2–5 at 750–800°C. At about 850°C and above the permeability is 1 and the ferromagnetic material becomes paramagnetic.

This shows that the depth of penetration will increase with temperature for both materials. However, the effect will be more pronounced in the ferrous material.

When undergoing the magnetic to nonmagnetic state at the Curie temperature, these materials will absorb a considerable energy for the change. This change will also take a noticeable time which should be added to the heating time.

The depth will suddenly increase two times or more when the temperature crosses 760°C (Curie temperature).

Nonferrous and nonmagnetic steels will show a steady increase in the depth of penetration with increase in temperature.

Magnetic steels (~ 0.4% C) are used for hardening. The case depth required is of the order of 1.0–3.0 mm. Heating is done to 850–1000°C. The above discussion shows that at constant current and frequency, the heating rate will be very fast up to 800°C. After that, it will slow down due to the sudden increase in the depth of penetration. If the case depth is to be controlled with some precision, the current and time will have to be tightly controlled. There is little control over frequency (as it is decided by the H.F. power supply).

Nonferrous materials are heated for through heating. Here the depth is larger in the beginning and increases steadily with the temperature. Hence, the frequencies required for surface and through heating are different.

10.1.6 Choice of Frequency

The choice of proper frequency is very crucial in designing a successful and efficient induction heating operation.

We have seen that the skin effect or the depth of a directly heated zone depends primarily on the frequency f and on the relative magnetic permeability μ_γ. For paramagnetic work only the frequency is important. For ferromagnetic work both f and μ_γ

play decisive roles because μ_γ decreases with temperature and spreads the skin deeper.

Frequencies are classified in three groups as shown in Table 10.1. High frequency generators are designed for a particular frequency or a group of frequencies in a given range. Only a few "standard" frequencies are commercially available, which are indicated in the table.

TABLE 10.1 Common Frequencies and Their Applications

Group and Range (Hz)	Typical commercially available frequencies (Hz)	Applications	Generator
Line or mains and their multiples 50–150 Hz	50, 100, 150 (60), (120), (180), where mains at 60 Hz	Melting and through heating of large size.	Mains frequency doubler or tripler
Medium 400–50 kHz	400, 1000, 3000 4000, 10,000 20,000	Surface heating for skin depth 0.01 to 0.4 cm object size 0.5 to > 10 cm. Through heating of smaller sizes. Welding and brazing from 0.5 to 1.0 cm thickness	Motor generator, spark gap generator, solid state generator.
High or radio frequency (RF) 100 kHz onward	100, 150, 200, 250 450, 600, kHz 2.5, 10, 25 MHz	Through and surface heating of small (< 0.1 cm diameter) articles, brazing, vacuum processing, semiconductor and crystal processing, tube welding	Vacuum tube oscillators or R.F generators

The size (diameter or thickness) of the work and the operation to be carried out (e.g., through heating, surface heating, etc.) will determine the frequency range to be chosen.

For through heating, it is desirable to have a large (0.05–0.5 cm) depth of penetration (δ) which can be achieved by relatively low frequencies in the range 1–10 kHz for work diameters 1–30 cm.

For surface heating applications, the penetration depth depends on the case depth desired. These are generally in the range 0.025–0.3 cm. Hence, it is necessary to choose higher frequencies than those for through heating. For work between 1.0 to 3.0 cm, frequencies in the range 1–50 kHz are useful.

For through heating or skin heating of small (0.1–0.5 cm) works, R.F or very high frequencies are used, as the depth of penetration is very small.

It is again pointed out that the frequency actually used in a given range is decided by the size of the job. More precisely it depends on the r/δ ratio obtained where r is the job radius (cm). For through heating r/δ should be small, while for surface heating it should be large.

Even in a given frequency range the choice may be restricted, as a generator in that range may be already available at works. The ratio r/δ in such cases may increase (or occasionally decrease) the heating time.

Process planning usually indicates the desired heating time. The work material properties, the required temperature, and the depth dictate the surface loading desired (kW/cm^2). The generator power available and the frequency will then finalize the heating time.

Another way of estimating the required frequency for through heating is the concept of "critical frequency."

The size, thermal properties, the required temperature, and the desired time decide the thermal loading (kW/cm^2) required. The skin depth is decided by the frequency (and permeability for magnetic work). If the chosen frequency is very high, the skin will be very thin. A lot of power will be concentrated on the skin in a very short time (a few seconds). As the dissipation of power to the interior is slow (being limited by thermal diffusivity α), this power will overheat and melt the surface. Hence, there will be an upper limit for frequency.

The coil efficiency compares the useful power absorbed by the work in the form I^2R_w, to the total power loss in the coil (I^2R_c) and the work (I^2R_w), i.e.,

$$\text{Coil } \eta = \frac{R_w}{R_c + R_w} \tag{10.6}$$

Where R_w and R_c is the work and coil resistance (ohm), respectively. It has been shown theoretically that for a given work diameter D there is a minimum frequency below which the term I^2R_c, i.e., the coil loss increases and the efficiency decreases. Thus, there is a lower limit to the frequency.

The critical frequency f_c should therefore be between these two limits, i.e.,

$$3\times 10^6 \frac{\rho}{\mu_\gamma D_w{}^2} < f_c < 6\times 10^6 \frac{\rho_w}{\mu_\gamma D_w{}^2} \tag{10.7}$$

where

ρ_w = Resistivity of work (ohm)
μ_γ = Relative permeability of work
D_w = Work diameter (cm)

Within these limits any available frequency can be chosen. A little higher frequency will result in a somewhat reduced heating time.

EXAMPLE 10.3

Determine the frequency range for through heating aluminum bar 0.02 m diameter to 500°C. The average resistivity of aluminum is 5.83×10^{-8} ohm.m. What will be the frequency range if the bar diameter is 0.01 and 0.1 m?

Solution

According to Equation (10.7) the frequency range will be

$$3\times 10^6 \frac{\rho_w}{\mu_\gamma D_w{}^2} < f_c < 6\times 10^6 \frac{\rho_w}{\mu_\gamma D_w{}^2}$$

For aluminum $\mu_\gamma = 1.0$

$$\frac{3\times10^6\times5.83\times10^{-8}}{0.02^2} < f_c < \frac{6\times10^6\times5.83\times10^{-8}}{0.02^2}$$

$$400 < f_c < 800$$

Hence, the frequency should be in the range of 400–800 Hz.

Repeating the same calculations for $d = 0.01$ and 0.10 m diameters

for 0.01 m diameter $1700 < f_c < 3400 \simeq 2$ kHz

for 0.10 m diameter $17 < f_c < 34 \simeq 50$ Hz

Note: *The recommended frequencies are much more (~450 kHz) than the above calculated values. Perhaps this is because the concept does not involve the power required to get the temperature in desired time. The maximum frequencies should be multiplied by a factor 3–5.*

10.1.7 High Frequency Generators

Generation of high power (2–2000 kW), high frequency current is a highly specialized field restricted to electric and electronic engineering. It is not possible to discuss their design in this book. Available generators, their principle, power, and frequency limitations are briefly reviewed. It is comparatively rare to design and build a generator for a given job. Manufacturers usually offer standard generators from which a suitable generator is chosen. Frequently, a generator is already available to the user. By compromising on frequency, and power, and controlling the cycle time, an available generator can be successfully matched to the job. All generators have a capacitor bank to correct the load power factor which is very low (~ 0.2–0.6).

10.1.8 Mains Frequency Generators

As the name suggests, these generators operate at 50 Hz (mains frequency) or at its multiples, such as 150 Hz (static frequency generators). Mains frequency generators are available from a few to about 1000 kW with a single or three phase output. They are mainly used for through heating of 50–150 mm billets as the depth of penetration is high at low frequencies. These generators are also used for large scale (foundry) melting in core-type furnaces. As there is no special machinery or transformers, these generators are comparatively cheaper.

10.1.9 Spark Gap Generators

In these generators a high frequency current is generated by sparking between two electrodes. A mercury pool in a vessel is one of the electrodes. The other electrode is a carbon or tungsten rod over the mercury. There is an adjustable gap between the mercury surface and the lower end of the rod. Hydrogen is circulated above the pool to prevent the oxidation of mercury. A typical generator is shown schematically in Figure 10.3. There may be more than one gap depending on the power requirements. The gap vessels have a water-cooling jacket.

Spark gap generators do not produce a fixed frequency. The output is in the form of a series of damped waves in each mains cycle. There is a problem of controlling the gap and the electrode wear.

Spark gap generators have a simple construction. They are available from 2–20 kW range and produce a frequency 150–600 kHz. They are much cheaper than vacuum tube generators.

10.1.10 Motor Generators

As the name implies, these generators consist of two main parts. One part is an AC induction or synchronous motor running on mains frequency. This drives a high frequency generator. The construction is generally vertical with the drive motor at the bottom and the generator at the top. The generator has a DC exciter generator. The output windings are on the stator. The rotor consists of a stack of electrical steel stampings with slots. The output frequency is given by the product of rotor slots and speed. Extensive cooling arrangements are required.

These are fixed frequency machines with frequencies from 1–10 kHz and power from 20–400 kW. Output voltages are

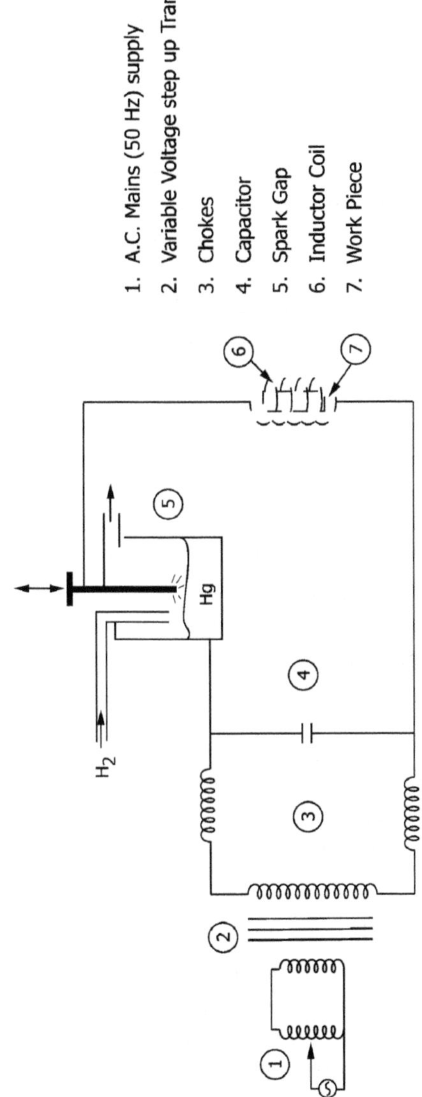

Figure 10.3 Spark gap generator.

dual, i.e., 400/800, 600/1200 V, etc. These are very sturdy machines generally requiring no elaborate maintenance. The generator and load are matched by a step-down transformer and tank circuit. The generator efficiency is about 90% at full load.

Previously, motor generators were the only commercial source for medium frequency operations. They are now superceded by solid state generators.

10.1.11 Solid State Generators

These generators are relatively new to the family of HF power sources. The development of switching rectifiers (SCR, thyristors) that can handle large currents ($\sim 10^3$–10^4 amp) has made it possible to build large HF power sources that are required for commercial purposes. Typical switching times of thyristors are a few microseconds. They are called "solid state generators" as they do not have any rotating parts or vacuum tubes.

The main constructional features of these power supplies are shown in Figure 10.4. The power is derived from 50 Hz, 3 phase AC mains and rectified by a suitable rectifying circuit (full wave, bridge, etc.). The rectifier may contain thyristors to obtain a variable voltage DC.

The DC supply passes through an inverter where it is switched at a high frequency producing a square wave switched DC (i.e., nonsinusoidal AC). There are three main types of inverters. They are:

1. Half wave series load inverter
2. Reactor inverter
3. Full bridge inverter

The circuit details of these inverters are not discussed here. The square wave can be shaped to resemble a sine wave. It is then fed to a tank circuit which acts as a buffer between the inverter and the coil. The tank matches the load and generator impedances, corrects the power factor, and produces a resonant condition to derive maximum power. There is usually an output transformer to lower the coil voltage. The load (coil) is connected to the tank output. There may be an extra correction capacitor in parallel with the coil to suit individual load demands.

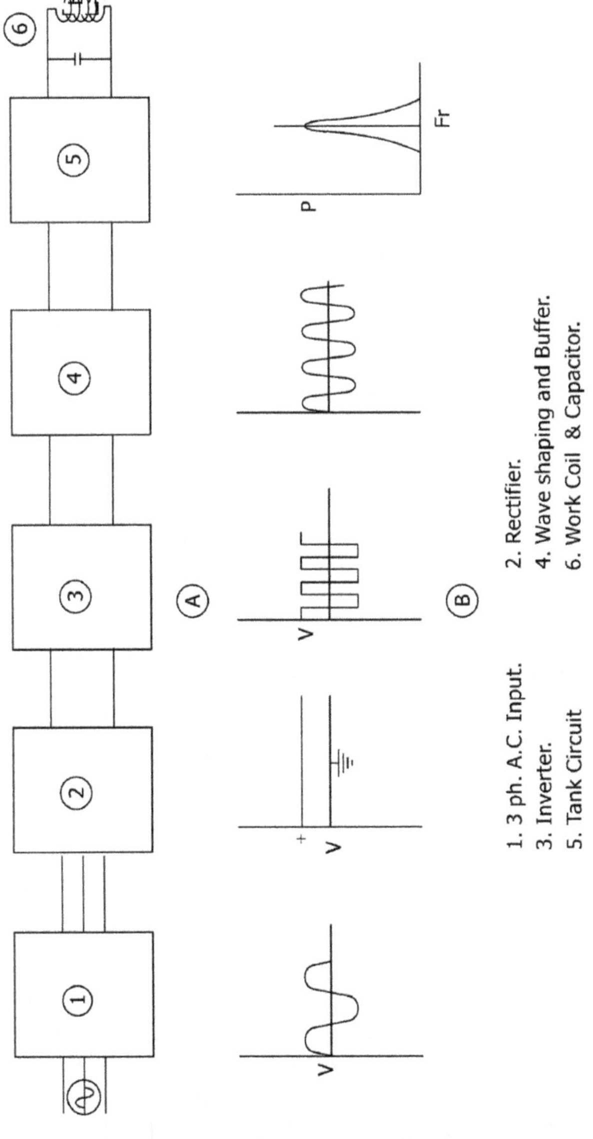

Figure 10.4 Solid state H.F. generator.

There are many protective and control facilities not shown in the figure. The output frequency is adjusted (often automatically) by adjusting the thyristor switching frequency.

In recent years, these power supplies have been fast replacing motor generators and spark gap units.

10.1.12 Some Features of Solid State Generators

1. These generators are available in the low frequency range (1–40 kHz) and medium frequency range (80–200 kHz). Higher frequencies are being offered with continued developments in thyristors and circuitry.
2. Power capacities are commonly available in the range of 10–250 kW, but units up to 2 MW are also being developed.
3. The typical generator efficiency is 0.8–0.9.
4. They have no moving parts or delicate vacuum tubes, resulting in reduced maintenance and noise problems.
5. The frequency and output voltage are adjustable, hence establishment of coil coupling, resonance, and a good power factor is easy and often automatic by taking feedback from the coil current.

10.1.13 Radio Frequency (RF) Power Generators

These generators are used in the range 20 kHz to several MHz. They are based on vacuum tube (triode) oscillators. The schematic construction of a typical RF generator is shown in Figure 10.5.

The input is a three-phase AC at 50 Hz. This passes through a switch gear and is converted to a higher voltage (7.5 to 20 kV) by a suitable HV transformer. It is then rectified to HV DC and fed to the oscillator. The main component is a vacuum tube which is a specially designed triode that can handle large power outputs. Considerable heat is produced in the tube and there are elaborate water-cooling arrangements. Oscillator circuits are of many designs. The main difference in the various designs is the method of grid control. The grid control voltage is derived from the tank circuit.

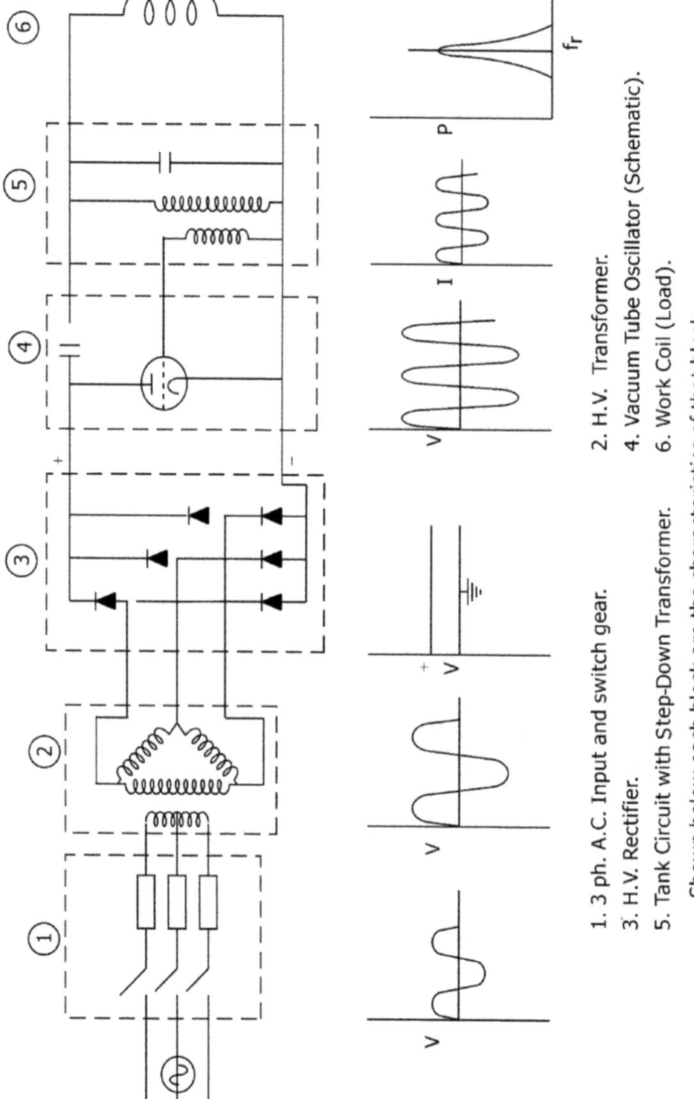

1. 3 ph. A.C. Input and switch gear.
2. H.V. Transformer.
3. H.V. Rectifier.
4. Vacuum Tube Oscillator (Schematic).
5. Tank Circuit with Step-Down Transformer.
6. Work Coil (Load).

Shown below each block are the characteristics of that block.

Figure 10.5 Vacuum Tube R.F. Generator.

The oscillator output is fed to the tank circuit, which contains a capacitor and a series inductor. The load matching is achieved by varying the frequency by grid control.

Commercial RF generators are available in 1–200 kW power capacities.

10.1.14 Features of RF Generators and Heating

1. Maximum power delivery, i.e., resonance, is established by an automatically varying frequency.
2. The heat generation in work is confined to a very thin surface layer, e.g., at 450 kHz and 800°C, the depth of penetration is about 8×10^{-6} cm. Hence, these generators are useful for through heating of very small (few mm diameter) works and for case hardening of shallow case depths.
3. The power density obtained is very high, e.g., for 1 mm diameter bar 2.5–15.0 kW/cm^2 can be developed, hence, heating time is very short.
4. RF gives better coupling at large air gaps. It is therefore suitable for semiconductor processing such as zone refining, doping, and single crystal growing.
5. Good coupling at large gaps and a high power capacity with a single or few coil turns makes RF generators suitable for welding and brazing.
6. RF generators are the costliest. Their maintenance is a costly and skilled job.
7. Extensive water cooling of tubes is necessary. The cooling has to continue for some time after the job is over.
8. Generator efficiency is 50–75% while the coil efficiency is 70–80%.
9. As their frequencies are near broadcast bands, RF generators produce electromagnetic interference and may require extensive shielding.

10.1.15 Generator and Coil Matching

Induction heating loads are highly reactive, i.e., their inductance is very high and have a lagging power factor at $\cos\phi = 0.1$–0.2. Because of its highly inductive nature, the coil is not able to draw maximum power from the generator. Hence, from the power utilization point of view, induction heating is very inefficient.

A simplified induction heating system is shown in Figure 10.6(A). The generator has a rated voltage V_G and an internal impedance Z_G. Thus, the voltage available at its terminals is E_G. The supply line from the generator to coil has an impedance Z_L and the voltage available at the coil is E_C. The coil draws a current I_C. If the supply line is very short, Z_L can be neglected. (This is why the generator is always located as close to the coil as possible.)

Electrically, maximum power is drawn when $Z_G = Z_C$, i.e., the generator and coil impedances are equal. There is generally no control over the generator impedance and Z_G is constant. The coil impedance consists of the active component R_C and the reactive component X_C (which is highly inductive). Thus, only X_C is adjustable by adding more inductance and capacitance to the coil. Usually, these additional components are located in a separate compartment known as the "tank circuit" as shown in Figure 10.6(B). As both the inductance and capacitance do not draw any real power, their addition does not change the coil current. The lagging power factor is corrected by the leading power factor of the capacitor.

Maximum power (for a given coil) is derived from the generator when the reactive power in Z_G and Z_C (along with tank impedance Z_T) is zero or very small. This condition is called resonance. It is established when the frequency, inductance, and capacitance in the circuit satisfy the condition.

$$f = \frac{1}{2\pi\sqrt{LC}} \tag{10.8}$$

At resonance, the load becomes purely resistive and the power factor also becomes unity. The frequency under these conditions is called the "resonant frequency f_r."

In HF generators such as motor generators or line frequency generators, the frequency is fixed and resonance is obtained by adjusting L and C in the tank circuit.

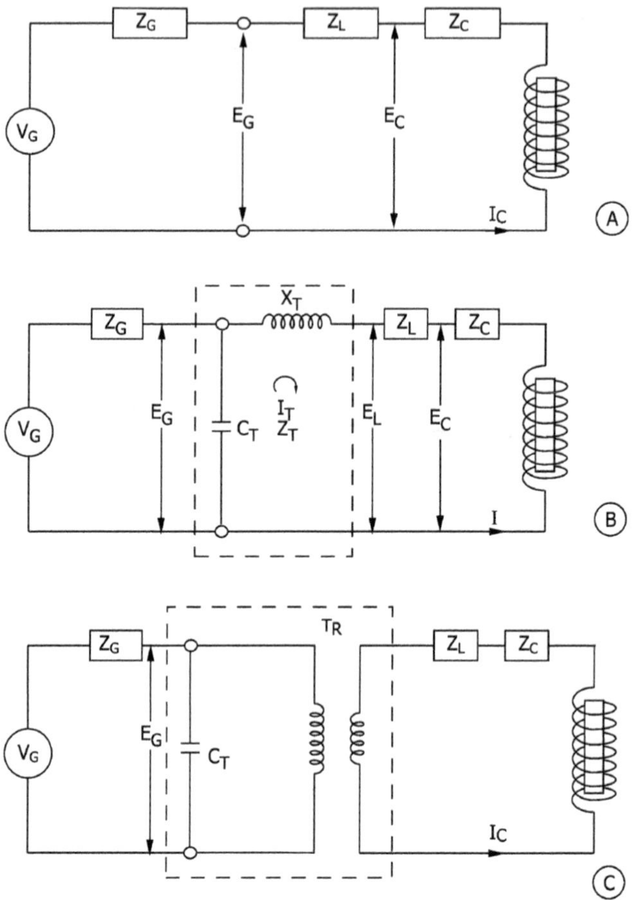

Figure 10.6 The tank circuit and load matching device.

In solid state (inverter) generators and RF generators the frequency is variable and resonance is obtained via frequency adjustment. The tank circuit contains fixed C and L components. In many instances an extra capacitor is connected near the coil to suit individual requirements.

Figure 10.6(B) shows tank circuit components in parallel with the coil. However, resonance is also obtained in series circuits as used for RF generators.

Besides adjustments in the impedance and correction of the power factor, a third possible adjustment is that of the coil voltage EC. Most high frequency generators have a fixed voltage.

The coil volts per turn, volt amperes, etc., determine the power requirements, and work size. For a given work, a certain number of turns are dictated by the available generator voltage. It is not physically possible to accommodate this number of turns and keep the same ampere turns by increasing the current. This can be achieved by using an HF step-down transformer as shown in Figure 10.6(C). Inversely, it is also possible that the calculated or required turns are few and the generator voltage is high, necessitating the use of a step-down transformer.

These transformers are water-cooled and are of the air or magnetic core type. They have a multiple turn primary with taps and a single turn secondary. The primary voltage is 380–400 V. The output voltages are 12, 15, 20, 25, 30, 40 V at 12,000 amp.

10.1.16 Thermal Requirements

The electrical energy supplied to the coil is expended in the following three components:

1. Heating the work to the required temperature and to the required depth, in the required time (P_1 watts).
2. Loss of heat from the work due to radiation, convection, and conduction (P_2 watts). These two types of heat form the "thermal requirement" of the work (P_T watts). There is a third component of heat.
3. The $I_C R_C$ or the "copper loss" in the coil, which is an unavoidable loss.

In this section we will present the procedure for establishing the "thermal requirement" of a given work. For simplicity, we will assume a cylindrical metal bar (billet) as the work to be through-heated to a temperature t_2.

Let

t_1 = Surrounding temperature (27°C)
C = Specific heat of work (kJ/kg°C)
ρ = Density (kg/m^3)

M = Weight of the work (kg)
d_W = Diameter (m)
ℓ_W = Length (m)
A = Surface area (m^2)
ϵ = Emissivity of the work
K = Stefen Boltzman constant = 5.67×10^{-8} w/m^2

The heat absorbed by the work in heating from t_1 to t_2 °C will be

$$\begin{aligned}\Delta H &= cM(t_2 - t_1)\ \text{J} \\ &= cM\Delta t\end{aligned} \tag{10.9}$$

The heat radiated by the surface

$$P = K\epsilon\, A\, [T_2 - T_1]\ \text{W} \tag{10.10}$$

where T_1 and T_2 are absolute temperature.

If heating takes place in time $\Delta\tau$, the total power for heating including radiation loss is

$$P_T = cM\frac{\Delta t}{\Delta\tau} + K\epsilon\ A\left[T_2^{\,4} - T_1^{\,4}\right]\ \text{W} \tag{10.11}$$

When the required temperature (t_2) is achieved all the power supplied will raise the temperature to an ultimate value T_u(K) and the extra power will be radiated. Assuming the power supplied remains constant

$$P_T = K\epsilon\ A\left[T_u^{\,4} - T_1^{\,4}\right]\ \text{W} \tag{10.12}$$

$$T_u = \sqrt[4]{\frac{P}{K\epsilon A} + T_1^{\,4}} \tag{10.13}$$

Substituting in Equation (10.11)

$$\frac{\Delta t}{\Delta\tau} = \frac{K\epsilon\ A}{cM}\left[T_1^{\,4} - T_2^{\,4}\right]$$

Integrating this between T_1 and T_2 gives

$$\tau = \frac{cM}{2K\epsilon\ AT_u^{\ 3}} \begin{bmatrix} \tan^{-1}\frac{T_2}{T_u} - \tan^{-1}\frac{T_1}{T_u} + \frac{1}{2} \\ \times \ln\frac{(T_u + T_2)(T_u + T_1)}{(T_u - T_2)(T_u + T_1)} \end{bmatrix} \tag{10.14}$$

This is the time required for heating from t_1 to t_2 °C with a given power input P_T.

Equation 10.9, Equation 10.10, Equation 10.11, and Equation 10.12 are useful in calculating the thermal requirements or power at its minimum value. The power so calculated will be that required for equilibrium or slow rate. The time calculated according to this power input will be much more than that required by the process. It is also assumed that the loss is only by the radiation, which may not be true.

It will therefore be necessary to increase the power input. If we plot t and T and τ for several multiples of P_T the results are shown in Figure 10.7.

The figure shows that the time of heating (τ), decreases with increasing power. After a certain power the rate of time t decreases and there is not much improvement. Hence, the calculated power P_T should be increased by a multiple of 2–6 depending on similar calculations or judgment.

The heating time for surface hardening of magnetic steel is determined in a similar way. The modification required is in estimating the skin depth and the heat loss from the skin to the interior, as well as radiation loss.

10.1.17 Design of the Coil

There are two aspects of coil design for induction heating.

1. Calculation of the number of coil turns, the coil current (primary current), etc., i.e., the electrical design of the coil.
2. The size and shape of the coil, cooling requirements, etc., which is included in the physical design of the coil.

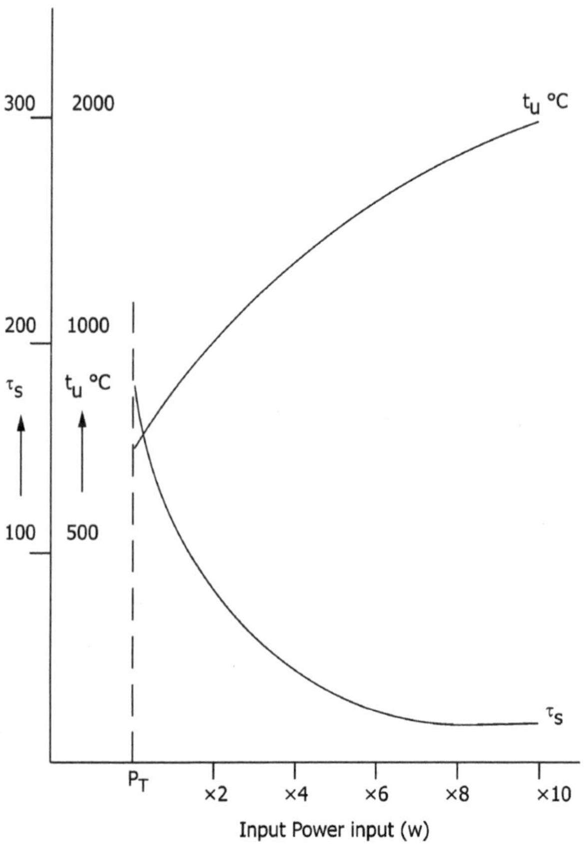

Figure 10.7 Heating time (τ_s) and ultimate temperature (t_u°C) vs. power input P_t.

10.1.18 Electrical Design of Coil

This design can be carried out by a number of theoretical and approximate or semitheoretical methods. Only two methods are discussed here for the design of a straight helical coil.

On passing a high frequency current through the coil, a magnetic field is created inside. Let this field have a strength H (amp turns/m). This field will induce eddy currents in the work as shown in Figure 10.8. The eddy currents will produce their own field, which will (vectorially) reinforce the original

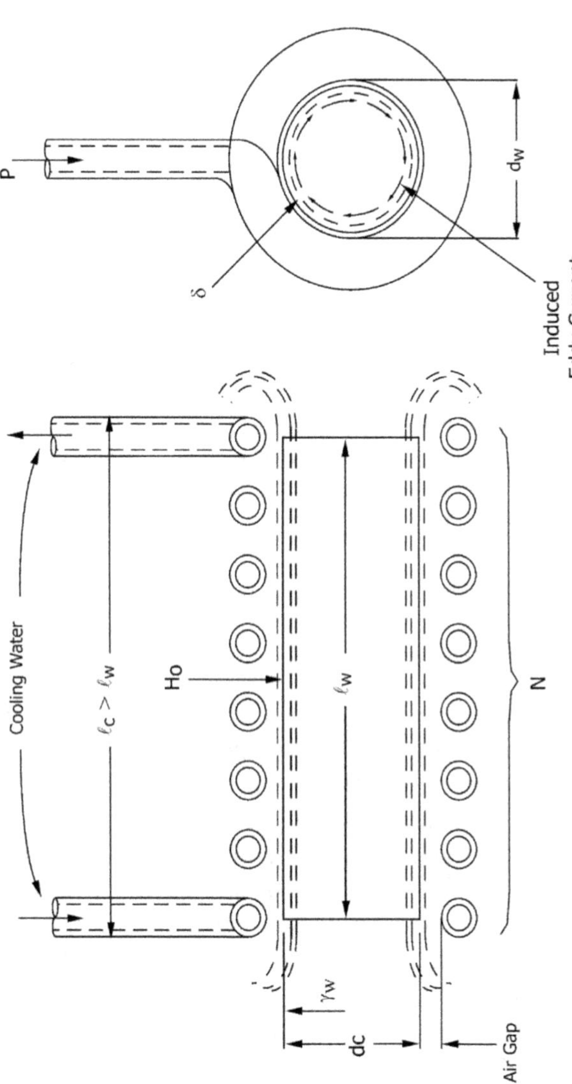

Figure 10.8 Work and coil parameters.

field H. Assume that this secondary field also has a strength H so that the total strength of the field outside the surface of the work is $2H = H_o$.

The power developed per meter length in the work due to the field is given by

$$P = 8\pi H_o \rho \times \frac{\gamma_w}{\delta_w} \times K \text{ w/m} \tag{10.15}$$

where

ρ = Resistivity (ohm.cm) of the work
γ_w = Radius of work (cm)
δ_w = Depth of penetration
K = Factor (0–1.0) depending on γ/δ_w

For $\gamma_w/\delta_w > 10$ $K = 1.0$.

The work power P(w) per meter of coil length is calculated as discussed in the last section. Thus, H_o can be calculated from the above equation. It is again pointed out that the energy conversion takes place as follows:

Electrical H.F power to coil	→	Variable magnetic field in work	→	Induced eddy current in work	→	P(Heat)

Considering the cooling water flow through the coil, an arbitrary choice is made about the size of the tube for the coil. Depending on the work length, frequently chosen tube internal diameters are 6–12 mm and the wall thickness is about $2 \times \delta_w$. If ℓ_w is the work length, the coil length (ℓ_c) chosen is slightly longer (10–20 mm). The spacing of the turns in length ℓ_w is adjusted to occupy maximum turns without the coils touching each other. This means that the tube occupies 70–80% of the coil length (see Figure 10.8).

The coil's internal diameter is taken as $d_c = d_w$ + air gap where d_c and d_w are the coil and work diameters, respectively.

For a coil to work better in a coupling, the air gap should be the minimum possible, consistent with the requirement of work centering and coil-work short circuiting. Normal air gaps are 25–50 mm depending on the work size and the frequency.

This decides (though arbitrarily) the number of turns that can be accommodated in a given coil length ℓ_w and, hence, turns/m can be calculated. If N is the number of turns

$$\frac{N}{\ell_w} = \text{Turns per meter}$$

Similarly the coil external diameter d_w and the tube length and the wall thickness is also known and the coil resistance R_c can be calculated.

As H_o, i.e., the ampere turns per meter of coil, are known, the coil current (I_c) can be calculated from

$$\left.\begin{aligned} I_c N &= H_o \times \ell \\ I_c &= \frac{H_o \ell}{N} \end{aligned}\right\} \tag{10.16}$$

Calculations of $I_c^2 R_c$ will give the copper loss (W).

The total power will then be

$$P = \text{Thermal power} + \text{copper loss}$$

Similarly, cooling-water requirements can be calculated to remove heat from copper loss.

The coil voltage is fixed by the available generator voltage and the transformer.

The method of calculation will be clear in the following solved problem.

10.1.19 Equivalent Circuit Method of Coil Design

At the beginning of this section, it was stated that induction heating works on the transformer principle. This concept is developed further and a design theory for coils and other electrical parameters is put forward by Baker et al. and presented in a compact scheme by Davis (See the *Bibliography* for Chapter 10).

The evolution of the circuit is shown in Figure 10.9(A), Figure 10.9(B). The final equivalent circuit referred to the primary is shown in Figure 10.9(C). The method is outlined briefly here to understand the approach. First, the required suitable frequency f is chosen. Expressions are then developed

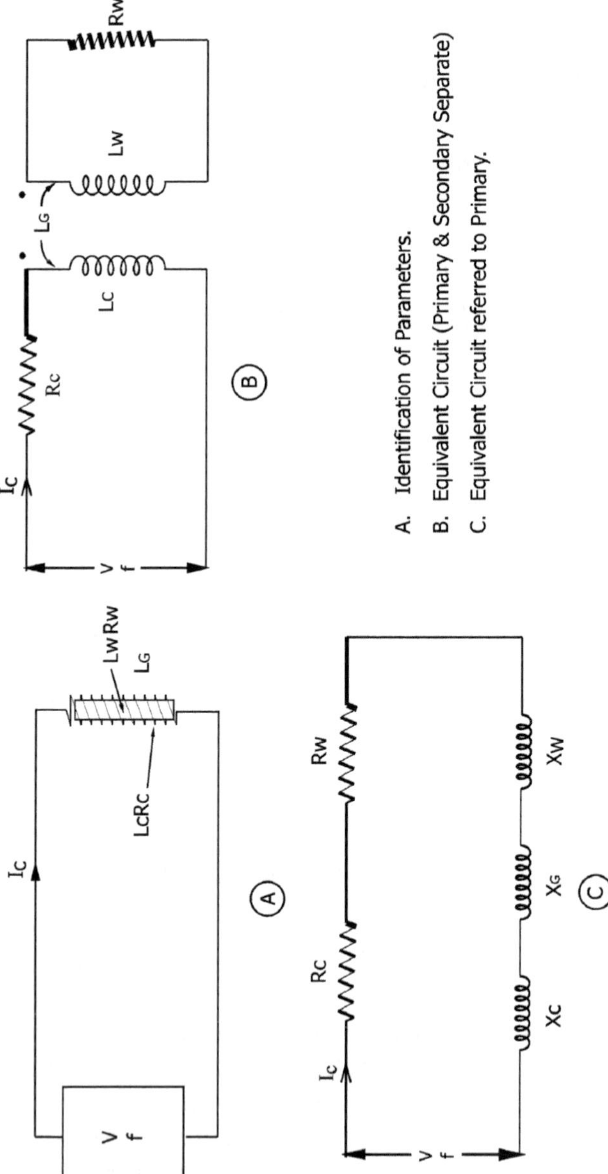

Figure 10.9 Development of equivalent circuit for induction heating.

for the five parameters, namely, coil resistance R_c, coil inductance X_c, gap reactance X_G, work resistance R_w, and work inductance X_w (see Figure 10.9(C)). The expressions involve their physical properties, a common factor K_1, and known dimensions. The only unknown quantity is the number of coil turns N. This quantity (N) is included in the common factor K_1 as N^2.

The thermal power P is calculated separately as outlined in the previous section. Similarly, the depth of penetration δ and the ratios d_w/δ_w are calculated. The following expressions are determined successively:

$$\text{Coil efficiency } \eta = \frac{R_w}{R_c + R_w} \tag{10.17}$$

$$\text{Impedance } Z = \sqrt{((R_c + R_w)^2 + (X_c + X_G + X_w)^2)} \tag{10.18}$$

$$\text{Coil power factor } \cos\phi = \frac{R_w + R_c}{Z} \tag{10.19}$$

$$\text{Coil volt ampere } (VA) = \frac{P}{\eta \times \cos\phi} \tag{10.20}$$

The next step involves calculating the factor Z/N^2. The factor K is of the form $K = AN^2$ where A is a numerical multiplier already calculated. Similarly Z is also known. Hence,

$$\frac{Z}{N^2} = \frac{Z}{K/A} \tag{10.21}$$

$$\text{Coil volts/turn } (V/N) = \sqrt{Coil\ VA \times \frac{Z}{N^2}} \tag{10.22}$$

$$\text{Coil ampere turns } (AT) = I_c N$$

$$= \sqrt{\frac{Coil\ VA}{Z/N^2}} \quad (\text{as } \frac{V}{Z} = I_c) \tag{10.23}$$

A suitable coil voltage is chosen. This will depend on the available generator voltage and available transformer tapping. Hence, V is known.

$$N = \frac{V}{V/N} \tag{10.24}$$

The accommodation of this number of turns in the practical coil length is checked and a suitable size of copper tubing is chosen. The wall thickness is about 2δ or a little less.

$$\text{Coil Current } I_c = \frac{AT}{N} \tag{10.25}$$

Once the tube size and length are known, the copper loss can be calculated from I_c and R_c from Equation 10.26.

$$W_c = I_c^2 \times R_c \text{ W} \tag{10.26}$$

To this we add the heat radiated from the work and absorbed by the coil (W_w).

The total heat to be removed by water circulation is $W_c + W_w$. The tube bore is known and the cooling-water requirement can be calculated.

This is an elegant method leading to calculations of many other parameters besides the number of turns.

We will adopt this method for calculations in the more elaborate designs worked out in second volume.

10.1.20 Physical Design of Coils

The shape of the coil depends on the geometry of the work. For simple cylindrical works to be surface heated or through heated, the coil has a simple helical shape as shown in Figure 10.1. The coil surrounds the work cylinder over the whole surface to be heated. If the work is long and coil power available is limited, a small coil is used and (usually) the work is moved through it. This is called "scanning heating." The coil may have a single turn or many turns.

There is an air gap between the coil and the work. The gap serves as the coupling medium. If the gap is too wide, the coupling is weak, resulting in inefficient heating. If it is too narrow, it will get heated by the heat radiated by the work. A narrow gap will make work centering very critical and there is a possibility of coil and work short-circuit. Thus the width of the gap, defined as coil diameter minus the work diameter, has to be correctly chosen. The gap is generally 25–60 mm. Thus

$$d_c - d_w \cong 25 - 50 \text{ mm} \tag{10.27}$$

The pattern of heat generaton in the work is governed by the "proximity" effect. It states that the induced current (and therefore the heat) in the work is concentrated in the coil shadow on the work. The shape of the shadow depends on the size and shape of the coil and the air gap as shown in Figure 10.10. Thus a round coil will produce a sharp shadow at a small gap. The shadow will broaden as the gap increases. A square coil will produce a flat shadow.

The proximity effect provides a clue about the design of coil shape for various work shapes. By proper shaping, the heat can be concentrated where desired. Some examples of coil shapes are shown in Figure 10.11.

For brazing applications, heat is concentrated a little away from the joint by avoiding the coil turns facing the joint.

If the heat is to be concentrated at the center of the work, the coil pitch is small at the center and spreads toward the ends. Similarly, the coil is made slightly convex (from the inside).

By making coil pitch smaller at the edges and wide at the center the heat can be concentrated at the edges. This effect can be enhanced by making the coil slightly concave from the inside.

A work having sharp corners (e.g., a square bar) shows heat concentration at the corners. This can be avoided by giving the coil ample radius at the corners.

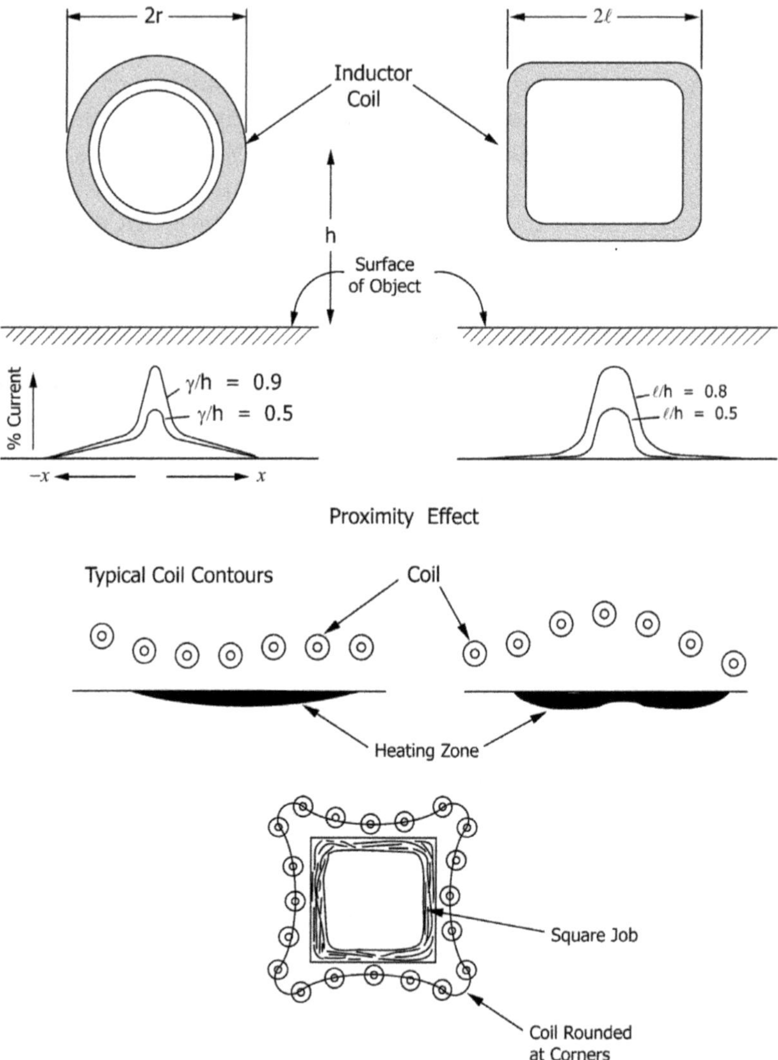

Figure 10.10 Proximity effect and typical coil contour.

In order to assure uniform heating even with a few variations in the work location and edge effects, the coil is usually a little longer than the work, i.e., $\ell_c > \ell_w$ by about 10–25 mm.

In many coils used for surface hardening the quenchant (usually water) is supplied simultaneously so that the quenching

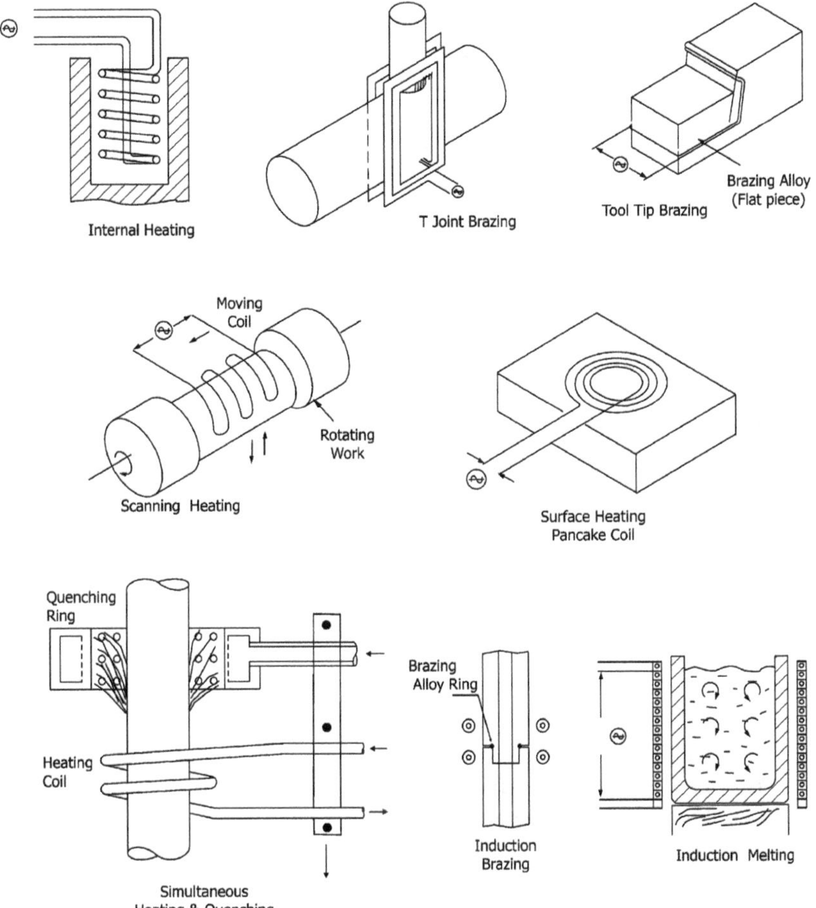

Figure 10.11 Proximity effect and typical coil contour.

occurs immediately after heating. In scanning heating the quenching ring and the heating ring move together.

Coil shape design and production of coils is a specialized technique. Only a few examples are cited above.

Coils are made from round or square copper pipes. This keeps the coil resistance low. Cooling water is circulated through the pipe to remove the heat due to the copper loss. The wall thickness of the tube is equal to or a little larger than the depth of penetration δ.

According to the "left hand rule" a current carrying conductor experiences a force F, which is at a right angle to the current I and magnetic field H. The induction coil and the work experience this force in both radial and longitudinal directions. The longitudinal force tends to push the work out of the coil. The radial force tends to make the gap uneven with a risk of a short-circuit. It is necessary to firmly locate and support both the work and the coil. The force is more noticeable on pancake coils for heating flat surfaces.

The flow requirement of cooling water is established from the coil's copper losses and radiation losses. The cooling takes place by forced convection. This mode of heat transfer is covered in detail in Chapter 3. The pressure drop in water due to flow through the coil is determined by a relation of the type.

$$\Delta P = 0.01613 f \frac{L \rho Q^2}{d^x} \tag{10.28}$$

where

ΔP = Pressure drop (m)
f = Friction factor (tabulated)
L = Length of the tube (m)
ρ = Density of water (kg/m^3)
d = Pipe bore (mm)
x = Exponent (4.87–5.0)

The coil contains many bends. Equivalent lengths for all bends are to be considered when calculating L. The above equation indicates that the tube bore should be as large as space and the number of turns permit. This equation will lead to the pump estimation. The circulating water should be clean and filtered.

EXAMPLE 10.4

A length of 8 cm at one end of a 0.3% carbon steel rod is to be heated for upsetting. The bar diameter is 1.0 cm and total length 25.0 cm. The temperature desired is 1000°C. Considering the demand, it is decided that induction heating will be used for the job.

Now determine:

1. The thermal power required
2. Suitable frequency
3. Approximate heating time
4. Coil turns
5. Coil current
6. Pipe size for coil
7. Copper loss
8. Total power required and the recommended generator

Data

1. Specific heat of steel (c) = 0.56 kJ/kg°C
2. Density of steel = 7.85 g/cm^3
3. Resistivity of steel (average) = 66.4×10^{-6} ohm.cm
4. Emissivity of steel = 0.8
5. Relative permeability of steel (30 – 760°C) = 15
6. Resistivity of copper = 2.2×10^{-6} ohm.cm
7. Ambient temperature = 27°C

Solution

Volume of steel to be heated

$$= \frac{\pi \times 1^2}{4} \times 6.3 \text{ cm}^3$$

Weight of heated portion

$$= 6.3 \times 7.85 = 50 \text{ gm } (0.05 \text{ kg})$$

Area of the radiating surface

$$= \text{Area of the curved surface} + \text{Area of one end}$$

$$= \pi \times 1 \times 8 + \frac{\pi \times 1^2}{4}$$

$$= 26 \text{ cm}^2$$

Heat required for heating from 27 to 1000°C

$$= \text{weight} \times \text{sp} \cdot \text{heat} \times \text{ temp difference}$$

$$= 50 \times 0.56(1000 - 27)$$

$$= 27.2\,\text{kJ}$$

Heat radiated at 1000°C

$$= 5.67 \times 10^{-8} \times 26 \times 10^{-4} \times 0.8(1273^4 - 300^4)$$

$$\simeq 300 \text{ w}$$

For quick heating we consider a power input

$P = 4 \times 300 = 1200$ w

The work is to be heated in a magnetic condition up to 760°C and in a nonmagnetic condition from 760–1000°C. Relative permeability $\mu r = 0.15$. The work diameter is small and frequency 10 kHz is suitable for heating in the whole temperature range.

Average depth of penetration up to 760°C

$$\delta_{760} = 5030\sqrt{\frac{66.4 \times 10^{-6}}{15 \times 10^4}}$$

$$= 0.11 \text{ cm (1.1 mm)}$$

Depth of penetration from 760–1000°C

$$\delta_{1000} = 5030\sqrt{\frac{66.4 \times 10^{-6}}{1 \times 10^4}}$$

$$= 0.41 \text{ cm (4.1 mm)}$$

$$= \frac{\gamma}{\delta_{760}} = 4.5\ F = 1.0$$

$$= \frac{\gamma}{\delta_{1000}} = 1.22\ F = 0.3$$

For heating up to 760°C

$$Tu_{760} = \sqrt[4]{\frac{P}{K\epsilon A} + T_1}$$

$$= \sqrt[4]{\frac{1.2 \times 10^3}{5.67 \times 10^{-8} \times 0.8 \times 26 \times 10^4} + 300^4}$$

$$= 1860 \text{ K } (1587°\text{C})$$

$$\frac{cM}{K\epsilon ATu^3} = \frac{0.56 \times 10^3 \times 0.05}{5.67 \times 10^{-8} \times 0.8 \times 26 \times 10^4 \times 6.43 \times 10^9} = 37$$

$$\tau_{760} = 37 \begin{bmatrix} \tan^{-1}\frac{1033}{1860} - \tan^{-1}\frac{300}{1860} + \frac{2.3}{2} \\ \times \log\frac{(1033 + 1860)(1860 - 300)}{(1860 - 1033)(1860 + 300)} \end{bmatrix}$$

$$= 37(0.507 - 0.16 + 0.630) = 34 \text{ sec}$$

For heating from 760 to 1000°C. It is assumed that power input is kept constant at 1.2 kW.

$$\tau_{1000} = 37 \begin{bmatrix} \tan^{-1}\frac{1273}{1860} - \tan^{-1}\frac{1033}{1860} + 1.5 \\ \times \log\frac{(1273 + 1860)(1860 - 1033)}{(1860 - 1273)(1860 + 1033)} \end{bmatrix}$$

$$= 37(0.600 - 0.506 + 0.192) = 11 \text{ sec}$$

Some time will be taken by the magnetic transformation at 760°C. As these calculations are beyond the scope, the time is arbitrarily estimated at 5 sec.

Thus, the total heating time at 1.2 kW is

τ = 34 + 11 + 5 = 50 sec (~ 1.0 min)

The length of the heated part is 0.08 m.
Power per meter of work

$$= \frac{1200}{0.08} = 1.5 \times 10^4 \text{ w/m}$$

Up to 760°C $\gamma/\delta = 4.5$ and $F = 1.0$
Power per meter

$$P = 8\pi H_o^2 \rho \times \frac{\gamma}{\delta} \times F$$

$$1.5 \times 10^4 = 8\pi H_o^2 66.4 \times 10^{-8} \times 4.5 \times 1$$

$$H_o = \sqrt{\frac{1.5 \times 10^4}{8\pi \times 66.4 \times 10^{-8} \times 4.5}}$$

$$= 14.14 \times 10^3 \text{ AT/m}$$

From 760–1000°C $\gamma/\delta = 1.22$ and $F = 0.3$
The power input is the same

$$H_o = \sqrt{\frac{1.5 \times 10^4}{8\pi \times 66.4 \times 10^{-8} \times 1.22 \times 0.3}}$$

$$= 49 \times 10^3 \text{ AT/m}$$

These AT are much more than those required to be heated up to 760°C, hence, power will have to be reduced to about 300 w to maintain the same AT

Consider $H_o = 20 \times 10^3$ AT/m throughout

Assume that we use a 6 mm internal diameter copper pipe for the coil. The wall thickness should be about the depth of penetration for copper (δ_c). The resistivity of copper is 2.2×10^{-6} ohm.cm

$$\delta c = 5030\sqrt{\frac{2.2 \times 10^{-6}}{1 \times 10^4}} = 7444 \times 10^{-5} = 0.07 \text{ cm } (0.7\,\text{mm})$$

Outer diameter of the copper pipe (d_p)

$= 6.0 + 2 \times 0.07 \simeq 7.6$ mm.

About 8–9 turns can be accommodated in a length 8–8.5 cms

$$\text{Coil current} = \frac{AT}{m} \times \ell_c / N = \frac{20 \times 10^3 \times 0.08}{8}$$

$$I_c = 200 \text{ Amp}$$

Coil diameter

$$d_c = d_w + \text{Air gap}$$

$$= 10 + 50 = 60 \text{ mm (6.0 cm)}$$

$$\text{Pipe length} = \text{Length of one turn} \times \text{Number of turns}$$

$$= \pi \times 6 \times 9 = 170 \text{ cm}$$

Cross-sectional area of pipe

$$= \pi/4(0.76^2 - 0.6^2) = 0.171 \text{ cm}^2$$

Resistance of coil at $\rho = 2.2 \times 10^{-6}$ ohm.cm

$$R_c = \frac{2.2 \times 10^{-6} \times 170}{0.171} = 2.19 \times 10^{-3} \text{ ohm}$$

Copper loss in coil

$= I_c^2 R_c = 200^2 \times 2.19 \times 10^{-3} = 87$ W

Generator power

= Work power + Copper loss = 1200 + 87 = 1.3 kW

To this we will have to add transmission loss, etc. To heat faster, the work power may be increased to 1500–2000 W.

A step-down transformer and a tank circuit will be required. For 10 kHz a solid state generator is recommended.

10.2 DIELECTRIC HEATING

10.2.1 Introduction

Like induction heating, dielectric or capacitive heating is a process in which heat is generated inside a dielectric (insulating material) by subjecting it to high frequency currents. As the name suggests, the process is applicable only to nonconducting materials.

10.2.2 Principles of Dielectric Heating

A dielectric material does not conduct electric current because it does not have free or conduction electrons. The ions or molecules have their electron shells firmly held and have a random orientation as shown in Figure 10.12(A). They are also randomly oriented.

If such a material is held between two metal (conducting) plates and subjected to DC voltage, the molecules, ions, or dipoles are oriented as shown in Figure 10.12(B). Some energy is absorbed in the orientation process. If the polarity is reversed (Figure 10.12(C)) the molecules turn about and assume a reorientation, again releasing and absorbing energy.

When subjected to an alternating potential, the molecules undergo orientation at the frequency of the applied voltage. Thus, they absorb and release energy in the form of heat in each cycle. They are heated by this energy. This is the principle of internal heat generation in dielectrics.

10.2.3 Review of Related Electric Properties

The electrical representation of the process is shown in Figure 10.13. A perfect capacitor has a current which leads the voltage by 90°. Generally dielectrics are not perfect. They have a resistance associated with them. Thus, a practical capacitor has a resistance that can be considered in series or parallel with the dielectric. The current then leads by an angle ϕ which is less than 90° by an angle δ. This angle δ is called dielectric loss angle. It is a fundamental parameter and is usually quoted as "tan δ."

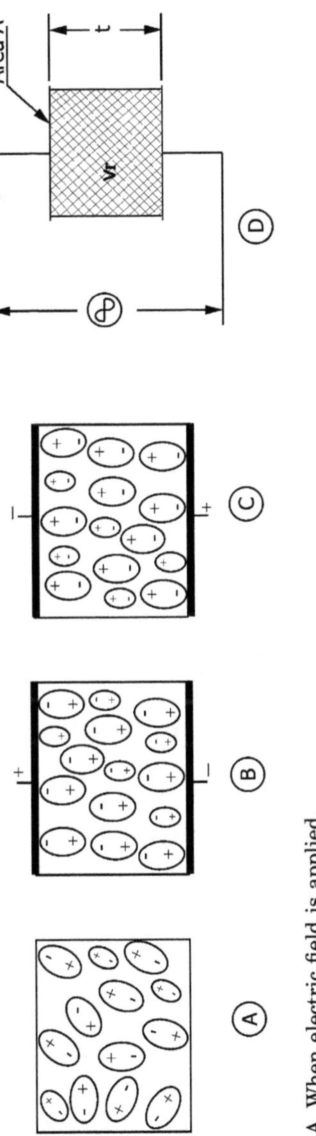

A. When electric field is applied.
B. Electric field applied from top to bottom.
C. Electric field applied from bottom to top.
D. Alternating source applied to a plate.

Figure 10.12 Molecules in electric field.

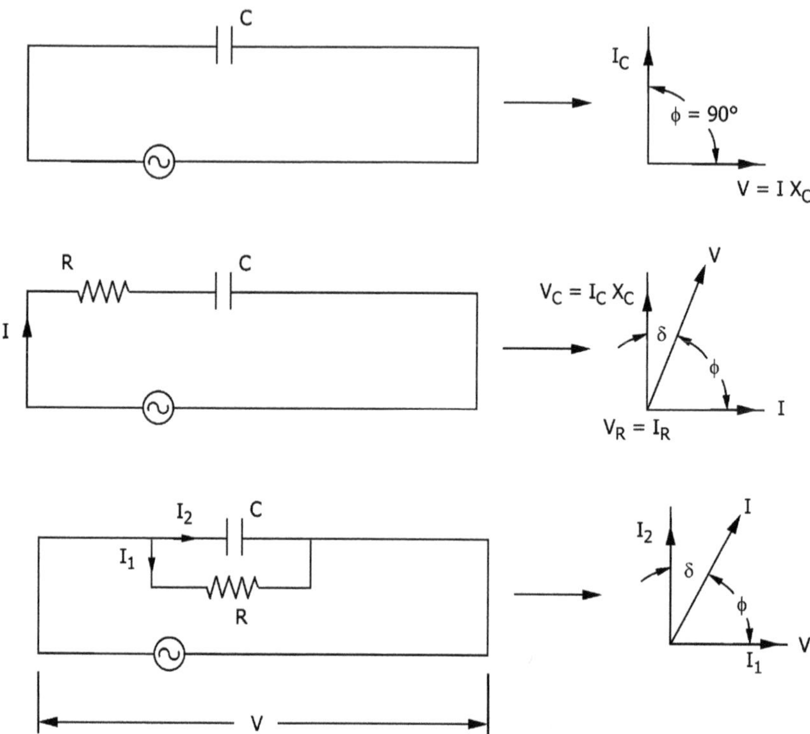

Figure 10.13 Dielectric (capacitor) as circuit element.

Air is taken as a practical standard dielectric. A measure of ability of a medium to orient or polarize an application of a voltage is called permittivity (ϵ). The permittivity of air (ϵ_o) is 8.854 (~ 9.0) × 10^{-12} F/m. The permittivity of other dielectrics (ϵ) is related to ϵ_o by relative permittivity ϵ_r given by,

$$\epsilon_r = \frac{\epsilon}{\epsilon_o} \tag{10.29}$$

All dielectrics fail when subjected to a very high voltage. The failure is by puncture or chemical change. The maximum voltage to which a dielectric can withstand for unit thickness is called its dielectric strength. The permittivity and dielectric strength depend on the temperature, frequency, moisture content, and purity of the material. Dielectric properties of some selected materials are given in Table 10.2.

TABLE 10.2 Dielectric Properties of Selected Materials

Material	Dielectric constant ϵ_r	Dielectric strength kV/mm	Loss factor $\epsilon_r \tan \delta$
Air	1.0	3.0	—
Natural Rubber	3.0	20.0	0.07
Polyethylene	2.5	18.0	0.0005
Polystyrene	2.3–2.6	19–28	0.0001–0.0005
Bakelite	6.5–7.0	12–16	0.030
Wood	2.3–3.6	Depends on moisture	0.03–0.1
Glass	3.5–1.6	1.2–6.0	0.002–0.004

The capacitance C of a capacitor is its ability to store a charge Q when subjected to a voltage V. It is measured in farads.

$$C = \frac{V}{Q} \tag{10.30}$$

A farad is a very large unit. It is generally quoted in μF (= 10^{-6} F) or picofarad pF (=10^{-12} F) etc.

In AC circuits capacitance is represented by reactance

$$X_c = \frac{1}{2\pi fc} \text{ (ohm)} \tag{10.31}$$

where f is the frequency.

The power stored in a capacitor of capacity C and Voltage V is

$$W = \frac{1}{2} CV^2 \text{ (joules)}$$

The dielectric loss and voltage gradient V are related by

$$P = kV^2 f \epsilon \tan \delta \tag{10.32}$$

where k is a coefficient accounting for the nonuniform field at the plate edges and the physical state (dry, wet, etc.) of the dielectric.

The terms $\epsilon \tan\delta$ are determined by the dielectric. Only the applied voltage V and the frequency f can be decided externally. To produce maximum heat, the voltage and frequency should be as high as possible.

The highest voltage that can be safely applied is determined by the breakdown voltage of the dielectric. Generally used voltage gradients are 70–150 V/mm.

The heating arrangement used in dielectric heating is similar to that shown in Figure 10.12(D). The dielectric work is sandwiched between the two electrode plates. To assure a uniform field at the edges, the plates are larger than the work. The whole sandwich is firmly held under pressure.

This arrangement forms a parallel plate capacitor. The capacitance of this is given by

$$C = \frac{\epsilon_o \epsilon_r A}{t} \tag{10.33}$$

where A is the area (m^2) and t is the thickness (m).

The unit of C is farad.

If there is a gap between the work and the plate, the capacitance is

$$C = \frac{\epsilon_o \epsilon_r A}{t - (d - d/er)} \tag{10.34}$$

where d is the thickness of the dielectric so that $t - d$ is the thickness of the air gap.

There are many other capacitor configurations possible.

10.2.4 Some Noteworthy Points about Dielectric Heating

1. The heat produced is uniform throughout because it is produced locally in all the molecules. The heating does not depend on thermal conductivity. In fact all dielectrics are generally bad thermal conductors and their heating by an external heat source is difficult. In induction heating the heat is produced only in a surface layer and conducted away.

2. Dielectric heating can be applied only to insulating materials. Metals, because they are electrical conductors, cannot be heated by dielectric heating. Thus, this is the only process for heating runners, glass, ceramics, polymers, etc.
3. The temperature obtained is limited to about 400°C. This is due to the inherent power factor and the risk of heating the plates and conductors, etc. In induction heating, there is no theoretical temperature limit.
4. As there is no "coil" only relatively simple shapes (mainly flats or cylinders) can be heated. In induction heating the coil can be shaped to suit the work. Moreover, it is away from the work due to the air gap. The capacitor is difficult to shape according to the work shape.
5. Heat production is rapid—usually a few seconds.
6. Local inhomogeneities in the dielectric such as bubbles, gaps, impurities, or foreign masses will produce nonuniform heating. For heating a charge of large area, it is advisable to use several small plates connected in parallel. This arrangement will also produce uniform potential all over the plate area.
7. The frequencies used are in the range 10–40 MHz and are obtained from vacuum tube oscillators similar to those used in induction heating. The relative permittivity V_r (dielectric constant) and the power factor tan d depend on frequency and temperature. When a known frequency is used, the term v tan d increases nonlinearly as the dielectric heats up.

Due to the capacitive nature of the process the power factor is leading. A variable inductor is used in series or parallel to correct the power factor and establish the resonance. As the power factor changes with temperature, an automatic tuning inductor is frequently employed.

10.2.5 Applications of Dielectric Heating

1. For lamination and veneering of wood. The laminates are glued. Dielectric heating heats the glue and sets it. Several sheets can be glued in one operation.

2. Dielectric heating can be used for drying food grains, vegetables, cloth, and many other nonmetallic products. Because of the simultaneous transfer of heat and moisture, careful design is required.
3. Joining of thermoplastics to each other and to different materials like glass and wood, sealing of plastic containers, and hermetic sealing.
4. Sterilization of nonmetallic objects like glass, ceramics, rubbers, and foods.
5. Vulcanizing and joining of rubbers.
6. Adhesive bonding, book binding, paper coating.
7. Paint and varnish drying.

10.3 MICROWAVE HEATING

Microwave heating was accidently discovered and developed in the last 30 years. In principle, microwave heating is similar to dielectric heating discussed in the last section.

10.3.1 Nature and Generation of Microwaves

Microwaves are electromagnetic waves in the frequency range 10^{11} to 10^{14} Hz, i.e., 1–100 GHz. The wavelength at these frequencies is 10^{-3} to 0.3 m (1–300 mm). Note that these wavelengths are comparable to the dimensions of the work to be heated and the container in which the heating takes place. Microwaves are used in communication and radar systems. Their high power generation (1–500 kW) was developed toward the end of World War II. Microwave heating was a secondary development but has now become a common household heating process. It is also increasingly used in the analytical, biological, medical, and industrial fields. At present, microwaves are mainly used in food-related fields, for reasons which will be clear in subsequent discussion.

Microwave generation is a specialized branch of electronic engineering and is beyond our scope. Basically, a 50 Hz supply is converted to microwave frequencies by using a vacuum tube called a “magnetron.” It consists of a hollow cylindrical metal block with precision-made circular cavities on the inner side. Electrons from a cathode are given circular paths in the cavities. Their frequency increases to GHz level by resonance.

Because of their large wavelength, generated microwaves are carried from the magnetron to the workspace (cabinet) via a precision dimensioned metal pipe called "wave guide" (usually having a rectangular section) or through a coaxial cable.

The magnetron generates waves in bursts, keeping the average power at a low level, and hence, does not require elaborate cooling arrangements.

Household microwave ovens are rated at about 1.0 kW. Larger commercial units are also in use. There is usually a level controller to adjust the duty cycle (on/off time ratio) and a timer to adjust the heating time.

Household ovens have a frequency of about 2.45 GHz. This frequency is chosen because it matches with the natural vibration frequency of water molecules. Higher and lower frequencies are available.

Exposure to microwaves is dangerous. It is avoided by carrying out heating in a closed cabinet. The door is interlocked with the power switch so that heating is not turned on when the door is open.

General features of the oven are shown in Figure 10.14.

10.3.2 Heat Generation by Microwaves

Microwave heating is mainly used for heating various foods, which contain water as one of the constituents. The water content may be as free H_2O molecules or combined with the molecules of other constituents such as fats and proteins. There can also be dissolved salts, alcohols, and other chemicals.

Water molecules will consist of a positively charged cluster of hydrogen and oxygen atoms surrounded by a negatively charged electron cloud. (This is a very crude description but is sufficient for our purpose.) The size of the hydrogen molecule is about 10^{-12} m. Similarly, the molecules of other constituents will have positive and negative charged regions.

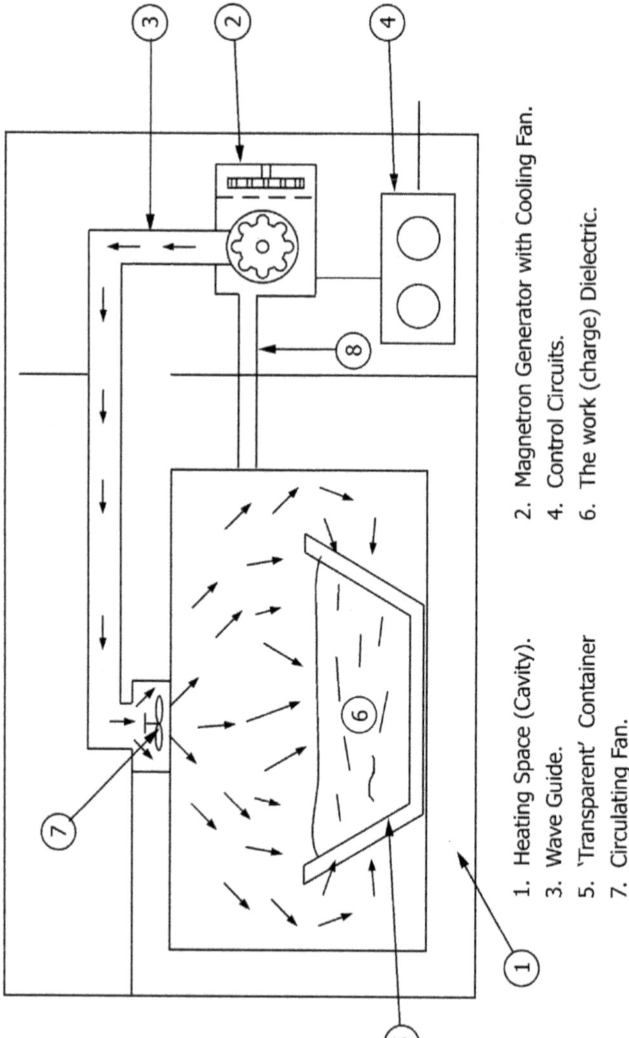

Figure 10.14 Typical microwave oven.

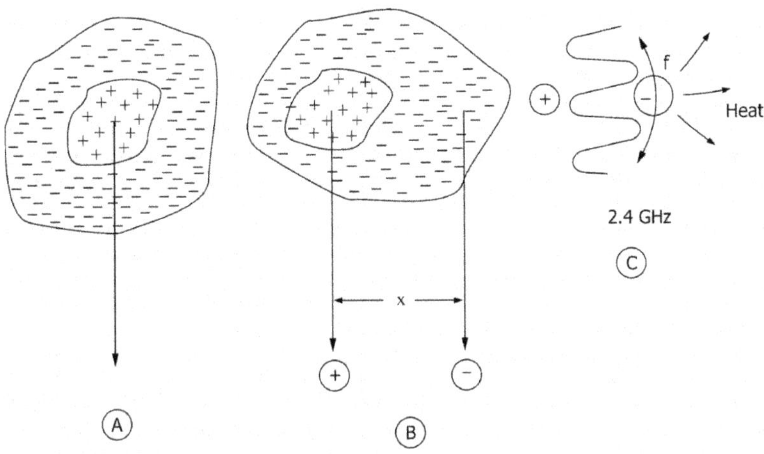

A. Non Polar Molecule.
B. Polar Molecule.
x - Dipole moment
C. Incident Microwaves make Polar Molecule to oscillate at Frequency f, and heat is generated.

Figure 10.15 Molecules and microwaves.

In some molecules, the center of gravity of positive and negative charges coincides (see Figure 10.15(A)) and there is no net moment. These molecules (e.g., H_2, N_2, C_xH_y) are called nonpolar dielectrics. These molecules are not affected by microwaves.

In water molecules (and many other molecules of "food"), the centers of gravity are separated by a finite distance x. These are called polar dielectrics. We are mainly interested in the water (H_2O) molecule, as it is a major common constituent of all foods (see Figure 10.15(B)). The exact molecular structure is not important and a polar molecule can be depicted as shown.

When subjected to microwaves, the water molecule starts vibrating and absorbs energy, which appears as heat as shown in Figure 10.15(C). When subjected to the electric field of the microwaves, the molecule tries to orient itself in the direction of field. As the field alternates at microwave frequency, the molecule undergoes changes in orientation at this frequency. Energy is absorbed from the field and reappears as heat. This is the principle of microwave heating. (Note that this is similar to the polarization in dielectric heating). The molecule requires some time to undergo the vibration cycle. This time is called "relaxation time." This places an upper limit to the frequency (the critical frequency) of vibration, i.e., frequency of the microwave. For water, this limit is 17 GHz. At a greater frequency the molecule cannot keep vibrating synchronously.

The extent to which a dielectric (e.g., water) undergoes polarization or orientation on the application of an electric field is called its permittivity ϵ_r (see ref. [10.2]). Permittivity depends on frequency, composition, and temperature.

Another factor related to heat generation by microwaves is "tan δ" or the "loss factor." This is related to the portion of entering (wave) energy converted to heat. A minor component of heat appears from the resistive heating (I^2R) due to the conductivity of the dielectric. The loss factor also depends on the composition and temperature. The maximum loss occurs at the critical frequency f_c.

A short list of dielectric loss and permittivity values of some materials is given in Table 10.3.

In the previous paragraph, we have discussed the behavior of polar molecules in microwave fields. To sum up, these molecules (or their dispersions) absorb microwaves and get heated. The temperature to which these materials get heated will depend on their permittivity (ϵ_r), their loss factor (tan δ), and the energy input, i.e., energy of the field.

Materials like glass, ceramics, and plastics are transparent to microwaves if they do not contain metal impurities (hence, the special "microwavable ware" in the market).

Large metal surfaces reflect microwaves as they have very little resistance. They do not heat and can be used for making the oven. Foils, metalized glassware, and ribbons heat up due to their higher resistance.

The work or the dielectric to be heated has some reflectivity and does not absorb all the incident waves.

Table 10.3 Dielectric Constant, Loss Factor, and Specific Heat of Selected Materials

Material	Specific Heat kJ/kg°C	Dielectric Constant Permittivity ϵ_r	tan δ@2.5 GHz
Carrots	3.80	65.0	0.43
Cabbage	3.93	62.0	0.41
Peas	3.30	64.0	0.50
Potatoes	3.35	38.0	0.31
Tomatoes	3.97	61.0	0.40
Onion	—	63.0	0.42
Spinach	3.40	62.0	0.80
Ham	2.51	45.0	0.53
Fish	3.35	—	0.57
Beef	—	40.0	0.30
Bread dough	—	20.0	0.20
Polyethylene	2.30	2.25	0.0003
Pyrex ware	0.84	4.0	0.0001
Silicon carbide	—	30.0	0.36
Water	4.18	80.3	0.005

10.3.3 Heat Produced in Microwave Heating

The energy absorbed w, in a volume V (cm^3) of a dielectric in a microwave field of frequency f and field strength E(V/cm) is given by

$$w = 0.57 \times 10^{-12}\, V f E^2\, \epsilon_r \tan \delta \text{ w/cm}^3 \tag{10.35}$$

where ϵ_r is the permittivity and tan δ is the loss factor.

The field strength E is usually not known. The microwave generators rated power $P(w)$ is known, as well as the design frequency f (GHz).

The rated power is calculated from heating the maximum quantity of a dielectric of known ϵ_r and tan δ (e.g., water) that can be heated to a fixed temperature. The time $\Delta\tau$ for attaining the desired temperature is noted.

The heat developed q (J) is given by

$$q = m \times c \times (T_2 - T_1) \text{ J} \tag{10.36}$$

where

m = Mass (gm)
ρ = Density (gm/cm^3) of the dielectric
T_1 = Initial temperature (°C)
T_2 = Final temperature (°C)
c = Specific heat J/gm°C

The volume will be $V = m/\rho$ cc

The power absorbed

$$w = q\ \Delta\tau \tag{10.37}$$

substituting in Equation (10.37)

$$w = 0.57 \times 10^{-12} \times (m/\rho) \times f\, E^2 \tan\delta$$

$$= m \times c \times (T_2 - T_1)\, \Delta\tau$$

These equations can be used to calculate the power output of the oven (W_M). These are the following possibilities:

1. $W_M > W$, the given load can be heated to the required temperature (T_2).
2. $W_M = W$, the power output and power absorption are just balanced.
3. $W_M < W$, the load cannot be heated as the oven output is limited, as compared to the absorption capacity (ϵ_r and tan δ) of the charge.

Foods contain more than one component. Each constituent has its own ϵ_r and tan δ values, hence, their heating rates will be different.

Dielectric constant, loss factor, and specific heat of selected materials is given in Table 10.3.

10.3.4 Some Peculiarities of Microwave Heating

Compared to other radiation heating processes, microwave heating shows some uncommon features.

1. The heat produced is deep inside the dielectric due to the high values of permittivity and the loss factor.
2. Each polar molecule vibrates in the field, hence, heating is thorough and uniform.

3. There is no heat transfer from external sources as in a furnace. There is no coil as in induction heating and no capacitor plates as in dielectric heating.
4. The container used for the work remains relatively cool.
5. The application is restricted to dielectrics based on water and is therefore (at least presently) applicable to food processing only.
6. Heating is clean, i.e., there is no contamination from gases, containers, etc.
7. The temperature of heating is restricted to about 250–300°C in domestic ovens.
8. Microwave ovens are often confused with conventional ovens. Heating in the latter is by radiation from the outside. Hence, they can be used for baking. Microwaves do not "bake;" they just heat from the inside.
9. Microwave ovens have a glass door. Glass is transparent to microwaves. To prevent the microwaves from escaping through glass, it is laminated with a fine metal mesh having "quarter wave" openings.
10. The microwave generator delivers the rated power in the programmed time and at the set level.

    The power absorbed by the dielectric depends on its above-discussed property. Hence, the efficiency of heating is difficult to judge.

    In all other modes of heating the required power is demanded by the load or the work, and is delivered accordingly.
11. Heating in a microwave field is decided by two factors.

    I. The maximum power that can be generated and delivered by the design of the source.

    II. The power that can be absorbed by the charge. This depends on the dielectric properties (ϵ_r, tan δ) of the charge. Whichever is greater decides the heating (temperature) attainable.

The (solved) examples will make this limitation clear.

The discussion in previous sections is centered on the application of microwaves for heating water-containing or water-based dielectrics. The common major use of this method is for heating foods and beverages in domestic and commercial fields. However, there are certain other applications that are possible and being developed. Some of these are listed in the following:

1. The analytical and physical properties of many substances in the food and related industries can be measured by using microwave techniques. Commercial instruments based on microwave measurements are already in the market.
2. Some minerals and compounds of barium, titanium, silicon, and carbon absorb microwaves. These materials can be heated to very high temperatures (~ 1800–2200°C) in a microwave field. This can be used in the synthesis of carbides, silicides, nitrides, titanates, etc. (see Figure 10.16). Recent research has proved the possibility of such proceses.
3. Thin metals are intensly heated in microwave fields. This heating is similar to induction heating. The depth of penetration in microwave fields is a few microns. It should be possible to heat thin wires, and foils by microwaves. The technique can be useful in microelectronic and solid state device fabrication.

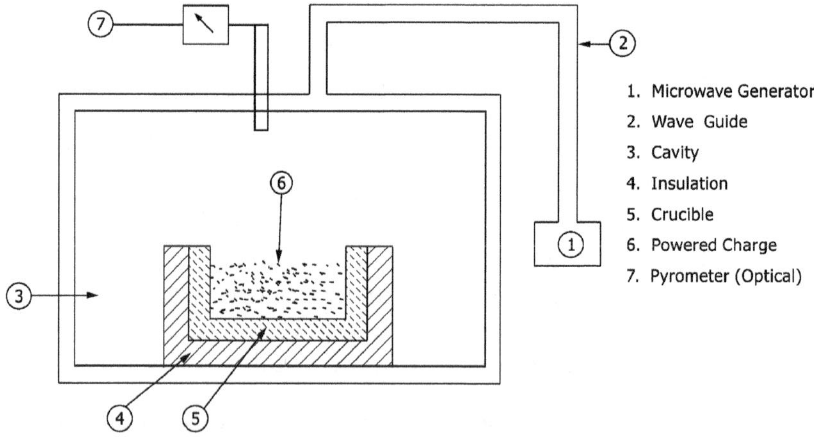

Figure 10.16 Microwave-assisted synthesis.

EXAMPLE 10.5

A domestic microwave oven has a rated output of 600 W at 2450 MHz.

Experiments were conducted by heating various quantities of water initially at 30°C. The temperature of water was measured after heating for different lengths of time. The results are shown in Figure 10.17.

Water has following related properties (assumed constant):

$$\text{Density } \rho = 0.986 \text{ g/cm}^3$$
$$\text{Specific heat } C = 4.178 \text{ J/g °C}$$
$$\text{Dielectric constant } \epsilon = 76.7$$
$$\tan \delta = 0.15$$

Now determine:
Field strength for water.
Power absorbed.
Comment on the result.

Solution

Consider the basic equation for microwave heating.

$$\frac{q}{V} = 0.556 \times 10^{-12} \times f E^2 \; \epsilon \tan \delta$$

Substituting for f, ϵ, and $\tan \delta$ in L.H.S.

$$= 0.556 \times 10^{-12} \times E^2 \times 76.7 \times 0.15$$

$$= 1.6 \times 10^{-2}$$

The field strength E (V/cm^2) is not known.

The experimental data show that 200 cm (~ 200 g) of water was heated to 80°C in 110 sec.

Hence,

$$\frac{q}{V} = \frac{200 \times 0.986 \times 4.178 \times (80 - 30)}{200}$$

$$= 210 \text{ J/cm}^3$$

$$= 1.6 \times 10^{-2} \text{ as calculated above.}$$

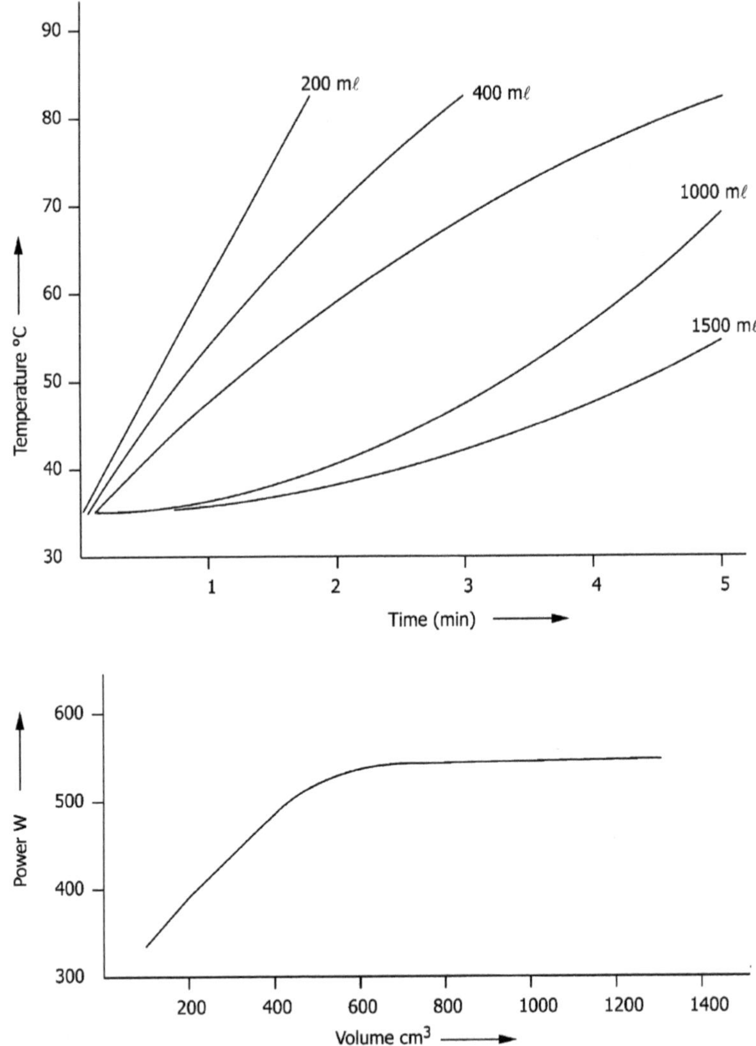

Figure 10.17 Microwave heating of water (Example 10.5).

Solving for E

$$E = \sqrt{\frac{210}{16 \times 10^{-2}}}$$

$$= 110 \text{ V/cm}^2$$

Power absorbed by 200 cm water in 110 sec is

$$W = \frac{J}{t} = \frac{200 \times 210}{110}$$

$$= 380 \text{ W}$$

Similar calculations for larger volumes of water produces the following results

Volume (cm^3)	Power to heat to 80°C
200	380
400	450
600	530
1000	530 (68°C)
1500	530 (55°C)

The above results show that the oven capacity (~ 600 W) limits the maximum quantity of water that can be heated to 80°C in less than 5 min and to about 600 cm^3.

For quantities about 200–400 cm, the power absorbed is limited by the electrical constants (ϵ, tan δ) and of boiling at 100°C.

Chapter 11

Concentrated Heat Sources

CONTENTS

In the last chapter we have discussed heating by electromagnetic radiation which "induces" heat in the object. In the last two decades two heat sources were developed which create heat in the object on incidence by focusing. The first such source is "laser" which is essentially a light beam which may or may not be in the visible range.

The second source is "a beam of fast moving electrons" which can be regarded either as a high energy particle beam or an electromagnetic wave.

Both beams are nonthermal but when focused and stopped by the object, develop a spot on the surface. This spot is of a very small size, but has an extremely high energy density.

This incident energy creates high heat flux which results in heating, melting, or evaporation of the material under the spot.

Both the sources mentioned above are commercially developed and used in a variety of very interesting applications. The development and application of these sources is currently on the frontier of heating technology. This chapter is a primer to the understanding of the peculiarities and capabilities of these "concentrated heat sources." A general idea about these sources as compared to the other traditional heating sources will be clear from Figure 11.1. This figure shows that an electric arc or plasma are also powerful concentrated sources. Both are used in applications involving melting, such as large scale melting in

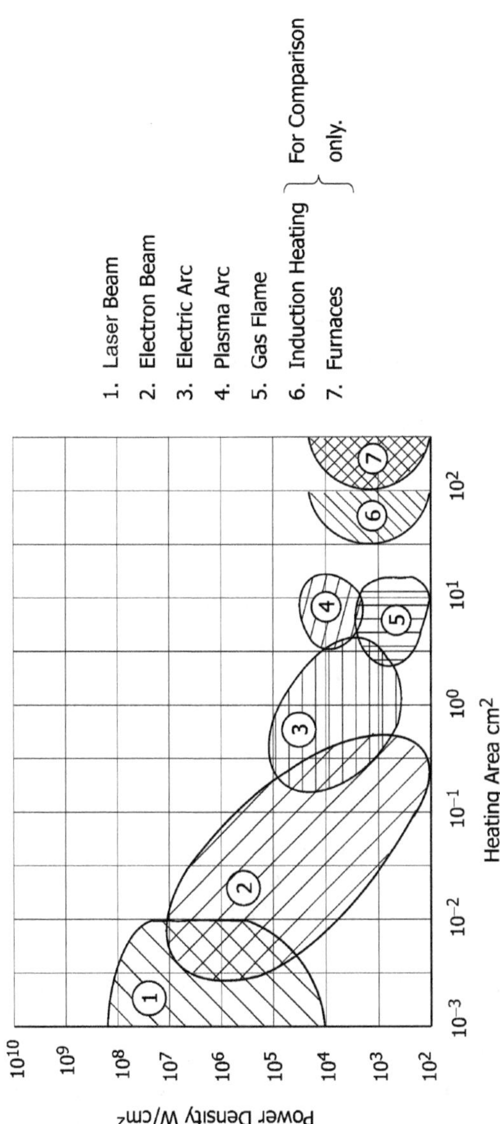

Figure 11.1 Heating area and energy density for some concentrated heating processes.

foundry or small scale melting in welding. Both operations are beyond the scope of this book and, hence, are not discussed here.

For both sources the theoretical treatment of source-material reaction discussed in this section applies equally well.

Extensive information on laser and electron beams as sources for material processing is given in the books listed in the *Bibliography*.

11.1 LASER

11.1.1 Introduction

The term laser is derived from "light amplification by stimulated emission of radiation." Broadly speaking, it is an ordered beam of monochromatic light. The theoretical possibility of such a beam was described by A. Einstein (~ 1917) and the first working laser was produced by T. Maiman in 1960. This first device was based on a doped ruby single crystal and had a very low power (mW).

In subsequent years, rapid developments in technology and worldwide research led to the development of a wide variety of laser generators with beam powers from a few watts to several kilowatts.

Lasers were soon realized as potential concentrated energy (CE) devices that can be commercially used for material processing. A focused beam of laser radiation can develop extremely high energy (heat) concentration ($\sim 10^4 - 10^8$ W/cm^2) on a small area ($\sim 10^{-2}$ cm^2) on the work surface. A large number of application areas were discovered to which new ideas are being constantly added. The use of laser as a "tool" is being exploited for processing all types of materials. Because of its delicacy and precision, it is also increasingly used in surgical operations.

Besides the above-mentioned application areas laser is being used in communication, surveying, printing metrology, analytical chemistry, holography, and several other fields. However, these applications are based on properties such as monochromaticity, coherence, and low divergence, and are beyond the scope of this book.

A very large number of atoms and molecules display lasing (i.e., emission of monochromatic, stimulated, and coherent radiation) when properly excited. Consequently a variety of commercial lasers are available. They are broadly classified as solid state lasers, molecular lasers, semiconductor lasers, excimer lasers, gas lasers, dye lasers, and so on.

In this book we are interested in so-called "power lasers" which can be used for heating, i.e., having a power output of the order of 10^4–10^8 w/cm^2 for sufficient time.

Presently, only two lasers are found to be suitable for such power applications. One is a solid state laser called Nd-YAG laser, which can give an output up to 1.0 kW. The other is a gas (molecular) laser based on CO_2 and is called the CO_2 laser. It is the most powerful laser presently available and can give an output up to 20 kW. For very small power requirements the ruby laser is used, which has an output of a few mW.

11.1.2 Generation of Laser Beam

The physics of the generation of laser beams by atoms and molecules is a very complex phenomena and a specialized branch of physics. In this book we are concerned only about the properties and application of laser beams for material processing (heating). The details of atomic and molecular processes leading to laser generation (lasing) are not important for these applications. What follows is a general and simplified account. For theoretical detail treatment, consult specialized literature listed in the *Bibliography*.

The electrons in atoms and molecules occupy certain energy levels or bands which are associated with a definite energy E. When the atom is unexcited, lowest energy levels are occupied. The amount of electrons that can occupy a given level is fixed and decided by the laws of quantum mechanics. It is expected that the reader is familiar with the basic atomic theory.

Atoms can be excited by the application of high voltage, high temperature, and many other techniques. On excitation the electrons acquire higher energy and jump to an appropriate higher energy level.

On removal of excitation the excited electrons in the higher level go back to their original lower level position. In doing so they get rid of the excess energy by emitting photons or light and other radiation.

If the energy in the higher level is E_2 (electron volts) and that in the lower level is E_1 the frequency ν (and wavelength λ) of the emitted radiation is given by the well-known relation:

$$\nu = \frac{E_2 - E_1}{h} = \frac{C}{\lambda} \tag{11.1}$$

where C is the velocity of light ($\sim 3 \times 10^{10}$ cm/sec).

In laser parlance the process of excitation of electrons to a higher level is called “pumping.” The return journey from a higher level to a lower one is called “decay.” The decay may take place in one step from E_2 to E_1 or a number of steps (such as $E_2 \rightarrow E' \rightarrow E'' \ldots E_1$) of successively lower energy.

According to the quantum theory the energy levels are discrete (definite) for a given atom/molecule. Hence, for satisfying the equation all the three variables E_1, E_2, and ν must be compatible for radiation emission. In the intermediate steps of decay, if there is no compatibility, that decay is emissionless.

For lesser generation, the lasing medium is selected so that it has a three- or four-level decay in which one intermediate state satisfies the conditions of the equation.

Besides the selected decay levels discussed above, the excitation and decay occurs at many levels and the resulting general light emission is called “fluorescence.”

Consider a substance having a four-level electronic structure with energy increasing from level 1 to 4 (Figure 11.2). In an unexcited state, level 1 is fully occupied by electrons. On excitation the electrons in level 1 will be pumped to level 4 creating vacancies in (the otherwise full) level 1. This is called “population inversion.”

On removal of excitation the electrons in level 4 will decay to level 3. This decay is emissionless. A further transition is from level 3 to 2 where the decrease in energy satisfies Equation (11.1) and a radiation of frequency ν will be emitted. The electron will then decay to level 1 without emission and thus return back to its original position.

The photons (radiation) emitted during the transition from level 3 to 2 will stimulate other electrons (undergoing the same $3 \rightarrow 2$ transition) to emit radiation of the same frequency ν.

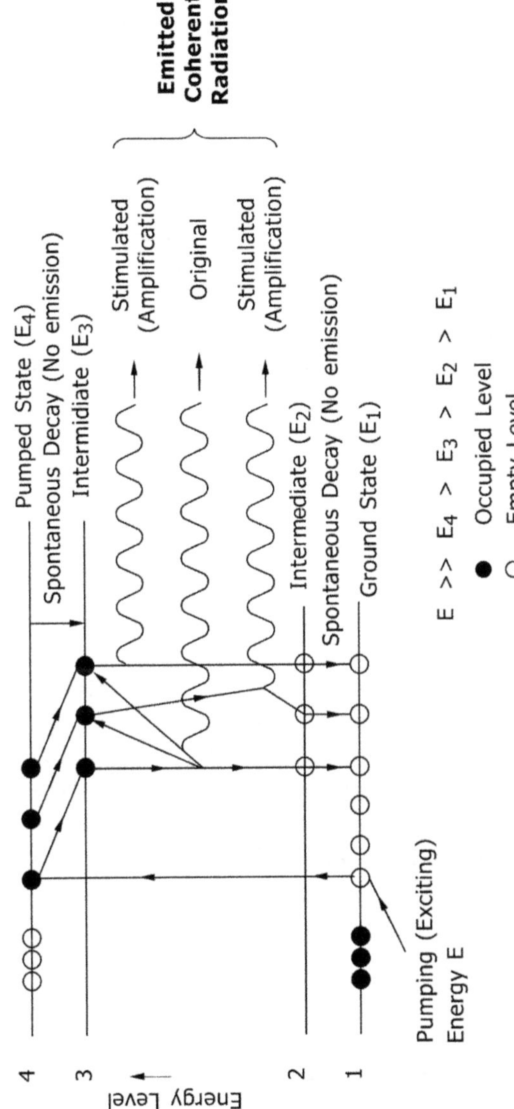

Figure 11.2 Laser emission by stimulation amplification.

This emission is not only of the same frequency (ν) but is also in the same direction and in phase with the original emission. This process is called "stimulation" and is the beginning of the laser beam formation.

Continued excitation (pumping) reinforces the beam, increasing its intensity.

A laser cavity is made by enclosing the lasing substance in the space between two mirrors (Figure 11.3) at two ends. One mirror is fully reflecting and the other is partially transmitting. The mirrors are perfectly parallel to each other.

When the radiation beam of stimulated radiation strikes the fully reflecting mirror it is reflected, i.e., turned back toward the cavity. On its journey to the second mirror it stimulates more electrons to transition $3 \rightarrow 2$ and emits radiation of the same frequency, same direction, and same phase. The beam intensity thus increases further. This is called "cavity gain."

On striking the semitransparent mirror, some radiation leaves the cavity as a coherent laser beam. The remaining radiation is reflected back and continues to travel to the fully reflecting mirror with increasing intensity.

The beam thus keeps traveling between the two mirrors and builds a resonance due to the particular distance (L) between the mirrors.

This is a simplified account of laser generation. The pumping action in some solid state generators is achieved by using high energy photons emitted by a powerful flash light (see ruby and Nd-YAG lasers in the next sections). In gas lasers the pumping is achieved by high voltage discharge.

Other radiation not compatible with the stimulated radiation escapes the cavity, and adds to the cavity loss.

Note that the input energy in a laser is that which is supplied to the pumping source.

11.1.3 Noteworthy Points about Lasers

Laser is a clean source of energy. There are no combustion gases. It is a beam of photons having no mass and charge. They do not exert any pressure on the work.

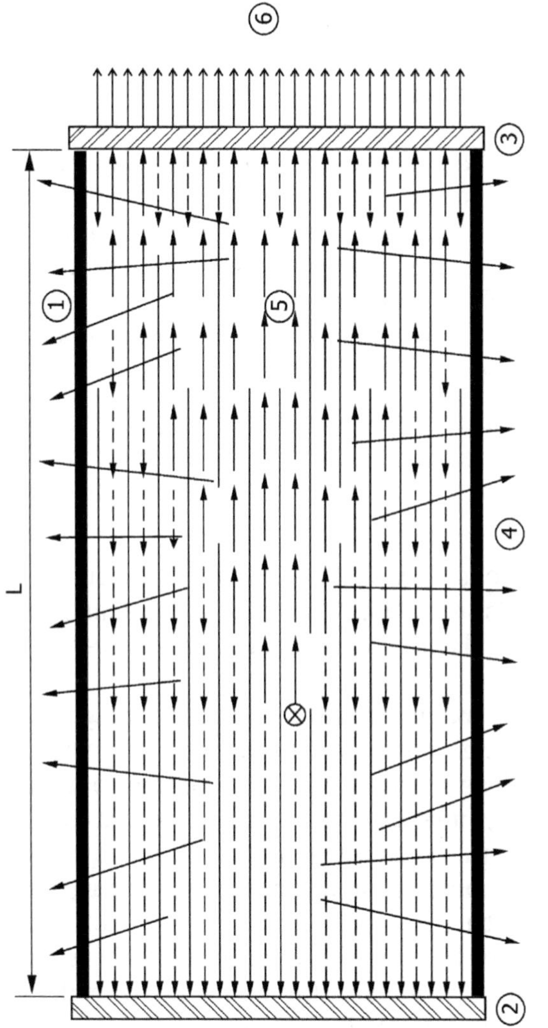

1. Laser Cavity containing Lasing Medium (Length L).
2. Fully Reflecting Mirror.
3. Semi-transperent Mirror.
4. Stray Incoherent Radiation.
5. Stimulated & Amplified Coherent Radium.
6. Emerging Laser Beam.

ⓧ Excited (Pumped) Atom/Molecule.

Figure 11.3 Building up of laser beam by stimulation and cavity resonance.

Laser is unaffected by air or any gas surrounding the work. It is affected only by plasma (ionized gas). Heating by laser does not require a closed surrounding like a furnace. Being a beam of light (mostly infrared), radiation can be transmitted over considerable distances by using a pipe (wave guide) and mirrors. Low power beams can be transmitted by using a relatively flexible optical cable. Hence, the laser source (generator) and the work can be separated.

By using a microscope and related optics the beam can be located precisely on the spot which is to be heated. A coordinated controlled work stage in combination with a microscope will enable the spot to be successively located at different sites on the work surface. The focused spot has a radius of a fraction of a millimeter which enables heating of a very small spot (~0.01 mm^2). Heating is thus confined to a very small area. This makes the following operations possible:

1. Welding of fine wires to each other (e.g., thermocouple junctions) or to integrated circuits or components.
2. Drilling (piercing or burning) of very small (10–500 μm) diameter holes.

The heating time is very short (~ msec). It depends on the work size and the pulse duration. This short time span restricts the heat affected zone (HAZ), hence, there are no internal thermal stresses.

The power density (w/cm^2) of a laser beam is very high (10^4–10^8 w/cm^2), hence, a very high power is concentrated on a very small area. At these power levels virtually any known material (metals, ceramics, glass, or polymers) can be melted or vaporized. This is achieved by controlling the beam power and the duration of heating.

Light particles (photons) do not have any electric charge. Their interaction with the work atoms or molecules of the work does not involve the completion of an electric circuit (in contrast to electron beams). Similarly, the heating does not involve setting up an internal electromagnetic field as in induction or dielectric heating. The heat evolution is purely by the impact of photons with the free electrons and atoms or molecules of the work. It is because of this reason that a laser beam heats any material.

This property makes lasers useful for processing any material or any combination of materials (e.g., dissimilar metal joining, ceramic cutting, drilling, etc.)

Cutting, drilling, or similar operations are carried out by a light beam used as a tool, and material removal is by precise melting and evaporation. In conventional machining, material removal is achieved by the impact of a hard tool on the work, i.e., by shearing off the material, piece by piece. The friction involves considerable unwanted heating. Usually we have to use a cutting fluid to reduce the friction and remove the heat. Laser machining does not require any coolant.

Heating rates achieved by a laser beam are very high and the spot is precisely located where required. Machining is therefore very fast and precise. There is very little swarf. The cut edge or welded joint is smooth and clean. Generally no post heating (machining) operations are required.

Laser machining does not require costly fixtures. The only facility required is a good programmable coordinate table with the required degrees of freedom. A programmable coordinate table makes it easy to change the work-piece or the site.

Laser operation is quiet and no gases are evolved, hence it is eco-friendly.

The laser beam and its profile can be precisely controlled and manipulated by techniques such as Q switching, pumping control, filters, lenses, TEM control, etc.

More than one laser generator (of the same wavelength and type) can be optically coupled to obtain higher power levels.

11.1.4 Limitations of Lasers

Laser heating is not a bulk heating process like furnace heating. The heat production is limited to a very small spot. Hence, laser processing is restricted to very small areas. Thus, laser is useful for micro-machining or micro-heating.

Due to the peculiar nature of the spot and work material interaction laser beams cannot penetrate deep (like a drill). The depth of penetration is usually restricted to a few mm.

To keep the laser beam focused on the work surface it is usually necessary to isolate the work table from the beam generator.

Laser optical components like mirrors and lenses do deteriorate with time. They are costly to repair or replace. Laser maintenance can be costly and is a specialized job.

Holes drilled by laser have a tapered shape. This taper is more pronounced for deep holes. Similarly, the hole edge of the top surface has a splashed edge due to melt flushing.

Laser radiation (especially focused) is damaging to the eyes and skin. Proper precautions, as recommended by the manufacturer, must be observed.

LASERS FOR MATERIAL PROCESSING

11.1.5 CO_2 Laser

This is a molecular gas laser and is the most powerful laser presently available.

The lasing medium is CO_2 gas in which lasing occurs due to changes in vibrational energy levels of CO_2 molecules. The lasing wavelength is in far infrared at 10.6 μm. The lasing medium consists of a mixture of CO_2, N_2, and He in the proportion 2:2:3. Addition of N_2 and He helps lasing of CO_2 and gives higher output and efficiency. Addition of He also helps to remove heat as it has a better conductivity.

The pumping is achieved by a high voltage 1–3 kV and high frequency (1–3 kHz) DC discharge.

The construction of a typical CO_2 laser is shown in Figure 11.4. This is a fast, axial flow-type generator. There are transverse flow designs which can give higher power outputs.

The lasing action occurs in a glass tube resonator. The tube has a fully reflecting mirror at one end and a partially transmitting window at the other end. Both the window and the mirror have a large radius and concave shape. The tube has a row of alternate positive and negative electrodes connected to a high voltage pulsing DC supply and controller.

The gas mixture gets considerably heated during lasing. Some CO_2 is decomposed to CO and O_2. To remove the heat

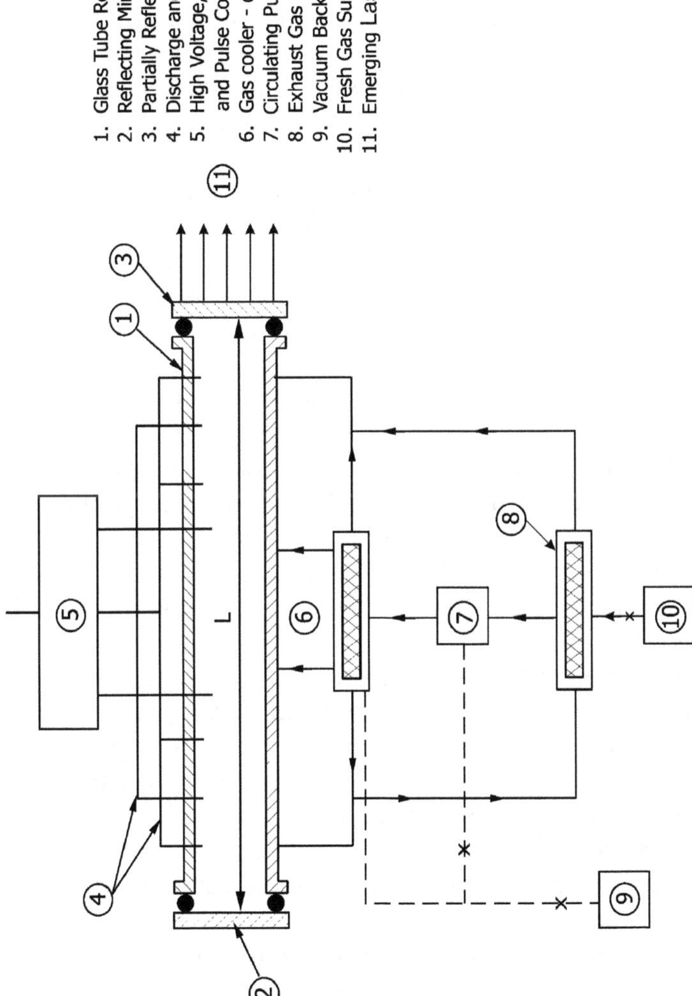
1. Glass Tube Resonator Cavity (Length L).
2. Reflecting Mirror.
3. Partially Reflecting/Transmitting Mirror.
4. Discharge and Ignition Electrodes.
5. High Voltage, High Frequency Power Supply and Pulse Control.
6. Gas cooler - distributor.
7. Circulating Pump (Roots Blower).
8. Exhaust Gas Cooler.
9. Vacuum Backing Pump.
10. Fresh Gas Supply.
11. Emerging Laser Beam.
1
2
3
4
5
6
7
8
9
10
11
L

Figure 11.4 CO_2 laser.

and the contaminants the gas is drawn off and cooled by a recirculating arrangement shown in Figure 11.4. Fresh gas is added when required. The axial gas flow rate is about 300 m/sec. The flow is maintained by a roots blower. The pressure inside the tube is less than the atmospheric pressure.

The output of CO_2 lasers increases with tube length and gas pressure. There is a limit to the tube length. Lasers containing several tubes in parallel are constructed to give a tube length of several meters; however, they are cumbersome to construct and keep in alignment. Lasers having a transverse flow are now available, giving an output of more than 20 J. The gas pressure in a transversely excited atmospheric (TEA) pressure type CO_2 laser is nearly atmospheric (~ 10^5 Pa). The discharge pulse frequency is 10–100 kHz.

Commercial CO_2 lasers have an output of up to 20 kW with a maximum efficiency of 20 kW.

CO_2 lasers have a wavelength of 10.6 μm with considerable line width. As these wavelengths are in the far infrared region almost all the optical materials normally used, such as quartz and glass, are unsuitable. The mirror, window, and all other optical components in the beam manipulation system are made of materials such as KCl, NaCl, Ga, As, Ge, Zn, Se. These are costly materials. Mirrors are of the gold-surfaced metallic type with water cooling. The optical components deteriorate and require replacement or resurfacing.

Low power (3–100 W) CO_2 lasers have a sealed tube, i.e., no gas recirculation. These are not very useful as a heat source.

Fast axial and transverse flow CO_2 lasers are the main types used today as power lasers in material processing. They are generally used in continuous mode but can be pulsed in the 1–10 μs range. The beam of CO_2 lasers can give various TEM modes though the fundamental (TEM_{00}) mode is more common. CO_2 lasers can be Q switched. The output can be controlled to produce a step or ramp pattern. The gas tube (resonator) may have external cooling.

High power CO_2 lasers used for machining have a gas-assist system to supply a reactive (air, oxygen) or inert (argon, helium) gas flow on the work surface at the hot spot.

There are other gas lasers based on CO, N_2, NO_2, H_2O, etc. The CO_2 laser dominates the fields.

Typical Properties of Commercial CO_2 Laser

Wavelength	10.6 μm with several lines
Transverse modes	Multimode. TEM_{00} (Gaussian) is commonly used
Output power (continuous)	1–10 kW commonly available
Pulsed output	10^{-3}–10^{-2} J
Gas pressure	10^4–10^5 Pa
Gas consumption	0.2–2.0 m^3/h
Flow	Fast axial or transverse
Beam diameters	5–25 mm
Efficiency	10–20%
Beam divergence	1–3 mrad

11.1.6 Nd-YAG Lasers

This is a solid state laser based on garnet (a mineral) doped with the rare earth element neodymium (Nd). The garnet used has a formula $Y_3 Al_2 O_{12}$ and is called yttrium-aluminum garnet, a synthetically grown single crystal having a diameter of 3.5–16 mm and a length of 45–250 mm. About 1–1.5 Nd is uniformly dispersed in it (doping) while the crystal is grown. The ends of the rods are highly polished and made exactly parallel.

The pumping medium used is intense white light produced by xenon flash lights or fluorescent discharge tubes. The output of these "pumps" can be controlled by varying the input energy and supply frequency.

The flash tube and the laser rod (medium) are enclosed in an internally reflecting container. There are various designs. The design shown in Figure 11.5 uses an elliptically shaped enclosure with the flash tube and the laser rod situated at the foci. This improves the pickup of the pumping light. The enclosure may be air- or water-cooled.

The pulse width and frequency is controlled by an external power and frequency source for the flash light. Recent versions of Nd-YAG laser use solid state diodes in place of flash lights. Similar

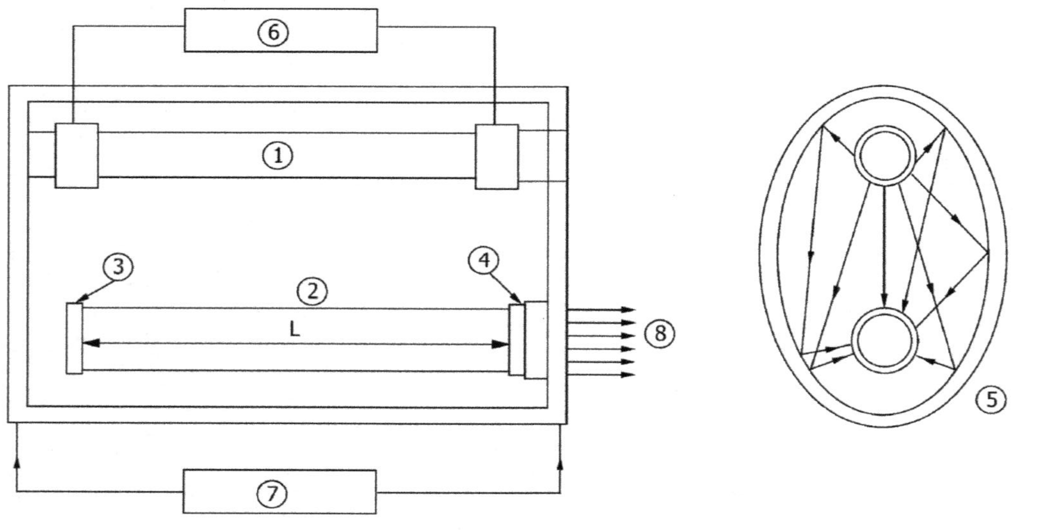

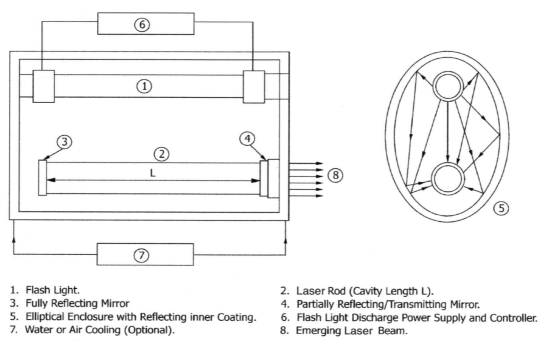

1. Flash Light.
2. Laser Rod (Cavity Length L).
3. Fully Reflecting Mirror
4. Partially Reflecting/Transmitting Mirror.
5. Elliptical Enclosure with Reflecting inner Coating.
6. Flash Light Discharge Power Supply and Controller.
7. Water or Air Cooling (Optional).
8. Emerging Laser Beam.

Figure 11.5 Solid state laser, e.g., ruby or Nd-YAG type.

lasers with less power output are neodymium yttrium glass (Nd-glass) and yttrium lithium fluoride (Nd YLF) lasers.

Nd-YAG lasers can be operated in both pulsed and continuous modes. The operation is usually in TEM_{00} mode. The laser can be Q switched.

The mirror and window at the end of the rod are made of quartz with suitable metal reflective coatings.

The laser rod and mirrors deteriorate due to internal defects, thermal fatigue, and atmospheric corrosion. Defective rods require replacement. In most cases mirrors can be recoated. The flash lights used have a typical life time of 10^6 cycles. In some (older) designs separate mirrors are not used. The polished rod ends are given a reflective coating.

Typical Properties of Commercial Nd-YAG Laser

Wavelength	1.06 μm continuous or pulsed
Transverse mode	TEM_{00}
Pulse energy	30 mJ–100 J
Pulse duration	0.2–20 msec
Pulse repetition rate	10^{-3}–10^3 Hz
Efficiency	2–5%
Continuous operation	10–30 W common–up to 100 w available
Q switched operation	200 mJ–3 J
Q repetition frequency	1–50 kHz
Pulse duration	Few milliseconds

11.1.7 Ruby Lasers

This was the first successful laser produced about 42 years ago and is still finding commercial applications.

In construction, the ruby laser is similar to the Nd-YAG laser described earlier. The laser medium or cavity is made from synthetically grown ruby single crystals. Ruby is $\alpha\,Al_2O_3$ or corundum. At the time of crystal growing, about 0.5% chromium is added and uniformly distributed. The lasing and stimulation takes place in the chromium atoms. Pumping is achieved by flash lamps around the ruby rods which are 45–250 mm long and have a diameter of 5–15 mm. Ruby is hard, mechanically strong, and has a good stability against radiation damage. Irregularities in crystal structure and uneven distribution of Cr and other impurities lower the efficiency.

Ruby lasers emit in the visible red region and have a wavelength of 0.6943 μm. They have a pulse output with a repetition frequency ~ 1.0/sec. The pulse energy is ~ d few J and duration ~ 1 msec. This laser can be Q switched to give a peak power up to 10^6 watts for 10^{-9} sec.

11.1.8 Longitudinal Modes of Laser Beam

A laser beam is a result of atomic/molecular stimulation and the resonance of the light waves so produced in the cavity. This process is already reviewed in the preceding section. Ideally, the resonance should occur only at one wavelength or frequency giving a single mode, monochromatic output (Figure 11.6) at wavelength λ_o.

In practice, the resonance occurs at many wavelengths due to the cavity dimensions and broad energy levels among which lasing and electron transfer occurs. The output beam is thus multimodal, i.e., it consists of many wavelengths centered about the main mode λ_o.

Some of the modes are absorbed by the cavity losses and in reflection at the mirrors. Those escaping in the beam are symmetrically situated about the main wavelength. The output of a laser is thus a beam having a predominant wavelength (or frequency) and a line width $\Delta\lambda$ or $\Delta\nu$.

The line width of a given laser depends on the cavity dimensions, material of mirrors, electronic structure of the lasing medium, and defects or impurities in it.

Numerically the line width can be defined as the ratio $\Delta\lambda/\lambda_o$ or $\Delta\gamma/\gamma_o$. For example, a low pressure CO_2 laser beam has a dominant wavelength 10.6×10^{-6} m and a line width 60×10^6 Hz.

The line width will be

$$\Delta\nu/C/\lambda_o \tag{11.2}$$

where C is the velocity of light (3×10^{10} cm/sec).

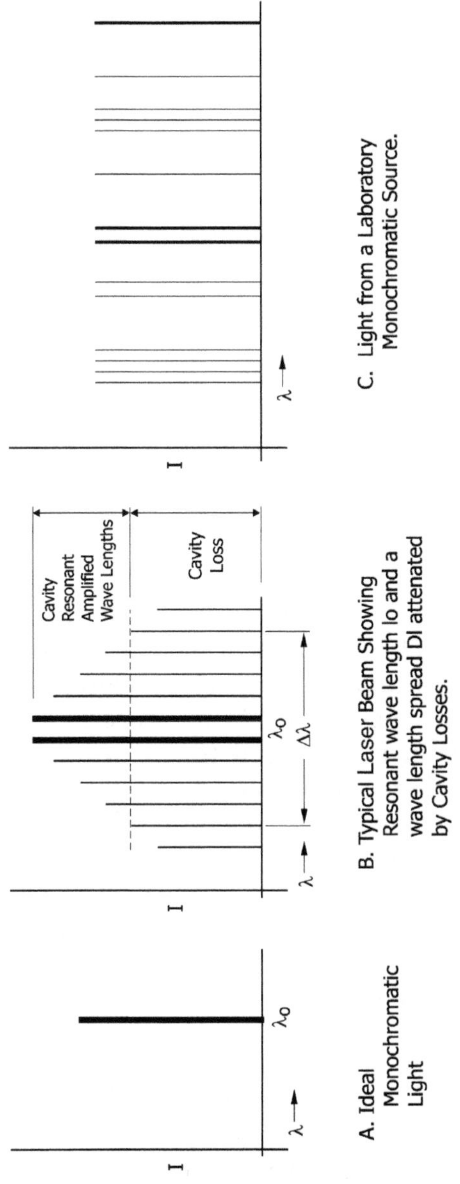

Figure 11.6 Wavelength distribution.

$$\text{Line width} = \frac{60 \times 10^6}{3 \times 10^{10}/10.6 \times 10^{-4}}$$

$$= 2.12 \times 10^{-7}$$

A Nd-YUG laser has a wavelength of 1.06×10^{-6} m and a line width of 1.2×10^{11} Hz.

Thus, the numerical line width is about 4.2×10^{-3} showing much broadening compared to a CO_2 laser.

The line width can be controlled to some extent on laboratory laser setups.

The fluorescent emission has a very broad spectrum (Figure 11.6) showing a number of strong or weak lines as there is no gain due to resonance.

The above calculations show that even with line broadening, the laser beam is still highly monochromatic.

11.1.9 Focusing Properties of Lasers

A TEM_{00} laser beam emerging from the window of the generator is a coherent and collimated beam having essentially a single wavelength *O*. For practical purposes, the beam can be assumed to have a diameter (aperture) equal to that of the window at the emergence point (see Figure 11.7) and a divergence θ. This beam is conveyed to the work by a suitable arrangement of mirrors and focused on the work surface. The focusing is done by a suitable converging lens having a diameter D and a focal length F.

Normally such a beam, when focused by a converging was (convex lens or combination lenses), should be thought of as coming to a point focus, i.e., having an infinitely small area. In practice, the beam focuses on a spot having a significant radius (r_f) and area.

Whenever a collimated beam passes through an opening (aperture) it bends (diffracts) and acquires a divergence of angle θ. If there are a number of apertures, diffraction will occur at each aperture and the beam will have a diameter larger than the original. The typical divergence of a laser beam is 1–5 mrad. Such a

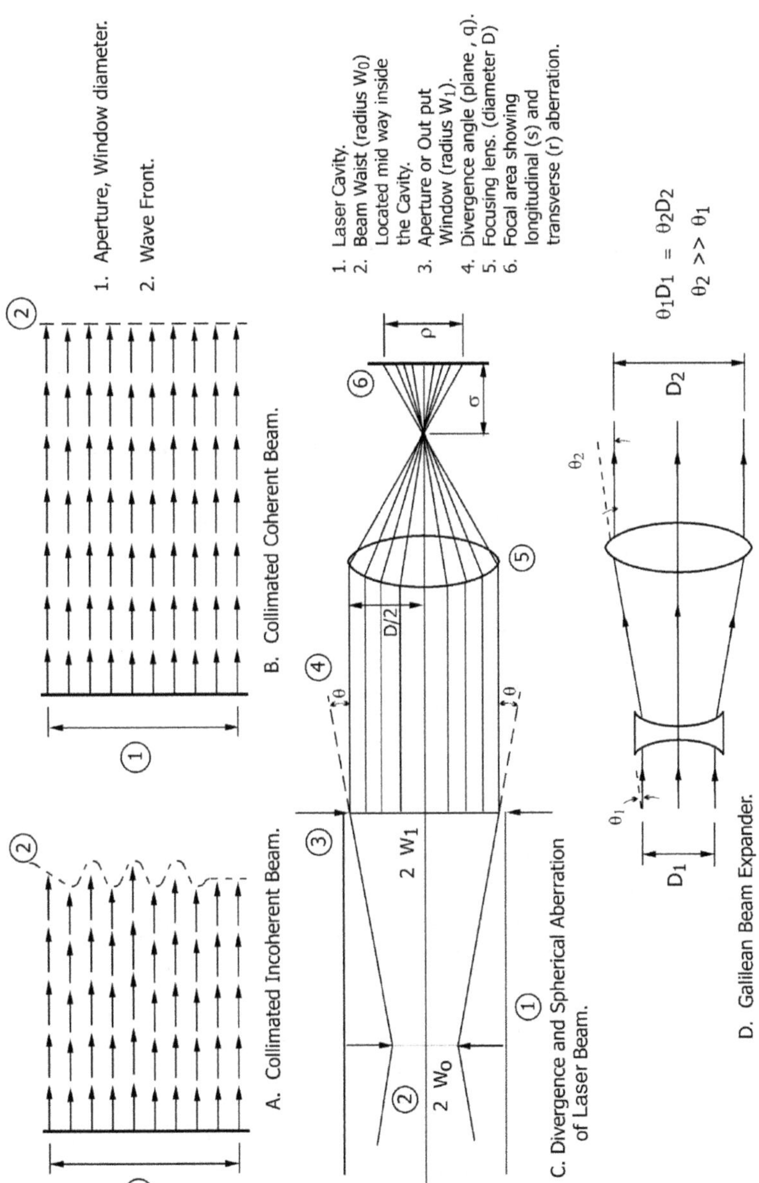
1
2
1. Aperture, Window diameter.
2. Wave Front.
A. Collimated Incoherent Beam.
B. Collimated Coherent Beam.
1
2
3
4
5
6
2 Wo
2 W1
D/2
θ
σ
ρ
1. Laser Cavity.
2. Beam Waist (radius W0) Located mid way inside the Cavity.
3. Aperture or Out put Window (radius W1).
4. Divergence angle (plane , q).
5. Focusing lens. (diameter D)
6. Focal area showing longitudinal (s) and transverse (r) aberration.
C. Divergence and Spherical Aberration of Laser Beam.
θ1
θ2
D1
D2
θ1D1 = θ2D2
θ2 >> θ1
D. Galilean Beam Expander.

Figure 11.7 Some characteristics of laser beams.

divergent beam focuses on a spot having a minimum radius r_o given by

$$r_o = F\theta \tag{11.3}$$

where F is the focal length of the focusing (convergent) lens. This shows that the spot diameter increases with focal length and angle of divergence.

The angle of divergence θ is related to the wavelength l and the lens (aperture) diameter D by the equation (assuming that the beam fills the aperture)

$$\theta = \frac{1.22\lambda}{D} \text{ rad} \tag{11.4}$$

As the wavelength is fixed, θ can be decreased by increasing the aperature (lens and beam size) D. The original beam diameter is fixed by the laser generator window. The beam can be expanded by using beam expanders. These are Galilean telescopes (used reversely) and give a magnification (increase in diameter) from 5 to 20. The divergence angle will be reduced by about one order.

Substituting for θ in Equation (11.3)

$$r_o = \frac{1.22F\lambda}{D} \tag{11.5}$$

Thus, by using a beam expander and a lens of large diameter the spot radius can be decreased to a limit. Also note that large diameter lenses will have a longer focal length F which will increase the spot size. It is therefore necessary to strike a balance between F and D.

Another basic limitation on spot size arises from a phenomenon called "spherical aberration" (Figure 11.7).

Spherical aberration arises from the fact that the rays of light coming from different parts of a lens get focused at different spots on the axis (axial aberration) and on the focal plane (lateral aberration). Thus, spherical aberration will always give a focal spot and not a point.

Spherical aberration can be reduced by using special aspheric lenses; however, applicable optical materials and their properties put a limit to the possible correction.

In summary, the focus of a laser beam produces a spot having a significant diameter. For many other reasons the practical "spot" is always larger than that calculated theoretically. A TEM_{00} beam has a Gaussian distribution throughout and the focal spot also inherits this. So, the power level is very high on the central spot and diminishes toward the edge. Consequently the material reaction varies from the spot center to the edge. The beam edges are trimmed by the aperture, which allows only about 86% energy to pass through.

11.1.10 Collimation

Scientific and industrial applications of radiation (light) usually involves focusing of a well-defined beam by a focusing lens system.

An ordinary source of light (like an incandescent bulb) emits radiation all around, i.e., through a solid angle of 4π steradians. It is virtually impossible to collect all this light and channel it into a beam.

The process of obtaining a well-defined beam of light, consisting of parallel rays is called "collimation." Most of the apparatus using an ordinary bulb catch a small portion of the radiation by a collimator consisting of a suitable lens and a long tube. This arrangement creates a parallel beam having divergence (~ few degrees).

A laser beam is inherently collimated, i.e., it consists of parallel rays requiring no further collimation. The beam has a diameter equal to that of the window (mirror) through which it emerges. There is very little divergence (~ 10^{-3} rad). This beam is easy to focus on relatively small areas (~ 10^{-1} mm) creating a high energy density.

The divergence can be further reduced by using a beam expander. This is a Galilean telescope used in a reverse manner so that the light enters the eyepiece and exits through the objective. The beam diameter can be increased by a factor of 5–20.

This makes it possible to use a large diameter focusing lens (Figure 11.7).

11.1.11 Coherence

This property is also inherent like collimation in a laser beam. A collimated beam from an ordinary light source consists of parallel rays but the wave fronts of individual rays are not synchronized with each other and the collective wave front is not well defined (Figure 11.7).

In a laser beam the individual wavefronts are perfectly synchronized to present a straight and collective wavefront. This synchronization is inherent and persists in both space and time. It can be imagined as the perfectly drilled marching of a group of soldiers showing synchronization at all steps at all times.

The main application of coherence is in measurement systems using laser, and in holography. They involve splitting the beam into two parts, passing them through different paths, and then recombining them to form a well-defined pattern.

Coherence is not treated in detail as it has no major significance in laser heating.

11.1.12 Depth of Focus

The laser beam coming out of the focusing lens converges to a focal spot (see Figure 11.8) of radius r_f and diverges beyond. Assume that the intensity I_o at the focal spot has a Gaussian distribution, then before and beyond the spot the intensity will remain Gaussian but will continuously decrease with distance z from it.

The depth of focus (z_f) is the distance between the focal plane and two symmetrical planes before and beyond. The intensity of the beam on these planes has a chosen ratio with that on the focal plane, (e.g., $I_z/I_o = 0.5$, etc.). Thus with respect to the focal plane at $z = 0$, the depth of focus will have two values $= I_z$.

If the radius at focal spot is r_o then its radius r_z at a (chosen) distance z is given by

$$r_z = r_o\left\{1+\frac{(\lambda z)^2}{\left[\pi r_o^{\,2}\right]}\right\}^{1/2} \tag{11.6}$$

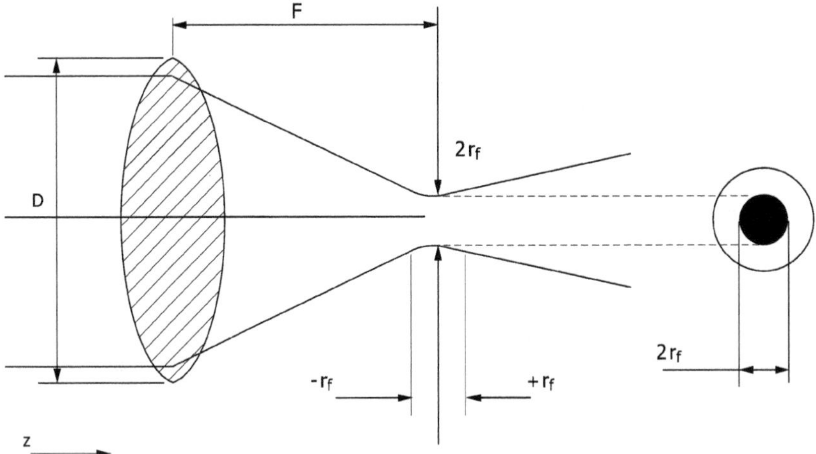

Figure 11.8 Concept of focal spot diameter and depth of focus of a laser beam.

where λ is the wavelength.

If $r_z/r_o = \sqrt{2}$

$$z = \frac{\pi r_o^2}{\lambda} \tag{11.7}$$

This value of z is the theoretical "depth of focus" and is called "Rayleigh range."

For practical purposes, it is more convenient to assume a r_z/r_o ratio of 5–10%. The calculations show that the form of Equation (11.7) remains the same except for a numerical factor (k) so that

$$z = k\frac{\pi r_o^2}{\lambda} \tag{11.8}$$

Where

$$r_o = \frac{F\lambda}{\pi D/2}$$

Substituting for r_o in Equation (11.8) gives

$$z_p = k\frac{F^2\lambda}{D^2} \tag{11.9}$$

Equation (11.9) relates the depth of focus to the focusing lens parameters F and D. It shows that for a large value of z, the focal length F should be large and the diameter D small.

A basic lens equation is

$$\frac{1}{F} = (n-1)\left[\frac{1}{R_1} - \frac{1}{R_2}\right] \tag{11.10}$$

where n is the relative refraction coefficient and R_1 and R_2 the curvature of the two faces of the lens.

For a given optical material n is constant. To obtain a large value of F, R_1 and R_2 must be large, i.e., the lens diameter D should be large. This shows that the requirement of small D and large F is self-contradictory. The depth of focus is therefore fixed when D is fixed by the beam diameter.

When a laser beam is incident on a material surface it is desirable that the beam is exactly focused on it. This will give maximum incident power (and hence efficiency). The material surface evenness, focusing accuracy, and machine vibrations are some of the factors which make exact focusing impossible. The concept of the depth of focus makes it possible to estimate the reduction in incident energy if the exact focus is missed.

In laser drilling and some other applications the depth and shape of the heat is affected and the machined zone changes with the position of the focus with respect to the surface. It is necessary to intentionally defocus the beam to achieve correct machining. The depth of focus is useful in beam defocusing as it gives an estimate of the incident spot size and energy.

EXAMPLE 11.1

A germanium lens used on a CO_2 laser has a clear aperture of 18.7 mm and a focal length 16.95 mm. The wavelength of the laser is 10.6 μm. Assume that the aperture is completely filled by the beam.

Calculate

1. The theoretical depth of focus.
2. The depth of focus at a 10% increase in focal diameter.

Solution

We first calculate the focal spot radius r_o by using the relation

$$r_o = \frac{F\lambda}{\pi D/2}$$

$F = 16.95$ mm, $\lambda = 10.6 \times 10^{-3}$ mm, $D = 18.7$ mm

$$r_o = \frac{16.95 \times 10.6 \times 10^{-3}}{\pi \times 18.7/2} = 0.0612 \text{ mm}$$

Theoretical depth of focus z

$$z = \frac{\pi r_o^2}{\lambda}$$

$$= \frac{\pi \times 0.0612^2}{10.6 \times 10^{-3}}$$

$$z = 1.11 \text{ mm}$$

For 10% increase in r_o

$$\frac{r_z}{r_o} = 1.1$$

From equation (11.6)

$$\frac{r_z}{r_o} = \left\{1 + \left[\frac{(\lambda z)}{\pi r_o^2}\right]^2\right\}^{1/2} = 1.1$$

$$\left[\frac{(\lambda_z)}{\pi r_o^2}\right]^2 = \sqrt{1.21 - 1}$$

$$z = \frac{0.458 \times \pi \times 0.0612^2}{10.6 \times 10^{-3}}$$

$$= 0.508 \text{ mm}$$

EXAMPLE 11.2

A laser beam of wavelength 1.06 μm has a diameter 3.0 mm when coming out of the generator.

The beam is expanded in a beam expander having a magnification by a factor of 10.

It is proposed to use a quartz lens having a transmission of 85%. Now determine:

1. The angle of diversion of unexpanded or raw beam
2. The diversion angle after expansion
3. The focal spot diameter for the raw beam
4. Focal spot diameter for expanded beam

Solution

The raw beam has a diameter of 300 mm and the wavelength 1.06 μm.

The angle of diversion is

$$\theta_r = \frac{1.22\lambda}{D}$$

$$= \frac{1.22 \times 1.06 \times 10^{-3}}{3.0}$$

$$= 4.3 \times 10^{-4} \text{ rad } (0.43 \text{ mrad})$$

$$= 246.4 \times 10^{-4} \text{ degrees}$$

$$\theta_1 = 3'26''$$

On expansion the beam diameter is $3 \times 10 = 30$ mm
The angle of diversion of the expanded beam is

$$\theta_2 = \frac{1.22 \times 1.06 \times 10^{-3}}{30}$$

$$= 4.3 \times 10^{-5} \text{ rad } (0.043 \text{ mrad})$$

Comparing θ_1 and θ_2 shows that the diversion is decreased by one order after expansion.

The raw beam has a diameter of 3.0 mm. Assume that the diameter of the quartz lens used is 4.3 mm and its focal length is 8.1 mm.

The radius of the focal spot will be

$$\begin{aligned} r_{01} &= F\theta \\ &= 8.1 \times 0.43 \times 10^{-3} \\ &= 3.48\,\mu\text{m} \end{aligned}$$

The expanded beam has a diameter of 30 mm. Assume that the lens used has a diameter of 31.5 mm and a focal length of 40 mm. The spot radius will be

$$\begin{aligned} r_{02} &= F\theta \\ &= 40 \times 4.3 \times 10^{-3} \\ &= 1.72\,\mu\text{m} \end{aligned}$$

This shows that the expanded beam gives a fine focus (hence, more energy density).

EXAMPLE 11.3

A laser beam has 0.1 cm focal spot and a power density q_o (W/cm²).

A steel plate is to be treated by this beam for cutting. Determine the time required for cutting a thickness h (< 0.2 cm) and the cutting rate.

Density of steel 7.8 g/cm³

Latent heat of evaporation 7 × 30 J

Solution

We will first determine the heat required for cutting by considering a thickness of 0.1 cm under the spot.

Area and volume of steel 0.1 cm thick and diameter 0.1 cm is

$$A = \pi \times 0.05^2 = 7.8 \times 10^{-3}\ \text{cm}^2$$

$$V = \pi \times 0.05^2 \times 0.1 = 7.85 \times 10^{-4}\ \text{cm}^3$$

At a density $\rho = 7.8$ g/cm^3

$$\text{Weight} = 7.85 \times 10^{-4} \times 7.8$$

$$= 6.1 \times 10^{-3}\,\text{g/mm thickness}$$

Latent heat of vaporization $L_B = 7 \times 10^3$ J/g

Heat required for vaporization of material having thickness h cm is

$$J = 6.1 \times 10^{-3} \times 7 \times 10^3 \times h$$

$$= 42.7 \times h$$

If the cutting time is t (sec), the heat given is

$$= q_o \times \text{area} \times \text{time}$$

$$= q_o \times 7.8 \times 10^{-3} \times t\ \text{J}$$

$$= 42.7 \times h\ \tau \text{ as calculated above.}$$

The time required is

$$t = \frac{42.7}{7.8 \times 10^{-3}} \frac{h}{q_o}\ \text{sec}$$

$$= 5.47 \times 10^3\, h/q_o\ \text{sec}$$

In time t a length equal to one spot diameter (0.1 cm) is cut. For continuous cutting the speed per min will be

$$= \frac{60 \times 0.1}{t}\ \text{cm/min}$$

11.1.13 Transverse Modes in Lasers

In the previous section we have discussed the generation and characteristics of longitudinal modes in a laser cavity.

There are other complex modes that occur in the cavity due to a variety of reasons. Broadly speaking, these are due to the limited transverse dimension of cavity (e.g., diameter), the curvature of mirrors, diffraction of electromagnetic waves, etc. These modes are called transverse electromagnetic (TEM) and affect the outgoing beam, both spatially and temporally (time-wise). They produce complex patterns of dark and bright shades (Figure 11.9). The result (from our point of view) is a symmetrical but nonuniform power distribution through the beam cross-section. The patterns are classified and numbered.

The simple TEM mode which is useful for our purpose is the TEM_{00} mode, in which the intensity and, hence, power has a Gaussian distribution as shown in Figure 11.9. It has a central spot of maximum intensity which decreases exponentially in the radial direction. The beam radius is defined as the distance from the center to a point at which the intensity is reduced to 0.135 ($1/e^2$) of its maximum value at the center so that the power (area under the $I - r$ curve) is 86.5% of the total area.

Some other TEM patterns are also shown in Figure 11.9. These are not used in material processing and are therefore not discussed further.

One useful aspect of the TEM_{00} mode is that it retains its Gaussian pattern throughout. The beam emerging from the generator and the final focal spot are both Gaussian though the beam diameter is different. The transverse mode of other TEM patterns change as they progress and undergo reflections or refraction in the beam manipulation and focusing systems.

Commercial lasers used in material processing (power lasers) are adjusted for the TEM_{00} mode.

11.1.14 Temporal Characteristics of Lasers

Laser generators have two distinct features. They are pulsed or continuous.

Pulsed lasers have a periodic output in short bursts or pulses. The pulse duration is in the 10^{-3}–10^{-6} sec range. The pulses are repeated after a time interval called the "pulse repetition rate" which has a wide range depending on the type of laser and mode of operation. Repetition rates from 0.1

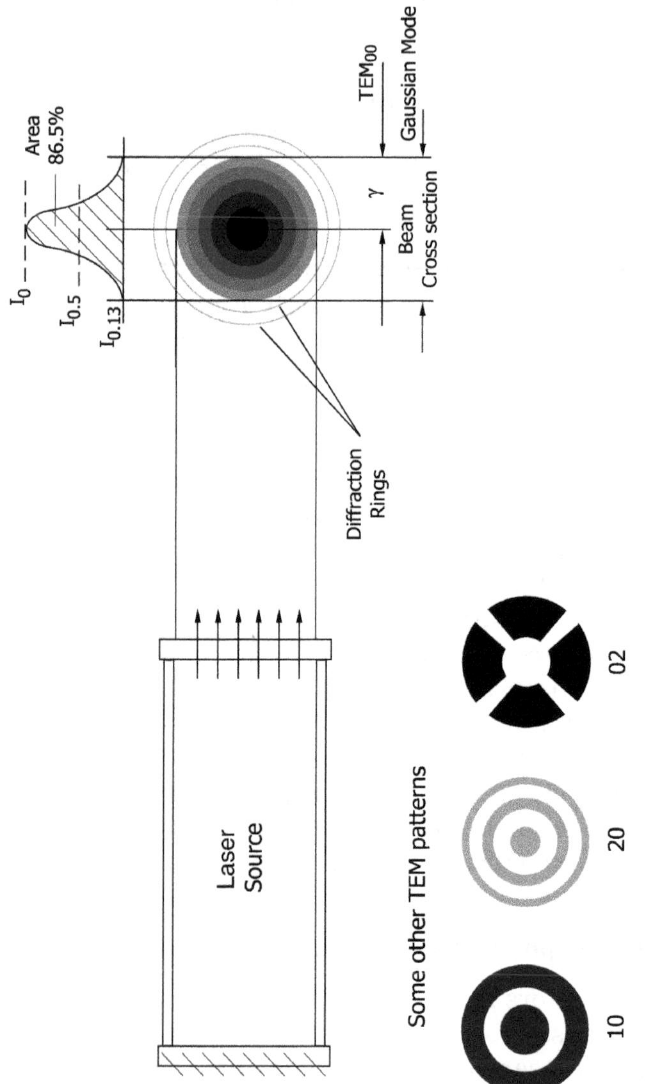

Figure 11.9 Transverse modes of laser beams.

to 10^3 Hz are possible. Solid state lasers, like ruby and Nd-YAG, which are pumped by flash lamps show a pulsed output.

A typical uncontrolled pulsed output is shown in Figure 11.10(A). The output consists of equally spaced spikes. The energy (intensity) increases to a maximum and then slowly decays. This is called a "free running pulse." This type of pulse is obtained by controlling the frequency of pumping lights or discharge circuits.

Total energy (J) in a pulse will be the area under the $I - \tau$ curve. For small generators it is of the order of 10^{-3} J. The power in the pulse will be average energy/sec (W), i.e., I_{av}/τ (Figure 11.10(B)). It is generally in the range of 1.0–30 W. Pulse characteristics are measured by an oscilloscope trace or by a suitable calorimeter. All the reported values of energy and power are for TEM_{00} mode.

Pulse discharge can be controlled by adjusting the pumping frequency and cavity optics.

A controlled pulse is shown in Figure 11.10(B). The energy increases quickly to a peak value and decreases to a plateau. If the pulse repetition rate is fast, we get a quasi or semi-continuous pulse.

In heating applications the pulse represents a "dose" of energy incident on the material surface. The intensity of the dose is in watts/unit area (e.g., W/cm^2), applied for the pulse duration time τ sec. It is important to choose a proper pulse power to achieve the proper temperature at the required site. A lower power will not raise the temperature to the required level. A higher dose will produce a very high temperature which will spoil the operation. We will discuss this point in more detail in subsequent sections.

11.1.15 Q Switching of the Laser Beam

This is a very interesting way of changing the pulse shape as shown in Figure 11.10(C).

The cavity in a Q switched laser contains a device called the Q switch. There are many types of Q switches such as a rotating prism or mirror, electro-optic or electroacoustic devices. These devices are switchable reflectors i.e., they will totally

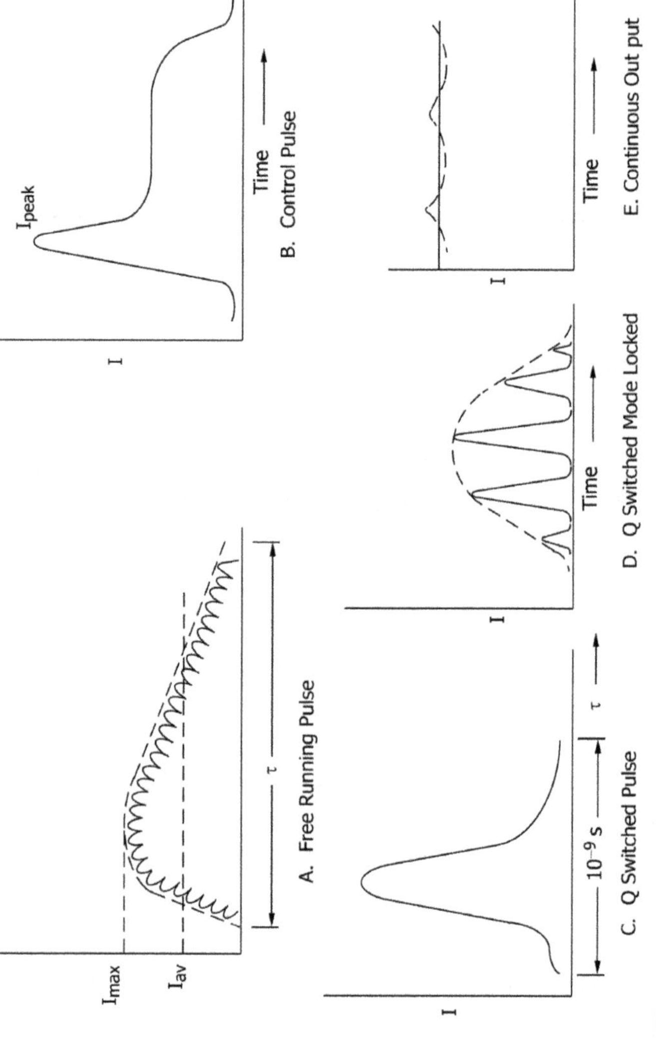

Figure 11.10 Typical laser outputs.

reflect the beam or allow it to pass through when a signal is applied. The position of the Q switch in a cavity is shown in Figure 11.10(B).

In the beginning, the switch is turned on along with the pumping source. No laser beam is produced as there is no resonant cavity. The pump keeps on exciting the electrons to a higher level resulting in a buildup of a large population in the higher level.

On switching off the Q switch, a resonant cavity is suddenly available. A powerful pulse of very high peak intensity but very short duration (~ 10^{-9} sec) emerges. The shape of a Q-switched pulse is shown in Figure 11.10(C). Q switching thus produces very powerful pulses (~ 10^8 W) of short duration. The pulse repetition can be ~ 10^3 Hz.

Q switching is very useful for precision cutting, drilling, and welding of thin sheets, fine wires, etc.

All the three lasers of interest to us, viz. the ruby, Nd-YAG, and CO_2 can be Q switched.

Another variety of a Q-switched pulse is the mode-locked pulse shown in Figure 11.10(D). It has roughly the same overall pulse shape but is made of a number of synchronized subpulses. Mode locking arises from multimode (frequencies) generation of longitudinal waves. Mode locking is achieved by generating a broad line width and adjusting the cavity losses.

Regular Q-switched and mode-locked Q–switched pulses are sometimes simultaneously generated.

A technique somewhat similar to Q switching is called "cavity dumping." Here the laser is continuously pumped and the beam is "stored" in the cavity (similar to charging a capacitor). On a signal the stored beam is dumped out in one large energy pulse of short duration.

Continuous Output

Some lasers like CO_2 and Nd-YAG give a continuous output as shown in Figure 11.10(E). The amplified output shows a wavy or spiked trace. In fact, CO_2 laser is almost always operated in CW (continuous wave) mode. The output power is 10–10^4 W. There is no control as in a pulsed output. The power incident on a material surface is controlled by moving the work or the laser beam to adjust the dwell time. Nd-YAG and ruby lasers rated as lower (10–100 W) are generally pulse operated.

11.1.16 Application of Lasers for Material Processing

For material processing the important attributes of laser are:

1. Generation of very high temperature on a very small area of the material surface in a very short time.
2. Absence of any protective atmosphere.
3. Practically no heat affected zone.
4. The laser cannot penetrate deeper than about 2.0 mm at the maximum. It is more suitable and efficient for thin work.

These attributes make it possible to heat any material to any desired temperature up to its boiling, burning, or charring point.

Consider a material heated from room temperature T_o (~ 25°C) to progressively higher temperatures. For our analysis, we will divide all materials into two categories.

1. Materials like metals, ceramics, glass, etc., which will progressively melt, boil, and evaporate.
2. Materials which do not melt, but char, burn, or disintegrate at higher temperatures, like wood, textile, and rubber.

When heated, materials absorb heat according to their specific and latent heats. The temperature-heat content characteristics for the two categories are shown in Figure 11.11.

For the first category, the heat absorbed from T_o to and above T_b will be (per unit mass)

$H_1 = C_S(T_m - T_o)$ up to melting point (T_m°C)

$H_2 = L_m$ latent heat (L_m) during melting

$H_3 = C_l(T_b - T_m)$ up to boiling point (T_m°C)

$H_4 = L_b$ latent heat (L_b) during boiling

$H_5 = C_v(T_f - T_b)$ up to final temperature (T_f)

ΣH = Total heat required.

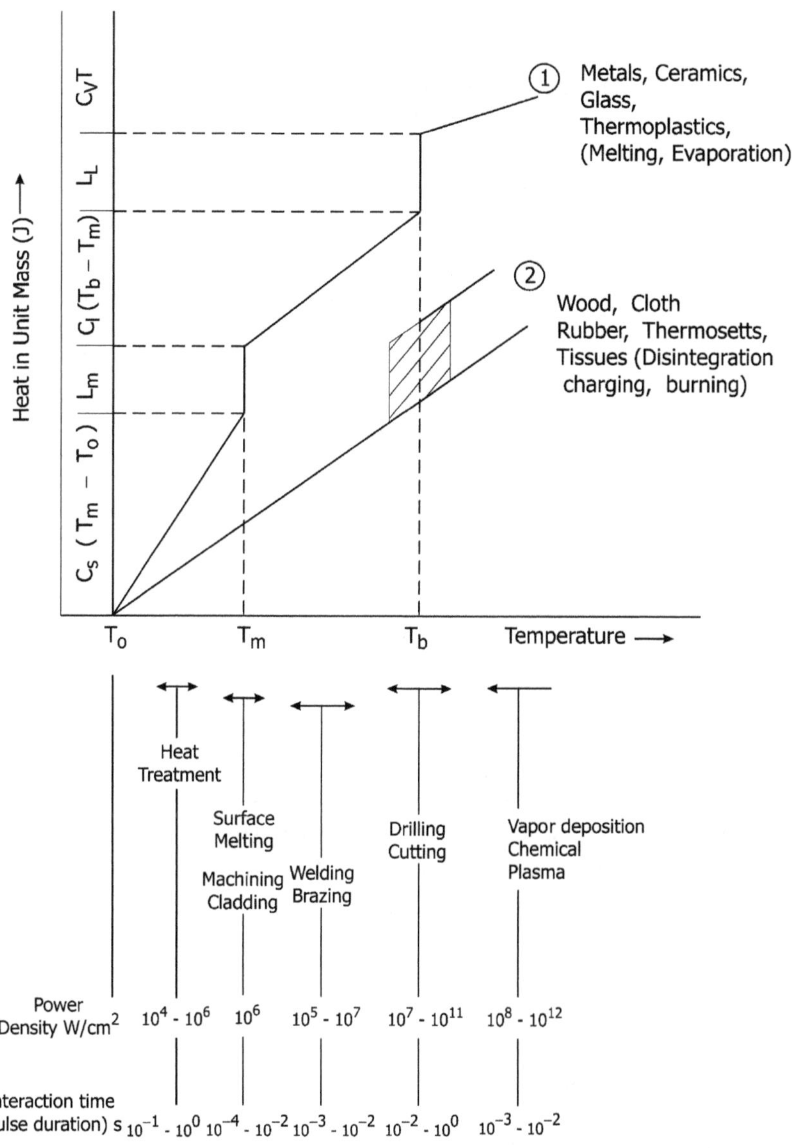

Figure 11.11 Energy, power pulse duration for laser applications.

For the second category the heat required will be $H = C_s (T_F - T_o)$.

Depending on the process to be carried out, the material will be required to be heated to any temperature from T_o to T_f by supplying the necessary heat at the process location.

If the heat supplied is less than that required, the process will be incomplete or unsuccessful. If it is much more than required, the material will melt or boil and evaporate.

It is therefore necessary to achieve a balance between heat requirement and supply. In laser heating there is practically no control on the energy of the beam. Control can be exerted by adjusting the time of laser application. In a pulsed laser, time is controlled by pulse duration. For a continuous laser, time is adjusted by moving the process material.

Typical application areas for different temperatures, power densities required, and interaction time are indicated in Figure 11.11.

Consider heating of 0.4% carbon steel having the following average properties:

Density	7.67 g/cm^3
Melting point (T_m)	1500°C
Boiling point (T_b)	2860°C
Specific heat of Solid (C_s)	0.57 J/g°C
Latent heat of melting (L_m)	268 J/g
Specific heat of liquid (C_ℓ)	0.76 J/g°C
Latent heat of boiling (L_b)	6720 J/g

Heat required (J) per g for heating from 25 to 2860°C in various steps will be

$$H_1 = 0.57(1500 - 25) = 840$$

$$H_2 = L_m = 268$$

$$H_3 = 0.76(2860 - 1500) = 1033$$

$$H_4 = L_b = 6720$$

The above calculations show that heat required for heating to about 1000°C for processes such as heat treatment is about 556 J/g.

Heat required for melting, as in welding, is 840 + 268 = 1100 J/g.

Heat required for evaporation, as in cutting or drilling, is 840 + 268 + 1033 + 6720 = 8860 J/g, and the major part is that of the latent heat of boiling.

11.1.17 Laser-Material Interaction

A laser beam incident on and absorbed by a material, generates heat. This heat (at least in the beginning) is concentrated on the focal spot. Later it is transferred to the interior mostly by conduction.

The laser source may be a pulsed or a continuous one. If a single pulse is incident on the spot it transfers a single "dose" of energy (J) to the material during the pulse duration (10^{-3} sec). If required, more pulses can be given at an interval equal to the pulse repetition frequency.

In the continuous wave (CW) source the energy supply is continuous. The required power is transmitted by adjusting the time of exposure.

The transfer of energy to the material surface, its absorption, and transfer is governed by a very large number of variables. These can be divided into two categories.

1. Properties of the laser beam and the beam transfer and focusing optical train.
2. Properties of the work material, such as thermal conductivity, reflectivity, etc.

Many of the material properties are dependant on temperature and time.

The interplay of these two categories makes it virtually impossible to develop a satisfactory theoretical approach for laser material interactions that will be applicable in all situations. Our aim is to establish a method for estimating the temperature field in a laser-heated material for a given laser power or vice versa. Hence, the treatment of the interaction problem that follows is largely empirical. With some

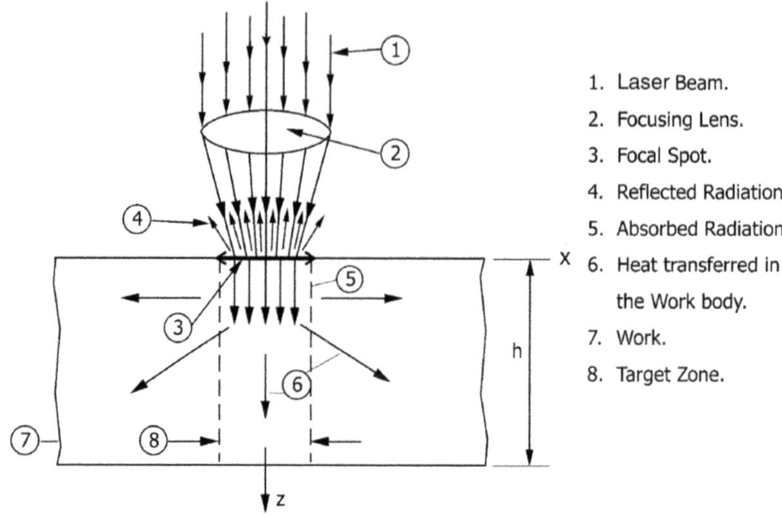

Figure 11.12 Energy transfer from laser beam to work.

modifications the treatment can also be applied to the electron beam (EB)-material problem.

The laser-material interaction–related zone is shown in Figure 11.12. The variables affecting the process can now be identified as follows:

1. Laser beam
   - Wavelength, frequency – λ, Hz
   - Mode – TEM_{00} or other
   - Emission – Continuous (CW) or pulse
   - Pulse duration – sec
   - Pulse repetition frequency – PRF Hz - Adjustable - Yes/No
   - Power – W
   - Q switching – Yes/No
   - Divergence angle
   - Beam diameter
2. Beam manipulation
   - Optics – Mirrors, lenses, material and absorption
   - Focal length – f (cm)

Depth of focus – cm
Interchangeable lenses – Yes/No
Focal spot radius – r_f(cm)
Monitoring microscope – Yes/No
Power measurement – Yes/No
Shutter and diaphragm – Yes/No

3. Material properties
   Type of material – Metal, polymer, ceramic
   Reflectivity and its dependance on surface finish, wavelength, temperature, etc.
   Thermal conductivity – K (W/cm°C)
   Specific Heat – C (J/g°C)
   Thermal diffusivity – α (cm²/sec)
   Density – ρ (g/cm³)
   Latent heat of fusion L_M – J/g
   Melting temperature T_M – °C
   Latent heat of fusion L_B – J/g
   Boiling temperature T_B – °C
   Thickness h – cm
   Processing required – cutting, welding, etc.
4. Machine related
   Work manipulation, axes, manual, automatic, programmable shielding, assisting gas control.

The behavior of materials on heating is shown in Figure 11.11.

On heating a metallic, ceramic, plastics, and glass materials will melt at the melting temperature and will absorb heat due to specific heat and temperature rise. At melting temperature T_m they will absorb the latent heat of fusion L_M.

If heating is continued further, the heat will be absorbed up to the boiling temperature T_B and on continued heating, the latent heat of boiling L_B will be absorbed. The vapor so produced will blow out with or without gas assistance.

The heat input to the material through the focal spot will have to be controlled to obtain the required processing temperature. Thus, for heat treatment the temperature required will be less than T_M. For welding it will be $>T_M$ and less than T_B. The depth to which the material gets heated depends on the thermal diffusivity and time.

Materials such as wood, paper, cloth, etc., do not melt; but they get charred or evaporated or burn on heating. For all types of cutting and drilling operations on both types of materials the temperature required is higher than the boiling or the burning temperature T_B.

Approximate temperature ranges, the power densities required, and observed interaction times are also shown in Figure 11.11.

A generalized commercial laser beam system is shown in Figure 11.13.

11.1.18 Reflectivity and Absorptivity

When radiation strikes a surface, a part of it is absorbed and the rest is reflected. The fraction of incident energy absorbed is called absorptivity (A). Reflectivity (R) is the fraction that is reflected. Hence,

$$R = (1 - A) \tag{11.11}$$

It is the absorbed fraction of incident energy that is useful for heating by laser. Hence, reflectivity represents a direct loss.

The reflectivity of a material surface is governed by many interrelated factors (see Figure 11.14). The surface condition is perhaps the most important. Clean, smooth, and polished surfaces are almost 100% reflective whereas oxidized, rough surfaces can have an absorptivity of about 70%. Very clean and polished surfaces are difficult to process by laser. It is sometimes necessary to coat these with antireflective coatings.

Reflectivity depends on the wavelength of the incident beam. Figure 11.14 (B) shows that the absorptivity is poor for the far infrared wavelength of a CO_2 laser (which is a commonly used power laser) and is better for a ruby laser (which does not have enough power!).

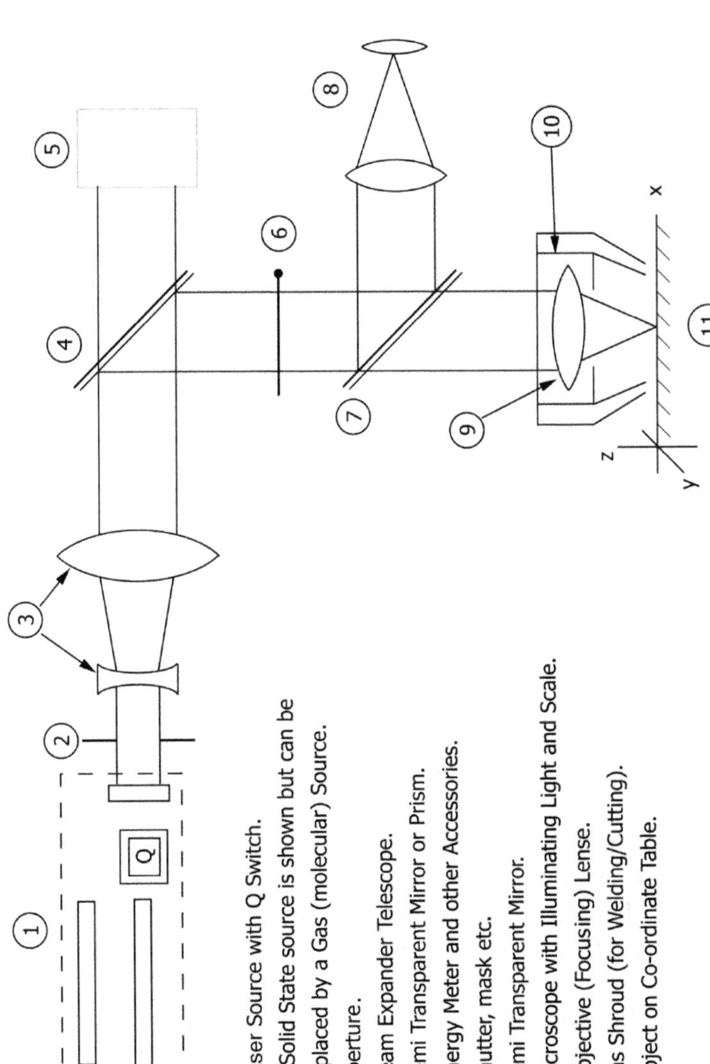
1
2
3
4
5
6
7
8
9
10
11
Q
x
y
z
1. Laser Source with Q Switch.
A Solid State source is shown but can be replaced by a Gas (molecular) Source.
2. Aperture.
3. Beam Expander Telescope.
4. Semi Transparent Mirror or Prism.
5. Energy Meter and other Accessories.
6. Shutter, mask etc.
7. Semi Transparent Mirror.
8. Microscope with Illuminating Light and Scale.
9. Objective (Focusing) Lense.
10. Gas Shroud (for Welding/Cutting).
11. Object on Co-ordinate Table.

Figure 11.13 A generalized laser beam system.

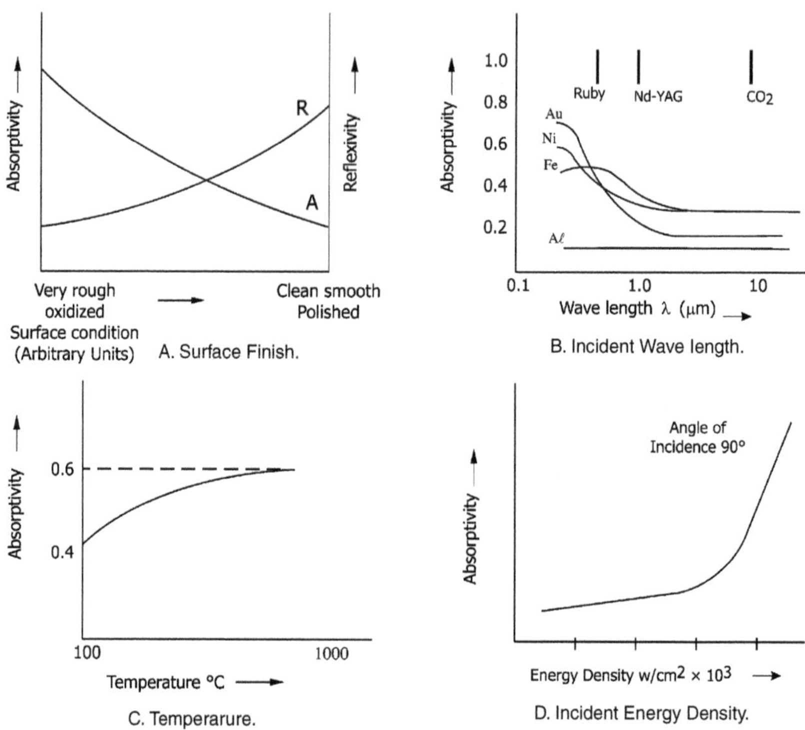

Figure 11.14 Absorptivity of metals, typical behavior.

Absorptivity which is low (~ 8%) at low temperature, increases to about 14–16% at higher temperatures. This has a direct effect on heating. The initial low heating rate due to high reflectivity increases quickly as the surface temperature rises. The angle of incidence also decides reflectivity. Best absorption is obtained when the angle of incidence is 90°.

Lastly, absorptivity increases with the energy density (w/cm^2) of the incident beam (Figure 11.14(D)). This may be due to faster temperature rise with a powerful beam.

Reliable data on reflectivity and absorptivity are scarce. The figures available may not be reproducible in each situation. For estimating, a reflectivity of 70–80% may be assumed after beginning the process. At the start, a 10% absorptivity appears to give reasonable estimates (applicable for very short, ≤ 1msec pulse durations).

When heating involves melting and evaporation, a higher apparent absorptivity (~ 50–70%) may be used.

11.1.19 Laser Penetration

Theoretical aspects of concentrated source-material reaction were already introduced in Section 4.7.

We have discussed the heating of the surface with time on exposure to a laser beam. We will now consider the penetration of heat along the vertical (z) axis. Here "penetration" is understood as the depth to which the material is heated to the desired temperature. It is assumed that the absorbed heat flows only in z direction and the related physical properties are constant.

If heating is to be done for surface treatment such as hardening, the desired temperature is less than T_M and higher or equal to the desired hardening or phase transformation temperature. For applications such as welding, the heating temperature is higher than T_M and lower than T_B. For drilling and cutting operations the temperature desired is higher than T_B so that the material evaporates through the full thickness. There is no single equation to predict the depth of penetration for all the processes given above.

For heating below the melting temperature T_M we can use Equation (4.33), Equation (4.36) to calculate the temperature at the desired depth z and time t, i.e., $T(z, t)$. This equation is strictly applicable for continuous laser or for pulsed laser in the time regime $t \le t_i$. Once the beam is discontinued (or the pulse is over), only the surface cools immediately. Due to the superheat ($T_M - T_P$) the heating still continues for some time. This heating is not calculated from the equation. Thus, the critical heating and cooling rates essential for heat treatment cannot be applicable over the full depth of penetration. If heating is continued for a longer time, heat will be lost by conduction outside the target zone.

For welding, it is essential to obtain a temperature $\ge T_M$ without evaporation. Melting absorbs additional heat as latent heat L_M. This heat will have to be supplied by a CW beam or from the superheat $(T_B - T_M)$ obtained from the pulse. Roughly, the penetration depth h(cm) obtained when the surface is heated to T_B can be estimated as

$$h = \frac{3.45}{T_M} \tag{11.12}$$

There is no satisfactory equation to predict the velocity of the solid-liquid boundary as the melting progresses. Penetration depth with lasers having power less than 800–1000 W is about 0.1 to 1.5 mm. Penetration increases with laser power. Thus laser welding in its normal form can be applied to microwelding (welding of small, thin materials). With lasers having power more than 1–5 kW, larger penetration can be obtained because of "keyholing." Depth up to 20 mm has been obtained.

Laser drilling or cutting involves vigorous vaporization. The vapor so produced is blown or burned by a suitable gas jet. This exposes a fresh surface for vaporization. Thus, material removal is very fast. As the evaporation front recedes from the focal spot, the intensity of beam decreases and evaporation stops due to inadequate energy supply. This restricts the cutting and drilling depths to a maximum of 1–3 mm depending on the laser power. Gas blowing with air or oxygen increases the depth.

Reflectivity (or absorptivity) greatly affects all laser heating processes. A suitable allowance for this must be made in estimating laser power requirements.

The time duration of heating is the most important control variable. A shorter time will produce lower temperature and penetration. A longer time will heat the surface to an undesirably higher temperature resulting in plasma generation and thermal shock. The heat affected zone (HAZ) will also be broad.

The penetration depth can be controlled by slightly defocusing the beam so that the focus spot is little above or below the material surface. Remember that the beam energy is maximum at the focus spot and will decrease away from it.

11.1.20 The Temperature Field

Laser heating applications involve a wide temperature range as discussed in Section 11.1.9. In this range the material undergoes a change of state from solid to vapor. The heat flow is from the surface to the interior.

There is no one equation that can be used to calculate the temperature field in the material. Available equations are quite complicated and are applicable under limited conditions only.

The laser energy input is of two types, continuous wave (CW) and pulse. Separate equations are involved for these two types of inputs. We treat the input as a point source having a uniform energy density and a circular shape of radius r_f. The heat flow in the material is assumed to be unidirectional (direction z, Figure 11.12) in the process time t sec. The change in thermal properties with temperature is neglected.

For a continuous point source the equation chosen is

$$T(z,t) = \frac{2q_o\sqrt{\alpha t}}{K}\left[ierfc\frac{z}{2\sqrt{\alpha t}} - ierfc\frac{\left(z^2 + r_f^{\ 2}\right)^{1/2}}{2\sqrt{\alpha t}}\right] \tag{11.13}$$

where

$T(z,t)$ = Temperature on axis at depth z and time t
q_o = Input power density W/cm^2
α = Thermal diffusivity cm^2/sec
K = Thermal conductivity w/cm°C
r_f = Radius of focal spot cm
t = Exposure time sec

(The *ierfc* functions in the brackets are integrated complementary error functions which are available in tabulated form.)

The temperature at surface $T(0,t)$ is given by

$$T(0,t) = \frac{2q_o\sqrt{\alpha t}}{\sqrt{\pi}K} = \frac{1.1284\,q_o\sqrt{\alpha t}}{K} \tag{11.14}$$

The power density required to obtain a certain temperature T at the surface is

$$q_o = \frac{0.856KT}{\sqrt{\alpha t}} \tag{11.15}$$

The limiting temperature at the surface when $t \to \infty$ is

$$T(0,t) = \frac{q_o r_f}{K} \tag{11.16}$$

If the input is a pulse we have to consider two time regimes. The first regime is from 0 to t_i where t_i is the pulse duration (S). In this time the pulse energy (J) is deposited on the surface and is heated to a temperature T. In the second regime which lasts from t_i to t the deposited energy heats the interior. There is no fresh energy input. We again assume that both regimes are of sufficiently small duration to assure an essentially unidirectional heat flow.

The temperature distribution in the first regime is given by

$$T(\infty,t) = \frac{2q_o\sqrt{\alpha t}}{K} ierfc \frac{z}{2\sqrt{\alpha t_i}} \tag{11.17}$$

Here $t \le t_i$.

At the end of the first regime ($t = t_i$) the surface has attained a temperature $T(0, t_1)$ which will decrease logarithmically in the second regime. The temperature distribution is now given by

$$T(z,t) = \frac{Q_o}{A\rho c(\pi\alpha t)^{1/2}} \exp\left[-\frac{z^2}{4\alpha t}\right] \tag{11.18}$$

where $t > t_i$ and

Q_o = Pulse energy (J)
ρ = Density (g/cm^3)
c = Specific heat (J/g°C)
A = Spot area = πr_f^2 (– cm)

The power density q_o to obtain a certain temperature at the surface $T(0, t_i)$ will be given by

$$q_o = \frac{0.884\,K\,T}{\sqrt{\alpha t_i}} \text{ W/cm}^2 \tag{11.19}$$

which is similar to Equation 11.17 as $t \to \infty$, $T \to T_o$, i.e., surrounding or initial temperature.

We can rearrange Equation 11.19 above to obtain the pulse time t_i to heat the surface to the required temperature.

$$t_i = \frac{0.79 K^2 T^2}{q_o^{\,2} \alpha} \tag{11.20}$$

In Equation (11.19) and Equation (11.20) we can substitute T_M or T_B in place of T to obtain the power density required for melting or boiling temperature at the surface.

In cutting and drilling operations we are interested in continuous evaporation from the surface and have to consider the latent heat of evaporation L_B(J/g). The power density for vigorous evaporation is given by

$$q_c^v = \rho L_B \left(\frac{\alpha}{t} \right)^{1/2} \tag{11.21}$$

A continuous laser power q is given in watts. For a pulsed laser what we know is the average or total energy in the pulse (Q_{av} or Q) and the pulse duration t_i (sec). This can be converted to watts as

$$q = \frac{Q_{av}}{t_i} \text{ W} \tag{11.22}$$

If r_f is the spot diameter (cm) the power density is

$$q_o = \frac{Q_{av}}{t_i \times \pi r_f^2} \text{ W/cm}^2 \tag{11.23}$$

11.2 ELECTRON BEAM HEATING

11.2.1 Introduction

An electron is a fundamental particle carrying a fixed negative charge (–e) and having a mass (m). It can be accelerated to tremendous velocities by an electromagnetic field. This increases the kinetic energy of the electron.

When a beam of electron (EB) so accelerated strikes a target matter, it loses kinetic energy, which is converted into heat in the target. This is the principle of electron beam heating.

Note that the beam itself is "cold" and the heat is created inside the target.

An electron beam can be precisely focused (like a beam of light) by an electromagnetic focusing coil which acts as a lens. The charge on the electron thus enables the electrons to be accelerated to high energy and also to concentrate the energy on a small spot by focusing.

The heat created is focused on a small spot ($d \leq 1.0$ mm) and an extremely high temperature is created locally. Thus the target spot can be melted or evaporated in a very short time (~ few msec). The bulk target material is essentially unaffected.

Due to the tremendous kinetic energy, the beam can penetrate the target to a considerable depth.

All the above-mentioned characteristics of the electron beam make it an ideal heat source for welding, small scale melting, or evaporation applications.

In the next sections we will discuss in detail the generation, control, properties, and applications of the electron beam (EB) as an industrial heat source.

11.2.2 Generation of Electron Beam

A general industrial setup for producing and controlling an electron beam is shown in Figure 11.15. The electrons are produced by indirectly heating a negatively charged cathode made from a material such as lanthanum hexaboride, which releases electrons on heating to 1400–1600°C. The heating filament is made of tungsten. The cathode heater assembly is sealed and has a life of 200 h. The released electrons are collected and directed forward by the cathode cup. A bias cup controls the electrons moving ahead. This acts like a grid in a triode vacuum tube.

An anode cup with a hole in the center is next. It is positively charged and accelerates the incident electron stream which passes through the hole. The emerging beam is shaped by a coil below the anode. The potential difference which acclerates the electrons to high energy is applied between the cathode (–ve) and anode (+ve) by a high voltage cable. A ceramic high voltage insulator electrically separates the anode and cathode. Depending on the design and application the voltage is between 80 and 150 kV and is variable to suit the requirements.

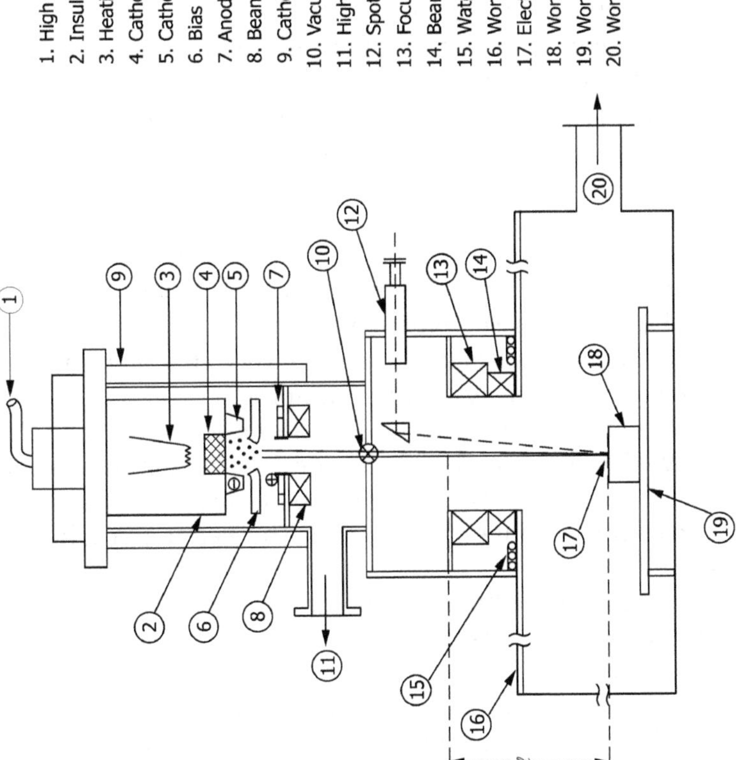

1. High voltage cable
2. Insulator
3. Heating Filament
4. Cathode (-ve)
5. Cathode Cup
6. Bias (grid)
7. Anode with hole
8. Beam shaping coil
9. Cathode cooling jacket
10. Vacuum isolation valve
11. High vacuum connection
12. Spot observation telescope
13. Focusing coil
14. Beam deflection coil
15. Water cooling
16. Working Chamber
17. Electron beam
18. Work
19. Work table and manipulator
20. Work chamber vacuum connection

Figure 11.15 Electron beam machine.

The whole arrangement, known as the "electron generator," is installed in a casing. There is vacuum connection on the side and an isolation vacuum valve at the bottom of the casing.

A chamber next to the electron gererator contains the focusing coil and deflection coil. The focusing coil focuses the beam on a small area on the work or target. It also adjusts the distance (ℓ) between the focusing coil and the target. This distance and the focus spot diameter are important process parameters.

The work is situated in a vacuum work chamber located below. There are arrangements such as a coordinate table, holding devices, etc., in the work chamber. A vacuum connection evacuates the chamber to about 10^{-2} to 10^{-4} torr.

The work can be remotely manipulated inside the chamber to bring the required location in the beam path.

Electrons on impact (at high velocities) on the target surface generate X-rays which can be hazardous, hence the chamber is protected by proper shields. The observation window is made from lead glass which is opaque to X-rays.

Electrons in the beam on impact with gas molecules in the air lose their energy. Hence, the whole process is carried out in vacuum. The generator compartment requires a higher vacuum (~ 10^{-6} torr), hence it is isolated from the work chamber and has a separate high vacuum system for pumping.

There is an optical observation system through which the spot and the beam action can be observed. The deflection coil deflects the beam about the central spot. Some machines have a beam tracking arrangement so that the beam can follow a programmed path.

The current in the high voltage circuit (cathode-anode) is a few milliamperes. Extensive water-cooling arrangements are made in the generator.

The work chamber can be of a small size (0.5 m^3) to a very large one (2–5 m^3) depending on the requirements.

An electron beam carries a current (due to the charge) and the circuit between the anode and the work has to be completed. This is achieved by grounding the work (0 V). It is then necessary that the work is conducting. This is why EB can only operate on metals.

11.2.3 Characteristics of EB

Charge (–e) on the electron = 1.6×10^{-19} coulombs (C)
Mass of the electron (me) = 9.1083×10^{-31} kg

If an electron (initially at rest) is placed in an electric field of intensity E (V/m), it will experience a force F which will cause it to move with a velocity U (m/sec) toward the positive.

The work done when an electron moves under a force F in a field having a voltage V is given by

$$\text{Word done} = eV \tag{11.24}$$

The kinetic energy acquired by the electron due to the work = $\frac{1}{2} m_e U^2$

Equating the work and energy

$$eV = \frac{1}{2} m_e U^2$$

$$U = \sqrt{\frac{2eV}{m_e}}$$

substituting for e and m_e, the values given above

$$U = 600\sqrt{v}\ \text{km/sec} \tag{11.25}$$

Thus, if the cathode-anode voltage is 80 kV

$$U = 600\sqrt{80 \times 10^3}$$

$$= 1.7 \times 10^5\ \text{km/sec}$$

$$\text{Kinetic energy } = \frac{1}{2} m_e U^2$$

$$= \frac{9.1083 \times 10^{-31} \times (1.7 \times 10^8)^2}{2}$$

$$= 1.315 \times 10^{15}\,\text{J}$$

It is difficult to establish a formula for the focal spot diameter of EB as a large number of variables are involved. It involves some machine parameters, and the data given by the manufacturer is useful for that particular machine. There are several measurement techniques that can be used to measure the actual diameter produced under given operating conditions. The spot diameters generally available are in the range of 10^{-3}–1.0 cm. The diameter decreases with increasing accelerating voltage. If the beam current (I) and accelerating voltage (V) are known, the total beam power is $I \times V$ and the power density

$$q = \frac{I \cdot V}{(\pi \times d^2/4)} \text{ w/cm}^2 \tag{11.26}$$

can be calculated. Typical power densities for EB are in the range of 10^5–10^8 w/cm^2.

High energy electrons penetrate a metal target to a considerable depth. The electrons collide with the free electrons and the ionic lattice. At each collision some energy is lost which is converted into heat. The number of collisions undergone before the electron comes to rest (very low kinetic energy) is a statistical phenomena.

Due to each collision the direction of the electron changes and the incident beam scatters inside the target. The depth(δ) to which the electrons penetrate is given by

$$\delta = 2.6 \times 10^{-12} V^2/\rho$$

where

V = Accelerating voltage and ρ the target density (g/cm^3)

Thus, if $V = 75$ kV and $\rho = 7.8$ g/cm^3

$$\delta = 2.6 \times 10^{-12} \times (75 \times 10^3)^2 \times 7.8 = 0.114 \text{ cm}$$

At 100 kV $\delta = 0.203$ cm

11.2.4 EB—Noteworthy Points

An electron beam can be continuous or pulsed. It is initiated by the application of current to the cathode and accelerating voltage between cathode and anode pulsing can be easily achieved by switching the power.

During their journey from the cathode to the target, the electrons will suffer energy loss if they collide with gas molecules in the air. To avoid this loss the entire apparatus has to be kept under vacuum. The need for vacuum (~ 10^{-4} to 10^{-6} torr) is one of the disadvantages of EB. Small machines have a single vacuum system. Larger machines have a high vacuum system for the electron beam generator. The work chamber, which is isolated from the generator, has a separate, large capacity, and medium vacuum system.

As mentioned before, an electron beam can operate only on metallic (conducting) targets.

The energy density in the spot is very high (~ 10^8 w/cm^2) and is easily controlled by the high voltage and the focusing currents. This enables the creation of very high temperatures at the spot. There is no temperature limit. Any metal or alloy can be melted or evaporated at the spot. Due to vacuum there is no oxidation at the target.

Due to the electric control, the beam can be deflected so that it can scan the required target area. It can be tracked or programmed to follow a particular path over the target surface.

Due to local heating of a small area in a very short time (~ 1.0 msec) the heat affected zone is very narrow or nonexistent.

Deep penetration, narrow heat affected zone, no temperature limit, easy control on the time and energy of the beam, and clean environment (no oxidation) make EB an ideal heating method for welding. Virtually any metal or dissimilar metals can be precision welded.

Depending on the work manipulating system and the size of the work chamber, very small or relatively large assemblies can be welded on production scale. Hence, EB is used for electronic welding (microwelding), automotive components, rocket components, etc.

EB is also used for special alloy melting, vacuum sputtering, drilling etc.

11.2.5 EB—Material Interaction

When an EB strikes the target it penetrates to a considerable depth until its kinetic energy is fully absorbed by the target lattice and free electrons. The main body of heat is created below the surface and acts essentially as an internal volume source of heat.

If the beam is pulsed, a certain "dose" of energy is deposited as heat which then spreads around in the target.

In the case of a continuous beam the penetration is deeper and may pierce the target. As the process time is short (0.1– 2.0 sec) the heat flow is mainly in the up-and-down stream of the beam.

The beam and the target reaction produces emissions which consist of target atoms, ions, electrons, and X-rays (Figure 11.16). Due to the intense heat and sublimation, plasma or vapor or small molten drops are also produced. In fact, by using an apropriate spectrometer these emissions can be used for chemical analysis and characterization of the target material. The hot spot also gives out thermal radiation.

The heat spreads up and melts the material. The expansion of the material and occluded gases produces vigorous currents. This causes spurting and the molten metal is ejected out. The heat that spreads downward causes melting mostly confined to a conical volume pointing downward.

The energy distribution in the focal spot is gaussian with a peak at the center.

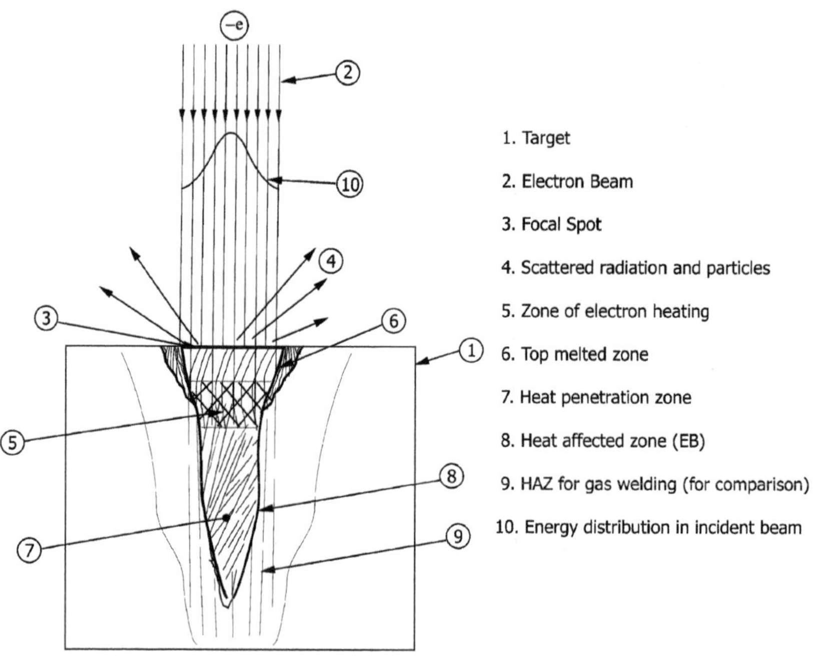

Figure 11.16 Electron beam-target interaction.

For deep penetration the beam is focused at a distance lower than the target surface. This is achieved by adjusting the current in the focusing coil. It changes the distance ℓ between the coil and the focal spot. Thus, this distance ℓ and the focusing coil current are critical parameters.

There is no exact analytical method of predicting the penetration depth, spread, and temperatures in the reaction zone. Control is established by the accelerating voltage, cathode current and the focusing coil currents, and pulse time. Like the laser beam, the target material has some reflectivity which reduces the efficiency.

11.2.6 Commercial EB Equipment

The specifications of commercial EB machines usually contain the following specifications:

Anode voltage kV and range of variation
Beam current mA
Pulse operation
Pulse repetition rate – Hz
Pulse duration and range – msec
Beam diameter – mm
Vacuum system – Pressure limits – mbar, torr
Pumping time
Beam deflection and scanning – mm^2
Focusing parameters – Lens to target distance
Beam viewing and tracking (optional)
Work chamber volume – m^3, length, width, and height
Coordinate table length, width, range, and programming
Gun positions – standard and optional
Work piece – Maximum dimensions accomodated
Additional manipulation axes
Total power requirement kW
Auxiliaries – Transformer
Water requirement – m^3/h
Operational data are supplied by the manufacturer to suit customers' requirements
Servicing and spare parts supplied by the manufacturer

Chapter 12

Vacuum Engineering

CONTENTS

12.1 INTRODUCTION

Processing of materials at low pressures is now a well-established industrial technique. It is used in a variety of manufacturing processes such as heat treatment, melting, degassing, brazing, electron beam welding and processing, surface coating, semi-conductor processing, and food processing.

"Vacuum" means the absence of air or any other atmosphere. Theoretically it is impossible to obtain such a totally atmosphere-free state. What we can achieve is lower and lower pressures, i.e., lesser and lesser atmosphere (in a confined vessel). Free space such as interstellar space also contains some atmosphere. It contains gases, ions, or electrons, etc., but some particles always exist.

Historically, vacuum was known about 150 years ago but it was a laboratory curiosity. It soon became apparent that materials displayed astonishing properties if processed in vacuum. Small vacuum chambers and furnaces were built in the early part of the 20th century. The only pump available for producing vacuum was the reciprocating pump, which had many limitations.

Two simultaneous developments at that time made it possible to build larger vacuum vessels. One development was that of a variety of pumps such as oil-sealed pumps, rotary-vane pumps, diffusion, and molecular pumps. These pumps had large capacities, and if used in series, could create very low pressures. A short description of these pumps and their pressure ranges is discussed in subsequent sections.

Another development was a variety of vacuum gauges. The height of a liquid (e.g., mercury) column was the first vacuum measuring device. It was very crude and could be used only to about 1/500 atmospheric pressure. This was refined in the Macleod gauge to measure up to 10^{-4} atmosphere. Subsequent developments brought about the Pirani gauge, thermocouple gauge, ionization gauge, and many others which have very low pressure ranges and are rugged, for use in industrial environments. These gauges are also discussed briefly.

Manufacturing techniques such as welding and sealing also helped the development of large vacuum vessels having very low or no leaks. Testing methods for leaks were also developed.

Today, vacuum chambers are available with capacities ranging from a few liters to several m^3. The vacuum that is routinely produced is 10^{-2}–10^{-7} atmospheres, while 10^{-9} –10^{-15} is also not uncommon. Vacuum is now a well-established, large scale industrial technique.

Though the use of vacuum in industries is common, the design and manufacture of vacuum systems is still a highly specialized field. There are many misconceptions about the production, measurement, and maintenance of vacuum.

It is therefore desirable to include a brief outline of vacuum technology in this book. Vacuum melting furnaces are not included. Only vacuums up to 10^{-5} atmospheres are discussed in some detail. For extensive treatment of vacuum engineering consult the *Bibliography*.

12.2 UNITS FOR VACUUM

The fundamental unit of vacuum is that of pressure, which is Pascal (Pa, N/m^2). However, vacuum is always quoted with reference to environmental or atmospheric pressure which is 10^5 Pa, or 1 bar, or 760 mm Hg. One bar is defined as 750 mm Hg. Hence, units commonly used for vacuum are

$$1 \text{ torr} = 1.0 \text{ mm Hg} = 10^3\ \mu\text{Hg (micron)} = 133.3 \text{ Pa}$$

$$\text{or} \quad 1 \text{ mbar} = 10^2 \text{ Pa}$$

For conversion

$$1 \text{ mbar} = 7.5 \times 10^{-1} \text{ torr} = 750 \text{ microns}$$

$$\text{or} \quad 1 \text{ torr} = 1.33 \text{ mbar} = 1000 \text{ microns}$$

As far as possible, we will use torr in our discussion as it appears more frequently in literature.

Thus, vacuum is quoted in 10^{-n} torr where n is a whole number.

Hence, 1 atmosphere is 7.5×10^{-1} torr. Subsequent lower pressures will be 10^{-1}, 10^{-2}, etc.

Low pressures (not much lower than atmosphere) are quoted in mm of water (kg/m^2) or mm of Hg.

Another unit we will be coming across when dealing with vacuum systems is "throughput" (Q). This is defined as the product of the pumping speed (S) and the inlet pressure (P), i.e.,

$$Q = P \times S$$

$$= \frac{dv}{dt}$$

Q can also be considered as the quantity of gas in pressure × volume units at a specified temperature. Q can be quoted in Pa × m^3 /sec, or more conventionally as torr × liter/sec.

Throughput will be useful in calculating pumping capacity, system leaks, and outgassing.

12.3 VACUUM PUMPS

Vacuum or low pressure is achieved in a container or vessel by removing the air or gases in it. The devices used for this are collectively called "vacuum pumps."

There are many types of vacuum pumps in use. Some of them are "true" pumps. They draw the gas out by suction, compress it, and discharge it into the atmosphere. These are mechanical pumps. They operate at high pressure but have limited displacement, pressure, and range.

Next are the blowers such as roots pumps. They operate at low pressure but have a large displacement. Principally, blowers only push the gases and are not pumps.

For low pressure in the vacuum range (10^{-3}–10^{-9} torr) diffusion and molecular pumps are used. They trap gas molecules physically or mechanically and push them ahead. These pumps are displacement or drag pumps and can operate at low pressures only. They always require a backing pump.

For higher vacuum (10^{-5}–10^{-12}) there are many devices such as ion pumps, and cryo pumps. They are based on various phenomena displayed by gas molecules at very low pressures.

A brief review of the operating principles and characteristics of some vacuum pumps for use in (760 to 10^{-5}) torr range are discussed in this section. For more information about these and other pumps refer to special books in the *Bibliography*, and manufacturers' literature.

12.3.1 Positive Displacement Pump

These are purely mechanical pumps. They create a vacuum inside a "cylinder" by a moving piston; gas is drawn in, compressed, and discharged into the atmosphere at a slightly higher pressure than the atmospheric one. Thus their discharge is always at atmospheric pressure. Hence these pumps are at the exit end of a vacuum pumping system. The discharge pressure is regulated by a spring loaded valve.

Though theoretically, the intake pressure of these pumps is supposed to be zero, in actuality, it is of the order of 10^{-1} to 10^{-3} torr due to leakage and other problems.

Commercial positive displacement pumps have a rotary piston. Two common designs are shown in Figure 12.1(A) and Figure 12.1(B).

The rotary pump in Figure 12.1(A) has a solid shaft rotating eccentrically in the cylinder. A ring with a sliding ring is mounted on the shaft and carries a slider. The contact point between the cylinder and the ring on one side and the slider on the other side divides the cylinder into two cavities. The slider also acts as the gas inlet. The gas is received, compressed, and discharged through the spring-loaded valve. As the gas heats during compression, the pump is cooled by oil around it. There may be additional cooling pipes to circulate the water and cool the oil.

Figure 12.1(B) shows a sliding vane pump. The eccentrically rotating shaft has a diametrical slot carrying two radial vanes. The inner ends of the vanes are connected by springs. The outer ends rotate in contact with the inner surface of the cylinder. Thus, two rotating cavities are created. Air or gas is

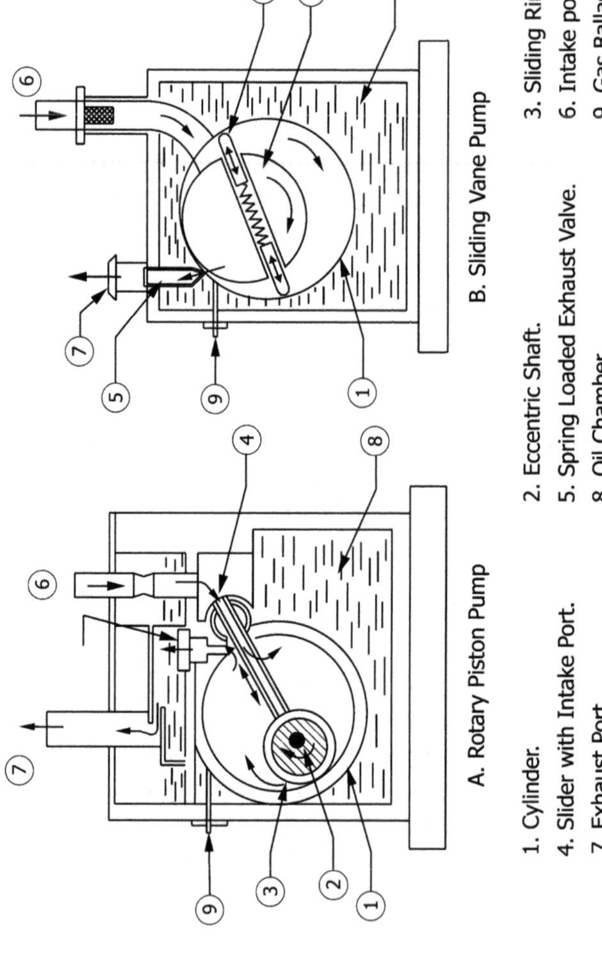

Figure 12.1 Positive displacement pump.

received via the intake port, compressed, and discharged through the spring-loaded valve. This pump is also oil-immersed.

Both pumps usually have a gas ballast facility. Through this, the measured quantity of air is injected into the compressed gas a little before discharge.

If the gas being pumped has moisture or other condensable vapors, these condense during compression due to reduction in volume and vapor saturation. The condensed vapors create corrosion problems and also contaminate the oil. The injected ballast prevents condensation, allowing vapors to escape with the exhaust. Gas ballasts raise the ultimate (minimum vacuum) pressure and thus reduce the pump capacity.

Limitations and Advantages of Mechanical Pumps

1. These are the only available pumps that can discharge the pumped air directly into the atmosphere.
2. They have a constant pumping speed from atmospheric to about 1.0 torr pressure. At lower pressures the speed drops rapidly, as shown in Figure 12.2. The lowest attainable pressure is about 10^{-3} torr.
3. At lower pressures, the pump cannot compress the drawn gas to discharge (atmospheric) pressure and thus pumping is effectively stopped.
4. Due to rubbing between various parts there is considerable friction and wear. Hence, the pump speed is limited to about 350–700 r/m.
5. The gas ballast solves the condensation problem but raises the ultimate pressure.
6. The ultimate vacuum (pressure) can be lowered by using two pumps in tandem. A large capacity fore pump draws air from the vessel, compresses it, and discharges it into the backing pump which compresses it still further and discharges it into the atmosphere. The pumping characteristics of this arrangement are shown in Figure 12.2.
7. These pumps are excellent for roughening, i.e., quickly reducing the pressure to about 1–10^{-2} torr so that other pumps (blower or diffusion) can take over for further pressure reduction.
8. Pumping capacities available are 10 to 1000 m^3/h.

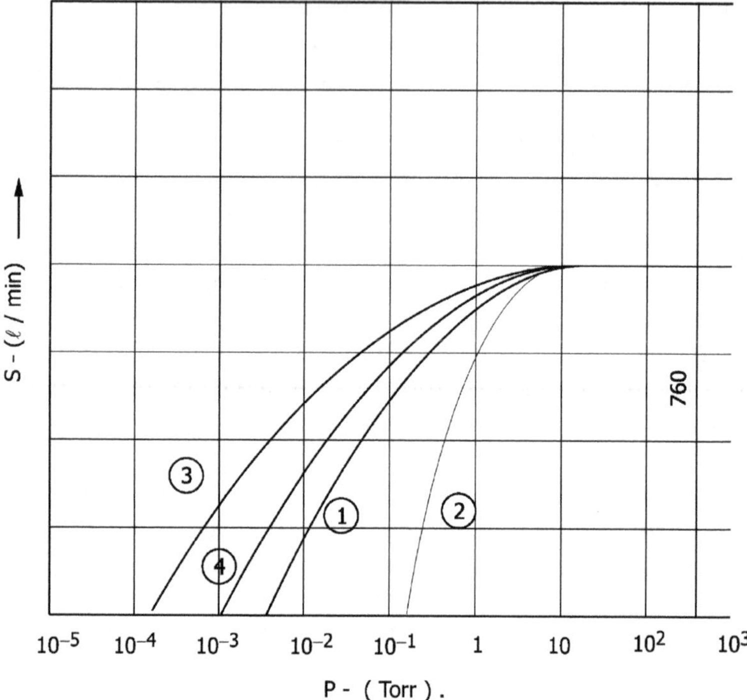

1. Single Stage without Gas Ballast
2. Single Stage with Gas Ballast
3. Two Pumps in Tandem without Ballast
4. Two Pumps in Tandem with Ballast

Figure 12.2 Pressure-speed characteristics of a typical positive displacement pump.

12.3.2 Roots Pump

This pump is a blower and is mainly used as a booster or enhancer for positive displacement pumps.

The construction of a typical roots pump is shown in Figure 12.3. Two counter-rotating rotors of approximately the same shape, rotate in a housing. The clearance between the rotors, and the rotors and housing is 75 to 300 microns. This clearance makes the rotation free (without contact) and hence

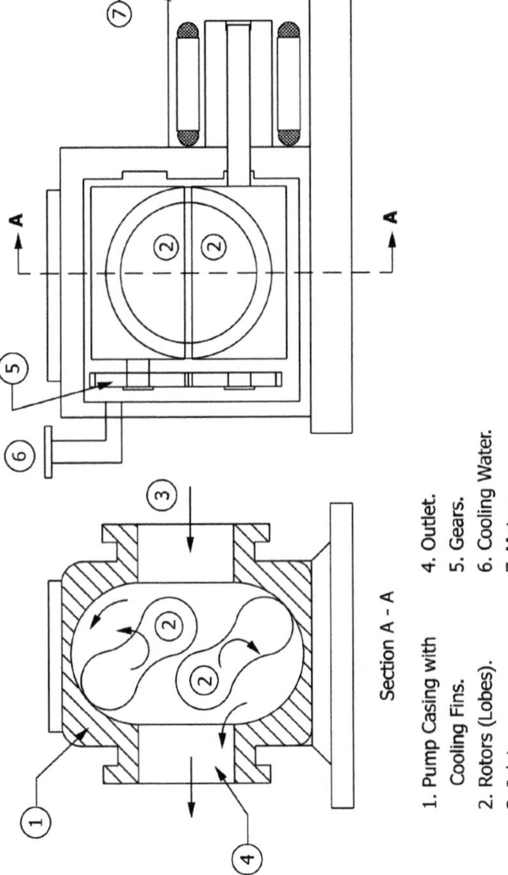

Figure 12.3 Roots pump (Blower).

r/m is possible without friction. The inlet and outlet ports are as shown. The outlet port is smaller than the inlet.

During rotation, gas molecules are trapped between the casing and the rotors and are transported toward the outlet. There is not much compression, hence, this is a low-pressure high-discharge pump. The pumping speed (ℓ/sec) depends on the lobe-casing volume and r/m. Some blowers have spiral-shaped rotors. The compression ratio is about 1:10. Blowers with a capacity of 500–2000 m^3/h are available.

Due to the peculiar manner of trapping and pushing the gas, a blower has different speeds for different gases.

Typical pressure-speed characteristics of roots blowers are shown in Figure 12.4.

Some peculiarities of roots pumps:

1. A roots pump is really a blower. It pushes a high volume of gas at low pressure. Hence, it cannot be used at high pressures, i.e., the initial stages (roughening) of vacuum.
2. The maximum permissible pressure difference is 50–60 torr. If connected to discharge directly to the atmosphere (760 torr) it draws out a large volume initially. When the pressure in the vessel drops, it cannot develop enough output pressure and develops considerable heat. It is therefore necessary to back the blower with a rotary pump.
3. Such a combination of the blower and rotary pump can be used in the 760–10^{-3} to 10^{-4} torr range. The ultimate pressure is about 5×10^{-5} torr.
4. The pumping speed depends on the r/m. The usually available speeds are 1000–3000 r/m but higher speeds are possible.
5. As there is considerable heat generation, the blowers have elaborate water cooling.
6. The size (rating) of the backing pump has an effect on the blower performance. Usually, a backing pump of about 1/10 pumping speed is used.

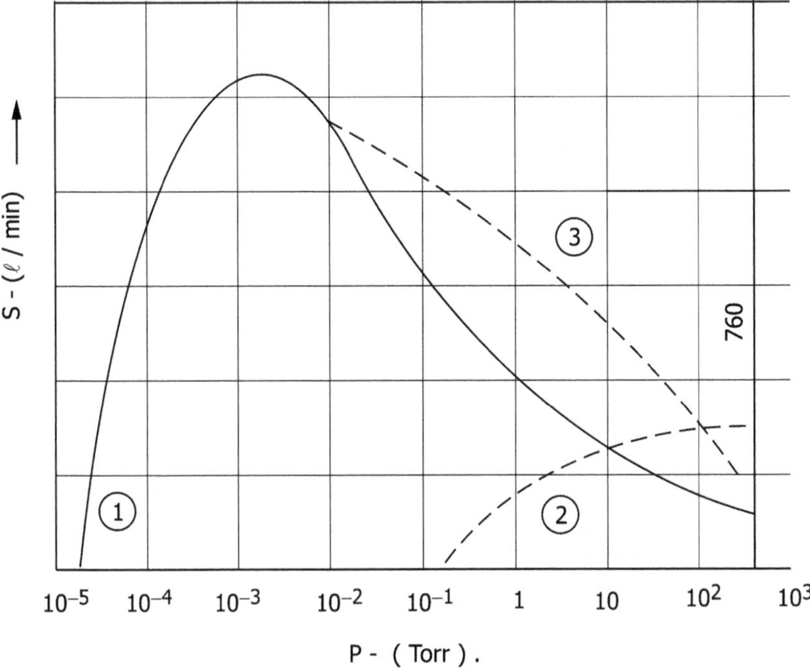

Figure 12.4 Pressure-speed characteristics of a roots pump.

7. Blowers have a maximum pumping speed of about 0.05–0.1 torr, below which the speed drops. At 0.001–0.01 torr the speed is 50–75% of the maximum.
8. Blowers are free from oil contamination.
9. Positive displacement pumps are useful up to a maximum of 1.0 torr while diffusion pumps can pump efficiently from 10^{-2} to 10^{-3} torr. Hence, there is a gap between 1 and 10^{-2} torr when a diffusion pump is connected directly to a rotary pump. Blowers are very useful to cover this gap. Hence, diffusion pump-blower-rotary pump is the best combination.

10. Roots pumps can be connected in series or parallel to handle large amounts of gases containing condensable vapors.

12.3.3 Diffusion Pumps

In these pumps the air molecules coming from the vessel are caught up or trapped by a stream of vapor and are carried along with it toward the exit (foreline). In principle, the pumping action is similar to an ejector pump.

The pumping medium, i.e., the vapor is produced at the base of the pump by boiling a suitable fluid. As the vapor carrying the gas descends, it condenses on the cold walls of the pump casing. The trapped air or gas molecules are released near the lower end and are drawn out by the suction from the foreline.

The vapor stream is produced in two or more stages by the jet assembly over the top of the boiler (Figure 12.5).

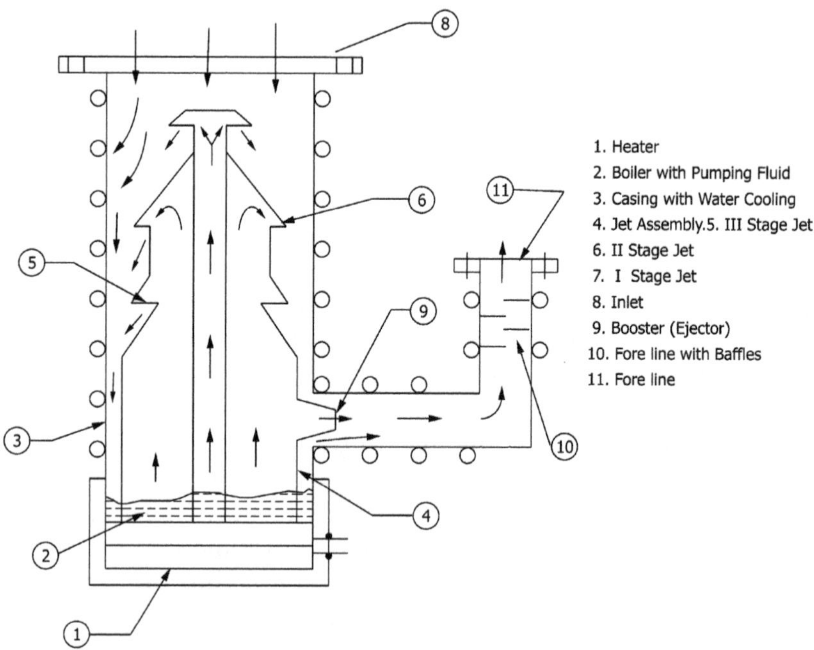

Figure 12.5 Oil diffusion pump.

A further enhancement in pumping is obtained by an ejector or booster in the foreline. The pump walls and foreline are cooled by water. This cooling removes the heat of condensation and reduces the back streaming of vapor to the vacuum vessel.

The pumping speed is determined by the area (diameter) of the inlet and what is called the H_o factor. This factor is 0.3–0.5. For air the pumping speed S is given by $S \approx 2D^2$. Diffusion pumps with diameters of 25–1200 mm and air pumping capacities from 10–100,000 (ℓ/sec) are available.

In earlier days mercury was used as the pumping fluid. It is still used in small laboratory-type pumps. Presently, all commercial pumps use synthetic oils and are sometimes called "oil diffusion pumps." Among many desirable properties of the pumping oils, the following are most important:

1. Low vapor pressure
2. High molecular weight
3. Low latent heat of evaporation
4. Noncorrosive and nontoxic

A number of synthetic oils are developed. They are proprietary formulations based on hydrocarbons, silicones, and other organic groups. The boiler has a sealed electric heater.

The oils may consist of a single compound or a number of compounds. In the latter case a "fractionating pump" is used. Different components vaporize at different pressures and rise to different jets.

Noteworthy points about diffusion pumps

1. Pumping characteristics of a typical diffusion pump (P–S curve) is shown in Figure 12.6. The figure shown is typical for pumping air. Note that at pressures higher than 10^{-3} torr, the speed drops down, and somewhere between 10^{-2}–5×10^{-3} it becomes practically zero. For pressures lower than 10^{-3} the speed is constant up to about 10^{-7} torr.
2. The maximum constant speed depends on the inlet size (dia D). Theoretically, there is no lower limit. In practical pumps, the ultimate or blank off pressure is

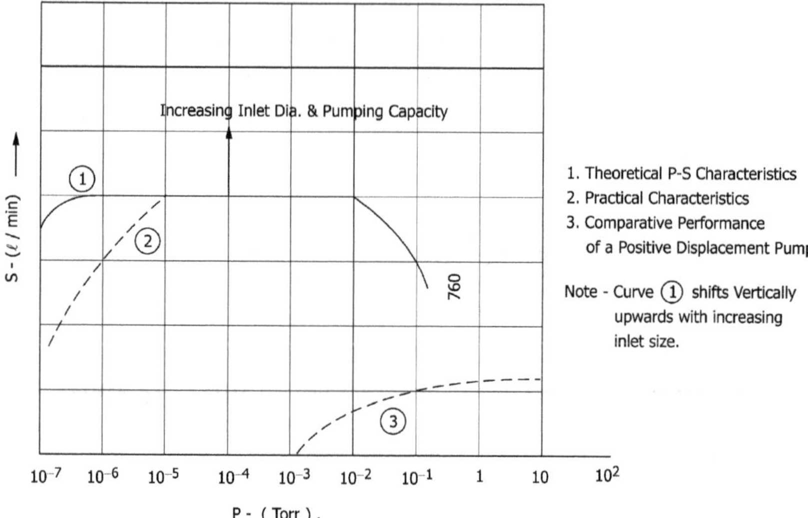

Figure 12.6 Typical P-S curve for a diffusion pump.

about 10^{-9} torr. The ultimate pressure depends on the back-streaming of oil from the pump to the vacuum vessel. The back-streaming is avoided by using a trap or baffle at the inlet to condense the oil vapor.

3. Diffusion pumps draw the gas with the vapor stream. The drawn gas is collected in the foreline where it acquires a higher pressure, which is called the "foreline pressure."
4. As the diffusion pump can pump only from about 10^{-3} torr, it is necessary to connect a mechanical pump between it and the atmosphere. This mechanical pump must have an intake pressure lower than the foreline pressure. If it is not so, air collected in the foreline cannot be drawn out.
5. Thus, there is a maximum limit to the foreline pressure. If pressure higher than this limit is reached, it will start affecting the pumping speed at the inlet of the diffusion pump. This limiting pressure is called the "fore pressure tolerance" of the diffusion pump. This is an important design parameter when selecting

the diffusion-mechanical pump combination. The fore pressure tolerance (at full load) is given by the manufacturer. It is usually between 15 and 50 × 10^{-2} torr. The backing pump must have an intake pressure lesser by at least one order.

6. Diffusion pumps require water-cooling. The water requirement as well as the power required for the oil boiler is given by the manufacturer. Diffusion pumps have no moving parts. They do have oil which may produce back-streaming vapors.
7. As the diffusion pumps can start pumping only at about 10^{-3} torr, it is best to use a roughening pump directly connected to the vessel. The diffusion pump is connected in line after the vessel pressure reaches 10^{-3} torr, as shown in Figure 12.6.
8. When the vessel is very large (≥5 m^3) a number of diffusion pumps are connected in parallel to achieve a quick pump-down time.

Diffusion pumps are mounted vertically. They are connected to the vessel via a short, large-diameter pipe or an elbow. The conductance of these must be taken into account when deciding the pump capacity. These connections often contain a baffle or trap to condense oil vapors back streaming into the vessel. The traps and baffles also have a conductance which significantly lowers the available pumping capacity.

12.3.4 Molecular Pumps

These pumps are based on the phenomenon of molecular impact in a low pressure regime. If a gas molecule strikes a surface, it rebounds in all possible directions that have no relation to the angle of incidence. At low pressures (< 10^{-3} torr) the molecules have a large mean free path. Thus, when molecules strike a surface, some of them rebound in a direction which is more or less parallel to the surface. If the surface is moving at a fast speed, these molecules will be carried along the surface, i.e., they will be pumped away.

There are two main types of molecular pumps based on this principle. The molecular drag pump uses a high speed rotating disk and a spiral pumping path. These pumps have many mechanical problems and are now superceded by turbo molecular pumps.

A typical turbo molecular pump is shown in Figure 12.7(A). It consists of several pairs of stator and rotor blades with angular grooves on their edges as shown in Figure 12.7(B). There are several configurations of this. The one shown in the figure has a central inlet with blade pairs on both sides, an outlet on one side, and an integral motor. The clearance between the blades is about 1.0 mm (much smaller than the mean free path ~10^2–10^3 m).

The rotor blades are rotated at 10,000–20,000 r/m by the motor. Gas molecules are drawn in the stator grooves and pumped by the rotor grooves. The number of grooves, their angle, the clearance, and the speed are critical design factors. Higher r/m increases the pumping speed and limitation appears to be the bearings. Pumps with capacities 1000–20,000 ℓ/min are available.

The pressure-speed characteristics of turbo molecular pumps are shown in Figure 12.8.

The pump starts pumping from about 10^{-2} torr and has a constant speed down to 10^{-9} torr. It is thus necessary to use a backing pump.

Turbo molecular pumps are oil free and are not affected by condensable vapors. They are better but costlier than diffusion pumps. Because of their very high speed they have mechanical problems. These pumps are mainly used in electron beam machines, solid state processing, etc., where vacuum higher than 10^{-5} torr is required.

12.4 PUMPING SYSTEM DESIGN

12.4.1 Selection of Vacuum Pumps

After reviewing various aspects of vacuum pumps and vacuum systems we can now tackle the main problem of the designer, i.e., how to select the pump or pumps required to

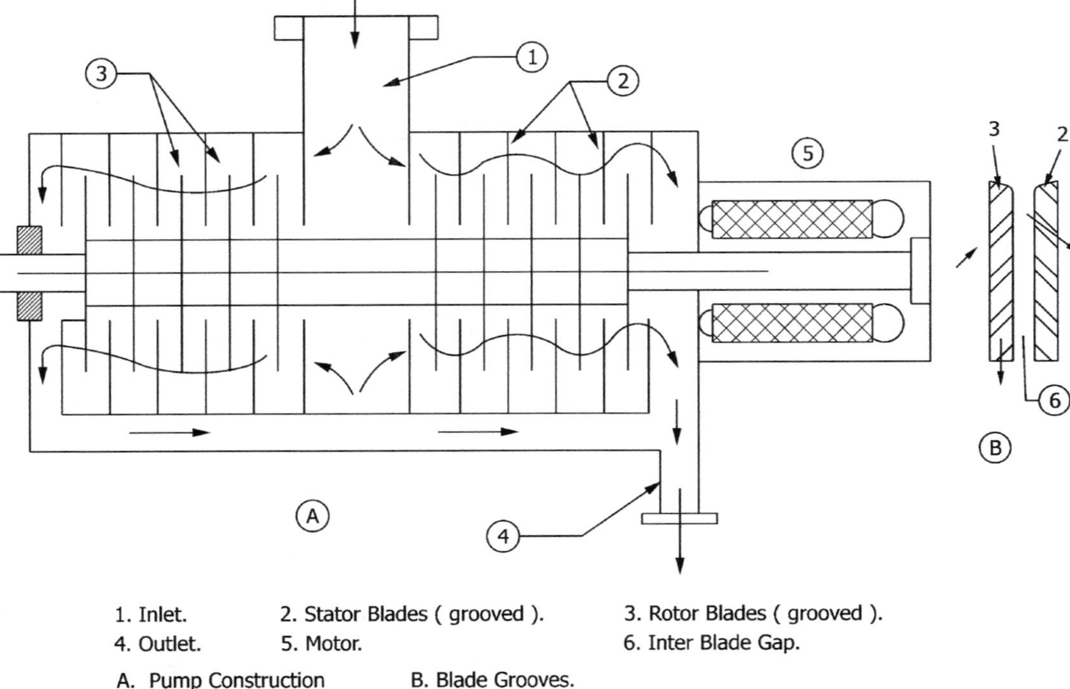

1. Inlet. 2. Stator Blades (grooved). 3. Rotor Blades (grooved).
4. Outlet. 5. Motor. 6. Inter Blade Gap.

A. Pump Construction B. Blade Grooves.

Figure 12.7 Turbo molecular pump.

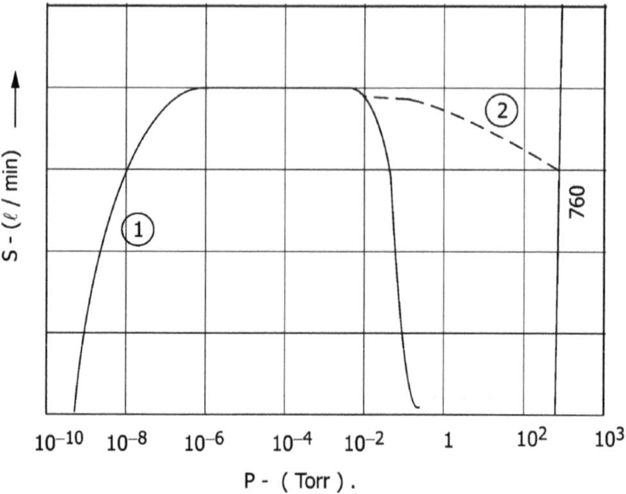

Figure 12.8 Pressure-speed characteristics of a turbo molecular pump.

achieve the given vacuum pressure in the given time and to sustain it during the given process time.

The various available pumps have their characteristic pressure range and pumping speed. They give optimum performance within these limits. The following are the general operational ranges of these pumps:

Mechanical pumps	760–0.1 torr
Roots pumps	$760–10^{-3}$ torr With input/output pressure difference of about 40 torr.
Diffusion pumps	$10^{-2}–10^{-7}$ torr
Turbo molecular pumps	$10^{-3}–10^{-9}$ torr with cryogenic trap or $10^{-3}–10^{-7}$ without traps.

The ultimate pressure of these pumps is usually much less than the ranges given above.

This tells us that with the exception of low vacuum of about 10^{-1} or 10^{-2} torr, no one pump can produce higher vacuum. Another important point is that only mechanical, i.e., positive displacement pumps, can pump out at atmospheric pressure. Hence, in virtually all the pump combinations, the mechanical pump will be the output or the last pump and will have to be used as the backing pump.

Consider the pumping speeds (ℓ/min). Following are the typical speeds of various available pumps:

Mechanical pumps	400–40,000	ℓ/min
Roots pumps	10,000–50,000	ℓ/min
Diffusion pumps	3,600–540,000	ℓ/min
Turbo molecular pumps	3,000–25,000	ℓ/min

This shows that pumps applicable for low pressures have high pumping speeds. This is fortuitous because at low pressures the volume of gas to be pumped out is large (PV = Constant). However, the pumps will have to be matched to their fore and backing pumps and the conductance of the piping.

The output of the mechanical pump can be increased by connecting two or more pumps in series (compound) or in parallel.

After we reach a pressure of about 10^{-3} torr there will be outgassing from the vessel, its furniture, and components. The gas load due to outgassing can be estimated from the available data but such estimates could be very different (generally lower) than the actual one. The pumps should be capable of taking up this load and maintaining the vacuum. Outgassing generally subsides or comes to equilibrium with pumping.

On further evacuation there will be an additional gas load due to the process. This load will arise at the process pressure and temperature, and can be estimated with reasonable accuracy. The process gas load will continue throughout the process time. The pumps will have to take up this load and maintain pressure at the level required by the process.

The normal leaks and permeation will also pose an additional load but this is generally negligible compared to the outgassing and process loads.

The piping used to connect the vessel and the pumps will have conductance, as discussed in Section 12.5. The size (diameter) of the plumbing will be dictated by the flange size of the chosen pump. The length will depend on the system layout but should be the shortest possible. For a given pipe or fitting, or their combination, the conductance will depend on the flow regime, i.e., turbulent, viscous, intermediate, or molecular. Conductance will lower the actual speed available at the input and, hence, will influence the pump selection.

12.4.2 Calculation of Pumping Speed

In the beginning, the whole system will be at atmospheric pressure (P_A). We will assume that the system contains dry air.

The first stage of evacuation will be roughening, i.e., to bring down the pressure (P_R) to 1–10^{-1} torr. A positive displacement pump is used for this. If a time t min is assumed for roughening,

$$\left.\begin{aligned} t &= \frac{V}{S_P} \ln \frac{P_A}{P_R} \quad \text{or} \\ t &= 2.3 \frac{V}{S_P} \log \frac{P_A}{P_R} \end{aligned}\right\} \tag{12.1}$$

where V = Effective volume of system (ℓ)

S_P = Pumping rate (ℓ/min)

This is modified by using the factor K for different pressure ranges as given in Table 12.1. We can also use this equation to determine the roughening pump speed S_P, for pump selection.

If the air contains substantial moisture, a gas ballast will have to be used, which will raise the P_R from 10^{-1} to 1.0 torr. The roughening pressure can be lowered to 10^{-2} and a higher pumping speed can be obtained by using a compound pump.

TABLE 12.1

Pressure Range, Torr	Service Factor K
760–100	1.0
110–10	1.25
$10–5 \times 10^{-1}$	1.5
$5 \times 10^{-1}–2 \times 10^{-4}$	2.0
$5 \times 10^{-2}–2 \times 10^{-4}$	4.0

The roughening pressure can be lowered further to 10^{-3} to 10^{-4} torr by using a roots pump as the forepump and backing it by a rotary displacement pump. In the absence of outgassing loads, the combination will have a fairly flat P–S characteristic. If outgassing exists, the blower capacity (S_B) will have to be matched to the gas load.

A roots blower can also be used to the evacuate directly to the atmosphere at the very beginning of evacuation. Blowers have large pumping speeds and can bring down the system pressure to about 100–10 torr in a very short time. After this, the blower could be backed by a mechanical pump. This arrangement will require elaborate plumbing.

Experience has shown that a blower works best when backed by a mechanical pump having a capacity of about 1/10 of the blower.

For pressures below 10^{-3} torr to about 10^{-8} torr, a diffusion pump will have to be backed by a blower or a mechanical pump. It is in this pressure range that most of the outgassing load will appear. The pump capacity should be adequate to remove the outgassing, process gas load, and maintain the system pressure at the desired level. Diffusion pumps have very large pumping speeds. To realize these speeds the foreline pressure will have to be at least one order below the maximum tolerable value. This will be supplied by the manufacturer. It will be about 0.18–0.30 torr. The gas flow in the pressure range of 10^{-3}–10^{-8} torr is in the molecular or intermediate range and the conductance of pipes and traps in the suction line must be considered.

Instead of a diffusion pump we can use a turbo molecular pump in the 10^{-3}–10^{-10} torr pressure range. The pump has a flat S–P characteristic and is oil-free. However, it has a low capacity and is expensive; hence, it is mainly used for special applications. It will have to be backed by a blower or a mechanical pump.

The layout of a typical vacuum system will be as shown in Figure 12.9.

The example that follows will make clear the process of pump selection.

12.5 CONDUCTANCE AND PUMPING SPEED

The vacuum chamber and the pumps are connected by pipes. There may be one or more pumps, and the connecting plumbing may be quite complex. Besides the pipes, it contains many other components such as traps, baffles, valves, and elbows.

The plumbing offers considerable resistance to the flow of gases and affects the pumping speeds. In vacuum parlance this resistance is called "conductance (C)." Each component has its own conductance. The total conductance of a system is calculated by considering their way of connection, i.e., in series, parallel, etc. This is similar to the calculation of the resistance of an electric circuit containing a number of resistors.

Consider a simple system as shown in Figure 12.9(A), where the vessel and pump are connected by a large, straight, circular pipe.

Let S_P = Pumping speed of pump ℓ/sec

S_E = Effective speed available at the vessel

The two speeds are different because of the conductance offered by the intermediate pipe which consumes some part of the pump speed. Thus, S_P and S_E are related by the following relation:

$$\frac{1}{S_E} = \frac{1}{C} + \frac{1}{S_P} \tag{12.2}$$

where C is the conductance of the pipe. The units for C are the same as S_E and S_P, i.e., ℓ/sec or m^3/sec, etc.

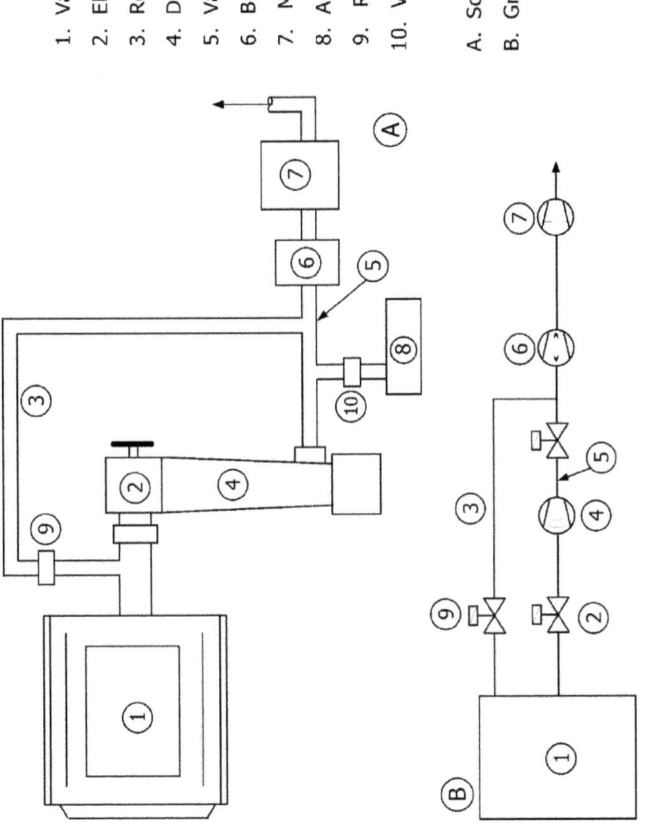

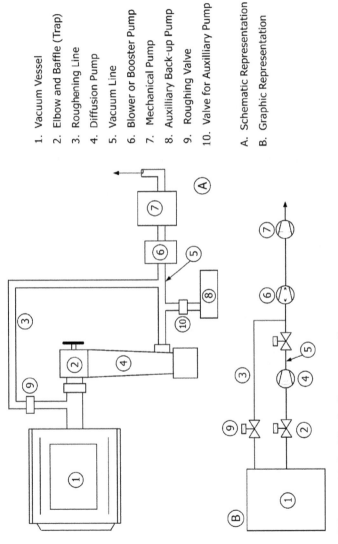

Figure 12.9 Typical vacuum line.

If there are more pipes connected in series, the total conductance C is given by

$$\frac{1}{C}=\frac{1}{C_1}+\frac{1}{C_2}+\frac{1}{C_3}+\cdots \tag{12.3}$$

where C_1, C_2, C_3, etc., are the conductances of individual pipes. If these are arranged in parallel,

$$C=C_1+C_2+C_3+\cdots \tag{12.4}$$

Let us consider the relative magnitudes of the quantities in Equation (12.3) and Equation (12.4).

If $C = S_P$ then $S_E = S_P/2$, hence, only 50% of the pump speed is available at the vessel. If $C < S_P$ lesser speed is available. If $C > S_P$ more speed is available. Thus, at $C = 4\ S_P$ 80% of the pump speed is available at the vessel. Note that 100% is never available at the vessel. (Unless of course when $1/C = 0$!). This also shows that if the conductance C is limiting, there is no use increasing the pump speed; i.e., a larger pump is not necessarily a solution.

The conductance of a pipe depends on the flow regime, i.e., viscous, intermediate, or molecular (see Table 12.2) and on the geometry of the pipe.

For a circular pipe of length L (cm) and diameter D (cm) carrying air at 20°C at average pressure P (torr), the conductance C is given by

$$C=180\left(\frac{D^4}{L}\right)P \tag{12.5}$$

The above equation applies for viscous regime only. As seen before viscous regime exists if $P \times D \le 0.5$ torr cm.

If $P \times D$ is between 0.5 and 0.005 the flow is in intermediate range and the conductance is given by

$$C=3.3\times10^{-2}\left(\frac{D^4}{L}\right)P+10\frac{D^3}{L} \tag{12.6}$$

Here, the first term is the conductance due to viscous flow, and the second term is for the molecular flow.

TABLE 12.2 Characteristics of Vacuum Systems

Characteristics	Vacuum			
	Low	Medium	High	Ultrahigh
Pressure Range (typical) mm Hg (torr)	760–1	$1–10^{-3}$	$10^{-3}–10^{-7}$	$10^{-7}–10^{-10}$
Number of molecules per m^3 ($N_{AV} = 10^{26}$)	$10^{25}–10^{22}$	$10^{22}–10^{19}$	$10^{19}–10^{13}$	10^{19} and less
Mean free path (cm). For air at 20°C (approximate for comparison only.)	4.5×10^{-3}	0.45	4.5×10^{3}	4.5×10 4.5×10^{6}
Viscosity and thermal conductivity	Independent of Pressure	Depends on mean free path/ Mol.dia.	Directly proportional to pressure	Both phenomenon practically absent
Gas Flow*	Viscous	Mixed	Molecular	Molecular to no flow at very low pressure

* This also depends on the pressure, viscosity and vessel, pipe dimensions.

For the molecular flow, where $P \times D \leq 0.005$ the conductance for air at 20°C is given by

$$C = 12.1\frac{D^3}{L} \quad \frac{\ell}{s} \tag{12.7}$$

This equation shows that in the molecular flow regime the conductance is independent of pressure. If $L \leq D$, the above equation can be written as

$$C = 9.1\, D^2 \quad \frac{\ell}{s} \tag{12.8}$$

Equation 12.8 can also be modified to take into account the end corrections which arise due to the entry aperture between the vessel and the tube.

Equation (12.7) and Equation (12.8) also show that to have a large conductance, i.e., more pump speed at the vessel, the connecting pipe should have a large diameter and short length.

Besides the simple tube connection between the pump and the vessel, another connection commonly used is a right angle elbow. The conductance of an elbow can be calculated from the above equations by substituting D' for D where

$$D' = (L_{ax} + 1.33\,D)$$

where L_{ax} = Axial length (cm) and
D = Diameter (cm) of the elbow

For more complex shapes and their conductance, consult specialist literature listed in the *Bibliography*.

12.6 BAFFLES AND TRAPS

These are devices to remove condensable vapors from the air or gas being pumped out. There are several sources of condensable vapors in a vacuum system.

The air always contains moisture or humidity. At the right combination of pressure and temperature, the gas is saturated and the vapor and moisture are condensed.

Various materials used in the construction of the vessel give out large amounts of water vapor at the outgassing stage. It is reported that the outgassing "gas" is about 75% water vapor.

The charge to be processed in vacuum also has absorbed water vapor which is released at the right pressure.

In a high or medium vacuum system the outgoing vapor enters the diffusion pump and condenses there along with the pumping fluid (usually oil). The oil gets contaminated, thus affecting the pumping rate.

Elastomers and seals give out hydrocarbon vapors during evacuation. These vapors also condense and contaminate the oil.

What is discussed above is the source of vapors going out from the vessel. One other source of vapor goes in to the system. This is the oil vapor escaping from the pump inlet into the vessel.

If the pumping system contains only a roots pump or a mechanical pump, the outgoing vapors can be removed by a gas ballast arrangement. The condensing vapors lower the ultimate pressure of the pump. Roots pumps are unaffected by the presence of vapors.

The diffusion pumps are the type mostly affected by the presence of vapors. It is therefore necessary to remove the vapors by condensation before they enter the diffusion pump. Traps and baffles are such devices. They are located between the vessel and the fore pump. Technically, baffles provide a zigzag path for the vapor so that it meets cold surfaces and condenses before entering the pump. Traps are intentionally cooled or refrigerated to condense the vapors. Both devices condense the vapors and the condensate remains in the device. It re-evaporates upon heating and enters the system. To minimize this, there is a baffle valve which isolates the vessel from the pump.

The arrangement of these and some typical baffle and trap designs are shown in Figure 12.10.

Traps are cooled by water circulation, dry ice, refrigeration, liquid nitrogen, or air depending upon the cooling requirements. The condensate, either liquid or solid, is removed (emptied) when the system is off. Alternately, the vapors are absorbed by a suitable molecular absorbant (e.g., activated charcoal or other synthetic material).

Theoretical calculations of the baffle and trap conductance is tedious. It is best to rely on manufacturers' data.

12.7 OUTGASSING

All materials, when exposed to low pressure, release gases. This phenomenon is called "outgassing." Either during manufacturing or in subsequent processing and handling, materials absorb gases such as water vapor, hydrogen, and oxygen.

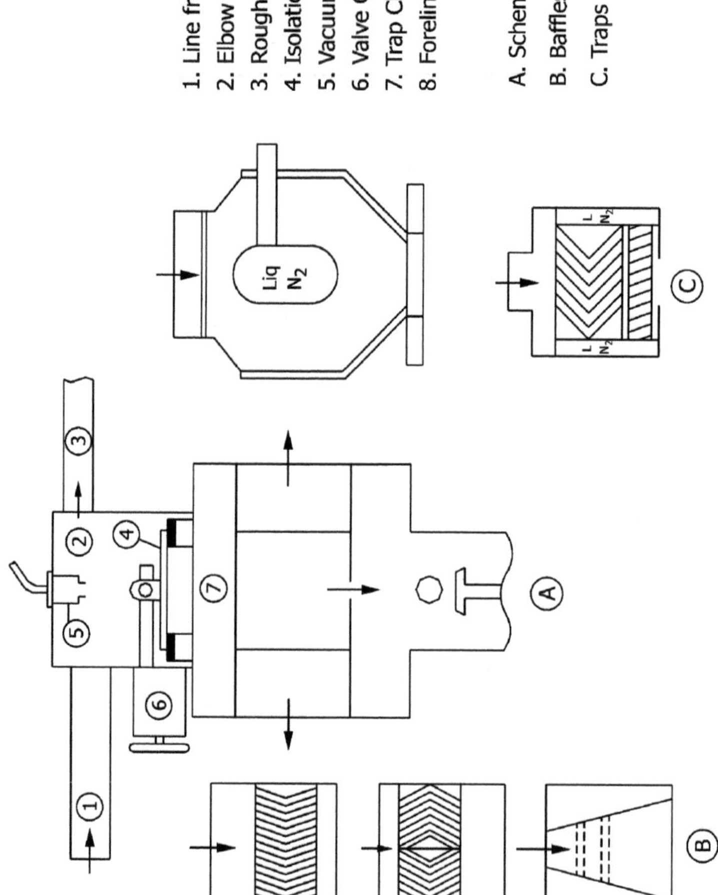

Figure 12.10 Vacuum foreline with traps.

These gases are linked up with the chemical structure of the materials.

If they are bonded with the surface layers, only then are they said to be "absorbed." On the other hand, they may be dissolved in the material and evenly distributed throughout. Some gases may be occluded in the surface defects (e.g., cracks) and some may be in the form of small bubbles. There are many other forms or mechanisms of absorption of gases by material surfaces. These are subjects of study in surface sciences. It should be remembered that these gases are not present because of surface uncleanliness. Otherwise, clean surfaces would also have absorbed gases (in fact, more). In the construction of vacuum furnaces and accessories we use metals, polymers and rubbers (gaskets), glasses (view ports), ceramics, etc., and all these materials have absorbed gases mainly on their surfaces.

On exposure to vacuum (and temperature) the absorbed gases are evolved. The pressure at which a material outgasses, and the rate of evolution varies from material to material, its processing history, and prior exposures. An elevated temperature (baking) helps outgassing. At elevated temperature the material outgasses at a higher pressure and at a faster rate.

One common fact about all materials is that the rate of outgassing first increases and then diminishes with time. Most materials outgas in 10^{-2}–0^{-4} torr pressure range. Evolution of absorbed gases is called "desorption."

Organic materials like elastomers and polymers give out additional gases due to chemical decomposition. They are generally blends of several chemicals with the main polymer. These chemicals evaporate or decompose at low pressure and elevated temperature.

On exposure to vacuum for a prolonged time, many materials release gases due to diffusion from internal layers or permeation of atmospheric gases through them.

All vacuum systems show leaks. These may arise from manufacturing defects (e.g., faulty welding, improper seal, design, etc.) or due to aging with use or need of maintenance. In a well-designed and maintained system, these leaks are taken care of by the pumping system.

Outgassing of many construction materials is extensively studied. Outgassing rates are measured in throughput per unit area, i.e., torr × liters/sec × cm^2 and are published in specialist literature. A few are quoted below.

Stainless steel, aluminum	1.5×10^{-9}
Mild steel	5×10^{-7}
Teflon	5×10^{-6}
Neoprene	2×10^{-4}
Porcelain	6×10^{-7}

The above quoted rates are for comparison only. Actual rates depend on surface condition and time in vacuum.

Many metals contain lead, sulfur, zinc, cadmium, and other low melting point impurities or alloying elements. These vaporize on exposure to vacuum. In general, alloys like stainless steel, mild steel, etc., have very high vapor pressure and do not add to outgassing.

Surface dirt, grease on the vessel, or charge outgas and further load the system.

In conclusion, outgassing poses an unpredictable and uncontrollable problem in the design of pumping systems. At about 10^{-2} the outgassing produces a gas load and raises the vessel pressure. The pumping system should be capable of handling this additional gas load at low pressure and stabilizing the vacuum to the desired level in reasonable time.

As compared to outgassing, vaporization and diffusion evolve vapors and gases at a very slow rate and the evolution continues at a constant rate for a very long (many days) time. Hence, they pose a permanent, steady but predictable problem. After outgassing, the pumping system comes to equilibrium with vaporization. This puts a limit on the ultimate (long term) vacuum that can be obtained in a given system.

The outgassing problem can be reduced by heating the vessel during the roughening stage. This is called "bake out." A temperature of 200–400°C is used. Due to the combination of heating and vacuum, the exposed surfaces outgas quickly. This reduces the outgassing load at lower pressure and the required vacuum is produced in a shorter time. Bake out is limited to surface outgassing (desorption) only. Surfaces once baked out will reabsorb gases if exposed to the environment.

EXAMPLE 12.1

A furnace having 2.0 m^3 volume is to be evacuated to 10^{-3} torr. Initially the furnace is filled with dry air at atmospheric pressure and 300 K.

Assume:

1. Constant furnace temperature (300 K)
2. Air contains 21.0% O_2 and 79% N_2 by volume
3. R = 287 J/kg K

Now determine:

1. The quantity of air initially present
2. The air remaining in the furnace after evacuation to 10, 1, and 10^{-3} torr

Initially,

$V = 2.0\ \text{m}^3,\ P_1 = 10^5\ \text{Pa},\quad T = 27°\text{C}$

$$P_1V = m_1RT$$

$$1 \times 10^5 \times 2 = \text{m}_1 \times 287 \times 300$$

$$\therefore m_1 = 2.3256\ \text{kg}$$

Finally,

$V = 2.0\ \text{m}^3,\ P_2 = 10\ \text{torr} = 1.333 \times 10^3\ \text{Pa},\ T = 27°\text{C}$

$$P_2V = m_2RT$$

$$1.333 \times 10^3 \times 2 = m_2 \times 287 \times 300$$

$$\therefore m_2 = 0.03\ \text{kg}$$

Repeating the calculations for

$P_3 = 1.0$ torr and $P_4 = 10^{-3}$ torr, we get

$m_3 = 0.003$ kg (1.0 torr)

$m_4 = 0.3 \times 10^{-5}$ kg (10^{-3} torr)

Comparing m_1 and m_2, the air removed in bringing the pressure from 1 atmosphere to 1/750 atm is

$$2.3256 - 0.03 = 2.2956 \text{ kg}$$

i.e., 98.7% of air is pumped out in the first stage leaving behind only 1.3%.

Subsequent stages will progressively pump out more air but the remainder will never be zero!

12.8 VACUUM PUMPING (PRESSURE–TIME RELATIONS)

Consider the evacuation of a vessel having a volume V m³ by a pump with a pumping speed S m³/sec.

Assume

1. There are no leaks or gas evolution (outgassing)
2. The pump speed is constant at all pressures
3. The resistance (conductance) of the connecting pipes and fitting is negligible

Gas pumped out in time dt at a vessel pressure
$dp = V\, dp/dt$
Gas removed by pump = S_P giving

$$\frac{dp}{P} = \frac{S}{V} dt \tag{12.9}$$

If in time t the pressure goes down from P_1 to P_2, the above equation on integration gives

$$t = \frac{V}{S} \text{ in } \frac{P_1}{P_2} \tag{12.10}$$

$$t = 2.3 \frac{V}{S} \log \frac{P_1}{P_2} \tag{12.11}$$

The above equation is applicable between 760–1.0 torr. In this range the gas flow is viscous and the outgassing is negligible. Mechanical pumps (rotary vane, blowers) used in this range show a fairly constant pumping speed.

The useful application range of this equation can be extended by including a factor *K* so that

$$t = 2.3K\frac{V}{S}\log\frac{P_1}{P_2}\ \text{sec} \tag{12.11}$$

The factor *K* is empirical and is usually supplied by the pump manufacturer as a service factor. Typical service factors are given in Table 12.1.

Mechanical pumps cannot be used at pressures below 1.0 torr as their pumping speed drops rapidly to zero at about 10^{-2} torr. The range can be extended to 10^{-4} by using two pumps in series such as a blower followed by a rotary pump or two blowers in series.

Equation (12.11) can be rearranged to the form

$$P_2 = P_1e^{-kt} \tag{12.12}$$

which will show that the pressure is reduced by a logarithmic decay and is zero only after infinite time, irrespective of the limitations of the pump.

The design problem for vacuum systems is how to reduce the pressure to a desired value in a reasonable pump down time and hold it there for the required process time.

Equation (12.12) is used when the pumping speed (S) is constant, i.e., the *S-P* characteristic is flat. Outside the constant S range, the curve slopes and we will have to use an equation based on constant throughput Q, ($= P \times S$).

$$t = \frac{V}{Q}(P_1 - P_2) \tag{12.13}$$

If this part of the characteristic curve is not a straight line, it is divided into small segments. The equation is separately applied to each segment and the time values are added to get the total time for evacuation in that range.

The use of these equations will be clear from the solution of the following problems.

When the outgassing load is constant in the molecular range and when there are no process gases, we can use the pumping speeds demanded by the required throughput for time calculation.

EXAMPLES

Data

Vessel–mild steel cylindrical vessel 1.2 m diameter and 1.5 m height. Initially the vessel contains dry air at 760 mm pressure and 30°C temperature.

Volume – 1.7 m^3 (1700 ℓ)

Surface Area – 5.65 $m^2 = 5.65 \times 10^4$ cm^2

Outgassing rate (0–10h)

Mild steel – 4.2×10^{-7} torr ℓ/sec cm^2

Elastomer – 2×10^{-4} torr ℓ/sec cm^2

Glass – 1.16×10^{-9} torr ℓ/sec cm^2

Gasket (elastomer) – length 0.4 m–400 cm

Exposed thickness – 0.15 cm

EXAMPLE 12.2

The above discussed vessel is to be evacuated to 1.0 torr in about 15 min. Calculate the pumping speed and select a suitable pump. Disregard outgassing and the conductance of the pipe between the pump and the vessel.

Solution

The pump has to discharge at atmospheric pressure, hence, a positive displacement or mechanical pump is required. The data mention "dry" air but some amount of moisture (at least humidity) is expected. Mechanical pumps can handle the moisture by introducing ballast. The pumping capacity will be reduced by ballast. Ultimate pressure without ballast will be about 10^{-2} torr. With gas ballast it will be about 10^{-1} torr (single stage). The vessel pressure required is 1.0 torr. Hence moisture, if present, will not be a problem.

The pumping speed-pressure characteristics (Figure 12.11) show a constant speed in the range 760–1.0 torr. Hence, we can apply Equation (12.1), i.e.,

$$t = \frac{V}{S} \text{ in } \frac{P_1}{P_2} \qquad \text{or}$$

$$t = K \times 2.3 \times \frac{V}{S} \log \frac{P_1}{P_2}$$

$V = 1700\ \ell$, $t = 15$ min $= 900$ sec, $P_1 = 760$ torr, $P_2 = 1.0$ torr
Substituting

$$S = \frac{V}{S} \times 2.3 \times K \log \frac{760}{1}$$

The pressure range is on the high side, hence, $K = 1.1$

$$S = \frac{1700}{15 \times 60} \times 2.3 \times 1.1 \times 2.88$$

$$= 13.76\ \ell/\text{sec} \quad \text{or}$$

$$= 826\ \ell/\text{min}$$

Nearest available pumps are (from manufacturers' data)

Single stage $S = 750\ \ell/\text{min}$

Two stage $S = 1000\ \ell/\text{min}$

Calculating evacuation times for both

$$t = \frac{1700}{750} \times 2.3 \times 1.1 \times 2.88$$

$$= 16.5 \text{ min}$$

$$t = \frac{1700}{1000} \times 2.3 \times 1.1 \times 2.88$$

$$= 12.4 \text{ min}$$

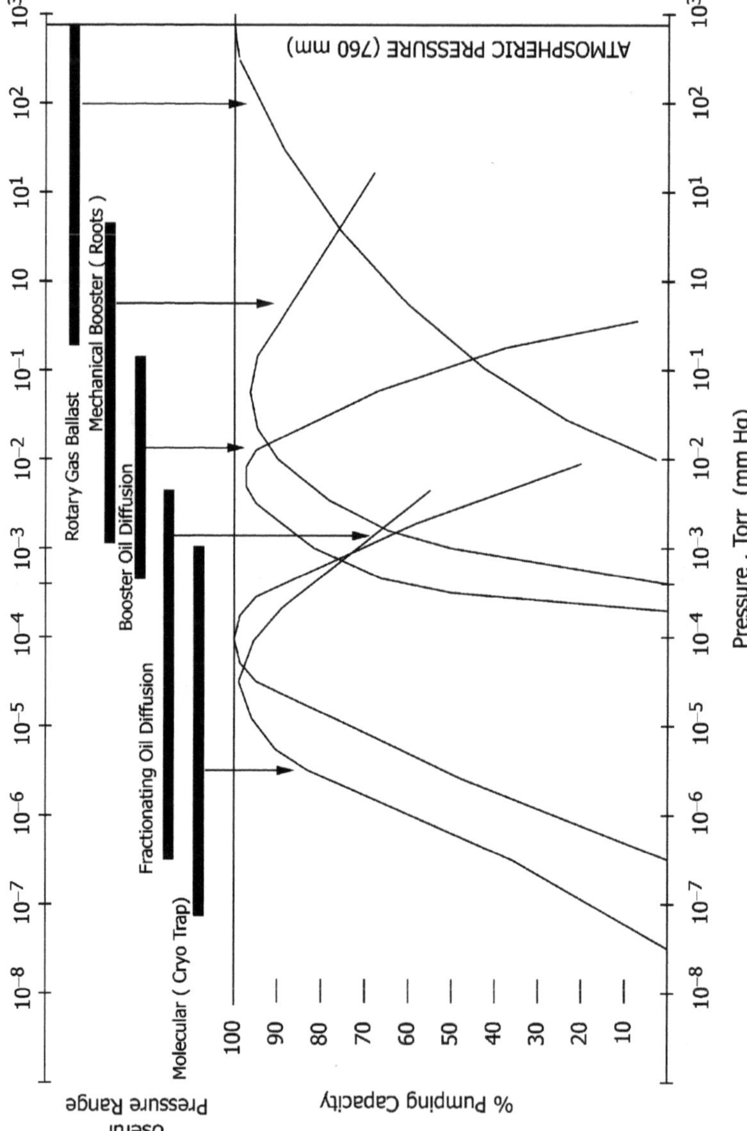

Figure 12.11 Typical pumping characteristics of vacuum pumps.

Hence, any of these pumps can be used. The single stage 750 ℓ/min will be cheaper.

EXAMPLE 12.3

The same vessel is connected to the pump by a pipe having diameter 30 mm and length 1.2 m. Calculate the conductance of the pipe and the pump capacity if the evacuation from 760 to 1.0 torr is to be done in 15 min.

Solution

In the given pressure range (760–1.0 torr), the gas flow will be molecular. Let us check this.

For molecular flow

$$D \times P > 5 \times 10^{-1} \text{ cm} \cdot \text{torr}$$

$$D = 3.0 \text{ cm}, \quad P = (760 + 1)/2 \sim 380 \text{ torr}$$

$$D \times P = 3 \times 380 \gg 5 \times 10^{-1}$$

Hence, the molecular flow is confirmed.

The conductance in this range is given by

$$C = \frac{184 \times D^4 \times P}{\ell} \frac{\ell}{S}$$

$$\ell = \text{length} = 120 \text{ cm}$$

$$C = \frac{184 \times 3^4 \times 380}{120}$$

$$= 4.72 \times 10^4 \, \ell/\text{sec}$$

If a pump of capacity P is chosen the effective speed P_E is given by

$$\frac{1}{P_E} = \frac{1}{P} + \frac{1}{C}$$

Substituting

$$\frac{1}{P_E} = \frac{1}{1000} + \frac{1}{4.72 \times 10^4}$$

The second term on LHS is very small and can be neglected. Hence, the pumping speed is unaffected by the pipe conductance.

EXAMPLE 12.4

The same vessel is to be evacuated to 10^{-3} torr in about 15 min. Choose a suitable pump. Calculate assuming there is no outgassing.

Solution

We have a choice of pumps in the given situation.

1. Rotary piston pumps with an ultimate pressure about 10^{-4} torr are available. These are two-stage pumps.
2. We can use a roots blower backed by a mechanical pump.

The rotary pump shows a constant speed (flat) behavior from 760 to about 10^{-2} torr and a constant throughput from 10^{-2} to 10^{-4} torr.

Time required for lowering pressure from 760 to 10^{-2} torr is

$$t = K\frac{V}{S} = 2.3 \log \frac{P_1}{P_2} \text{ sec}$$

In this range $K = 2$. Arbitrarily, we assume a pumping time 10 (< 15) min

$$= 2.0\frac{1700}{10}2.3 \log \frac{P_1}{P_2}$$

$$= 3816 \ \ell/\text{min}$$

Nearest pump available is a two-stage rotary pump with a pumping capacity 3333 ℓ/min and an ultimate pressure 10^{-4} torr.

For finding out the amount of time to pump from 10^{-2} to 10^{-3} torr, we can use the constant throughput equation

$$t = \frac{V}{Q}(P_1 - P_2)$$

$$\text{Throughput} = P \times S$$

$$= 760 \times 3333$$

$$= 25 \times 10^5 \text{ torr } \ell/\text{min}$$

$$t = \frac{1700}{25 \times 10^5}(10^{-2} - 10^{-3})$$

$$\simeq 6 \times 10^{-6} \text{ min}$$

This shows that the pressure reduction from 10^{-2} to 10^{-3} is practically instantaneous. This is so because there is practically no gas load (outgassing).

We can reduce the pump capacity by allowing 15 min for pressure reduction from 760 to 10^{-2} torr. This gives

$S = 2544\ \ell/\text{min}$

Nearest available pump is 1660 ℓ/min. For this, the pumping time is

$$t = \frac{1700}{1600} 2.3 \times 4.88$$

$$\simeq 23 \text{ min}$$

Let us consider a combination of a roots pump backed by a mechanical pump. Assume

1. The mechanical pump evacuates down from 760 to 10 torr.
2. The roots pump evacuates between 10 and 10^{-3} torr.

3. The maximum permissible pressure difference between the inlet and outlet of the roots pump is 60 torr.

We use a mechanical pump to reduce pressure from 760 to 10 torr in 10 min

$$S = 1.1 \times 2.3 \times \frac{1700}{1600} \times \log \times \frac{760}{10}$$

$$= 812 \ \ell/\text{min}$$

Nearest pump is a single-stage rotary piston pump of 1000 ℓ/min capacity and 2×10^{-2} ultimate pressure. The recalculated pumping time is

$t = 0.8$ min

EXAMPLE 12.5

The vessel in the previous problems is to be evacuated to 10^{-5} torr and maintained at that pressure for one hour.

The pumps to be used are a diffusion pump, backed by a roots pump, backed by a mechanical pump, which exhausts into the atmosphere.

The vessel and its components start outgassing at about 10^{-3} torr and contains for 1 h to 10^{-5} torr at a constant rate.

The diffusion pump is connected to the vessel via:

1. A straight pipe 50.0 cm long.
2. A right angle elbow having each leg 25.0 cm long.
3. A baffeled elbow with molecular transmission probability 0.35.

Now determine:

1. Pumps required.
2. Time to reach 10^{-5} torr from 10^{-3} torr.

Solution

As the evacuation process involves outgassing at pressures 10^{-3} and lower, we have to design the pumping requirement starting with the diffusion pump.

The surface area of the vessel is 5.65×10^4 cm^2 and the outgassing rate is 4.2×10^{-7} torr ℓ/sec·cm^2. Outgassing load from vessel is

$$\begin{aligned} &4.2 \times 10^{-7} \times 5.65 \times 10^4 \\ &= 2.37 \times 10^{-3} \text{ torr} \cdot \ell/\text{sec} \end{aligned} \tag{I}$$

The elastomer area

$$= \text{ length } \times \text{ exposed surface}$$

$$= 400 \times 0.15$$

$$= 60 \text{ cm}^2$$

Elastomer outgassing rate is 2×10^{-4} torr·ℓ/sec·cm^2. Hence, the outgassing load from elastomer is

$$\begin{aligned} &= 2 \times 10^{-4} \times 60 \\ &= 12.0 \times 10^{-3} \text{ torr} \cdot \ell/\text{sec} \end{aligned} \tag{II}$$

Total outgassing rate is

$$\begin{aligned} &= (2.3 + 12) \times 10^{-3} \\ &= 14.3 \times 10^{-3} \text{ torr} \cdot \ell/\text{sec} \end{aligned} \tag{III}$$

The outgassing starts at 10^{-3} torr and continues to 10^{-5} torr. The gas load at 10^{-3} torr

$$\begin{aligned} &= \frac{14.3 \times 10^{-3}}{10^{-3}} \\ &= 14.3\, \ell/\text{sec} \end{aligned} \tag{IV}$$

At 10^{-5} torr it is

$$\begin{aligned} &= \frac{14.3 \times 10^{-3}}{10^{-3}} \\ &= 14.3 \times 10^2\, \ell/\text{sec} \end{aligned} \tag{V}$$

The conductance of the elbow-trap will lower the pumping speed of the pump.

Let us choose a diffusion pump with a capacity (P_C), $2 \times 14.3 \times 10^2$

$$P_C \simeq 30 \times 10^2 \ \ell/\text{sec}$$

Manufacturers' literature shows an oil diffusion pump having a capacity 3000 ℓ/sec at 10^{-5} torr. It has three stages and an inlet diameter 250 mm.

Mean free path of air at pressure P is

$$\lambda = 4.86/P \text{ cm}$$

Hence,

at 10^{-5} torr, $\lambda = 4.86/10^{-5} = 4.86 \times 10^5$ cm

at 10^{-4} torr, $\lambda = 4.86/10^{-4} = 4.86 \times 10^4$ cm

at 10^{-3} torr, $\lambda = 4.86/10^{-3} = 4.86 \times 10^3$ cm

The ratio D/λ for various pressures is

at 10^{-5} torr, $D/\lambda = 4.86/10^{-5} = 4.86 \times 10^5$ cm

at 10^{-4} torr, $D/\lambda = 4.86/10^{-4} = 4.86 \times 10^4$ cm

at 10^{-3} torr, $D/\lambda = 4.86/10^{-3} = 4.86 \times 10^3$ cm

Hence, the flow will be molecular throughout.

If the diffusion pump is connected to the vessel by a circular pipe 250 mm dia. and (say) 500 mm long pipe, the conductance is given by

$$C = 12.1 \times D^3/L$$

Substituting,

$$C = 12.1 \times 25^3/50$$

$$= 3781 \ \ell/\text{sec}$$

The effective pumping speed P is given by

$$\frac{1}{P_E} = \frac{1}{P_C}\frac{1}{C}$$

$$P_E = 1673 \ \ell/\text{sec}$$

Actual required speed is 1430 ℓ/sec

Hence the chosen diffusion pump is adequate. We can, however, choose a slightly smaller pump.

The introduction of the pipe has reduced the chosen capacity by

1673/3000 ~ 56%

Let us consider a plain right-angle elbow connecting the pump and the vessel.

The conductance of the elbow is given by

$$C = \frac{12.1 \times D^3}{L_1 + L_2 + 1.33D}$$

where L_1 and L_2 are the lengths of each leg of the elbow and D is the diameter (both in cm).

Assume that each leg is 25.0 cm long, D = 25.0 cm

$$C = \frac{12.1 \times 25^3}{25 + 25 + 1.33 \times 25}$$

$$= 1180 \ \ell/\text{sec}$$

Substituting

$$\frac{1}{P_E} = \frac{1}{30 \times 10^2} + \frac{1}{11.8 \times 10^2}$$

$$P = 1297 \ \ell/\text{sec}$$

Thus, the pumping speed with elbow is short of the required speed 1470 ℓ/sec and we will have to modify the elbow, or choose a larger pump, or do away with the slightly reduced capacity.

Consider now an elbow with a baffle. This is a component of a complex shape and we will use the probabilistic method.

The molecular conductance of a given passage C is obtained from

$$C = C_O \times Pr$$

where

C_O = Conductance of aperture with an opening of area equal to that of the given passage
$= 3.64\ (T/M)^{1/2}\ A$

T = Absolute temperature (K)

M = Molecular weight of gas (g·mol^{-1})

Pr = Transmission probability

Pr is calculated and published in special books

Assume $T = (273 + 30) = 303$ K

$M = 28.98 \sim 29.0$ for air (g·mol^{-1})

$A = \pi \times (25^2)/4 = 490\ \text{cm}^2$

$C_O = 3.64 \times (303/28.98)^{1/2} \times 490$

$= 2018\ \ell/\text{sec}$

The probability Pr for a trapped elbow is about 0.35 (This value is taken from manufacturers' literature).

The conductance of the baffle elbow is

$$C = 5.765 \times 10^3 \times 0.35$$

$$= 5765\ \ell/\text{sec}$$

The pump speed with this elbow is

$$\frac{1}{P_E} = \frac{1}{P_C} + \frac{1}{C}$$

$$= \frac{1}{30 \times 10^2} + \frac{1}{11.8 \times 10^2}$$

$$P_E = 1202\ \ell/\text{sec}$$

This capacity is less than the minimum required speed, 1473. Hence, a larger pump is required when using a baffle/elbow combination. Next available pumps have 4000 and 5500 ℓ/sec speed. The above calculations will have to be repeated to check their suitability.

12.9 CALCULATION OF PUMPING TIME

For determining the pumping time from 10^{-3} to 10^{-5} torr, we will assume that the pump chosen has sufficient capacity to handle the required throughput through the connecting pipes, valves, baffles, etc.

It is estimated that the volume of the pump and connecting piping is of the order of the volume of the vessel. This shows that the total pumping volume is $1700 \times 2 \cong 3000$ ℓ.

The calculated outgassing load is 14.3×10^{-3} torr·ℓ/sec.

At 10^{-5} torr the speed is 1430 ℓ/sec, similarly at 10^{-4} torr it is 143 ℓ/sec and at 10^{-3} torr it is 14 ℓ/sec.

If we plot the required speed against the pressure on log × log plot, we get a straight line showing a speed decreasing with increasing pressure.

The usual pumping time equation

$$t = \frac{V}{S} \text{ in } \frac{P_1}{P_2}$$

can be applied if we split the pressure range (10^{-3} to 10^{-5} torr) into suitable steps. The smaller the chosen range, the more accurate will be the time estimate.

Let us split the pressure range as follows:

1. 10^{-3}–5×10^{-4} torr Average speed 37 ℓ/sec
2. 5×10^{-4}–10^{-4} torr Average speed 87 ℓ/sec
3. 10^{-4}–5×10^{-5} torr Average speed 457 ℓ/sec
4. 5×10^{-5}–10^{-5} torr Average speed 1000 ℓ/sec

Applying the equation to first range

$$t_1 = 2.3\frac{3000}{37}\log\frac{10^{-3}}{5\times10^{-4}}$$

$$= 56 \text{ sec}$$

Similarly

$t_2 = 55.3$ sec

$t_3 = 4.5$ sec

$t_4 = 4.8$ sec

Total time = 120.3 sec, say 2.0 min.

Thus the time to pump from 10^{-3} to 10^{-5} torr is about 2 min as calculated. We may apply a service factor of about 5 so that the total time is about 10 min.

Note: There are other methods of calculating the time more accurately. However, for a preliminary estimate, the above given method is adequate. The service factor is mainly dependent on the pump combination, conductance of plumbing, and experience.

EXAMPLE 12.6

The same vessel as in the previous problems is evacuated from atmospheric to 10^{-5} torr pressure by a diffusion pump 30×10^2 ℓ/sec with an effective speed 12×10^2 ℓ/sec.

The critical backing pressure (fore pressure tolerance) for this pump is 0.35 torr.

Now determine:

1. The backing pump(s) required
2. The time required

Solution

The outgassing load from 10^{-3} to 10^{-5} torr (as calculated in the previous example) is 14.3×10^{-3} torr$\cdot\ell$/sec.

The diffusion pump has a speed of 30×10^2 ℓ/sec. The maximum throughput of diffusion pumps is generally at 10^{-3} torr.

The maximum throughput of the chosen diffusion pump will be

$$3000 \times 10^{-3} = 3.0 \text{ torr}\cdot\ell/\text{sec}$$

The maximum fore pressure tolerance of this pump is 0.3 torr. Discharge at this pressure is $3/0.3 = 10$ ℓ/sec or 600 ℓ/min.

The backing pump chosen should have a capacity larger than 600 ℓ/min at 0.3 torr.

Let the required capacity be 700 to 800 ℓ/min.

If we choose a positive displacement pump (mechanical pump) the pumping capacity of medium pumps is limited to about 300–400 ℓ/min at 0.3 torr and rapidly increases at higher pressures exceeding 0.3 torr. Thus, a mechanical pump alone is not suitable. If we are required to use the gas ballast the speed decreases still further.

We will have to back the diffusion pump by a roots blower or booster. This booster will further have to be backed by a mechanical pump.

The final choice is to use a blower to reduce the pressure from $10–10^{-3}$ torr and a mechanical pump in the 10–760 torr range.

Manufacturers' literature shows a roots pump having a capacity 45 ℓ/sec at 0.3 torr which reduces to about 40 ℓ/sec at 10^{-2} torr. Hence it has an adequate backing capacity. Moreover, its capacity (speed) at 1–10 torr is also 35–45 ℓ/sec.

The throughput of blower at 8–10 torr is 8×40 to 10×40, i.e., 320–400 torr ℓ/sec.

Literature shows both a sliding vane and a rotary piston pump with speeds of 350–370 ℓ/min. These are single or double stage pumps. We choose a sliding vane pump having a speed 2500 ℓ/min or 41.6 ℓ/sec at 760 to 1.0 torr, and an ultimate pressure 10^{-2} torr.

The final choice of pumps is

$10^{-5} - 10^{-3}$ torr, diffusion pump 3000 ℓ/sec

$10^{-3} - 10$ torr, roots pump 45 ℓ/sec

$10 - 760$ torr, mechanical pump 41 ℓ/sec

The time to evacuate from 760 to 10 torr is

$$t_1 = 1.5 \times 2.3 \times \frac{1700}{45} \log \frac{760}{10}$$

$$= 276 \text{ sec}$$

Time for reducing pressure from 10–10^{-3} torr

$$t_2 = 1.5 \times 2.3 \times \frac{1700}{45} \log \frac{10}{10^{-3}}$$

$$= 780 \text{ sec}$$

Total time to reach 10^{-3} torr from 760 torr

$$= t_1 + t_2 = 276 + 780$$

$$= 1056 \text{ sec} \cong 18 \text{ min.}$$

Pumping from 10^{-3} to 10^{-5} is achieved by the diffusion pump which has a flat pressure speed characteristic showing constant speed. The time required is (as estimated in the previous example) 5–10 min.

Hence, time for pumping from atmospheric pressure to 10^{-5} torr is

= 23–28 min or about 30 min (1/2 h)

12.10 MEASUREMENT OF VACUUM

Gauges used for the measurement of low pressures are different from those used for normal pressures. Ordinary Bourdon tube gauges can be used from 760 to 0.5 torr, which is only the beginning of the vacuum range. Just as in the case of pumps, there is no one gauge useful for vacuum measurement at all ranges. Every gauge has its applicable pressure range. In many instances more than one gauge is used to cover different pressure ranges. There are three main types used in the 760–10^{-5} range.

1. Mechanical Gauges: 1.0–10^{-3} torr
2. Conductivity Gauges: 1.0–10^{-4} torr
3. Ionization Gauges: 10^{-3}–10^{-9} torr

The only gauge which directly measures the vacuum pressure is the MacLeod gauge. In this gauge, a gas sample of known volume, from the vacuum system, is isolated and compressed to a known volume. The difference in the pressure is measured on a liquid (mercury) coloumn. MacLeod gauges can be used from 10 to 10^{-6} torr. However, the instrument requires manual operation and is intermittent. It cannot be used for industrial applications. However, because it gives absolute pressure measurement, it is used for the calibration of other gauges. Hence, the MacLeod gauge is not discussed further.

12.10.1 Mechanical Gauges

This type of gauge is similar to the aneroid barometer. It contains a metal box or capsule with a very thin diaphragm on one side. The diaphragm is connected with fine linkage to a pointer. The capsule is evacuated to very low pressure (~ 10^{-8} torr). It is enclosed in a chamber (see Figure 12.12) which is open to the vacuum. Thus, the diaphragm is subjected to the vacuum pressure of the system from the outside and to its own vacuum from the inside. This pressure differential causes the diaphragm to deflect proportionally. The deflection is amplified mechanically by the linkage and exhibited by the pointer.

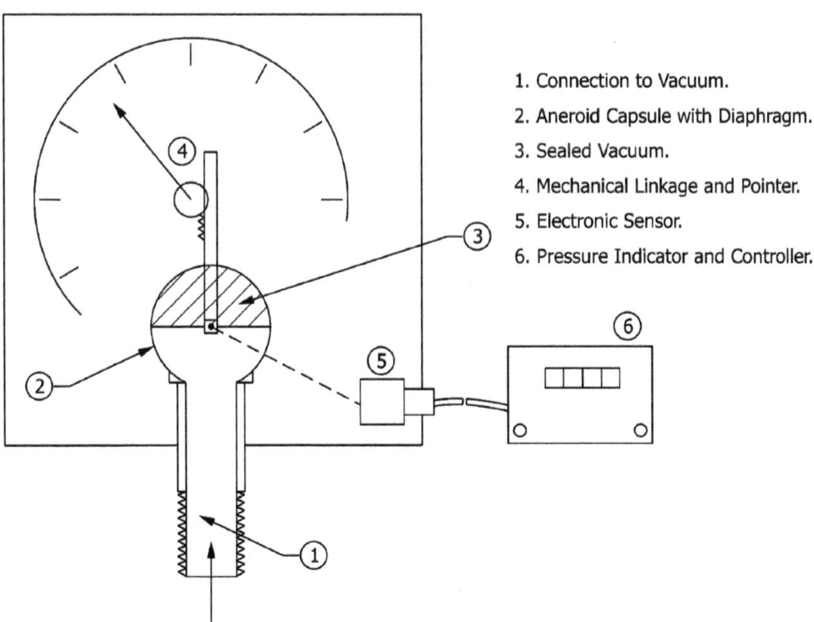

Figure 12.12 Mechanical and electro-mechanical vacuum gauge.

The linkage is replaced by a sensor and an electronic system in some designs.

Diaphragm gauges can be used in the range 10^{-1}–10^{-4} torr. Due to their all-metal construction they can be baked to 180°C for outgassing. These are simple and rugged devices independent of atmospheric pressure, and are sensitive. They are affected by temperature and aging. With electronic sensing, they can be used for an extended range up to 10^{-8} torr and remote control.

12.10.2 Conductivity Gauges

Thermal conductivity of a gas is a function of its density, i.e., pressure in the 1.0–10^{-5} torr pressure range. In gases, thermal conduction takes place because of the impact of molecules with the hot body (or heat source). As the pressure decreases, the number of molecules in a given volume decreases and their mean free path increases. This reduces the number of impacts and, hence, the conductivity (Figure 12.13(A)).

Conductivity gauges are based on the indirect measurement of conductivity of a hot filament placed in vacuum. There are two major types of conductivity gauges.

1. *Thermocouple Gauge*
   The construction of a thermocouple gauge is shown in Figure 12.13(B). A filament suspended at the ends is heated by an electric current. The filament is made of platinum or nickel alloy. A chromel constantan or similar base metal thermocouple made from thin wire is welded to the filament. The emf of the thermocouple depends on the filament temperature, which in turn depends on thermal conductivity (i.e., heat transfer by conduction). The measuring circuit contains the cold junction compensation. The envelope of filament and thermocouple is open to vacuum. As the pressure decreases, the filament temperature increases and is measured on an indicator calibrated in pressure.

   The useful range of a thermocouple gauge is 1.0–10^{-4} torr depending on the design. It is rugged and gives

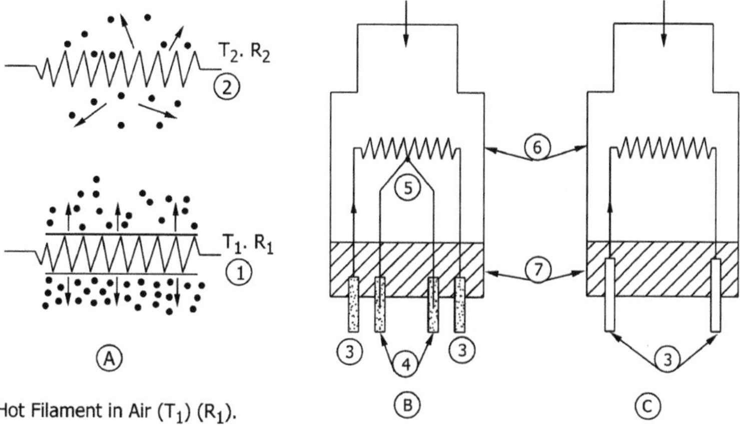

1. Hot Filament in Air (T_1) (R_1).
2. Same Filament in Vacuum (T_2) (R_2) Note $T_1 < T_2$, $R_2 > R_1$.
3. Filament (I , R , T).
4. Thermocouple.
5. Hot Junction Welded to Thermocouple.
6. Envelope (Glass).
7. Glass-metal Vacuum Seal.

Note - Measuring and Heating Circuits not shown.

A. Heat conduction in Air & Vacuum. B. Thermocouple Gauge. C. Resistance (Pirani) Gauge.

Figure 12.13 Thermal conductivity vacuum gauge.

continuous readings. Electrical variations in the filament current or aging affects the accuracy of the gauge.

2. *Resistance or Pirani Gauge*

In construction, the Pirani gauge is similar to the thermocouple gauge except that the envelope contains only the filament, as shown in Figure 12.13(C). The filament is heated at a constant voltage. The change in conductivity due to vacuum, is reflected in the change in resistance. By using a Wheatstone bridge the resistance is measured and interpreted in pressure. A similar compensating resistance is used in the adjacent leg of the bridge. Recent designs use a thermistor instead of the filament.

The general useful range of the Pirani gauge is 1.0–10^{-4} torr, which in some designs is extended to about 10^{-6} torr. This is a rugged instrument giving continuous reading (with a self-balancing bridge). Readings are affected by contamination with vapors.

12.10.3 Ionization Gauge

If an ionizing particle such as electron or proton strikes a gas molecule, the later is ionized. A positive ion and electrons are produced. These products can be collected on oppositely charged electrodes (cathode, anode). The resulting current will be proportional to the number of ions, which in turn, is proportional to the pressure. This proportionality is generally nonlinear.

In construction, ionization gauges are similar to electronic valves or vacuum tubes with one end open to the vacuum. The principle of ionization gauges is shown Figure 12.14(A).

There are many variations in the design. It is not possible to discuss all of these. Moreover, many designs incorporate some method of shaping the interelectrode field to enhance the capture of ions and develop a better current. Most of the advanced gauges are used for measuring vacuum of $> 10^{-9}$ torr and hence are beyond the scope of our discussion. One such gauge is the Bayard Alpert gauge, which is used in the ultra-high vacuum range (10^{-9}–10^{-12}).

Only two basic ionization gauges commonly used in the 10^{-5} range are described. Ionization gauges remove molecules from vacuum and hence, may act as a "pump" in the envelope.

Cold Cathode (Penning) Gauge

The construction of a Penning, or cold cathode, vacuum gauge is shown in Figure 12.14(B).

The tube is essentially a diode. The central electrode has a rectangular window shape and is positively charged to about 2 kV. Surrounding the anode is a rectangular (solid) cathode which acts as the electron generator. The envelope is open to vacuum. On the outside is a permanent magnet with its field at a right angle. The combined electric and magnetic fields make the ion path very long, thus increasing ionization and the resulting current.

The gauge is used in the range 10^{-4}–10^{-7} torr. The current produced is large enough and does not require amplification. The response is nonlinear. Due to the large degree of ionization there is a risk of contamination and pumping. Compared to other ionization gauges, Penning gauges are cheaper. They are

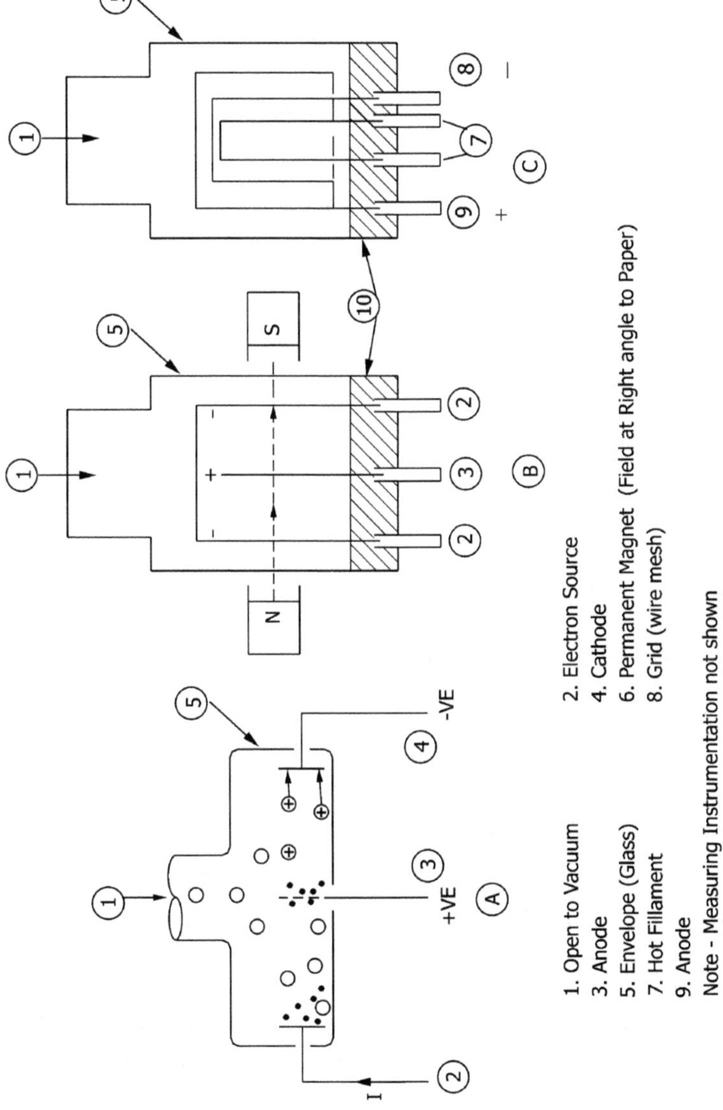

Figure 12.14 Ionization vacuum gauge.

bulky due to the permanent magnet. They are not very accurate but are good for general vacuum duty where high accuracy is not required.

Hot Cathode Gauge

These gauges are like a triode or tetrode with one or more grids generally placed between the anode and the cathode. The cathode is hot, i.e., thermionic, and is made of thoriated tungsten. It is heated to about 350–450°C. The general construction is shown in Figure 12.14(C). One end of the envelope is open to vacuum.

The hot filament releases electrons which ionize the gas molecules. The grid is at a negative potential and collects the positive ions. This creates a grid current. Similarly, the electrons are collected at the anode and create a current in the anode circuit. In another version, the grid is positively charged and the surrounding cathode is negatively charged. The ratio of the ion and electron currents is proportional to the pressure. The measuring circuits are not shown but the electron current is kept constant so that the ion current is directly proportional to the pressure. The gauge has different responses to different gases.

Hot cathode gauges are useful in the 10^{-3}–10^{-9} torr range. There is a risk of filament burnout and contamination. The gauge is not robust and generally not used for industrial applications.

"Alphatron" Ionization Gauge

A radioactive material continuously emits α, β, and γ radiation. The α radiation is helium nuclei with a charge of +2 and is least penetrating. They ionize a gas molecule on impact.

Alphatron vacuum gauges use a small mass of radioactive material as a source of α radiation. Thus, compared to the previously described ionization gauges, the Alphatron is a passive device.

The construction of an alphatron gauge is shown in Figure 12.15(A). The radiation source is a radium salt rolled into a gold foil. It has an activity of 1.5–100 micro curies and

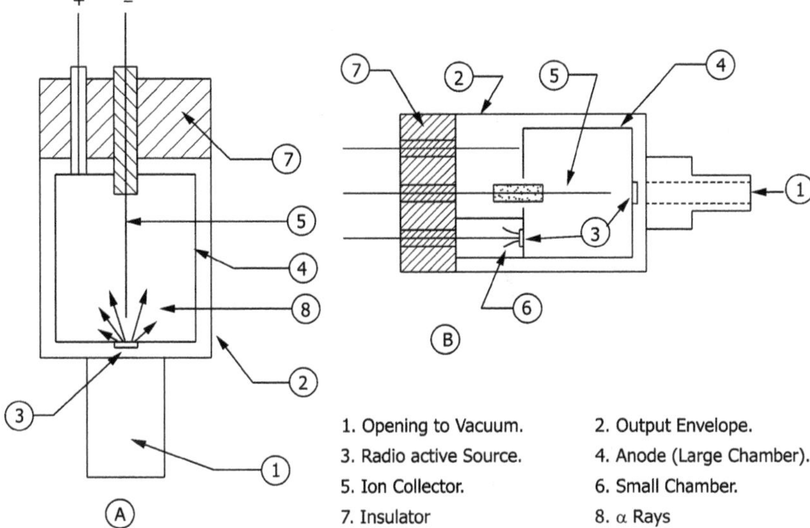

Figure 12.15 "Alphatron" ionization gauge.

so is quite safe. The source, in the form of a very small packet or capsule, is kept on a tray attached to the anode chamber. The cathode or ion collector is a wire centrally located and insulated from the cathode. One end of the envelope is open to vacuum. A single chamber gauge can measure vacuum in the 1.0–10^{-4} range. Figure 12.15(B) shows a double chamber gauge. The small chamber can measure in the 760–1.0 torr range. Thus the gauge can be used for continuous vacuum readings from the atmosphere to 10^{-4} torr.

Ionization produces a very small current (~10^{-12} A) and needs amplification. The output is linear. Though the gauge head is cheap and rugged, the instrumentation makes this gauge expensive.

Choice and Installation of Gauges

1. In a vacuum system, gauges are located in all the lines connecting the vessel and the various pumps.
2. The actual gauge to be installed at a point should cover the pressure range expected in that line. Thus the

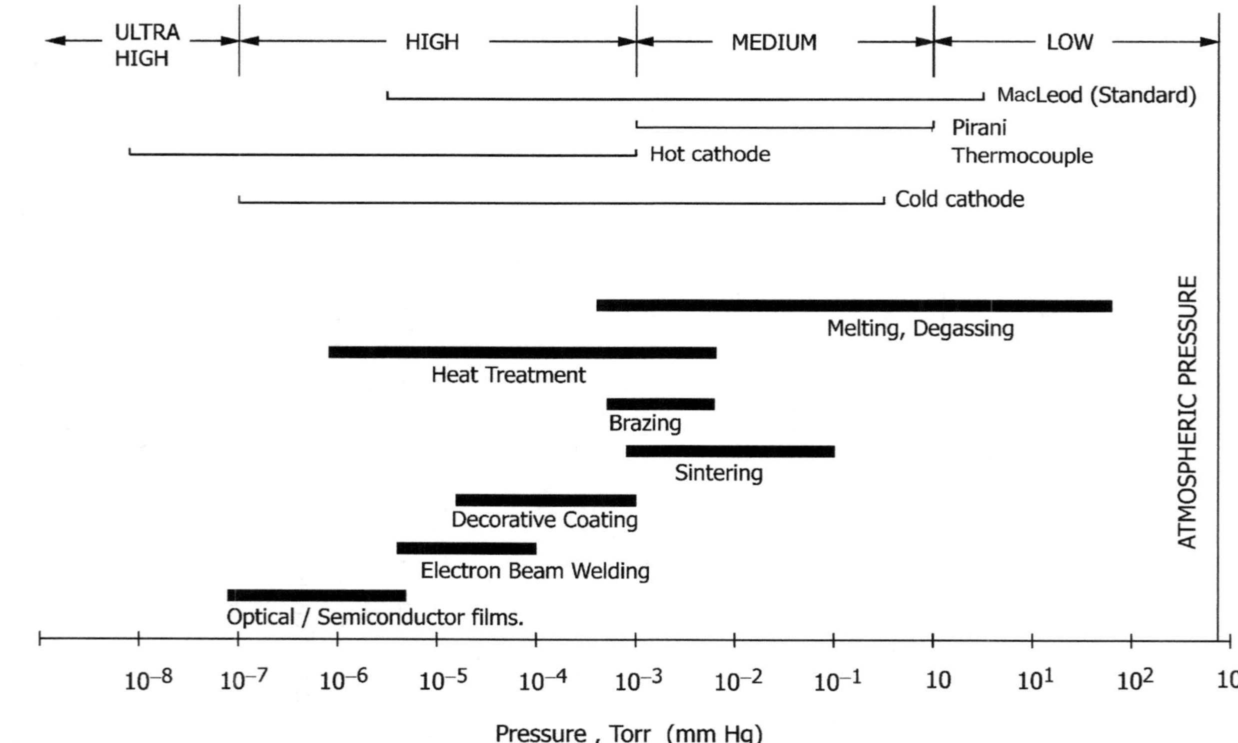

Figure 12.16 Vacuum, classification, gauges, and applications.

pressure will be lowest in or near the vessel. It will be higher in the line between the pumps and highest at the outlet to the atmosphere.

3. If some process is to take place in the vessel, locating the gauge in the vessel is avoided to guard against spattering and contamination. This also reduces the connection through the wall and attendant leakage.
4. For a small pressure difference ($1–10^{-1}$ torr) mechanical gauges are used.
5. A high vacuum gauge for a vessel is generally located in the pipe or elbow outside the vessel.
6. All gauge envelopes are miniature vacuum vessels and are affected by their own outgassing.
7. In large systems pressures are monitored and controlled from a central panel. This requires remote sensing gauges (ionization alphatron, thermocouple) and extensive instrumentation.

There are many other gauges available for higher vacuum ranges. These are not discussed here as they are beyond the arbitrarily chosen limit of 10^{-5} torr for this book.

Pressure-wise classification of vacuum, applicances, and gauges is shown in Figure 12.16.

Chapter 13

Protective Atmospheres

CONTENTS

13.1 INTRODUCTION

Both electric and fuel-heated furnaces contain an oxidizing atmosphere with free oxygen, moisture, and CO/CO_2.

Most metals and alloys are oxidized in such an atmosphere. For some metals (aluminum, stainless steels) the oxide is in the form of a tightly adhering skin, preventing further oxidation. Metals like carbon alloy steels, and brasses form a heavy and loose scale which drops off. Thus, their oxidation is continuous and represents significant waste.

Depending upon the gas composition and temperature, steels either lose carbon (decarburization) or gain carbon (carburization). Some metals, especially in the liquid state, pick up oxygen, hydrogen, and nitrogen, which affects their service properties.

In some treatments of steels, they are intentionally exposed to special atmospheres to obtain a surface layer containing higher concentrations of C and N, as in gas carburizing and carbonitriding. Some solid state semiconducting materials are doped (microalloyed) by exposing them to dope (alloy element) gas for a long time.

For temperatures more than about 1600°C, we use tungsten, molybdenum, or carbon resistors. These materials require an inert atmosphere or (in some cases) vacuum as a protective atmosphere.

Sintering of carbides requires a dry hydrogen atmosphere. Some brazing and welding processes (TIG) require inert gas or vacuum.

Thus, many heating processes require special atmospheres other than air. The furnace designer has to make modifications in the design to suit the furnace for these conditions. The actual manufacture of these atmospheres is a specialized field best left to experts. Some atmospheres are explosive and toxic. It is necessary to design the furnace for their safe handling and disposal.

The actual atmosphere used for a given process and its quality is usually specified by the customer's metallurgist. This again is a specialized field not directly connected to furnace design.

The preceding discussion is a general review of the need for special atmospheres, usually called "protective atmospheres." For more detailed information on these, refer to special literature.

13.2 MANUFACTURED ATMOSPHERE

These are prepared by partial combustion or dissociation of gases such as hydrocarbons and ammonia. They predominantly contain H_2 and CO (as reducing components), N_2 — through air (as inert component), and small amounts of H_2O, CO_2, and fuel.

The principal reactions are

Natural gas: $2CH_4 + O_2 + [N_2] \rightarrow 2CO + 4H_2 + [N_2] + \Delta H$

Liquefied petroleum gas (LPG) Propane:

$2C_3H_8 + 3O_2 + [N_2] \rightarrow 6CO + 8H_2 + [N_2] + \Delta H$

Ammonia: $2NH_3 \rightarrow N_2 + 3H_2$

The above reactions are carried out in metal retorts filled with catalysts such as iron and nickel pellets. The reaction products are controlled by controlling the temperature, flow rate, and immediate cooling of the products after they leave the reaction chamber. The catalyst enhances the speed of the reaction but does not change the products and their yields.

Protective atmospheres prepared from natural gas or LPG are of two types. Exothermic atmospheres are manufactured by partial or complete combustion in a furnace. A nozzle mixing burner with elaborate control of fuel/air ratio is used.

Endothermic atmospheres are prepared by heating an air-gas mixture in a metal retort which is externally heated. The reaction products are cooled as soon as they leave the retort. The retort is heated to 950–1000°C. These atmospheres contain more CO and H_2 than the exothermic gases.

Both exothermic and endothermic gases will contain small amounts of uncombusted fuels.

A "lean" nitrogen-based atmosphere is prepared by exothermic combustion of fuel gas. CO_2 and water vapor are removed from the combustion products, leaving a mixture of N_2, CO, and H_2.

Typical gas generators for exothermic, endothermic, and dissociated ammonia atmospheres are shown in Figure 13.1, Figure 13.2, and Figure 13.3.

13.3 PURE GAS ATMOSPHERES

13.3.1 Nitrogen

Pure and dry nitrogen offers a cheap source of neutral atmosphere as it does not react with most metals when in the molecular (N_2) state.

Nitrogen is manufactured from liquefaction of air which separates it from oxygen and argon. It is nonexplosive and has the same density as air. Moisture and other minor constituents are removed to produce dry and pure nitrogen.

It can also be manufactured from ammonia by partial combustion. If the combustion is carefully controlled, mixtures of N_2:H_2 with 99–20 N_2 can be produced. Pure nitrogen can also be mixed with other gases to obtain reactive mixtures. As no reactor is required, nitrogen-based atmospheres are cheap and easy to handle. The dew point of nitrogen is –60 to –68°C.

Nitrogen is used for sintering, brazing, and annealing. It is also used for purging furnaces (removal of air) before introducing other atmosphere. It is also used for packaging foods.

13.3.2 Hydrogen

This is the most powerful reducing atmosphere. Hydrogen is manufactured from natural gas or by electrolysis of water. The latter method produces oxygen as a by-product.

Hydrogen requires very careful drying to reduce the water content to 8–10 ppm, giving a dew point of –68°C. Oxygen is also removed to < 1 ppm. Thus, the gas has a very high H_2/H_2O ratio.

Hydrogen is available in cylinders as a compressed gas. Bulk hydrogen is supplied in liquid state by tankers.

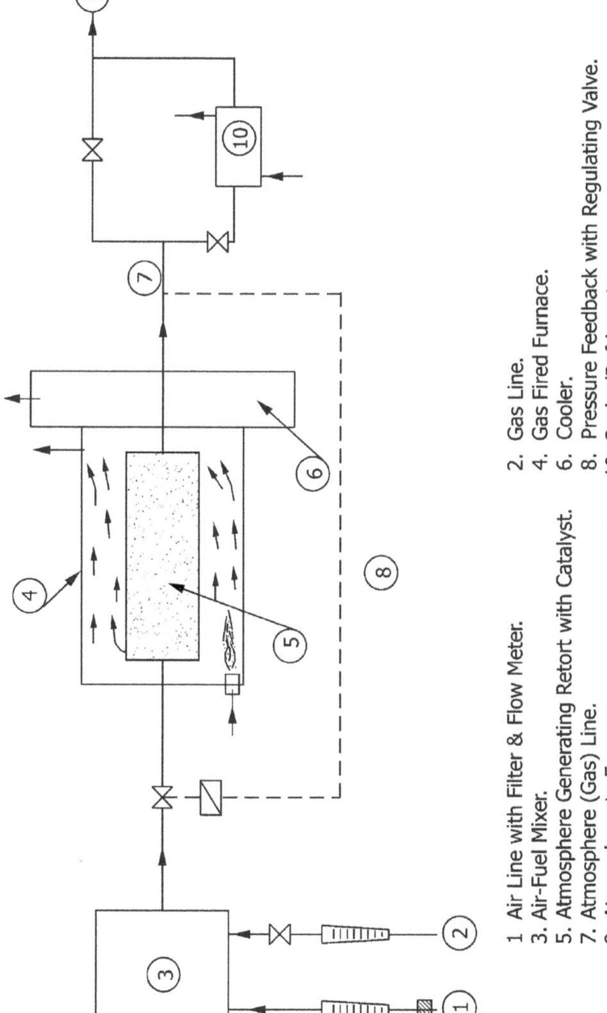

Figure 13.1 Endothermic atmosphere generator.

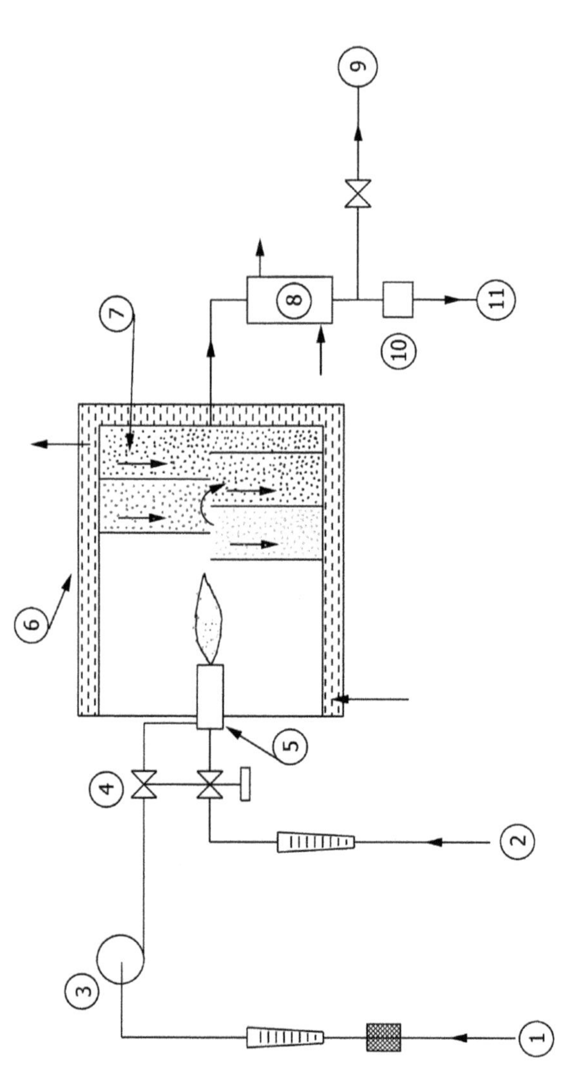

1. Air Inlet with Filter & Flow meter.
2. Fuel Gas Inlet with Flow meter.
3. Compresor / Blower.
4. Air-Fuel Control Valves.
5. Burner.
6. Water Jacketed Furnace Retort (900 °C).
7. Catalyst.
8. Atmosphere (gas) Cooler.
9. Vent / Bleed off.
10. Refrigeration (optional).
11. Atmosphere to Furnace.

Figure 13.2 Exothermic atmosphere generator.

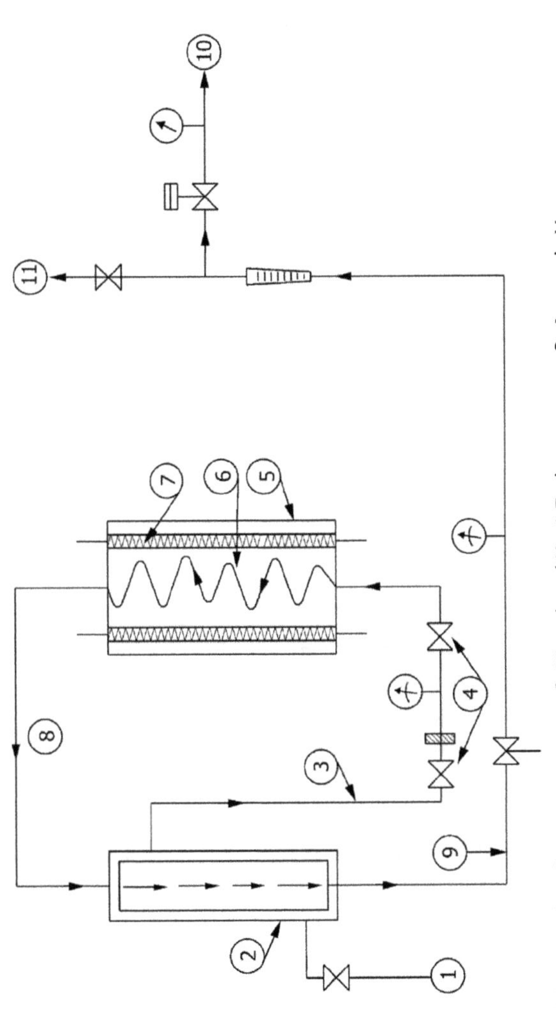

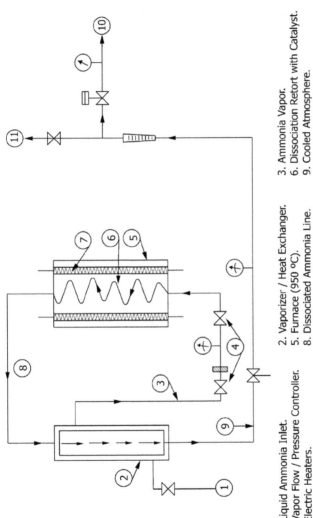

1. Liquid Ammonia Inlet. 2. Vaporizer / Heat Exchanger. 3. Ammonia Vapor.
4. Vapor Flow / Pressure Controller. 5. Furnace (950 ºC). 6. Dissociation Retort with Catalyst.
7. Electric Heaters. 8. Dissociated Ammonia Line. 9. Cooled Atmosphere.
10. Atmosphere Line to Furnace. 11. Bleed Off.

Figure 13.3 Dissociated ammonia atmosphere generator.

Hydrogen is explosive and requires very careful handling. It has good thermal conductivity.

Many metals, especially steels, absorb hydrogen, which leads to embrittlement and fracture. It is also a decarburizer and cannot be used for carbon steels. It is mainly used for annealing low carbon steels, electric and stainless steels, and for sintering of carbides.

13.3.3 Helium and Argon

These are true inert gases having absolutely no reactions with any metals or other materials (like ceramics, glasses, and plastics).

Helium is separated from natural gas, which contains helium 3–10% depending upon the source locality. Air contains about 1% argon from which it is separated by liquefaction. Both gases are costly as compared to other atmospheres. They are treated to remove oxygen and moisture to less than 0.1%.

Helium is lighter than air, while argon has a comparable density. These gases are safe and easy for transport and handling.

Because of their cost, they find use in special applications such as TIG welding, sintering, heat treatment of special high-temperature high-strength alloys, and nuclear materials.

Some commonly used protective atmospheres and their applications are given in Table 13.1.

13.4 HEATING OF PROTECTIVE ATMOSPHERE FURNACE

In atmosphere furnaces, it is vital to avoid mixing combustion products of fuel with the protective gas.

If a fuel-fired furnace is to be used for economic reasons, there are two options:

1. Use radiant tubes (see Section 6.12).
2. Use an inner metallic muffle for the process and an outer furnace for combusting fuel.

TABLE 13.1 Selected Protective Atmospheres and their Applications

Type of Atmosphere	Nominal Composition vol. %							Dew Point°C	Application	Fuel required m^3 fuel for 1000 m^3 atm
	N_2	CO	CO_2	H_2	CH_4					
Lean Exothermic	86.8	1.5	10.5	1.2	—			4.5	Bright annealing copper, sintering ferrites, purge gas.	120
Rich Exothermic	71.5	10.5	5.0	12.5	5			10	Bright annealing low C steel, silicon steels, steel/copper brazing, sintering.	155
Endothermic	40–45	20	0–0.5	34–40	0.5–1.0			–10 to +10	Hardening, carburizing with CH_4 added, sintering, brazing.	120–150
Dissociated Ammonia	25	—	—	75	—			–50 to +60	Brazing, sintering, bright annealing.	10 kg
Argon	—	—	—	—	—	A 100%	CO_2 and H_2O Removed	—	Welding, Ht. treatment of special steel semi-conductor processing.	—
Helium						He 100%				—
Vacuum	—	—	—	—	—	—	—	—	Sintering, degassing melting, brazing.	—
Nitrogen	>99.9	—	—	—	H_2O <0.001	—	–60		Neutral for annealing, etc, purging.	
Hydrogen	—	—	>99.9	—	H_2O <0.001	—	–68		Reducing, sintering.	

Source: *Metals Handbook* Vol. II "Heat Treating, Cleaning, Finishing," Eight Edition, American Society for Metal, Ohio 1964.

The heat transfer to the charge will be through the muffle walls. The atmosphere and the charge are thus effectively separated from combustion products.

The best alternative to the use of fuel combustion is electrical heating. There are no combustion products and better control on heating is possible.

The operating temperature range of most metallic and nonmetallic heating materials is lower in reducing gases. Thus, for gases containing CO, H_2 (dry or moist), and hydrocarbons, the maximum operating temperature for Fe-Cr-Al alloys is 800–1100°C and for Ni-Cr alloys, it is 750–1000°C. Silicon carbide heaters work well up to 1500°C. A more detailed discussion of these limitations is given in Chapter 8 and Chapter 9.

Thus, the electrical heaters will have to be operated at a lower maximum temperature if they are directly exposed to the atmosphere. If indirect heating through the metallic muffle is adopted, the maximum temperature for air can be used.

In both cases (fuel and electrical heating), indirect heating will lower the heat utilization. It will also depend on the maximum permissible temperature of the muffle alloy.

Some Relevant Information

Both continuous and batch-type furnaces use fans to obtain even circulation of the atmosphere. The fans are made from temperature-resistant alloys and are generally of the open-blade centrifugal type. The fan motor and reduction gear is located outside. The shafts have gas-tight seals (sometimes water-cooled) to avoid atmosphere leakage.

The interior of an atmosphere furnace should not have any sharp corners or partitions. These will hinder gas circulation and may keep trapped air.

13.5 DETERMINATION OF ATMOSPHERE CONSUMPTION

All types of atmosphere generators are manufactured in a range of capacities and are rated in m^3/h. The designer has to determine the expected consumption of gas for his furnace, and choose a matching generator from the available models.

The gas requirement depends on three main factors:

1. Type of furnace
2. Size of furnace
3. Operational stage of the process

13.5.1 Batch Type

In steady state operation this type will have a fixed volume. Though the furnace is designed to be gas-tight, there will be inevitable leaks through doors and other openings or cracks, or joints. The furnace interior is kept at a slightly higher pressure than the atmosphere to avoid ingress of air.

The leaks must be estimated and a compensating flow has to be maintained to ensure a proper atmosphere around the charge. In fact, a slightly higher flow rate will be required.

The leaks are not constant but will depend on the age and maintenance of the furnace.

In transient operations such as charging, quenching, and loading the furnace doors are open (at least partially). To ensure that air does not enter the furnace, a large flow will be required. The issuing gas will then burn at the door on a flame curtain and will stop air entry. This flow will depend on the size of the opening and the pressure difference.

At the beginning and the end of a campaign the furnace must be filled or emptied of the gas. A purging gas may be used to assist this process. These operations have to be carried out with minimum wastage of time and gas. Thus, a large flow of atmosphere and purging gas will be required.

Generally, the amount of gas required for filling and emptying the furnace is 8–10 times the volume of the furnace. The process is carried out in 15–20 min. The flame on the flame curtain gives a good indication of the presence of atmosphere inside.

Figure 13.4 and Figure 13.5 show a typical batch furnace with a protective atmosphere. Mechanical loading/unloading equipment will shorten the “open” time of the door and thus save gas.

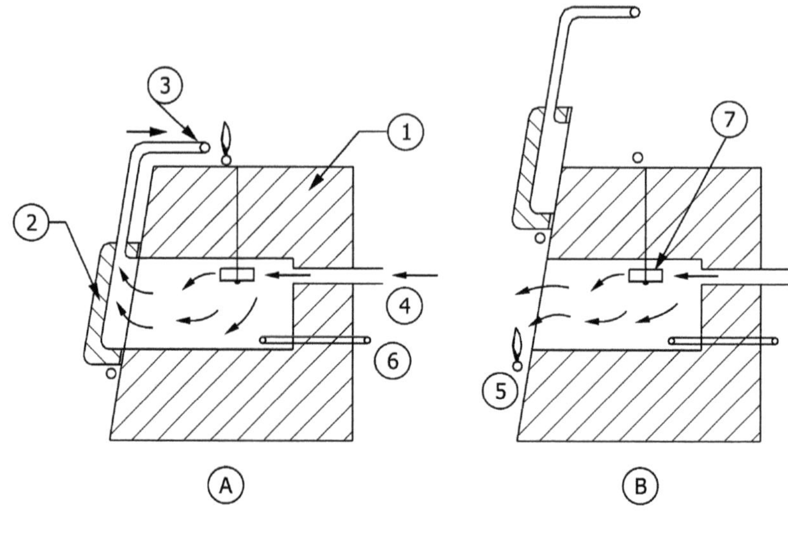

1. Furnace Muffle (Electric heaters not shown).
2. Sliding Door with Atmosphere Outlet.
3. Burner.
4. Atmosphere Inlet.
5. Flame Curtain.
6. Sampling Probe.
7. Fan (optional).

A. Door Closed. B. Door Opened.

Figure 13.4 Atmosphere circulation in a batch furnace.

13.5.2 Continuous Type

These furnaces have an open entry and exit, and the gas introduced at the center continuously flows toward these two openings. The flow rates will then depend on the size of the openings and the pressure difference. The escape of gas is unavoidable and will have to be compensated by keeping a proper inflow of the gas.

By hanging ceramic or glass fiber tape curtains at the openings (see Figure 13.6) the amount of escaping gas can be substantially decreased.

Continuous furnaces requiring protective atmospheres are usually made of metallic muffles. A well-maintained furnace has no leakages. The escaping gas is burned out at the flame curtain.

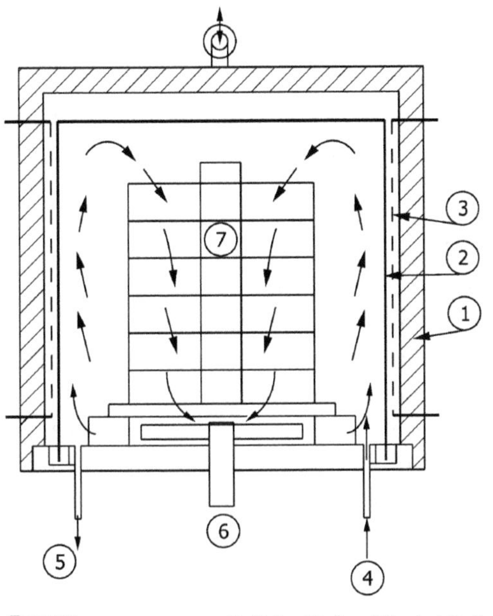

1. Furnace.
2. Bell with Sand Seal at Bottom.
3. Electric Heaters.
4. Atmosphere Inlet.
5. Atmosphere Outlet.
6. Circulating Fan.
7. Rack with Charge.

Figure 13.5 Bell-type furnace with protective atmosphere.

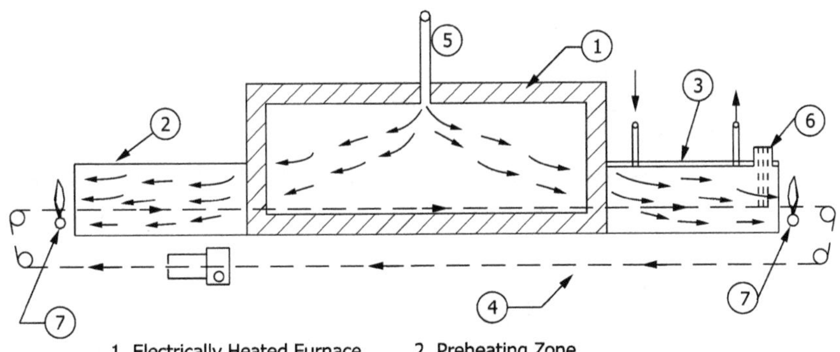

1. Electrically Heated Furnace.
2. Preheating Zone.
3. Water Jacketed Cooling Zone.
4. Continuous Belt.
5. Atmospheric Inlet.
6. Hanging Glass Tape or Metal Chain Curtains.
7. Flame Curtains.

Figure 13.6 Atmosphere circulation in a continuous furnace.

The gas flowing out of the exit is usually cool, as the exit section of a continuous furnace is also used as a cooling section by the use of a water jacket.

The gas flowing toward the loading or entry port is hot and is used for preheating (and dewaxing in sintering) the charge. Thus, if the entry port is larger than the exit, a substantial part of the flow can be directed toward the entry.

At the beginning or end of a campaign the furnace will have to be filled or flushed out. A high flow rate will be required, similar to batch furnaces.

The example following this section will show the method of calculating gas consumption.

It should be noted that there are no "standard" requirements. Calculations will give probable values. Actual flow rates will depend on the condition of the furnace, nature of the charge, time cycles, and operation practice.

The chosen generator should be capable of supplying adequate quantities of gas under all possible requirements.

EXAMPLE 13.1

The article shown in Figure 13.7(A) is to be carburized. There will be 100 pieces in a batch. The carburization process has the following requirements:

1. Depth of carburization = 1.5 mm
2. Original carbon content = 0.2%
3. Density of steel = 7.834 g/cm^3
4. Carburization temperature = 900°C
5. Carburization time = 8 h
6. Typical carbon gradient is as shown in Figure 13.7(B)
7. Carrier gas (endothermic) composition, 40% N_2, 20% CO, 40% H_2
8. Enriching (carburizing) gas = Propane (C_3H_8) added to carrier to 10% by volume
9. The case depth (mm) varies with time (h) according to the relation

   Case depth = $0.525t$ mm
10. The pressure inside the furnace = 25.0 Pa

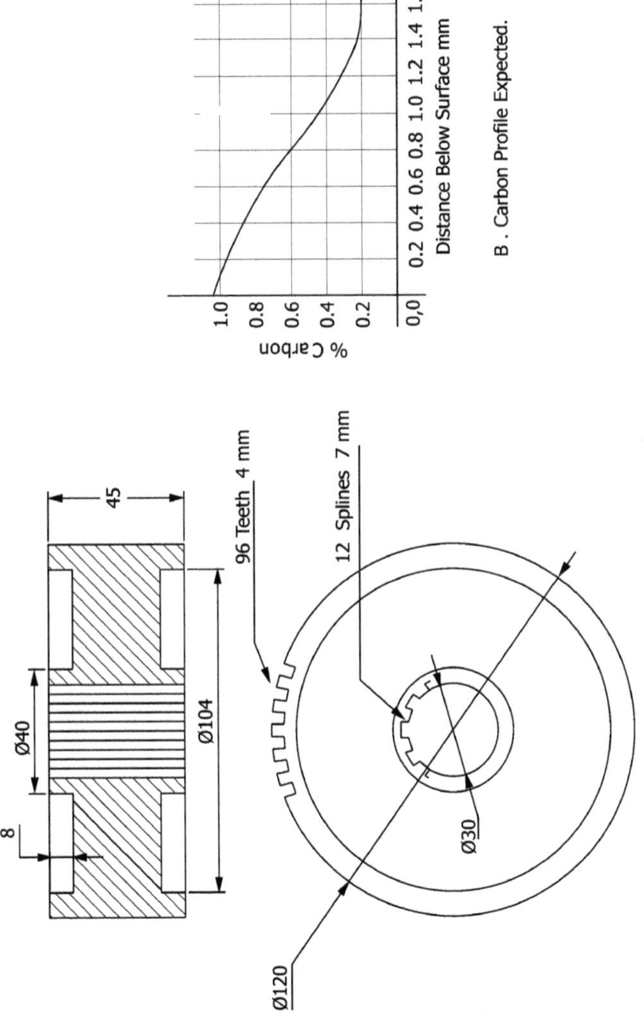

A . Article to be Carburized (Dimensions in mm.)

B . Carbon Profile Expected.

Figure 13.7 [Example 13.1]

Now determine the requirement of carrier and carburizing gases.

Solution

Assume that the whole article (gear) is carburized.

Estimated surface area

$$= 13 \times 10^4 \text{ mm}^2$$

Geometrical (gross area)

$$= 44 \times 10^3 \text{ mm}^2$$

This shows that the surface area is increased 300% by machining.

$$\text{Aprox. volume} = 4.8 \times 10^5 \text{ mm}^3$$

At a density 7.834 g/cm^3, weight of the article is 3.8 kg.

Considering that 100 gears are to be carburized in one lot, and leaving enough space for gas circulation, the volume of the furnace is

$$V_f = 0.25 \text{ m}^3 \text{ minimum}$$

Thus, for trial calculations the assumed furnace (muffle) size is $0.5 \times 0.5 \times 1.0$ m.

Carrier gas composition is 40% N_2, 20% CO, 40% H_2.

Density of these gases at 25°C is

$$N_2 = 0.5 \text{ kg/m}^3$$

$$CO = 0.26 \text{ kg/m}^3$$

$$H_2 = 0.038 \text{ kg/m}^3$$

As the density of the gas is proportional to their volume percentage, the gas density works out to be 0.69 kg/m^3.

Enriching gas contains

$$C_3H_8 = 10\%, \text{ Carrier gas} - 90\%$$

The composition of carburizing gas for 1 m^3 is

$C_3H_8 = 10\%$

$N_2 = 36\%$

$CO = 19\%$

$H_2 = 35\%$

Density of C_3H_8 is 1.967 kg/m^3. Hence, the density of the enriched carburizing gas work out to be gas 0.73 kg/m^3.

Carburizing components of the enriched gas are C_3H_8 and CO. Respective molecular weights (kg) moles are

C_3H_8 – 44, (82% C), CO – 28, (43% C)

Recalling that one kg mole of all gases occupy 22.4 m^3 at STP,

0.1 m^3 C_3H_8 contains 0.160 kg C

0.19 m^3 CO contains 0.106 kg C

Total carbon available from 1 m^3 of gas

$= 0.160 + 0.106 = 0.266$ kg

$= 266$ g/m^3

The surface area of carburized portion $= 13 \times 10^4$ mm^2.
Depth of case is 1.5 mm.
Volume of carburized portion

$= 13 \times 10^4 \times 1.5$

$= 20 \times 10^4$ mm^4/per article

At a density of 7.834 kg/cm^3, weight of the carburized material

$= 7.834 \times 200$

$\simeq 1.6$ kg

As there are 100 gears in one batch, total weight of the carburized material is

$= 1.6 \times 100$

$= 160$ kg

The surface carbon is about 1.1% and core carbon is 0.2%. The carburization profile shows a nonlinear form.

Mean carbon content (C_M) of carburized case (logarithmic) is

$$C_M = \frac{C_S - C_C}{2.3 \log C_S / C_C}$$

where C_S and C_C are carbon percentage at surface and core, respectively.

$$C_M = \frac{1.1 - 0.2}{2.3 \log 1.1/0.2}$$

$$= 0.53\%$$

Weight of carbon used for carburization

$$= 160 \times 0.053$$

$$= 8.48 \text{ kg.}$$

This is the total carbon demand.

Carbon available from the gas is 0.26 kg/m^3.

Gas required 8.48/0.26 = 32.6 m^3.

The rate controlling step in carburization is the absorption or diffusion of carbon in austenite, which is a slow process.

We can calculate the extent of diffusion if the carbon gradient, area, time, and temperature are known. The diffusion coefficient of carbon is of the order of 10^{-7} cm/sec. However, these calculations will not give information about the gas requirement.

Empirical equation for case depth with time for 900°C is

$$\text{case depth} = 0.525\sqrt{t} \text{ mm}$$

where t is the time in h.

A plot of case depth with time is shown in Figure 13.8. However, this equation is applicable only for $t > 2$ h.

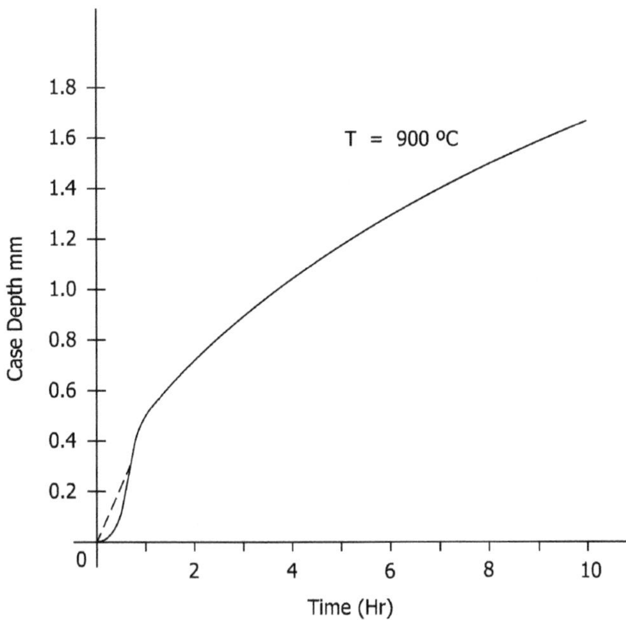

Figure 13.8 Case depth vs. time. Dotted curve $x = k\sqrt{t}$.

It shows that the rate is not constant with time. It is very slow in the beginning for $0 < t < 1$, it increases with time and then decreases for $t > 3–4$ h.

Thus we have to estimate the rate of gas supply to the surface. The following general remarks will have to be taken into account:

1. On exposure to a carbon containing atmosphere, the surface and a few atomic layers below will absorb carbon and will be saturated at a carbon content shown by the Fe–C diagram.
2. Carbon will then diffuse into the interior at a rate given by the case depth-time plot.
3. The surface will continue absorbing carbon at a rate equal to depletion and remain saturated.
4. Carbon atoms will move from the gas to the surface under the concentration gradient $C_g - C_s$, where C_g is the carbon in gas.

5. Enough gas must be supplied to keep C_g near the surface constant. This can be achieved by continuously supplying fresh gas or circulating gas in the furnace by using a fan.
6. According to the theory of adsorption all carbon atoms incident on the surface are not adsorbed and some are "reflected" back.

In conclusion, the gas to be supplied to the furnace must be in excess of the theoretical requirement.

Let us assume that an excess of about 200–300% gas gives satisfactory results.

Thus, carburizing gas requirement is

$$32.6 \times 3 \cong 100 \text{ m}^3$$

Though the carbon penetration rate changes with time, the gas will be supplied at a constant rate throughout.

Thus, for carburizing, the gas rate is

$$100/8 \sim 12\text{–}13 \text{ m}^3/\text{h (minimum)}$$

Let us now check the gas flow when the furnace door is open.

The furnace opening, i.e., the door, is assumed to be 0.5×0.5 m of which about 80% is opened when charging or downloading.

$$\text{Area of opening } A = 0.5 \times 0.5 \times 0.8 = 0.2 \text{ m}^2$$

The gas density is 0.7 kg/m^3. The gas flowing out will be the carrier gas and not the carburizing gas.

The pressure inside is 25 Pa gauge.

The velocity of escaping gas will be

$$v = \sqrt{\frac{2(P_f - P_A)}{\rho}} = \sqrt{\frac{2 \times 25}{0.7}} = 8.45 \text{ m/sec}$$

The volume of flow will be

$$Q = Av$$
$$= 0.2 \times 8.45$$
$$= 1.7 \text{ m/sec}$$
$$= 6085 \text{ m}^3/\text{h}$$

This requirement is much higher than the process requirement calculated earlier (12–13 m^3/h).

It can be reduced by:

1. Having a smaller door and longer furnace
2. Mechanically-assisted opening or closing for quick handling
3. Purging the furnace with a cheap gas such as air or nitrogen and opening or closing at lower (~250–400°C) temperatures to avoid oxidation of charge.
4. Providing a vestibule for loading and unloading

The operation cycle for carburizing will be approximately as shown in Figure 13.9.

EXAMPLE 13.2

A continuous sintering furnace has a 350 × 350 mm loading port and 350 × 250 mm unloading port.

The protective gas pressure in the furnace is 50 Pa gauge (5.165 mm H_2O). The furnace volume is 1.2 m^3. Density of gas is 0.3 kg/m^3.

Assume a contraction coefficient of 0.6.

Now determine:

1. The velocity and the volume rate of gas flow from the furnace ports
2. The volume flow rate if both the gates have flexible curtains having a hydraulic resistance of 25

Solution

The exit velocity of the gas is given by

$$= \sqrt{\frac{2(P_f - P_A)}{\rho}}$$

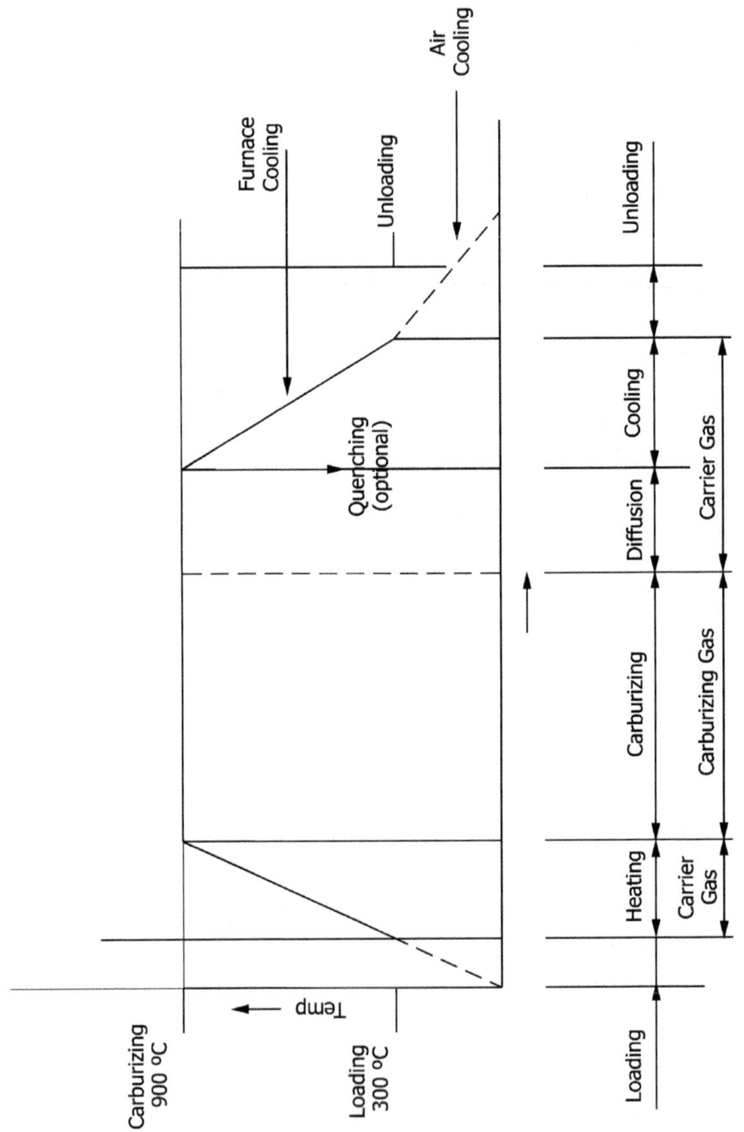

Figure 13.9 Suggested carburizing cycle.

where

P_f = Furnace pressure (Pa)

P_e = Atmosphere pressure (Pa)

ρ = Gas density kg/m^3 = 0.3

Given $P_f - P_e = 50$ Pa

Substituting,

$$= \sqrt{\frac{2 \times 50}{0.3}}$$

= 18.25 m/sec.

The volume flow rate will be

$Q = \Phi\, Av\mu^3$/sec.

For the loading port

$Q_1 = 0.6 \times 0.35 \times 0.35 \times 18.25$

= 1.34 m^3/sec.

For unloading port

$Q_2 = 0.6 \times 0.35 \times 0.25 \times 18.25$

= 0.96 m^3/sec.

Total flow rate = $Q_1 + Q_2$

= 1.34 + 0.96

= 2.30 m^3/sec.

In the presence of the curtain, the velocity will be given by

$$= \sqrt{\frac{2(P_f - P_A)}{\zeta\rho}}$$

where ζ resistance coefficient for the curtain = 25

Substituting,

$$= \sqrt{\frac{2 \times 50}{25 \times 0.3}}$$

= 3.65 m/sec.

For the loading port

$$Q_1 = 3.65 \times 0.35 \times 0.35$$
$$= 0.447 \text{ m}^3/\text{sec.}$$

For unloading port

$$Q_2 = 3.65 \times 0.35 \times 0.25$$
$$= 0.319 \text{ m}^3/\text{sec.}$$

Total discharge

$$= Q_1 + Q_2 = 0.447 + 0.319$$
$$= 0.766 \text{ m}^3/\text{sec.}$$

$$\text{Reduction in discharge due to curtain} = \frac{0.766}{2.30} = 33.3\%$$

13.6 INSTRUMENTATION FOR PROTECTIVE ATMOSPHERES

Protective atmospheres used for various processes contain a number of constituent gases. Thus the constituents of fuel-derived atmospheres are CO, CO_2, CH_4, and small amounts of other hydrocarbons. Atmospheres derived from ammonia contain N_2, H_2, and undecomposed NH_3. Gases like argon contain traces of oxygen and moisture which can be harmful to the charge. In fuel-derived gases the CO/CO_2 and H_2/H_2O are important variables. These ratios are critical in processes such as carburizing, hardening, etc.

When using protective atmospheres it is necessary that the key constituents should be controlled to achieve successful processing. In this section we will review some of the commonly used analytical techniques for measuring the important constituents.

13.6.1 Dew Point Measurement

Moisture enters protective atmospheres from a number of sources. The air used for combustion is generally humid. The combustion process itself will generate H_2O as the product of combustion of hydrogen. Reaction of atmosphere and the charge will also produce H_2O.

If a gas mixture contains moisture, its maximum amount will depend on the temperature.

At some temperature the gas gets saturated with moisture. More moisture than that required for saturation at that temperature condenses on a suitable (cold) surface as dew.

This phenomenon is discussed in more details in Appendix D. Standard charts for saturation against temperature are also available.

In dew point measurement, the moisture content is indirectly measured by determining the dew point of the gas.

In the standard method a vessel coated with silver is used as the condensing surface (Figure 13.10(A)). It is cooled by a cooling mixture such as ice or dry ice. The gas is circulated on the outer surface. The formation of dew is visually checked and the temperature of the dew formation is noted. Thus, the drier the gas the lower will be the dew point.

This method is very simple but it is periodic in operation and requires manual help. It is used for checking the moisture content by occasional sampling for quality control. It cannot be used in the described form for continuous or online control.

For continuous measurement, the basic technique used is the same, i.e., condensation on a cold bright surface, but elaborate arrangements are made for automatic heating and cooling of the surface, measurement of its temperature, and detection of the dew.

One such instrument is shown in Figure 13.10(B). The gas is circulated through a cell which contains the mirror (condensing surface). The mirror is heated or cooled by a solid-state Peltier device (see Chapter 14). The temperature of the mirror is measured by a resistance thermometer.

A light beam strikes the mirror and is reflected and detected on a photo diode. When moisture condenses on the mirror, its reflectivity is reduced, which reduces the strength of the reflected beam and, hence, the signal. The temperature of the mirror is the dew point. There are many modifications and design variations of the instrument available. They are called "optical hygrometers."

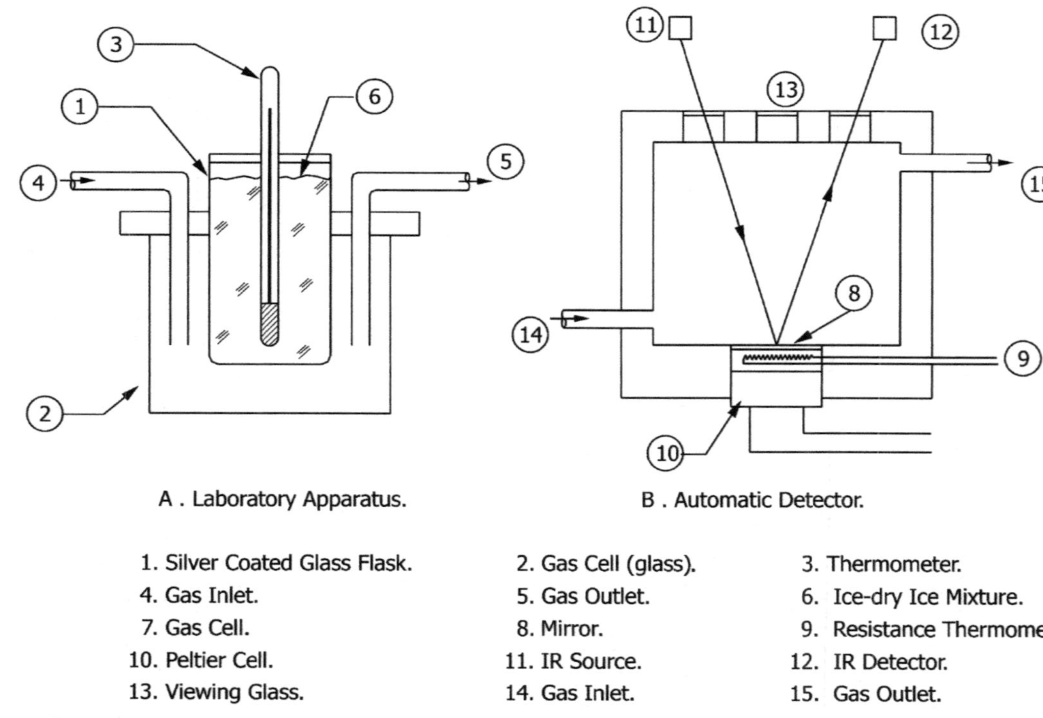

Figure 13.10 Dew point detector.

They give continuous readings, do not require manual attention, and can be used as control and recording. They are much more expensive than ordinary hygrometers.

13.6.2 Measurement of CO, CO_2, CH_4, and NH_3

These constituents are measured by the infrared (IR) absorption technique. This method is discussed in detail in Chapter 14. It basically contains an IR source such as a hot filament, an absorption cell for gas circulation, and an IR detector. The absorption bands of these gases are overlapping. It is therefore necessary to remove the interference by using a comparison cell containing a "standard" gas sample and appropriate optical filters. The detectors used are thermopiles, solid state detectors, or a gas-filled capacitive microphone.

A typical IR absorber detector is shown in Figure 13.11. At the top of the cell is an IR source which sends radiation through the cell below it. The gas sample circulates through the cell. The cell is gold-plated from inside to increase its reflectivity. The transmitted radiation is filtered by the filters located below.

In the lower part, there are two identical cells. One cell contains a nonabsorbing gas such as nitrogen. The other cell contains a standard gas sample. Both cells are called sensitizing cells.

Equal amounts of radiation coming out of the sample and filter above pass through the sensitizing cells.

Those passing through the nitrogen cells pass through without being absorbed. Those passing through the other cells are absorbed according to the absorption spectrum of the sensitizing gas.

The IR radiation passing through falls on the detector, producing a differential signal. This signal is processed and displayed as the percentage of that gas in the sample.

The effect of interfering gases is removed by a cell located near the filter. This cell contains a mixture of background gases that absorb radiation according to their absorption spectrum.

There are many other designs of this detector having their own special features. If a number of constituents are to be detected, each constituent requires a separate cell sensitized specially for that gas.

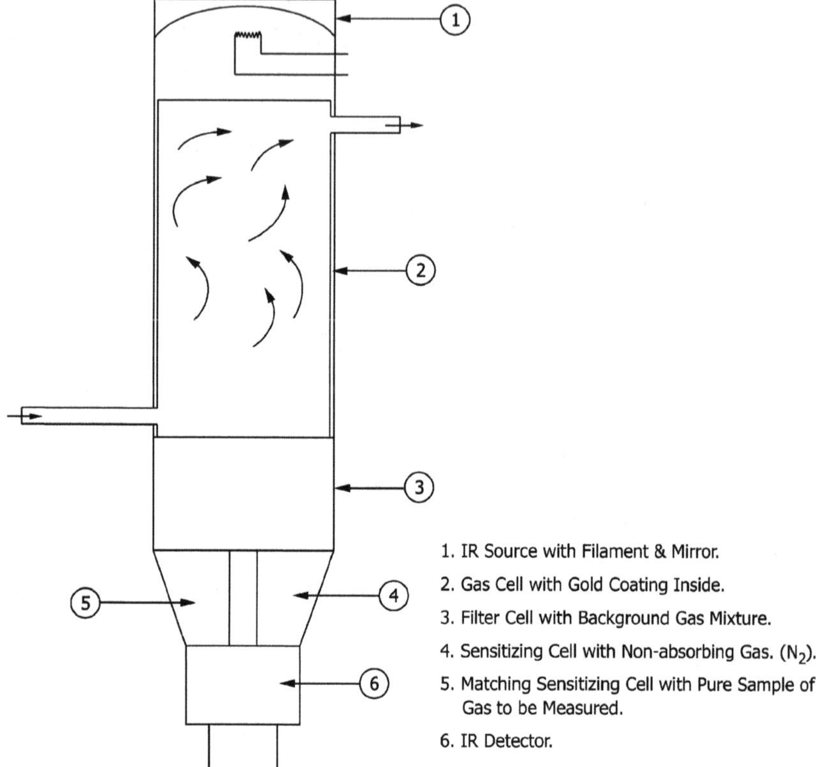

Figure 13.11 IR detector for Co, CO_2, CH_4, etc.

IR absorption instruments are used for quantitative measurement of CO, CO_2, NH_3, H_2O, SO_2, hydrocarbons, NO_2, etc. They cannot be used for pure diatomic gases such as N_2 or O_2 as they do not absorb IR radiation. Almost all the gases listed earlier can be measured in the 0.001–100% range. The instruments give continuous measurements. They are very costly and, due to complex electronic circuits on the measuring side, require expert maintenance. They can be used for remote sensing and control.

13.6.3 Detection of Oxygen

Oxygen is not normally a constituent of protective atmospheres. It may be present when excess air is used or when intentionally added to bring about oxidation.

Oxygen cannot be determined by IR absorption at it does not absorb radiation. The method for continuous measurement of oxygen is a relatively recent development.

The instrument used is shown in Figure 13.12. The cell consists of a porous zirconia (ZrO_2) tube which has a thin platinum plating on both the inner and outer surfaces. The inner surface of the tube is exposed to the gas to be measured. The outer surface of the tube is exposed to the atmosphere. Thus there is an oxygen concentration differential between the inner and outer tube surface. Oxygen molecules flow through the pores under this differential. An emf is impressed between the two surfaces. This creates a current which is

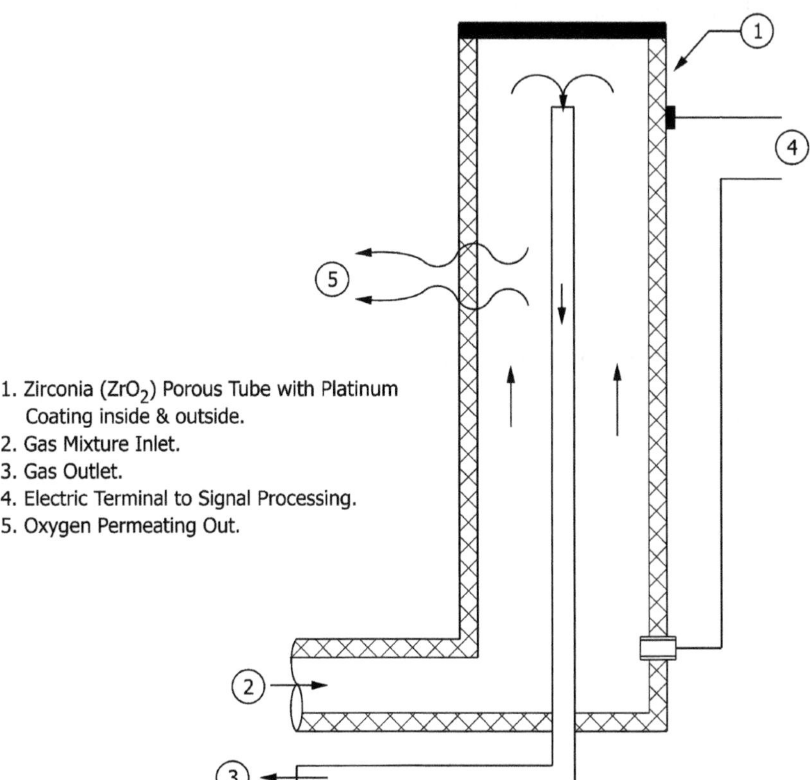

Figure 13.12 Instrument for measuring oxygen content in gas mixture.

proportional to the oxygen concentration. The current is amplified and displayed.

This is a cheap and rugged instrument. There are other types of instruments for oxygen measurement based on magnetic properties.

13.6.4 Selection of Analytical Instruments

We have seen that analytical instruments for measurement of constituent gases in a protective atmosphere are now available. Except for nitrogen, almost all other gases can be measured.

All these instruments are very costly; hence, it is necessary to critically assess their need in a given situation. In most cases it is sufficient to measure one critical variable (say dew point or CO/CO_2 ratio).

In large production units, these instruments can be used for automatic control of one or more atmosphere generators.

It is also necessary to determine the placement of detection probes. They can be located between the generator and the process (furnace), or in the process, or in the exhaust line. The best location is decided by the process needs. Thus for head treatment the sample tube is located near to the charge as possible.

The need for continuous detection and measurement can be avoided by absorbing an unrequired constituent (e.g., H_2O, O_2, CO_2) by using a molecular sieve. Thus only periodic monitoring is sufficient.

Chapter 14

Temperature Measurement

CONTENTS

14.1 INTRODUCTION

There are different objectives for heating an object — such as melting, heat treatment, joining, and processing. The only way to know whether it is heated properly is to measure its temperature. Thus, the temperature measurement and its control is of paramount importance in the design and operation of all types of heating apparatus.

Temperature measurement is arbitrarily divided into two branches — thermometry and pyrometry. The instruments used are called thermometers and pyrometers, respectively. Thermometery is concerned with temperatures below 350–400°C. Pyrometry measures temperatures that are beyond the range of conventional thermometers.

The temperature of an object is a relative property. When we say that a certain body is at a high temperature what we mean is that its temperature is higher than some other "cold" body.*

The standard scale used is degree celsius (°C) for which it is agreed that 100°C will be the boiling point of water. The scale space is divided into 100 equal parts, each part as 1°C.

Ideal gases expand on heating and contract on cooling. If such a gas is continuously cooled, at some low temperature its volume will be zero. This state is achieved at –273.15°C. This is taken as the zero point of the thermodynamic scale or the Kelvin scale (K). The value of a degree of temperature in °C and K is same, i.e., an interval of unit Kelvin is same as an interval of unit degree Celcius. However, in order to convert

* For uniformity, precision, and convenience we need a scale having international agreement.

from Kelvin to degree Celcius one must add 273.15. In some countries the Fahrenheit scale (°F) is still in use. However, °C and K scales are now universally adopted.

Except the 0°C and 100°C points on the celsius scale, many other points are standardized for the extension to higher and lower temperatures. For example, the melting point of gold is 1063°C, and the boiling point of oxygen is –182.97°C. The methods of calibration are also standardized.

The expansion of liquids (especially mercury) with temperature was used for construction thermometers. Mercury in glass thermometers are useful up to 325°C while mercury in steel tubes could be used up to about 600°C. The former, being fragile, is used mainly in laboratories and the latter has been superceded by other pyrometers.

The resistance of metals increases with temperature. This property is used for making electrical resistance thermometers. Usually a thin wire of platinum (Pt) is used as the resistance source. The change in resistance is measured by a precision Wheatstone bridge. These thermometers can be used in the range –258 to 900°C and are extremely precise. The resistance is related to temperature by a relation such as:

$$Rt = Ro(1 + At + Bt)^2 \tag{14.1}$$

Rt = Resistance of platinum at any temperature t°C
Ro = Resistance of platinum at the ice point 0 t
= True temperature
A and B = Constants related to each platinum thermometer/pyrometer

They are used for delicate thermocouples but due to their delicate construction, find little use in industry.

The change of color of a heated object with temperature is used to "judge" the temperature between 600–1300°C, but is not useful as an instrument.

The pyrometers and thermometers that find wide use in an industrial environment are based on thermoelectrical and radiation phenomena. These are considered in detail in the next sections along with some other types of minor instruments.

This introduction is an overall review of the science of temperature measurement. For reference and more details consult specialist literature[1,2] in the *Bibliography.*

14.2 THERMOCOUPLE PYROMETERS

These pyrometers, invented long ago, still dominate the field of temperature measurement. The basic principle behind thermocouple is known as the Seebeck effect which states that, "In a circuit made of two different metals, if one junction is kept at a temperature t_1, and the other at a lower temperature t_2 then an e.m.f. E is created in the circuit (Figure 14.1(A)) which is proportional to the temperature difference $(t_1 - t_2)$, i.e.,

$$E = K_s (t_1 - t_2) \tag{14.2}$$

where K_s is the proportionality constant known as the Seebeck coefficient.

The junction at the higher temperature (t_1) is called the hot junction and the other junction at the lower temperature is called the cold junction. A basic thermocouple pyrometer operating on the Seebeck effect is shown in Figure 14.1(B).

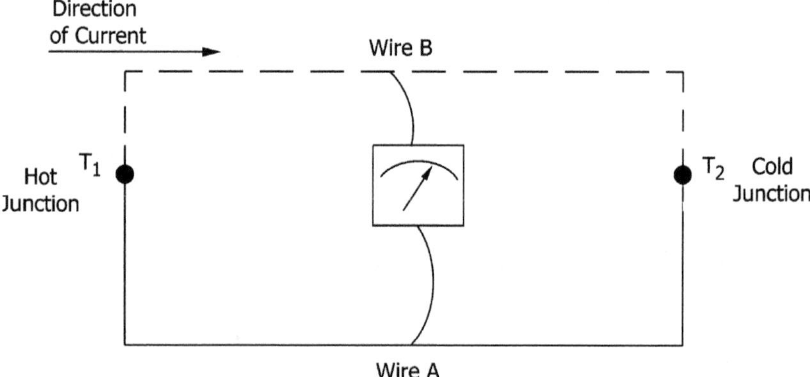

Figure 14.1 A Seebeck effect.

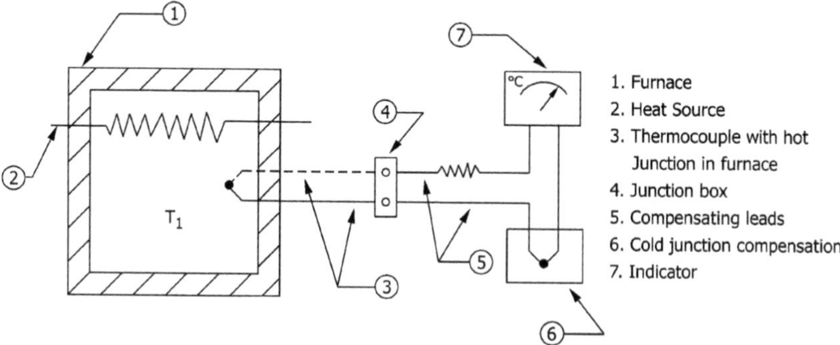

Figure 14.1 B Thermocouple pyrometer.

Two other effects are associated with the Seebeck effect.

1. Peltier effect states that the Seebeck effect is reversible. If in a circuit of two dissimilar metals a direct e.m.f. is impressed (say with a cell or battery) then one of the junctions becomes hot and the other cool. The Peltier effect is the principle on which thermo-electric refrigeration is based. At least up to now, these cooling devices cannot economically compete with mechanical refrigeration. They have their advantages and limited applications.
2. The Thompson effect states that if in a current carrying a conductor, two points are at dissimilar temperature, then heat is either evolved or absorbed by the conductor. The direction of heat reverses with the reversal of the direction of the current.

Both the Peltier and Thompson effects are of no use in pyrometry and hence, are not discussed in detail.

There are some practical corollaries which we can derive from the above three laws of thermoelectricity.

Corollary 1:

A thermoelectric circuit made of the same material will not produce any e.m.f. even though the two junctions are at different temperatures. The circuit must be made of two dissimilar metals.

Corollary 2:

If a circuit made of two dissimilar metals contains other conductors and if the junctions of these extra materials are at same temperature, the thermoelectric e.m.f. produced will not be affected by the presence of the extra conductors. It will be the function of the difference of hot and cold junction temperatures only.

Corollary 3:

Consider a circuit made in two parts from two dissimilar metals. Let the hot and cold junction temperatures in the first circuit be t_1 and t_2 and let those in the second circuit be T_2 and T_3. Also, let $t_1 > t_2 > t_3$.

The e.m.f. produced in the two circuits will be

$$E_1 = K_s(t_1 - t_2) \text{ and } E_2 = K_s(t_2 - t_3)$$

If we now consider one circuit of the same metals with junctions at t_1 and t_3 the e.m.f. produced will be

$$E_3 = K_s(t_1 - t_3) = E_1 + E_2 \tag{14.3}$$

Corollary 4:

In the Thompson heating and cooling of a conductor, both difference of temperature and a preestablished current are necessary. Only one condition cannot produce the effect.

14.3 PROPERTY REQUIREMENTS OF THERMOCOUPLE MATERIALS

1. The useful temperature range of a thermocouple should be a little less than their melting points. Like all metals, the thermocouple metals lose their mechanical rigidity at temperatures near melting. Generally, the alloys chosen can be used up to a temperature 100–150°C less than the melting temperature. If a thermocouple consists of two metals (or alloys), then the above criteria will apply with respect to the alloy with the lower melting point.

2. The chosen thermocouple should produce a large e.m.f. in the temperature range. A very small e.m.f. is difficult to measure with consistent accuracy.
3. As far as possible, the relation between the e.m.f. and the temperature should be linear. Such linear e.m.f. can be easily measured on a simple voltmeter. With electronic circuits it is possible to linearize a nonlinear relation but these instruments are costly.
4. The e.m.f. temperature relation should be continuous, i.e., without any gaps or jumps.
5. The thermocouple metals must be resistant to the effects of the atmosphere in the measurement environment (oven, furnace, etc.)
6. The thermocouple is usually made from wires. The alloys should have sufficient workability for the production of wires. Similarly, it should be possible to make these alloys with accurate composition and purity. Impurities, if not controlled, will affect consistency and reliability.

14.4 PRACTICAL THERMOCOUPLES

Considering the requirements of good thermocouple materials discussed in the previous section, a number of thermocouples have been evolved and standardized.

In earlier days the metals were manufactured by a few manufacturers. Hence they had proprietary names such as Chromel, Alumel, and Constantan. With the availability of ultra-purity constituent metals, sophisticated melting furnaces (e.g., vacuum induction melting), and fast and accurate analytical methods (e.g., spectrography), the number of manufacturers has increased. They have marketed similar thermocouples, differing only in some minor constituents.

Previously, thermocouples were classified in two categories: "noble" and "base metal." This classification was based on constituents. Base metal couples were of metals or alloys such as iron, copper, manganese, nickel, and chromium. The noble metal couples were made from platinum and its alloys.

Presently, all commercial thermocouples are classified by letter designation such as B, E, J, S, and T. This designation was suggested by ASTM and is now adopted all over the world. The designation does not depend on composition but by the E vs. t relation shown by the couple. Thus, even though the minor constituents are different, couples offering the same E, t relation belong to the same class.

Table 14.1 shows the classification of widely used commercial thermocouples and their useful temperature range.

The standardization of thermocouple wires is made by accurately determining the e.m.f. E and temperature "t" relation. For this the reference or "cold" junction is kept at 0°C. This reference junction temperature is internationally agreed upon. Hence, all the E and t data are reported with the tacit understanding that the reference junction is at 0°C.

The National Institute of Standards and Technology (U.S.) and manufacturers of metals have published such calibration data tables for all the standard thermocouples. The tables usually quote e.m.f. produced at every 1.0°C difference.

The data are fit into a standardized relation such as

$$\left.\begin{aligned} E &= a_0 + a_1 t + a_3 t^2 + \cdots \\ \text{or} \quad t &= a_0' + a_1' E' + a_2 E^2 + \cdots \end{aligned}\right\} \tag{14.4}$$

The values of constants a_0, a_1,.... or a_0', a_1', etc., are also accurately determined so that any interpolation can be done from the numerical tables.

The calibration is carried out by keeping the "measurement" or "hot" junction in a pure metal whose solidification temperature is standardized. The couple is dipped in a crucible containing the metal. It is then heated to a higher temperature so that the metal gets liquified. It is then cooled so that the solidification takes place at a constant temperature. The e.m.f. produced is recorded as a standard output.

There are internationally agreed primary and secondary calibration points.

Primary points are the boiling point (B.P.) of O_2, −182.97°C, B.P. of water, +100°C, solidification of silver and gold, 960.8°C and 1063.0°C, respectively, etc.

TABLE 14.1 Common Thermocouples, Classification, and Temperature Range

No.	Positive Wire (P)	Negative Wire (N)	Traditional Type	International Designation ASTM	Normal Temperature Range°C	Occasional Use Highest Temperature
1.	Pt_{70} Rh_{30}	Pt_{94} Rh_6	Noble	B	871–1705	1750
2.	Ni_{90} Cr_{10}	Ni_{45} Cu_{55} Constantan	Base	E	0–800	850
3.	Fe	Ni_{45} Cu_{55} Constantan	Base	J	0–700	760
4.	Ni_{90} Cr_{10} NICROSYL	Ni_{95} Mn_2 Al_2 Si_1 NISIL	Base	K	0–1200	1260
5.	$Ni_{84.5}$ Cr_{14} $Si_{1.5}$	$Ni_{95.5}$ $Si_{1.5}$ $Mn_{0.4}$	Base	N	0–1200	1260
6.	Pt_{87} Rh_{13}	Pt	Noble	R	0–1500	1700
7.	Pt_{90} Rh_{10}	Pt	Noble	S	0–1480	1700
8.	Cu	Ni_{45} Cu_{55}	Base	T	–184–+370	370
9.	W_{95} Re_5	W_{74} Re_{26}	—	Experimental Vacuum	1000–2700	2800
10.	W	W_{74} Re_{26}	—	Experimental Vacuum	1000–2600	2800
11.	Graphite with boron doping	Graphite	—	Vacuum	2000	2000

Secondary points are solidification temperatures of mercury, tin, aluminum, copper, nickel, tungsten, etc.

Further details of thermocouple calibration are beyond the scope of this book. Reference may be made to specialized books on pyrometers[1–4] listed in the *Bibliography*.

A graphical representation of e.m.f. temperature relationships for standardized thermocouples is shown in Figure 14.2. It can be seen that many thermocouples show a nonlinear relation. This nonlinearity can be corrected by using an indicator

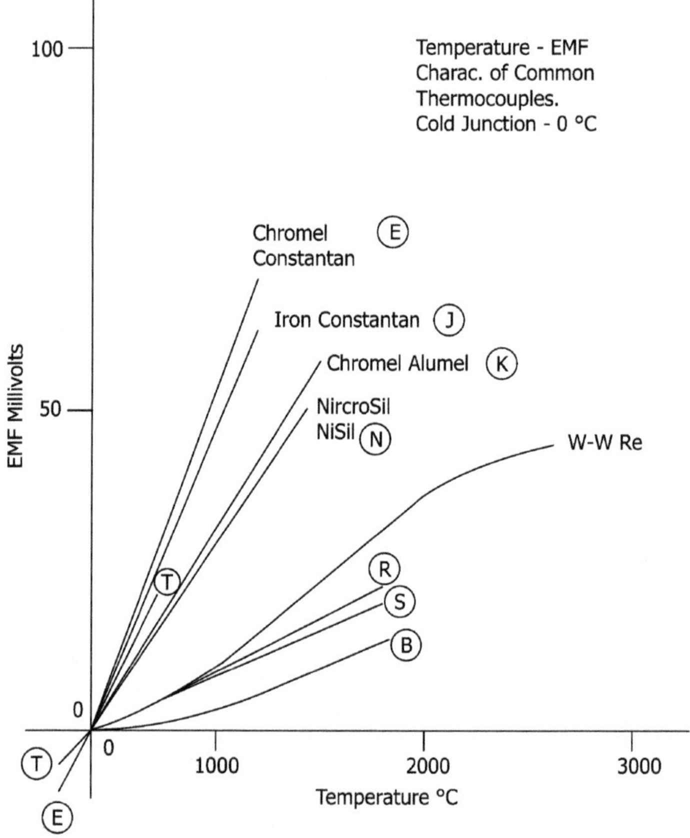

Figure 14.2 EMF-temperature relations for standard thermocouple.

with a specially ruled scale piecewise, or nowadays, electronically carrying out a piecewise linearization.

The hot junction of a thermocouple is made by twisting the wires together to form a mechanical joint. This joint is then welded to form an electrically homogeneous junction. This welding is done to form a bead. The welding is a delicate process, especially for thin wires. Precleaning and fluxing give a sound joint.

14.5 COLD JUNCTION COMPENSATION

The Seebeck effect, when applied to a standard thermocouple, requires that the temperature of one of the junctions (cold junction) must be kept constant.

In an industrial environment or in any practical application, the cold junction is located outside and exposed to atmospheric temperature. This temperature changes according to the time of the day, the season, and the surrounding industrial environment.

The standard calibration data for thermocouples are based on the cold junction temperature maintained at 0°C. In laboratories this condition is satisfied by keeping the junction in a mixture of melting ice and water. It is not practical or feasible to maintain the cold junction at 0°C in an industrial environment.

Consider in the first instance that the cold junction be kept at a constant nonzero temperature t_2 with the hot junction at a temperature t_1. The e.m.f. measured will be due to the difference $t_1 - t_2$. The standard e.m.f. from the tables is based on the difference $t_1 - t_0$. Thus,

$$\left.\begin{aligned} E_{\text{std}} &= K_s(t_1 - t_0) \\ E_{\text{mes}} &= K_s(t_1 - t_2) \end{aligned}\right\} \tag{14.5}$$

with the condition that $(t_1 - t_0) > (t_1 - t_2)$. Thus, the indicator temperature will be less than the standard value by an amount $t_2 - t_0$.

In practice, the temperature t_2 (environmental temperature) is not constant. Hence, the error $t_2 - t_0$ is also variable.

Devices or designs to remedy this source of error in temperature measurement are known as "cold junction compensation."

A number of such compensation methods are available.

1. If the accuracy of measurement required is not high (say = 10 to 15°C) the simplest and cheapest method is to forward the indicator needle to the assumed environment temperature. This will automatically add the assumed constant temperature to the measured temperature.
2. The indicator needle in some instruments is connected to a bimetallic spring. The spring expands or contracts on changes in environmental temperature. Thus, the needle moves forward or backward on the scale. This automatically compensates for the cold junction temperature. This method is better than the first method but still not accurate enough.
3. In some designs an artificial cold junction is kept in a miniature oven in the instrument. The temperature of this "oven" is kept higher than the maximum possible environmental temperature and is controlled by suitable circuits. This produces a constant error at all times. The required error voltage is automatically added to the measured voltage.
4. Some instruments use a thermistor in the thermocouple circuits. The resistance of the thermistor, and therefore the voltage, drop across it, changes with the environment temperature and compensates for the error.

There are other sophisticated methods of compensation. These are not discussed here.

For the furnace designer it is necessary to keep in mind two points.

1. For accurate temperature measurement some kind of cold junction compensation must be provided.
2. The compensation method to be selected will depend on the desired accuracy, reliability, and the cost.

14.6 COMPENSATING WIRES

Thermocouple wires are manufactured in two grades—standard and precision. Both grades are quite costly.

In most of the industrial furnaces and ovens, the temperature indicator or the controller is situated on a control panel which is at a considerable distance from the furnace. It is uneconomical to take the thermocouple wires over such distances. The thermocouple wires are terminated at the junction box outside the furnace.

From the junction box to the panel, other special but cheaper wires called compensating cables are used.

The temperature of the connection plate is generally in the range of 50 to 150°C and the reference junction in the instrument is in the range of 60 to 80°C. The compensating wires are so chosen that they produce the same e.m.f. as the thermocouple wires in the temperature range of 150 to 50°C so that no error results due to the termination of the thermocouple wires at the junction box (see corollary three above).

Compensation wires for different thermocouples are standardized and color coded. The color codes are not internationally standardized.

In the U.S. standardization, which is followed in many countries, the thermocouple wires have red jackets for all negative wires (N) and different colors for positive wires (P), e.g., J, K, T, E, N thermocouple positive wires are white, yellow, blue, purple, and orange colors. Extension or compensatory wires have different colors for positive and negative wires and the cable has usually a brown jacket.

It is vital to observe the correct polarity when connecting compensating cables to the thermocouple. A wrong connection (e.g., +ve thermocouple to –ve compensating cable) will result in considerable error.

14.7 CONSTRUCTION OF THERMOCOUPLES

Industrial thermocouples are constructed to withstand the rigors of the industrial environment for a long time. Bare thermocouples get corroded or oxidized by furnace gases and fail within a very short time.

The thermocouple wires are insulated from each other by using twin bore ceramic insulators usually made of alumina or mullite. The assembly is then inserted in a ceramic sheath in the shape of a one-end-closed tube. Many materials are used for the sheaths. Some common sheath materials are given in Table 14.2. When choosing a sheath, the temperature, environment, porosity, and thermal fatigue must be considered.

Metallic sheaths are also used either alone or as an outer cover for ceramic sheaths. If used alone, the hot junction is electrically insulated by a ceramic piece or powder. The assembled thermocouple can be up to 1 to 1.2 m long.

At the other end, the thermocouple is connected to terminals in a junction box. From these terminals compensating cables are taken out and connected to the indicator.

The construction of a typical thermocouple is shown in Figure 14.3. However, depending on the end use, many different designs are available.

TABLE 14.2 Thermocouple Sheath Materials

Sheath Material	Maximum Temperature°C
Carbon steel	550
Cast iron	700
Pure iron	700
18-8 Stainless steel	950
28 Cr stainless steel	1100
Chromel, Nichrome	1100
Incoloy, Inconel	1100
Fused silica	1050
Fire clay	1050
Silimanite, Mullite	1550
Silica, Alumina	1600
Silicon carbide	1650

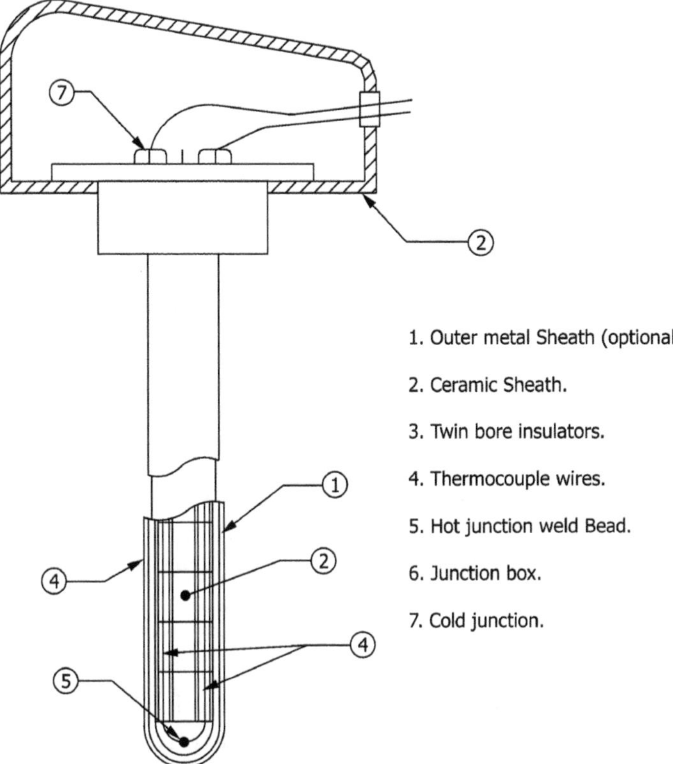

Figure 14.3 Construction of an industrial thermocouple.

14.8 SELECTION OF THERMOCOUPLES

The following points should be considered when selecting a thermocouple for a given application.

1. The normal temperature range and the maximum occasional temperature to which the thermocouple is to be used. This information will be useful in knowing which of the standard thermocouples are available for this range.
2. Operation of the furnace—continuous or batch type. If the latter is considered, what is the cycle time and how many batches are to be processed in a day or a week?

Thermocouples are prone to thermal fatigue. Their life is affected by the heating-cooling cycles. Base metal thermocouples of type E, J, K, N, and T are more affected than precious metal couples of B, N, R, and S.

3. The furnace atmosphere can be oxidizing, reducing, inert, or vacuum. The atmosphere to which different thermocouples can withstand are listed below.

Type	Atmosphere
K, E, N, R, S	Oxidizing and inert
J,T	Oxidizing and reducing
B	Oxidizing, inert, and vacuum

This shows that types J and T thermocouples can withstand both the oxidizing and reducing atmosphere, but their useful temperatures are only 760 and 370°C, respectively. Heat treatment furnaces generally have a reducing atmosphere and the temperatures are 800–1200°C. For this application, the cheapest choice will be type K but this will have a limited life due to corrosion and aging. Type N is the next choice for higher temperatures. It will have a better stability (physical and thermoelectrical) than K but will be somewhat costlier.

Type R and S (Pt-Pt, Rh) can be used in oxidizing and inert atmospheres up to 1480°C while type B can be used to 1700°C under similar conditions. These thermocouples will have a long life but cannot be used in reducing (heat treatment, sintering, etc.) atmospheres. They are much more costly than the base metal types.

Currently, there is a growing trend to avoid reducing atmospheres and use vacuum instead. Only B type thermocouples are suitable but they must be covered by a suitable impervious sheath and also by an outer metal sheath (if the temperature and atmosphere permits).

4. The response of a thermocouple to changes in temperature is also an important factor. A thermocouple will indicate the correct furnace or object temperature when it reaches that temperature. This is achieved by conduction of heat through the sheath and conduction

through the bead (not junction). Conduction is a slow process. The sheath is ceramic and has an inherently low conductivity. Additionally, if the thermocouple wires have a large diameter, they will take time to heat up.

Thus a thermocouple has a slow response. It can be improved by using thinner sheaths and wires but this will be at the cost of service life and rigidity. Note that this response factor has nothing to do with the controller/indicator response.

5. Thermocouples and the related control system are important components of the total cost of the furnace. The total expected life of the system and the cost of repairs, replacements, and calibration in this period must be given consideration in costing. The relative initial cost of precious metal couples is much more than base metals but they give better service for a long time.
6. Small furnaces usually require only one thermocouple, strategically placed in the furnaces. Precision furnaces, though small, may require more thermocouples to maintain the required temperature near the charge. Larger and multizone furnaces also require more thermocouples than one.

14.9 RADIATION PYROMETRY

14.9.1 Principle of Radiation

A body at a temperature higher than its surrounding emanates electromagnetic radiation. The intensity of this radiation is proportional to the fourth power of the absolute temperature of the radiating body. This is the Stephen Boltzman's law of radiation.

The radiation can be directly measured by using a sensor or it can be compared to the radiation of a body of known temperature. This is the principle of radiation pyrometry. There are several types of instruments incorporating radiation measurement. One immediate advantage of this method is that there is no contact between the measuring instrument with the hot body, and at least in principle, the body and the instrument can be far apart as long as they "see" each other.

We will come across "radiation" in detail when we discuss heat transfer (Chapter 3). Here only relevant facts are reviewed.

A hot body emits radiation all over the electromagnetic spectrum. The intensity of this radiation depends on the wavelength and the absolute temperature. At a given temperature the total energy emitted at all wavelengths is given by the Stephen Boltzman Law:

$$E_T = C_o(T)^4 \text{ W/m}^2 \tag{14.6}$$

where

E_T = Total energy emitted by unit surface (W/m²)

T = Absolute temperature of the emitting body (K)

C_o = Stephen Boltzman constant (5.669×10^{-8} W/m·k⁴)

The total energy will be the area under the E_λ plot against wavelength λ for the given temperature T as shown in Figure 14.4. Note that the radiant energy is spread all over

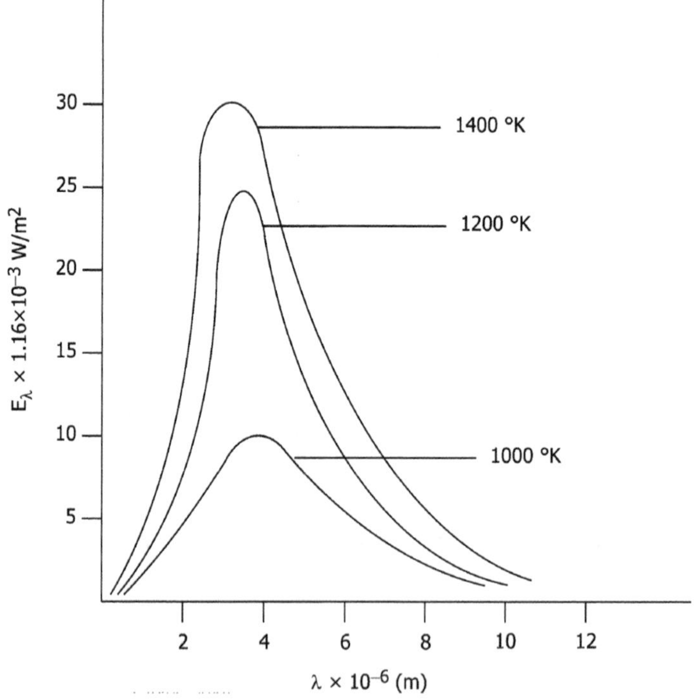

Figure 14.4 Radiation intensity of a black body.

the spectrum and the intensity is different for each wavelength. At the extreme ends, i.e., X-rays and far infrared and radio wavelengths, the intensity is very low. Most of the energy is between 0.5 to 10 μm region. An enlarged portion of the E_λ against λ part of Figure 14.4 is shown in Figure 14.5 at various temperatures.

The intensity E_λ at a given wavelength and temperature is given by Plank's radiation law.

$$E_\lambda = \frac{C_1\lambda^{-5}}{e^{C_2/C_0T} - 1} \text{ W/m}^3 \tag{14.7}$$

where

λ = Wavelength (μm)
T = Absolute temperature (K)
C_1 = Constant (3.69×10^{-16} W/m^6)
C_2 = Constant (1.44×10^{-2} m, K)
C_0 = Stephen Boltzmann's constant

Equation 14.6 and Equation 14.7 are for black body radiation. In practice we have to consider the emissivity ϵ of the radiating body. Hence the equations applied in practice are

$$E_T = \epsilon C_0 (T)^4 \text{ W/m}^2 \tag{14.8}$$

and

$$E_\lambda = \frac{C_1\lambda^{-5}}{e^{C_2/C_0T} - 1} \text{ W/m}^3 \tag{14.9}$$

Radiation pyrometers are based on Equation 14.8 and Equation 14.9.

14.9.2 Practical Problems

The application of Equation (14.8) and Equation (14.9) in practice raises several problems and puts limitations on radiation pyrometry. These are discussed below briefly.

1. The wavelength region having high intensity is between 0.1 to about 10 μm. In this region, 0.1 to 0.4

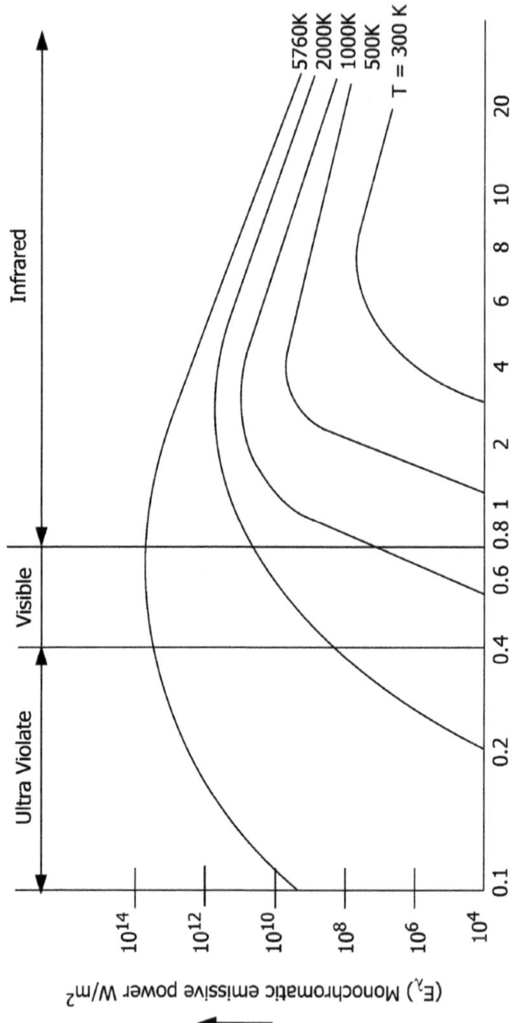

Figure 14.5 Variation of monochromatic emissive power (E_λ W/m^2) with wavelength (λ) and absolute temperature (K) for a black body.

is the ultraviolet region, 0.4 to 0.7 is the visible region, and 0.7 onward is the infrared region as shown in Figure 14.6. As the temperature of interest increases, the radiation becomes stronger toward shorter wavelengths. Hence, the practically applicable region for all temperatures is limited to approximately 0.5 to 8 μm, which lies in the visible (0.5–0.7 μm) and infrared (0.7 + 8.0 μm) region.

2. Any instrument built to sense the radiation will have to be enclosed to avoid dirt, dust, gases, etc., present in the industrial environment.

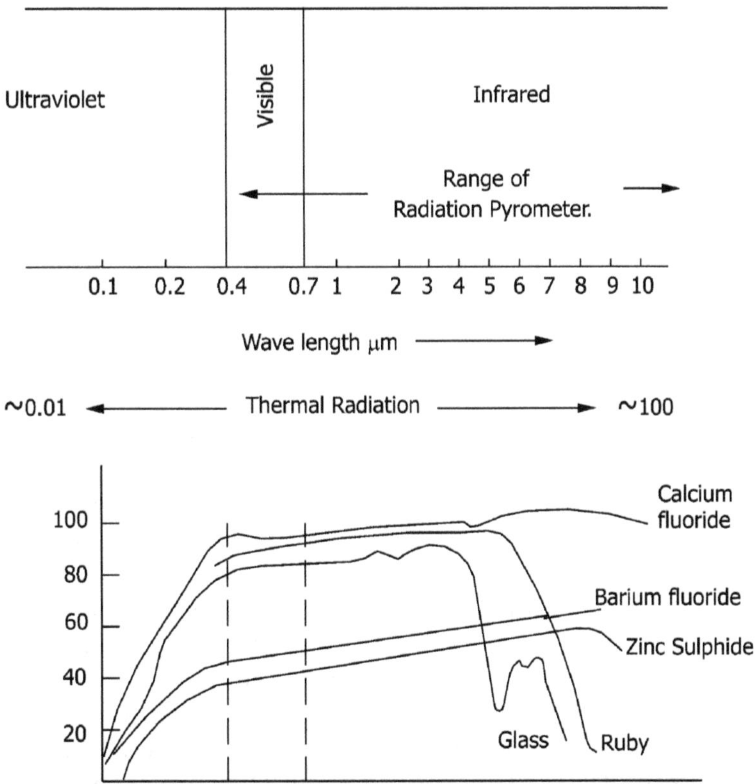

Figure 14.6 Transmission power of selected optical materials.

Such an enclosed instrument will have to be provided with windows and other optical components through which it can "see" the radiating body.

All the optical materials have their own characteristic transmissivity, i.e., they allow only particular wavelengths to pass through with sufficient intensity. For other wavelengths, they are opaque. Transmission power for many optical materials for various wavelengths is shown in Figure 14.6. It can be seen that ordinary glasses (crown glass, quartz, pyrex, ruby, etc.) have good transmission in ultraviolet and the visible region, but are opaque to infrared. Thus, glass windows are of no use at wavelengths higher than 2.5 μm. Barium fluoride and zinc sulphide windows have only about 60–80% transmission in the infrared and visible region. Calcium fluoride has good overall transmission. The last three materials can hardly be called glasses and their optical forms are very costly. Thus, the spectral region that can be seen gets limited by the window and the optical material chosen.

3. Radiation pyrometers will require a sensor which will sense the incident radiation and generate a measurable signal (usually electrical voltage).

   Radiation pyrometers traditionally used thermal detectors as sensors. Thermal detectors are usually thermopiles made of a number of thin gauge thermocouples connected in series. Their hot junctions form the radiation sensing surface. This face is coated black to form a black body. Thermopiles can detect signals (radiation) of all wavelengths and are ideal. However, they depend on being heated by radiation, and hence depend on heat transfer and are therefore slow. Further, their signal is very low (Figure 14.7(A)) and they require cold junction compensation.

   A number of semiconductor sensors are developed which depend on the quantum effect of incident radiation to generate a signal. They are based on materials like silicon, lead sulphide, indium antimonide, etc. (Figure 14.7(A)). Their response is practically

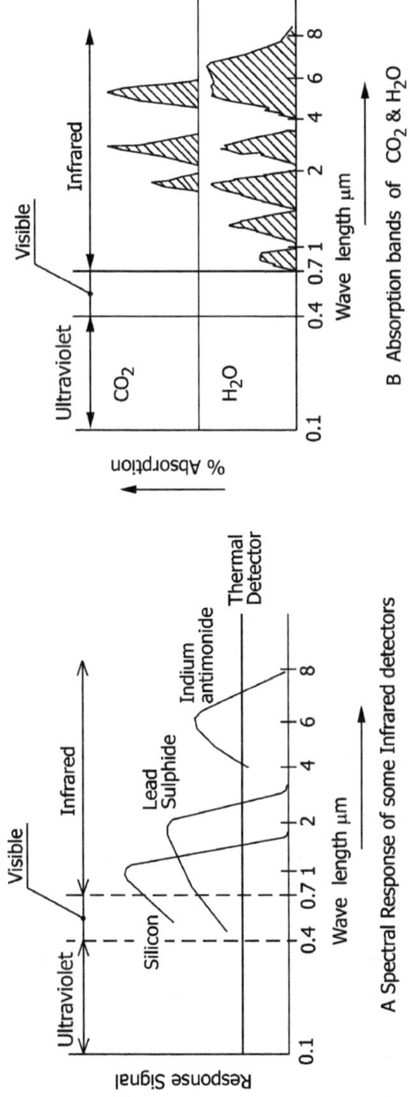

Figure 14.7 (A) Response of thermal detectors, (B) Radiation absorption by gases.

instantaneous but it is selective to wavelength. A "good" signal is obtained over a narrow band. Thus, silicon is suitable only around 0.8 to 0.9 μm and lead sulphide around 1.0 to 2.0 μm. Thus, the sensor used in an instrument also limits the wavelengths that can be sensed.

4. Radiation pyrometers do not require contact with the radiating body, hence there is a distance (sometimes considerable) between the two. The medium in-between is industrial atmosphere laden with dust, smoke and gases such as CO_2, water vapor, and reaction products.

   The dust particles scatter the radiation passing through them. CO_2 and water vapor selectively absorbs radiation as shown in Figure 14.7(B). Measurement of the radiation in the absorption bands should therefore be avoided. This again narrows the spectrum region of sensing. The absorption by these gases also depends on their concentration (partial pressure) in the air and the distance (optical path) between the object and the pyrometer.
5. It was mentioned when discussing the principles of radiation that the intensity of radiation of a nonblack body depends on its emissivity. Thus, for the same temperature there are two bodies with different temperature readings. The emissivity is a function of the surface condition, the wavelength, and temperature. Thus, radiation pyrometers will require some kind of emissivity correction. Many times, data available on emissivity are sufficient.

The above problems reduce the universality of radiation pyrometers. In practically all situations the pyrometer will have to be selected and tuned to the problem.

In real situations the placement of the sensing head with respect to the object is also to be assessed carefully to avoid background and external radiation affecting the incident beam.

14.10 DISAPPEARING FILAMENT PYROMETER

The tungsten filament of an electric bulb is a radiator. The intensity of radiation depends on the current flowing through it. The maximum temperature at the rated voltage is about 2800 to 3000°C. The minimum visible radiation is at 600°C. Hence we can obtain the radiation between 600 to 2800°C by changing the current.

The disappearing filament pyrometer uses the tungsten filament as a radiation source. Its radiation is compared with the radiation from the object. The object radiation is obtained in the background. The filament is in the foreground when seen through the eyepiece. By varying the current with a rheostat, the intensity of the filament radiation is so adjusted that at some current the filament disappears against the background. The intensity of the background and the filament is then equal.

The ammeter in the lamp circuits is calibrated in °C. Hence the temperature of the object is read on the ammeter.

The construction of a disappearing filament pyrometer is shown in Figure 14.8.

The whole instrument is built in the shape of a gun. A switch in the handle puts the lamp on when pressed. The handle also contains batteries. The ammeter is usually located at the back near the eyepiece. Note the focusing optics and the set of filters. The latter are to adjust for the shift of $E_T - \lambda$ graph with increasing temperature and the changes in emissivity.

The following points are noteworthy about this instrument:

1. It is very simple, portable, and easy to operate.
2. It operates in the visible region.
3. Personal error may arise due to defective eyesight, color blindness, etc.
4. Like other radiation pyrometers, this pyrometer also does not require contact with the body and hence is useful for melting furnaces, etc.

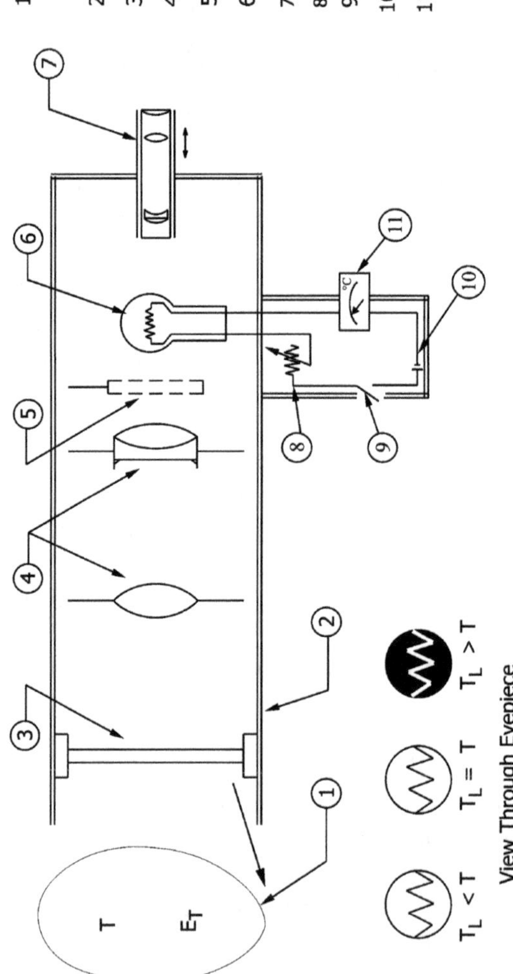

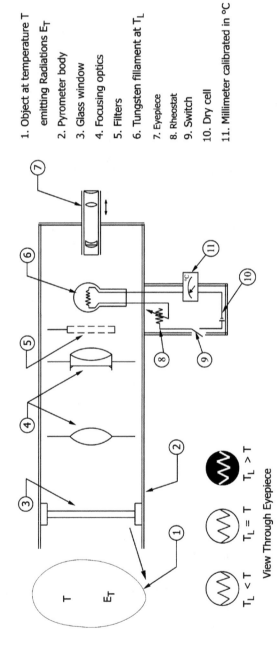

1. Object at temperature T emitting Radiations E_T
2. Pyrometer body
3. Glass window
4. Focusing optics
5. Filters
6. Tungsten fillament at T_L
7. Eyepiece
8. Rheostat
9. Switch
10. Dry cell
11. Millimeter calibrated in °C

Figure 14.8 Disappearing filament pyrometer.

5. The response is not quick, as the operator has to adjust the rheostat.
6. The measurements are not continuous.
7. It cannot be directly used for objects, such as the sun.

14.11 RADIATION PYROMETERS

The construction of a typical radiation pyrometer is shown in Figure 14.9. It generally consists of two parts — the sensing head and the electronic part. The sensing head is a sealed box and contains all the optics for sensing and sighting. The box has a window of a suitable optical material through which the radiation beam enters. There are suitable filters and diaphragms to condition the beam which is then separated into two parts. The central part is reflected by a mirror and taken to the sighting optics. Through this beam, the instrument is focused on the target or radiating body.

The surrounding beam is reflected and focused on the detector by a pair of concave convex mirrors. The signal, which is created at the detector, is carried by a cable to the electronics part. Here the signal is amplified. Again, control adjusts the signal for emissivity. The response of the detector is nonlinear and it is linearized and conditioned and fed to the indicator/controller/recorder. The indicator is usually digital.

The detector is thermal or semiconductor type. The construction of a thermal detector is shown in Figure 14.10. This detector requires a cold junction compensating device which is situated in the sensor box near the detector. In this type, the sighting optics can be situated directly behind the sensor as its center is open. The thermocouples used are of type K or N, i.e., base metal (e.g., chromel-alumel). They are made of very thin wires and their hot junctions are blackened to simulate a black body. Eight to ten thermocouples are situated radially and connected in series. This sensor can sense all the wavelengths passing through the windows (usually glass or quartz).

Semiconductor sensors have a response time of 0.01–0.025 sec. Depending upon the optics chosen, they can sense

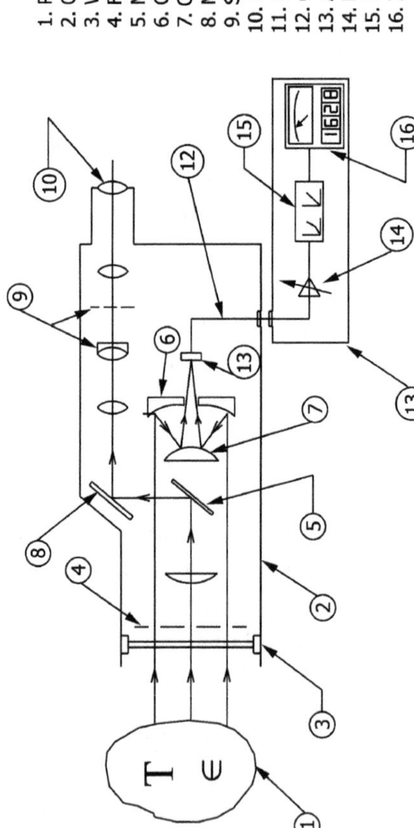

Figure 14.9 Monochromatic radiation pyrometer.

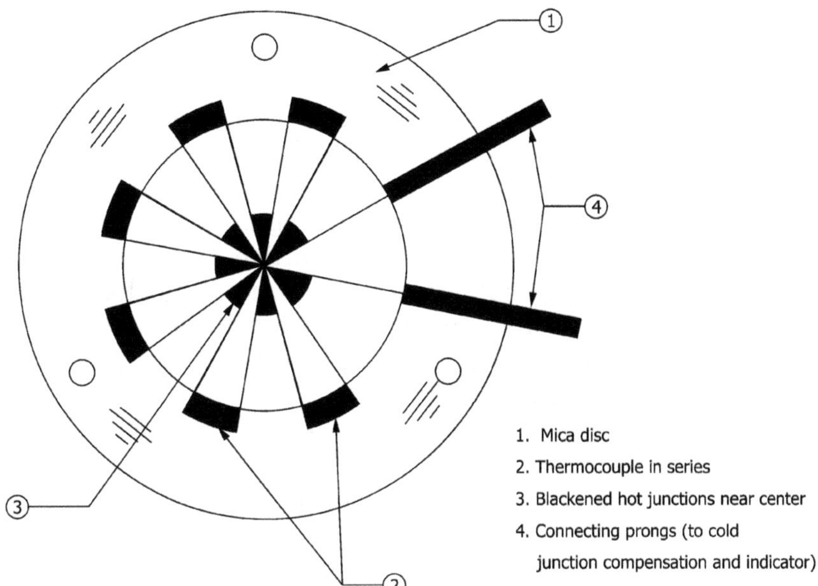

Figure 14.10 Thermal sensor.

0.65 to 10 μm radiation and can be used in the temperature range of 0–3500°C with an accuracy of 0.75% (full scale). The sensors may be sensing a very narrow wavelength band (e.g., 0.9, 1.0, 1.6 μm) or a broad band (8–14 μm).

Sometimes the sensing of the beam is obstructed and this causes incorrect readings. Such situations arise when

1. The object is too small so that the target area is incompletely filled.
2. Dust, smoke, or steam obscure the line of sight.
3. The windows are covered with dust or dirt and are difficult to clean.
4. Emissivity of the object changes due to surface conditions and composition.
5. Due to intermediate obstructions such as safety grills, furnace parts, and handling equipment the object cannot be observed fully.

The above difficulties can be solved in some situations by using a two-color pyrometer.

A two-color pyrometer senses two different wavelengths from the incident radiation. This can be achieved by using two separate detectors or a rotating filter disc as shown in Figure 14.11. A double detector device will sense the two wavelengths at the same time. A rotating filter disc will have a time lag in the two responses. If the two wavelengths are chosen properly they will be affected equally by the above-mentioned obstructions. Thus their effect will be cancelled if we take the ratio of the intensities. This ratio will then depend only on the temperature of the object.

There is a limit up to which the loss of signal due to obstructions will not affect the temperature reading. This limit depends on the design and measurement situation and

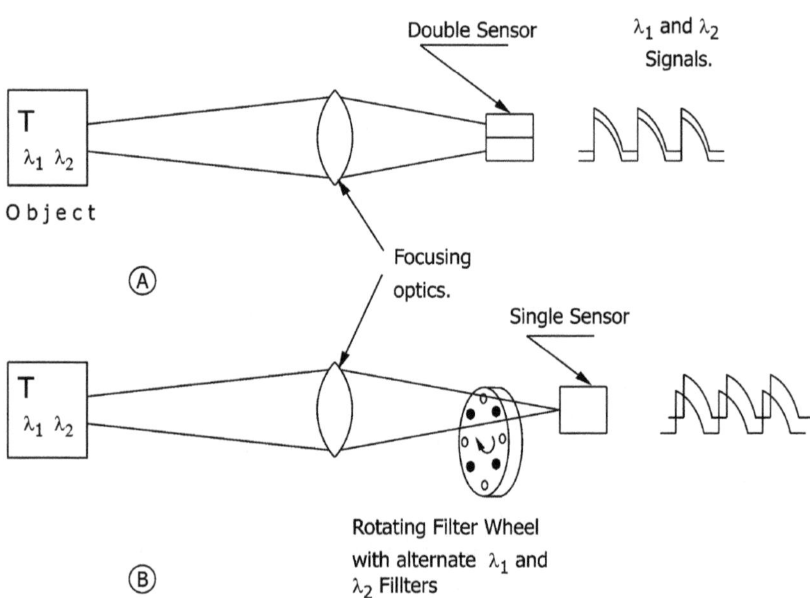

Figure 14.11 Principle of two-color pyrometer.

can be 5:1–25:1. The two wavelengths chosen are 0.7 and 1.8 μm and 1.55 and 1.68 μm. These instruments are used in heat treatment furnaces, foundries, kilns, ovens and the like. It is necessary to carefully assess the situation and locate the instrument to avoid reflected rays entering the beam.

14.11.1 Advantages of Radiation Pyrometers

1. There is no contact between the object and the sensing head.
2. Because there is no contact and the response is quick, the temperature of the moving bodies can be measured, e.g., ingot being rolled, flowing liquid metal, etc.
3. As long as the target area is fully covered, the temperature of small objects can be accurately measured.
4. There is (at least theoretically) no upper limit for temperature.
5. The instrument can be used for a wide range of materials during their processing such as rolling, extrusion, point drying, ovens for plastic, rubber, and paper.
6. Portable, battery operated, and handheld models are also available.

14.11.2 Limitations

1. The available optical and sensor materials pose a practical limit on the wavelengths that can be measured.
2. These pyrometers are not as universally applicable as thermocouples. Radiation pyrometers require careful assessment of the problem and choice of the suitable instrument.
3. Radiation pyrometers have a quick response and good accuracy but are costlier than thermocouples.
4. The temperature of objects with a thick scale cannot be measured accurately.
5. Emissivity correction is required.

14.12 MISCELLANEOUS TEMPERATURE-RELATED DEVICES

14.12.1 Temperature Indicating Colors

These are made from pure metals, eutectics, alloys, intermetallic compounds, and inorganic salts. They all have a calibrated single melting point or a very narrow solidus-liquidus gap. The calibration is equal to 1% and is available in the temperature range of 38–1371°C.

These materials are formulated in the form of sticks, liquids, pellets, stickers, and labels. They are used to mark the object to be heated (like a chalk) or are placed on the object in the furnace.

When the object surface acquires the calibrated temperature, these markings change their color or disappear, thus giving an indication of the temperature.

These markers are very useful to monitor furnace linings, furnace charge, and for calibrating the furnace pyrometer.

14.12.2 Bimetallic Devices

The basic temperature sensing element in these devices consists of a thin strip made of two different metals. One of the metals has a low coefficient of thermal expansion and the other has a high coefficient. The strip is made by corolling or electrodeposition.

When such a bimetallic strip is heated, it bends due to expansion. The angle of bending is proportional to the temperature (or to the temperature difference with the ambient).

The metals or alloys used for the strip must have the following properties to make a good instrument:

1. The difference between the expansion coefficients should be very large.
2. Bending should be proportional to the temperature over a wide range.
3. The expansion must be reversible and there should be no thermal fatigue and corrosion.
4. The strip should be available in various thicknesses.

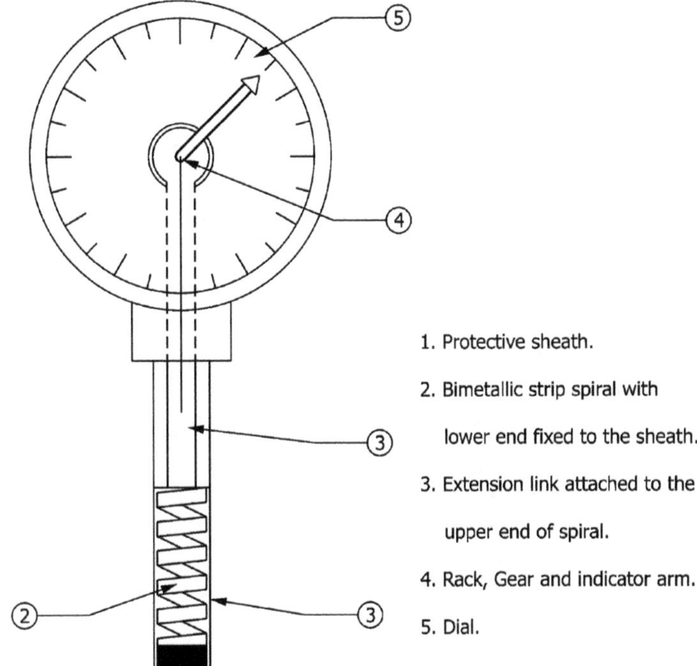

Figure 14.12 Bimetallic thermometer.

A number of bimetals are commercially available. The high expansion component is an alloy of iron, nickel, chromium, and manganese. Low expansion alloy is made from iron and nickel. They are applicable in the range of –30 to +300°C.

The construction of a typical bimetallic thermometer is shown in Figure 14.12.

The sheath encased strip is firmly fixed to the bottom of the sheath. The other end is fixed to a bar attached to a simple dial indicator. These thermometers are used for low temperature ovens and domestic apparatus. They are used as indicators and cheap and simple temperature controllers. They are of simple and rugged construction. Temperature indication and control can be achieved to 5°C in the range 80 to 350°C.

14.12.3 Bimetallic Energy Regulators

This is an interesting device which can be used for the temperature control of small and large furnaces and appliances.

The principle of operation of an energy regulator is shown in Figure 14.13.

A bimetallic strip has a resistance heater wound around it. One end of the strip is fixed. The free end has a contact attached to it. There is a movable contact opposite the contact on the strip. This contact is moved by a spring and knob. The movement of the knob adjusts the separation distance between the contacts.

When put in the ON position, the contacts touch each other. The strip heater is put on and the strip is heated. This causes bending and separation of contacts which cuts off the heater. The strip cools and straightens. Again the contacts touch and the heater is turned on. Thus, an ON-OFF cycle depends on initial contact separation which is adjusted by the knob. The ON-OFF cycle affects the outgoing supply similarly.

The power supply to the furnace is thus put on an ON-OFF cycle. This controls the energy supplied during a given time span. Hence, the name “energy regulator.”

Due to the cyclic supply the temperature of the furnace is indirectly controlled. The contacts are rated at 3–15 amps but can be supplemented by high rating electromagnetic contactors to control larger furnaces.

The following points about energy regulator are noteworthy:

1. The instrument has a simple construction and is very inexpensive.
2. It is completely independent of the furnace.
3. There is no temperature measuring component.
4. The temperature is controlled independently on a time cycle.
5. Precise control is not possible.
6. The degree of temperature control obtained for a given time cycle will depend on the thermal design and operation of the furnace.
7. A separate pyrometer will be required to measure the temperature.
8. There is no upper limit for temperature as far as the regulator is concerned.
9. Control can be obtained at 20–30°C about the desired value.

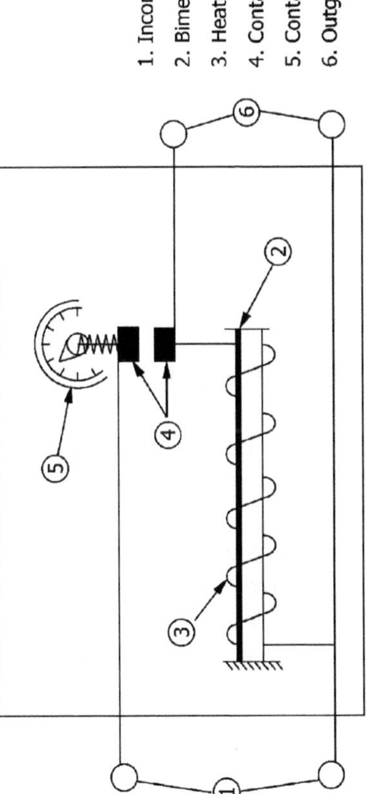

Figure 14.13 Bimetallic energy regulator.

14.12.4 Throwaway Tips

When measuring the temperature of molten metals like steel, cast iron, and nickel alloy the pyrometer is required to be dipped in the molten metal at temperatures exceeding 1500°C. The molten metals are extremely reactive and turbulent, which exposes the pyrometer to an extreme environment. This reduces its life.

To overcome this problem the pyrometers are made of throwaway tips or hot junctions. The tips are made of very thin Pt–Pt, Rh wires embedded in alumina powder and packed in thin metal or cardboard tubes. These tubes are plugged in the metal stem of the pyrometer. The tips (or the thermocouple) last for a very short time and then burn off. However, this time is sufficient to obtain a temperature reading.

Each tip cab be used only once, hence, the name "throwaway tip." However, they are quite inexpensive.

For cast iron and some steels, the solidus and liquidus temperatures can be correlated to the carbon and silicon content. Hence, with proper recording instruments, the throwaway tips can be used for determination of carbon, silicon, etc., in cast irons.

14.13 TEMPERATURE INDICATORS

In the previous sections we have reviewed the various types of sensors available for sensing the temperature, their advantages, and choice.

Every type of sensor will require some indicator to display the sensed temperature. In this section we will review the indicators commonly used and then the limitations.

Bimetallic sensors use a dial–gauge–type mechanical indicator working on a rack, a gear train, and a needle indicator moving on a graduated dial.

Mercury in steel tube thermometers works on pressure changes in the sealed tube due to the expansion of mercury. A simple burden tube indicator graduated in temperature scale is used for display, mainly used to measure low temperature. Measuring the temperature of flue gases, oil quenching tanks in heat treatment, and hard anodizing bath temperature are some of the examples.

The simplest indicator is a millivoltmeter with a dial big enough to read the temperature to the required accuracy. Some type of cold junction compensation is provided. As the indicator needle moves by a mechanical movement of a current carrying suspended coil, it does not have much power and sensitivity. Readings are usually restricted to 10°C. The instrument is delicate but inexpensive and simple. It is used for portable pyrometers.

Thermocouples produce a small e.m.f. proportional to the temperature. Depending upon the object condition it is constantly charging. The best instrument to measure small voltages is the potentiometer. Here the test voltage is balanced against a known comparable voltage. Balancing is done by moving a contact against a slide wire. The exact balance is obtained when there is no current. Thus this is a null method of measurement and hence very sensitive and accurate. However, the manual process of balancing takes time.

To overcome the time problem and to obtain quick and continuous readings, automatic balancing potentiometers were developed. Here the thermocouple e.m.f. is amplified and used to drive a servo motor to move the moving contact for balancing. At the null point the servo motor automatically stops. Thus, the pointer continuously moves from one balance point to the next. As the pointer is motor-driven its movement is powerful and can be used to drive the pen of a chart recorder. It can also be used to actuate the controller.

Self-balancing indicators are quite rugged. They are very sensitive and accurate and can be easily adapted to changes in the measuring range. They have a complicated construction, are very costly, and are not portable. They require an external pyrometer for their working.

The Wheatstone bridge used with resistance pyrometers can be made self-balancing.

Currently, the indicators used on all types of pyrometers are digital. They display the temperature in numbers, thus dispensing with the dial and pointer.

In digital indicators the analog voltage signal from the thermocouple is converted into a digital signal (numerical signal) and displayed on a counter. The details of their working are beyond the scope of this chapter.

These instruments are very fast and accurate. They require very little external power and due to developments in integrated circuits, they are very small in size. Hence they can be used on portable as well as panel instruments. Their indicating numbers are well-lighted so that there is no difficulty in reading. They are very fast acting, accurate, and sensitive.

The working details of all indicators discussed above are unimportant for the furnace designer. Enough information is given to make a choice for projected furnaces.

14.14 TEMPERATURE CONTROLLERS

For many heating operations, simple indication is not enough (e.g., heat treatment). It is necessary to control the temperature within certain limits or to control the heating and cooling rates, etc. For processes like zone melting or semiconductor, the temperature has to be maintained virtually constant, or within a very narrow band for a long time.

Where temperatures involved are not high (< 400°C) simple control instruments based on bimetallic switching action are sufficient. These controllers are called thermostats. They can maintain the temperature within a range of 10 to 20°C.

For higher temperatures, narrow temperature variations, and longer times, electronic controllers are used. It is not possible to discuss their design or circuits here as they belong to a specialized field of electronics and control engineering.

We will review their control actions broadly so that the furnace designer can make a choice of suitable controller.

Typical time-temperature cycles of a furnace are shown in Figure 14.14(A). The cycles can be divided timewise in three sections. Section I is heating from initial temperature to set temperature *Ts*, the second part (section II) shows the maintenance of *Ts* for a long (process) time. Lastly, section III

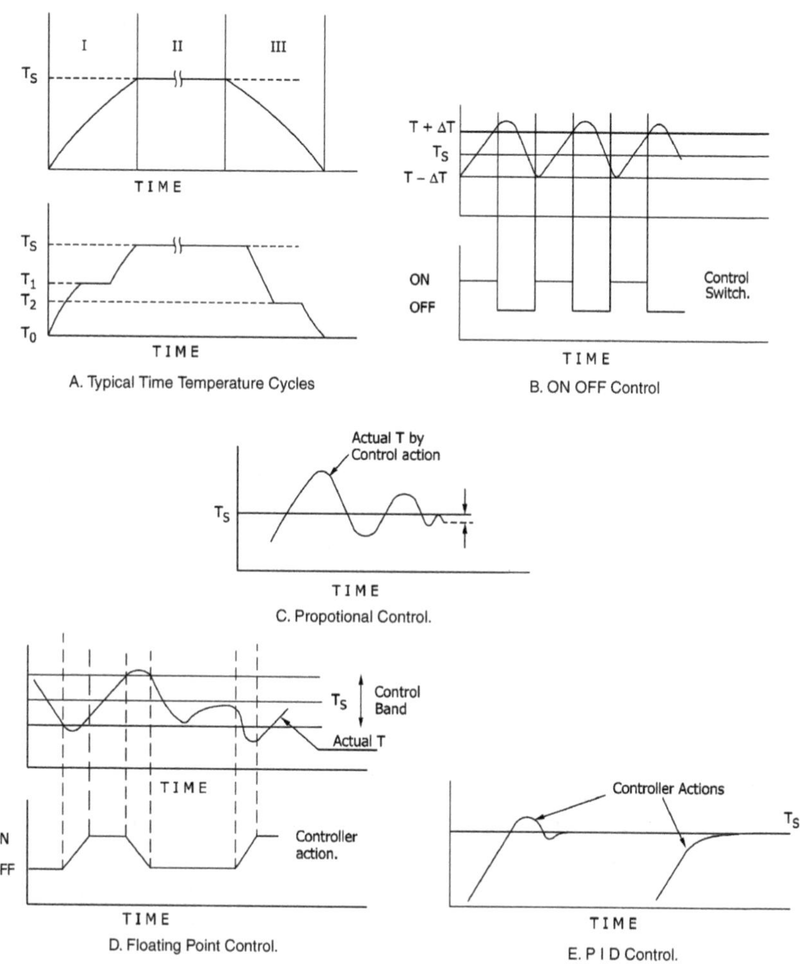

Figure 14.14 Temperature Control Cycles.

shows cooling. The second cycle shows that heating and cooling are done in steps:

$$T_0 \rightarrow T_1 \rightarrow Ts \text{ for heating and } T_s \rightarrow T_2 \rightarrow T_0 \text{ for cooling.}$$

These are the cycles desired.

The first controller cycle called On-Off, controls bands on the lower side $Ts - \Delta T$. The width of the band ΔT is adjustable.

It is also possible to have different widths on high and low sides.

The signal from the sensors is compared with the control band. If it lies out of the band, a switch operates to put the heater on or off as necessary. As only two control actions, on or off are available, the temperature is maintained fluctuating within the band. To obtain better control there may be two or more switches in series. Thus, for the off signal, the first switch will then turn off the second heater as required. This is the simplest controller and is suitable for large furnaces and wide control limits. The thermal mass of the furnace gives a slow control action.

Figure 14.14(C) shows the action of the proportional controller. On sensing the deviation of actual temperature from the set point Ts, the controller will take remedial action but will overshoot. This will initiate the next action to reduce the overshoot. These actions will continue giving temperature oscillations until the temperature is stabilized a little above or below the set point. This controller is better than the on-off control. The final deviation, i.e., the set-off, can be controlled or the oscillations can be reduced but cannot be removed. Proportional controllers are suitable for furnaces having low thermal mass and fast heating rates.

The next control option is the floating control. The controller senses the temperature against the allowable variation band. If the temperature is outside the band, the control device (motor, valve transformer tap, etc.) takes control action and, brings the temperature within the band. Once within the band, the controller action stops until an out-of-band signal is received. Thus, the controlled temperature oscillates within the band.

Mathematical operations on the controlled signal, with respect to time, can be performed electronically or numerically on a computer. Thus, its differential co-efficient dT/dt will give information about the slope of the controlled variable. Similarly, its integration $\int T\,dt$ will give its rate of approach to the set point and the direction of approach. Controllers that can perform differential and integral operations are called "derivative" and "integral" controllers. In actual control, these control actions are not used singly but in combination with proportional controllers. This gives proportional and integral controllers, (P,I) and PID controllers offering proportional, integral, and derivative actions. Normally, controllers can be set for any combination. The control response of PID controllers is shown in Figure 14.14(D) and Figure 14.14(E). PID controllers are useful when there are sudden changes in temperature. For example, the door of a furnace opened in midcycle (say for loading, etc.) and the temperature is drastically lowered due to loss of condition. These controllers are very costly. They are also used for precision furnaces.

All the controllers discussed above produce an output signal of the desired type. This signal is required to be amplified and matched to the controlling device.

The controlling devices are relays, contactors, motorized valves, solenoid valves, etc. They require considerable power for their operation.

When specifying a controller, the designer must carefully consider the following points.

1. The operation cycle — continuous or batch.
2. Allowable variations in set temperature, i.e., width of the control band.
3. Duration of the cycle.
4. Possible interference in operation.
5. The thermal inertia of the furnaces as a whole, both with and without load.

Remembering that simply specifying a sophisticated controller does not guarantee required temperature control. It has to be matched with the furnace and the load.

Chapter 15

Miscellany and Further

CONTENTS

15.1 INTRODUCTION

In this chapter we will collect a few loose ends before closing our discussions on heating.

A number of furnaces were introduced in earlier chapters. Here we will consider some special designs used for heat treatments, sintering, drying, and so on. Each design poses interesting problems for the designer.

Waste incineration as a combustion process was considered in detail in Chapter 5 and Chapter 6. Practical incineration design involves problems of material handling and waste gas and ash disposals. Three possible designs are presented next. They are for large scale (municipal), medium scale (industrial), and small scale (domestic) incinerators. The number of available and useful incinerators is very large. However, practically every situation requires a special design or modification depending on the type and quantity of waste available, and of course the local regulations.

Many large scale furnaces and heating installations can derive economic benefit by reusing the waste heat of the combustion gases. Heat exchangers which recover part of the heat from waste gases to heat combustion air are the relatively simple, cross-flow type. In some cases a lot of heat is taken up by the furnace walls or heating conductors. Though heat recovery from such applications is not economical, it is necessary to cool the parts to keep their temperature at an acceptable level. Such cooling is also done by heat exchangers.

It is in this context that a limited review of heat exchangers and simplified design calculations are presented next. For more details it will be necessary to consult references for Chapter 3 and Chapter 4, given in the *Bibliography*.

Drying is a process usually referred to as removal of moisture. The temperatures involved are low (80–250°). Drying is used in many diverse processes, each with its own peculiarities. Consequently, the variety of dryers available is large. Some typical dryers are briefly discussed.

Baking, like drying, is a low temperature process. It is mainly used in the food industry but has applications in many fields. The purpose of baking may be to make the article from dough (bread, cakes), to give crispness, to increase the taste or appearance, to improve bonding, and so on. Some representative baking ovens are presented to bring out their design features.

Fans are used in many furnaces and ovens. They are mainly used for the circulation of gases and supply air to the burners. Fan-induced air flow is also used for convective cooling. Considering these applications, a short and limited introduction to fans is also included. This should be helpful to the designer in choosing a proper fan for a given application from manufacturers' literature.

A short postscript on recently available materials appears at the end. The use of most of these materials in heating is not yet well established. However, they are sure to play an important role in the future.

15.2 SOME TYPICAL FURNACES

15.2.1 Rotating Hearth Furnace

This is a continuous type of furnace in which the hearth is made in the form of a flat, circular ring. It forms the bottom of a cylindrical furnace. The hearth rotates around its center while the furnace (i.e., the walls and top) is stationary. The hearth is rotated by a gear drive which may be located at the center or the outside. A radial partition wall separates the charging and discharging doors. A number of tangential burners are fixed on the upper side of the outer walls. Combustion gases exit through a port located opposite the discharge door. In large furnaces the hearth moves on a circular rail track. The construction is outlined in Figure 15.1.

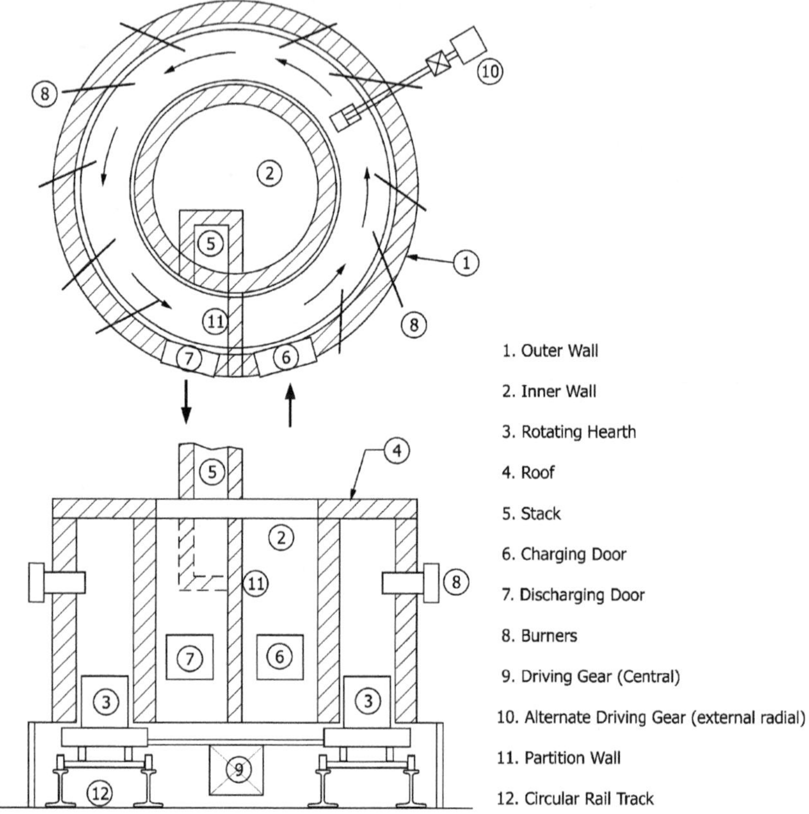

Figure 15.1 Rotating hearth furnace.

The following features make this type of furnace an attractive proposal for many applications:

1. The circular, rotating hearth makes these furnaces very convenient and compact preheating units for process applications such as forging or extrusion.
2. Large-sized furnaces have a hearth width up to 5.0 m so that medium to heavy-sized work pieces can be easily accommodated.
3. Variable rotational speeds and indexing makes the operation flexible. A heating cycle to suit any material can be easily adjusted.

4. It is possible to build smaller units having hearth widths up to 0.5 m. These small units can be used for light jobs such as heating for press working, intermediate annealing, and even drying or baking.
5. Large furnaces are usually oil and gas-heated. Smaller furnaces can use electricity or LPG.
6. As the charging and discharging doors are adjacent, one operator can carry out both the operations.
7. Linear continuous furnaces usually require some means of moving the work pieces through. In rotary furnaces no such mechanisms are necessary.
8. Instead of the hearth, the roof can be made as rotating. This will enable work pieces to be hung on the roof. They will receive uniform heat from all sides.
9. Some sealing arrangement will be required between the moving hearth and stationary walls to prevent the ingress of cold air or escape of hot gases.
10. The number, position, and capacity of the burners is an important design factor. This will help to adjust the desired heating and soaking periods.
11. The hearth can be divided into a number of zones by using hanging (chain) curtains.
12. If the burners are well above the workpieces, the major heating mode is radiation.

15.2.2 Automatic Integral Quench Furnace

Hardening heat treatment of steels requires heating the work to a certain temperature and quenching it in a suitable medium of a much lower temperature. The practical hardening cycle involves heating to hardening (austenizing) temperature at a certain rate and then holding it (soaking) at that temperature for a certain time. The heating rate and soaking time depends on the steel composition and size of the work. The quenching medium is usually oil.

Figure 15.2 shows a modern furnace in which all the operational sequences are carried out automatically. A typical cycle may be Loading → Heating → Soaking → Quenching → Unloading. The quenching tank and quenching operation is part of the system, hence the name “integral quench” furnace.

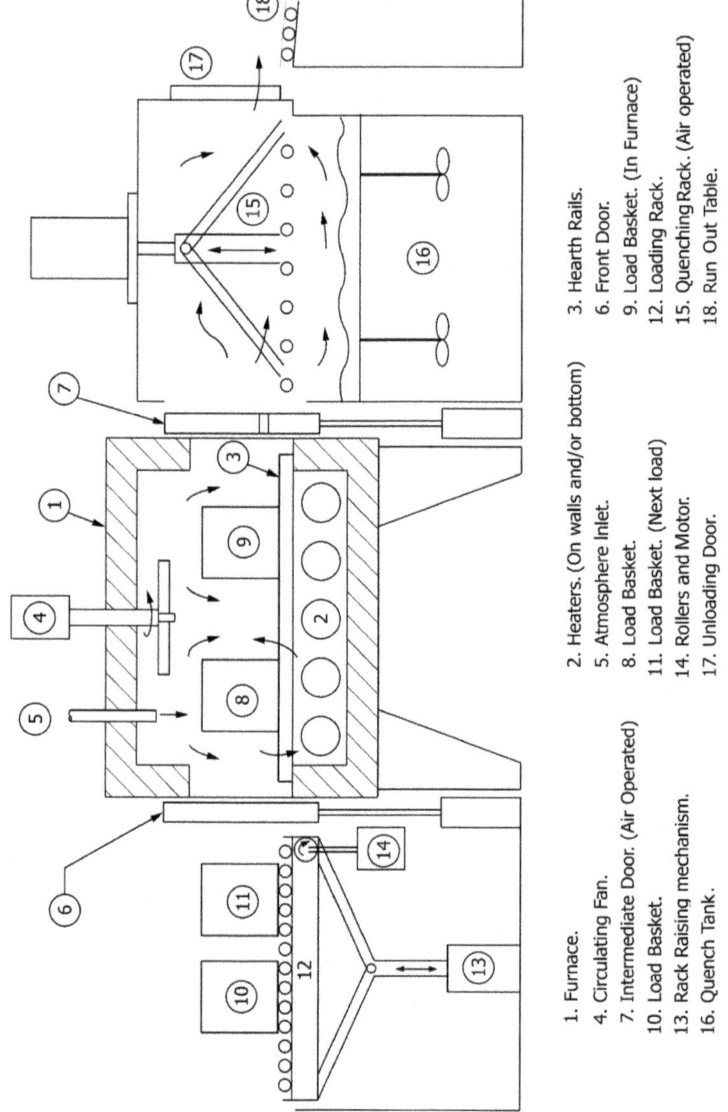

Figure 15.2 Automatic integral quench furnace.

The central part of the system is the furnace. It may be electrically or radiant-tube heated. A protective atmosphere is introduced at the top and circulated by a fan. The hearth has alloy rails on which the load rests. The furnace has two air-operated doors, one in front and the other in the back walls.

A complex charging/loading/pushing mechanism is located in front. The loading rolls and rails are accurately aligned with the hearth rails. The load charging/pushing is accomplished by forked racks which can be raised or lowered by an air-operated mechanism. The prepared (loaded) work baskets are kept on the rollers. An electric motor moves the rocks and baskets into the furnace chamber.

A quenching system is located next to the back door. After heating and soaking, the load baskets are pushed out on the air-operated quench frame. A quench tank is located below the frame. The frame and tank are located in a steel quenching chamber.

On receiving the hot load, the frame lowers it into the quenching tank. Thus the quenching operation is achieved without loss of heat or time. After the predetermined time in the tank, the frame and load are raised and discharged (pushed) over to the run out table through the unloading door.

The protective atmosphere is discharged through the intermediate door and occupies the space above the quench tank, before burning at the flame curtain at the discharging door. This arrangement protects the hot load for a short time before quenching. A flame curtain is also provided at the front door. A matching gas generator provides the required protective atmosphere.

The whole operating cycle, temperatures, loading, discharging, etc., is automatically sequenced so that no operator is necessary. Control is exercised from a separate station which can be located away from the furnace.

The quenching tank has a large capacity and is provided with circulators and heat exchanges to obtain uniform temperature and circulation. The typical furnace size is 600 × 900 × 450 mm with a quench tank holding 3000 l of oil. Larger units are available. These furnaces are extremely suitable for production jobs.

15.2.3 Vacuum Gas Furnace

Heat treatment of high alloy steels such as all types of tool steels presents some peculiar problems. These steels are machined or shaped in an annealed (soft) condition. The shapes are quite complicated and have varying thicknesses. Typical examples are dies, press tools, forging tools, cutting tools, etc. After machining they are hardened and tempered. Hardening requires thorough heating to temperatures in the order of 1000–1200°C followed by quenching (fast cooling) by a suitable coolant (quenchant). Tempering requires reheating of quenched articles to 600–1000°C, at a certain rate and then slow cooling at a certain rate. To avoid oxidation both operations require heating in a protective atmosphere. Quenching poses problems of distortion and fracture.

If quenching is carried out by using a cool inert gas, the cooling rate is less severe than that obtained by oils or water. This reduces or often eliminates the problem of distortion or cracking.

Vacuum-gas furnaces are designed for using a vacuum atmosphere for heating and an inert gas such as dry, pure nitrogen for quenching or slow cooling. As both vacuum and inert gas are used in the same enclosure, they pose interesting design problems. Heating is usually done by indirect electrical heating. Some vacuum tempering designs may be (externally) heated by gas, as the temperatures are low.

A typical vacuum-gas hardening furnace is shown in Figure 15.3.

The furnace is of the cold-wall vacuum-type having a horizontal cylindrical shape. The main body ① has a cooling jacket. The door ④ is also water-cooled and has a gasket to provide a vacuum-tight seal. Heating is achieved by indirect electrical heaters ③ placed radially. Either graphite or molybdenum heating elements are used. Both types require a low voltage, high current supply.

The work ② is usually kept in a basket and is supported by a combination of molybdenum and graphite supports. The vessel is evacuated through the vacuum port ⑤. The vacuum system consists of a diffusion pump backed by a roots blower and a mechanical pump.

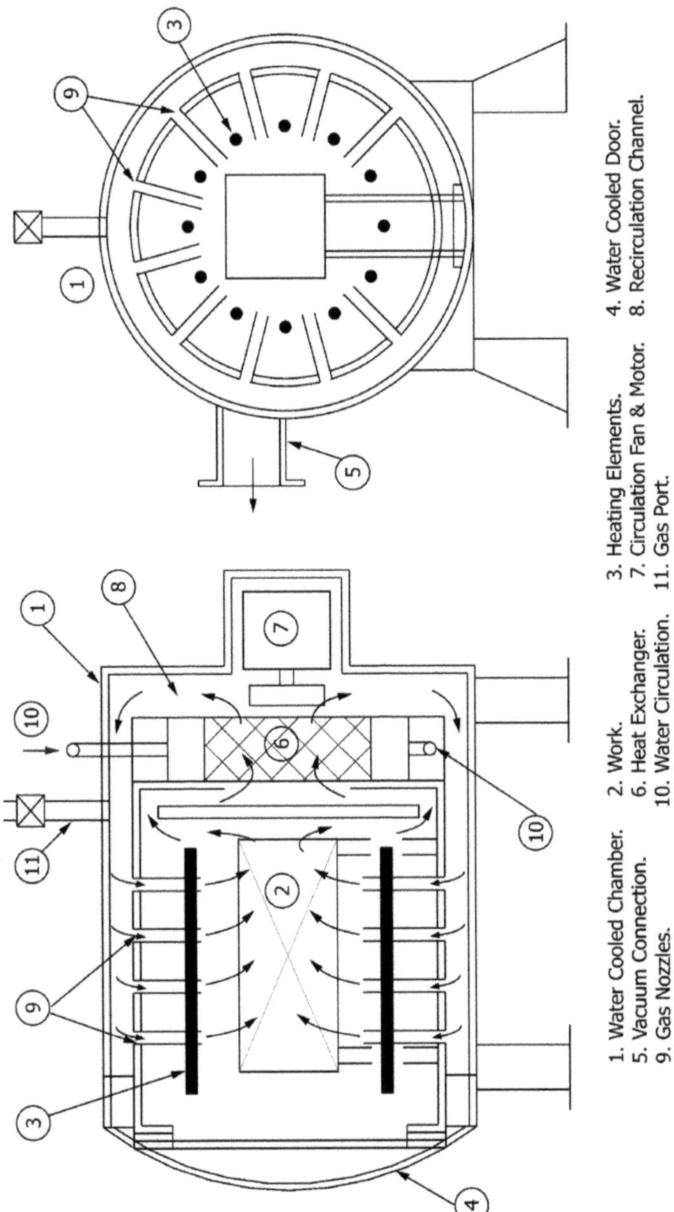

Figure 15.3 Vacuum-gas heat treatment furnace (For description see text).

For the quenching or tempering operation, dry pure nitrogen is supplied to the evacuated chamber through the opening until the desired pressure (2–3 Bar) is reached. The cool nitrogen enters the chamber radially through graphite nozzles, ⑨ and surrounds the work from all sides so as to obtain an even cooling rate.

The gas coming out of the nozzles gets heated on contact with the work and the hot furnace interior. It is axially drawn out by a fan ⑦ through the compact heat exchanger ⑥. Its temperature is brought down to less than 150°C and is recirculated via channel ⑨ to the nozzles and the chamber. The path of gas circulation is shown by arrows. The heat exchanger fluid is water and is admitted through the passage. The circulation goes on until the whole chamber cools to about 150°C and then the door is opened. This safeguards the graphite/molybdenum heating elements.

The same or a similar furnace is used for a subsequent tempering operation. Here, the temperature is lower and a nitrogen atmosphere is kept throughout the tempering period.

It can be seen from the figure that the design of the furnace is quite complicated. It requires several auxilliary systems and a complex sequencing and control system.

Specifications of a commercial furnace of this type are given below.

Work space	800 × 800 × 1200 mm
Overall size	2500 × 3200 mm long
Max. temperature	1350°C
Temperature uniformity in work space	5°C
Heating power	200 kW–3 ph
Fan motor	60 kW
Vacuum	10^{-2}–10^{-5} mbar
Leakage rate	$< 1 \times 10^{-3}$ mbar/sec
Weight of charge	1000 kg max
Quenching gas	Dry, pure N_2
Gas pressure during hardening	1.5–6 bar

For one quenching

Gas consumption	10 m^3
Water consumption	20 m^3/h quenching

15.2.4 Linear Continuous Furnaces

As the name suggests these furnaces have a straightthrough movement of charge. The through movement may be achieved either by pushing work loaded baskets through or by placing the work on a continuous, moving mesh belt.

A typical linear continuous furnace is shown in Figure 15.4.

The furnace consists of three main zones—preheating, heating, and cooling. The preheating zone operates at a low temperature (100–500°C) and is used for drying, dewaxing, or general preheating. This is followed by the heating zone at the required process temperature (900–1500°C). Heating on both zones is achieved by resistance elements, metallic or nonmetallic. The heating zone may be nominally divided into three parts. The first part will bring the work to the desired temperature (heating). The second part will maintain the temperature for some time (soaking). This can be achieved by varying the power input to the zones. The third or end part of the furnace will have no heating elements and will act as a slow cooling zone.

Many processes such as sintering and brazing, require a protective atmosphere which is usually reducing. Such reducing atmospheres are harmful to most of the resistance heating material. To avoid a reaction between the atmosphere and the heaters, a metallic alloy inner muffle is used to separate the atmosphere and work from the heaters. Small sections and short furnaces may use a ceramic inner muffle.

For achieving the required properties in the work, a certain temperature-time combination is required in the heating zones. The temperature or its gradient in the zones is maintained by a suitable temperature control system. The time is adjusted by adjusting the belt speed. With these two controls the furnace can be adjusted to process a

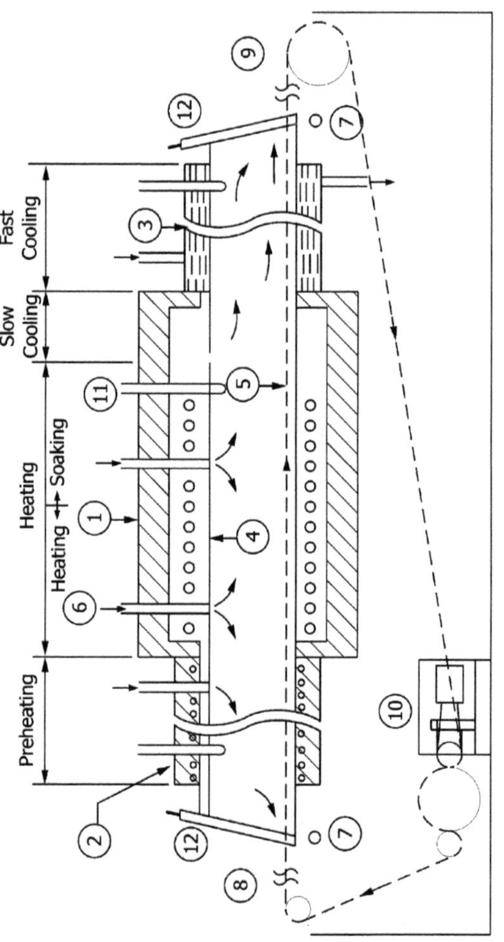

1. Heating Section with Resistance Heating Elements (High Temperature 900 - 1500 °C).
2. Preheating Section with (or without) Low Temperature Heating Elements.
3. Water Jacketed final Cooling Section.
4. Alloy Muffle (through out).
5. Alloy mesh Belt.
6. Protective Atmosphere inlets.
7. Flame Curtain Burners.
8. Charging (Loading) Section.
9. Discharge (Unloading) Section.
10. Belt Drive.
11. Thermocouple Wells.
12. Front and Back Doors (Air operated).

Figure 15.4 Linear continuous furnace (Mesh Belt type).

variety of materials. In a pusher pipe furnace the control is obtained by timing the introduction of a new work basket in the furnace.

The same basic design can be used for baking operations. The temperature required is of the order of 250–300°C. The cooling zone is usually not required. Baking evolves water vapor and carbon dioxide; hence the heating is best done in a stainless steel inner muffle. Embedded and sealed heating elements are used.

A very large variety of these furnaces are commercially available. Molybdenum, tungsten, or graphite heating elements with hydrogen as the protective atmosphere are available up to 2000°C.

The load capacity of mesh belts is a limiting factor. Commercially, Ni-Cr, Ni-Cu, monel, inconel, or stainless steel belts are available, which can be used up to 1100°C. For higher temperatures only pusher-type designs are possible with roller hearths.

The protective atmosphere is generally introduced in the central zone. It flows toward the doors where the chain and flame curtains are situated. A circulating fan may be used.

15.3 INCINERATORS

The production and accumulation of garbage in urban areas all over the world is assuming alarming proportions. Toxic and lethal by-products from chemical industries are polluting rivers and beaches and have already posed health problems. Incineration or burning off is one of the attractive (and some times the only) solutions.

The incineration process has been studied and developed in highly industrialized countries. However, it is rarely used in undeveloped or developed countries. It is with the desire to create an awareness of the technology of incineration that a limited review is taken in this chapter.

The nature and properties of "garbage" as a fuel are already discussed in Chapter 5.

The following points need consideration when designing an incinerator:

1. **Types of Garbage**
   Garbage can be classified on the basis of its origin.

   I. *Municipal garbage*—This is collected from streets, bins, beaches, and public places.
   II. *Establishment garbage*—This garbage originates from hospitals, restaurants, small nonchemical industries, etc.
   III. *Kitchen or household garbage*—From individual homes or residential buildings.
   IV. *Special garbage*—From chemical industries, slaughterhouses, food processing plants, markets, etc.

There can be many more types depending upon local industries and population.

2. **Composition of Garbage**
   Like other fuels (such as oil and gas) garbage does not have a single composition. The constituents depend on origin, geographical location, seasonal consumption trends of local populations, and many other factors. Thus, there are wide limits within which the composition varies. For example, the moisture content in municipal waste can vary between 10–40% depending on the season.

3. **Combustion problems**
   Due to variations in composition, the combustion of garbage cannot be carried out in a single stage and requires additional external fuel like gas or oil. Combustion is therefore invariably carried out in two stages.

   In the first stage (~ 600–800°C), the moisture is converted to steam and readily combustible constituents such as grass and paper are burnt. The steam and unburnt (pyrolized) gases flow to the next stage.

   Additional air and fuel are supplied to the gas in the next stage. Here complete combustion takes place. The waste gas is now essentially nontoxic and consists of CO_2, N_2, H_2O, etc. The temperature is 800–1000°C.

4. Waste gas from large incinerators usually require additional treatment such as cleaning, dust precipitation, and cooling.

Depending upon the operational practice, it may be possible to recover some of the waste heat by using a recuperater or waste heat boiler.

A large amount of ash is collected below the grates or in the post incineration chamber. This arises from metals and glass in the garbage. It is used for road building, ballasts, etc.

To make the incineration more easy and economical, it is necessary to separate large pieces of metal, glass, and stones, etc., from the garbage. This can be done mechanically or manually.

Three representative incinerator designs are discussed next. Note that there are many possible variations depending on the type and amount (T/day) to be incinerated, local available fuel, legal restrictions on waste gas constituents, water availability, and location and transport facilities.

15.3.1 Large Scale Municipal Incinerator

The typical municipal garbage has a nominal composition.

Dry dust, leaves, and wood	4–44%
Food, vegetable, and organic matter	8–30%
Paper cardboard	20–30%
Metals	5–10%
Rags, ropes, and textiles	2–10%
Plastics	1–6%
Unclassified	1–10%
Moisture	5–35%

The range is extremely wide. The volume of garbage is very large ($1–3 \times 10^3$ T/day).

A typical municipal incinerator is shown in Figure 15.5. Due to high volume, the loading is carried out by mechanical means such as a grab crane. The garbage is charged on the inclined, moving grate. Burners over and below the grate raise the temperature to 600–800°C. The moisture is evaporated and dry combustible matter starts burning. Combustion is

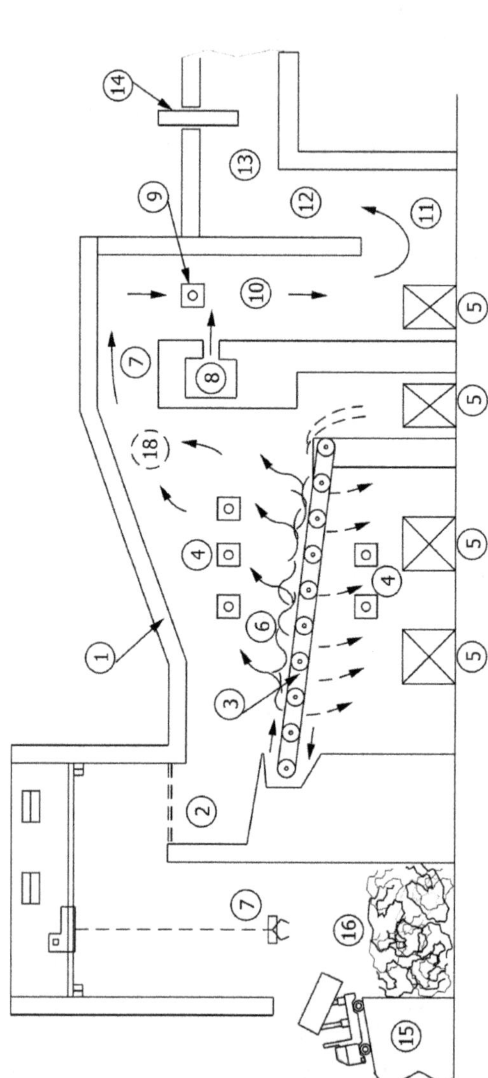

1. Refractory Walls & Roof.
2. Charging Port with Door.
3. Incineration Grate.
4. Burners above and below Grate.
5. Ash Pits and Doors.
6. Burning Garbage or Waste on Grate.
7. Combustion Gas Outlet Port.
8. Air Supply through Partition Wall.
9. Burner.
10. Air-Gas Mixing Chamber.
11. Port to Combustion Chamber.
12. Gas Combustion Chamber.
13. Waste Gas Flue to Stack, etc.
14. Damper.
15. Dumping Ramp.
16. Accumulated Garbage in Pit.
17. Garbage Lifting and Charging Gantry.
18. Possible Location of Waste Heat Boiler Tubes.

Figure 15.5 Typical municipal garbage incinerator.

completed as the mass reaches the end of the grate and is removed through the doors at the bottom.

Gases and steam evolved from combustion pass to a post-incineration chamber. It is supplied with excess air and the temperature is raised to 1000–1200°C by the auxiliary burner. The combustion is completed as the hot gases pass forward.

The gas pressure is controlled by a damper in the waste gas flue. Gas cleaning and dust precipitation is done before the gas is released into the atmosphere through the stock.

Incineration reduces the garbage volume by more than 90%. Sorted out glass, metals, plastics, and ash are saleable by-products.

15.3.2 Medium or Small Scale Incinerator

In these designs the volume required to be handled is not large (200–2000 kg/day), hence the charging is done manually. A U-type design is shown in Figure 15.6. The basic design is similar to that in Figure 15.5. The U construction has made the incinerator compact and hence, suitable for small or medium establishments. Gas cleaning and precipitation is usually unnecessary unless the charge contains hazardous chemicals. A tall stack is adequate.

Cremation of human bodies and incineration of dead animals and surgical waste are carried out in this type of incineration. They usually operate on electricity, gas, or light oil such as diesel.

15.3.3 Domestic or Office Incinerator

Domestic garbage contains moisture, food, and vegetable matter as the main constituents, along with small amounts of paper, plastics, glass, and metals.

Office garbage mainly contains all types of paper as the major constituent. Small amounts of chemicals, rubber, plastics, glass, and metals are the minor constituents.

A small hospital or clinic's garbage contains bandages, dressings, and cotton swabs in large amounts. Plastic syringes, bottles, tubes, tissue paper, organic matter such as dried blood, tissues, and discarded medicines constitute the rest. There is also a lot of moisture.

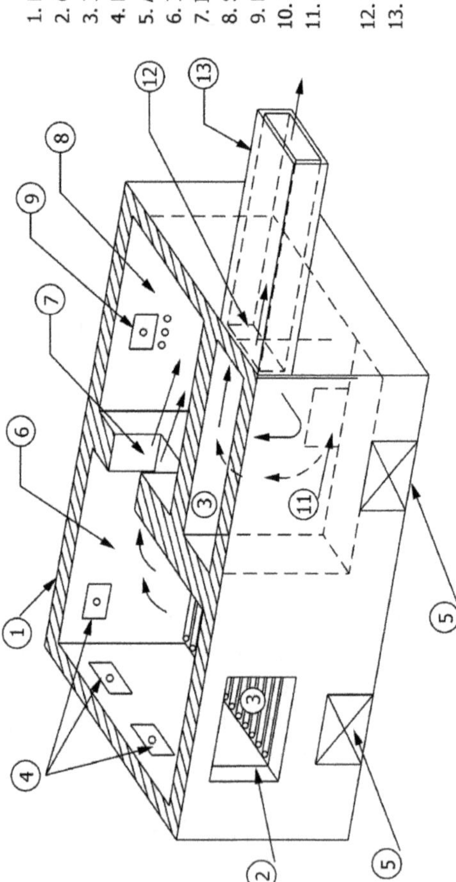

Figure 15.6 Small scale incinerator.

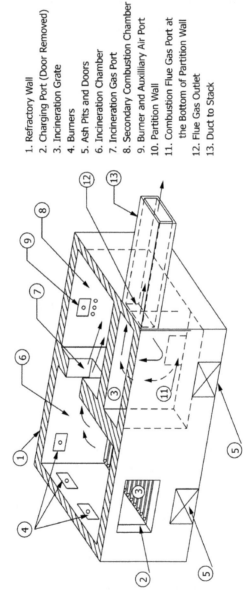

The volume is small (5–50 kg/day). The main requirements are a compact design, easy operation, and clean waste gas.

A typical incinerator of this type is shown in Figure 15.7. Outwardly the incinerator looks like a vertical-loading domestic washing machine. The incineration chamber is like a basket with a grill or grate at the bottom. A part of the wall is perforated for the gas to exit. There is a gas burner (or electric heater) for primary heating. The combustion gases pass through a secondary chamber that is the bottom part of the stock. An auxiliary burner located here completes the combustion. Air is drawn in through grills at the bottom.

Ash is collected in a drawer located below the grate.

15.4 HEAT EXCHANGERS

Heat exchangers are accessories used in heating/cooling processes to transfer waste heat from one fluid to other. Thus it involves two fluids, one hot and the other cool. Both fluids may be gases or liquids or one gas and the other liquid.

The hot fluid flows through a tube heating the tube surface, while the cold fluid flows over the tube and picks up heat from the hot fluid via the tube surface. Thus, the heat is exchanged between the fluids. It is also possible to achieve the same result by having the hot fluid flow over the tube while the cold fluid flows through.

Heat exchangers find wide applications in chemical and mechanical engineering. Almost all internal combustion engines (except small ones) require a separate heat exchanger (radiator) to cool the engine coolant. Similarly, air conditioning and heating require heat exchangers. Chemical industries use them to heat or cool process products and intermediates.

Combustion furnaces produce large amounts of flue gases which leave the furnace at a high temperature. If directly released into the atmosphere, considerable heat is wasted. If passed through an exchanger, at least some of the heat can be recovered by heating fresh air for combustion. It is also possible to obtain hot water or steam from flue gas heat exchangers

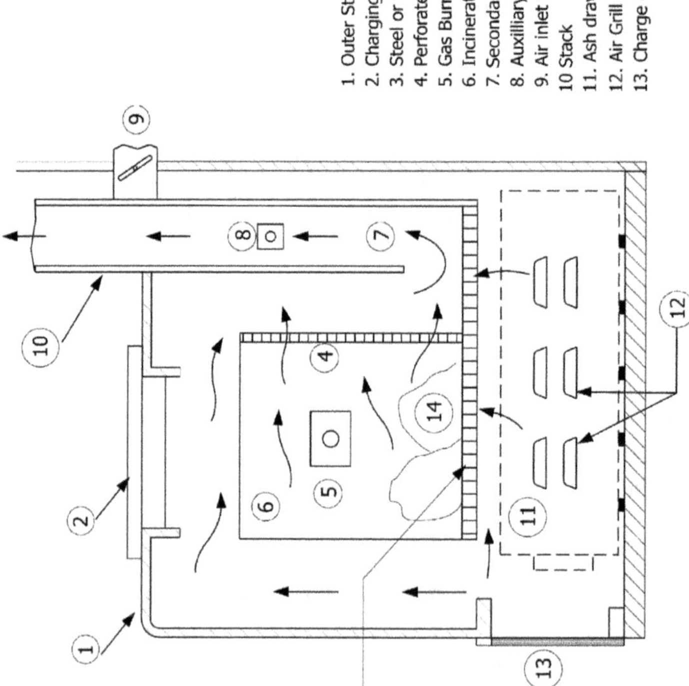

Figure 15.7 Domestic incinerator.

called "recuperators." Large induction heating plants use considerable quantities of water for cooling. The hot water is usually passed through an exchanger, cooled, and then reused. Hardening/quenching operations heat up the quenching media (oil or water). It is necessary to cool these in between quenching batches. Thus, there are many heating processes in which the use of a heat exchanger is necessary and economical.

Modern vacuum gas quenching furnaces heat the metal to the hardening temperature under vacuum. Quenching is achieved by introducing a cool gas (usually nitrogen) in the furnace. The gas picks up heat from the charge. A compact heat exchanger situated "inside" the furnace cools the gas and recirculates it over the charge. This is done to conserve the costly gas.

Another class of heat exchangers used in large furnaces is called "regenerators." These are operated in a cyclic fashion. For a certain period the flue gas is passed through them, so that the regenerator body (usually refractory) absorbs heat. In the next period, cold air is passed through the heated brick work to pick up heat. This cycle is repeated.

The furnaces and equipment discussed in this book are not likely to use regenerators. Hence, these are not discussed further.

15.4.1 Classification of Heat Exchangers

There are a vast number of heat exchanger designs and many schemes for their classification. As our interest in these devices is limited, we will adopt a classification based on the flow mode of the two heat exchanging media.

The simplest heat exchanger (Figure 15.8) is a tube in a tube heat exchanger. The hot fluid flows through the inner tube and the cold fluid through the outer tube. The temperatures are t_1 and t_2, $(t_1 > t_2)$. Heat transfer from the hot fluid to the cold takes place through the wall of the inner tube. The outer tube is insulated to prevent heat loss. In the figure, both fluids flow in the same direction, i.e., "parallel flow." The fluids can flow in opposite directions, i.e., "counter flow." The change in the temperatures of the fluids, as they pass through the

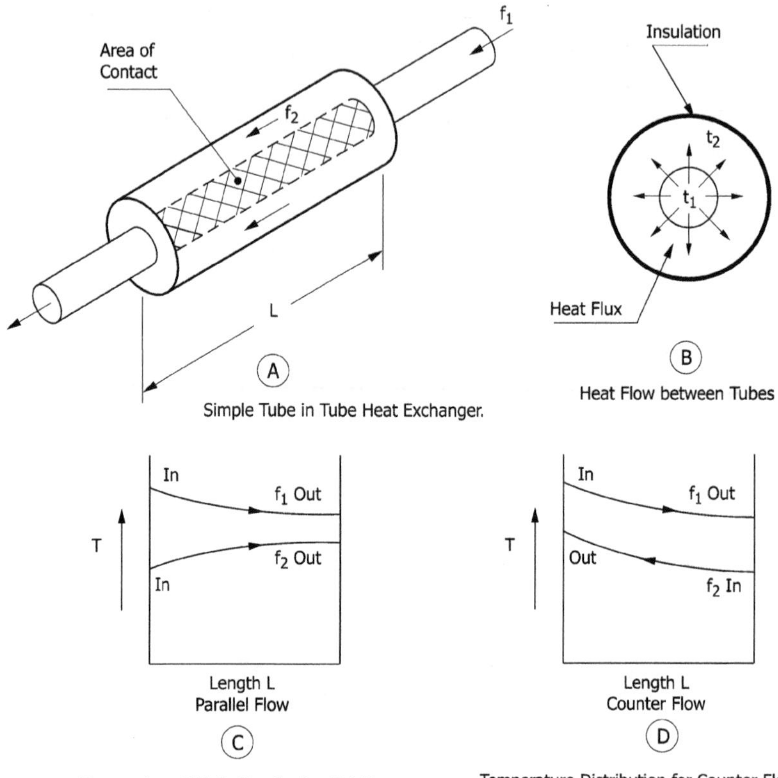

Figure 15.8 Simple heat exchanger.

exchange, are also shown in the figure. An important factor involved in the heat transfer between the fluids is the area of contact or length of contact. In a simple tube exchanger this area is very small and hence this type of exchanger has limited applications. The heat transfer coefficient in this and all other types of exchangers will depend upon the flow conditions, i.e., the Reynolds numbers of the two fluids. Usually the flow is laminar or transitional (Re 3×10^3 to 5×10^4). To increase the thermal contact surface, the hot fluid is given a number of passes through or inside the outer tube. This gives rise to shell and tube exchangers (Figure 15.9). Some shell and

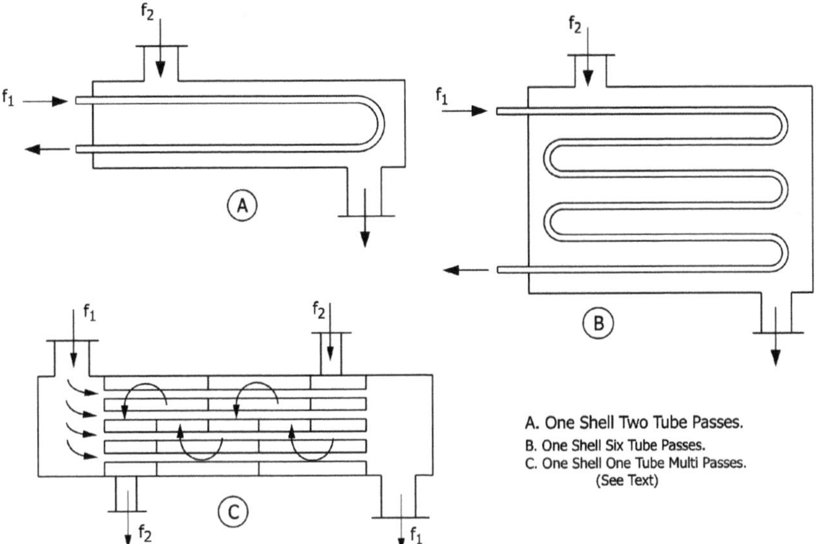

Figure 15.9 Some typical shell/tube heat exchangers.

tube exchangers are shown in the figure. The two fluids are called the "tube side" fluid (f_1) and the "shell side" fluid (f_2). Figure 15.9(A) shows a single-shell two pass-design while Figure 15.9(B) shows a single-shell six-pass exchanger. Figure 15.9(C) shows a single-pass shell, single-pass, multi-tube design. Here the fluid passes through the shell only once but there are four or more parallel tubes. Note the baffles which help to circulate the shell side fluid around the tubes. The temperature distribution of the fluids through the exchangers remains essentially similar to that shown in the figure. These types of exchangers are useful in recovering or dissipating heat from coolants or quenchants.

Besides parallel and counter-flow arrangements a third common scheme uses "cross" flow. Here one fluid (usually hot) flows through the tubes while the second fluid flows over the tubes at a right angle to the tube axis. A single tube cross-flow arrangement is shown in Figure 15.10. If the flow conditions of the outer fluid (f_2) are kept suitable (laminar or transient)

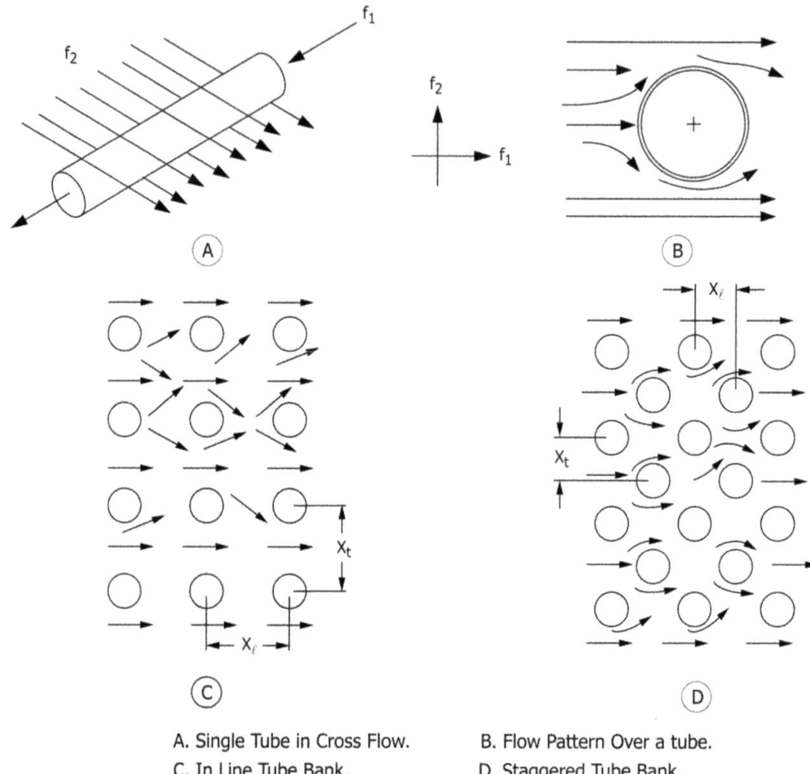

A. Single Tube in Cross Flow. B. Flow Pattern Over a tube.
C. In Line Tube Bank. D. Staggered Tube Bank.

Figure 15.10 Cross-flow heat exchangers.

heat is transferred via the tube wall by conduction and convection as the fluid-wall contact is good.

Single tube cross-flow heat exchangers are of little practical use as the surface area of the tube is very limited.

Practical designs of this type use tube banks as shown in Figure 15.10(C), Figure 15.10(D). Two types of tube arrangements are used. Figure 15.10(C) shows an "inline" arrangement while Figure 15.10(D) shows a "staggered" arrangement. In both types the tubes are placed at regular longitudinal and cross pitches x_ℓ and x_t. These pitches are related to the tube diameter d, such as $x_\ell = p \times d$ and $x_t = q \times d$. The tubes are welded to the header plates. These exchangers are extensively used for recovery of heat from waste gases. A typical cross-flow heat exchanger for recovery of heat (flue gas recuperator) is shown in Figure 15.11.

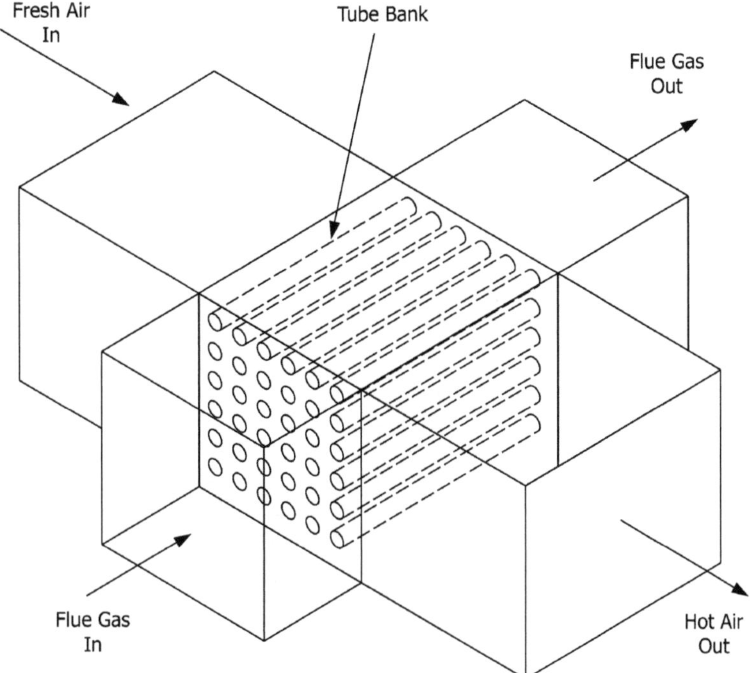

Figure 15.11 Typical cross-flow flue gas recuperator.

Heat exchangers of interest to us are constructed from welded steel. For temperatures higher than about 500°C, special heat-resistant alloys are used. For large furnaces and high temperatures the exchangers are constructed from refractory bricks and special shapes.

In some applications there are severe restrictions on the size and weight of the exchangers. Some special furnaces (e.g., vacuum and gas quench), all automotive applications, and small air conditioners have such requirements. In these cases "compact heat exchangers" are used. We have seen that the surface available for the exchange is a key factor in the exchanger design. The tube surface area in the exchange is increased by using a finned tube. The fins may be on the internal, external, or both surfaces as shown in Figure 15.12(A), Figure 15.12(B), Figure 15.12(C), and Figure 15.12(D).

To achieve further improvement in the heat transfer coefficient the construction is made from aluminium or copper.

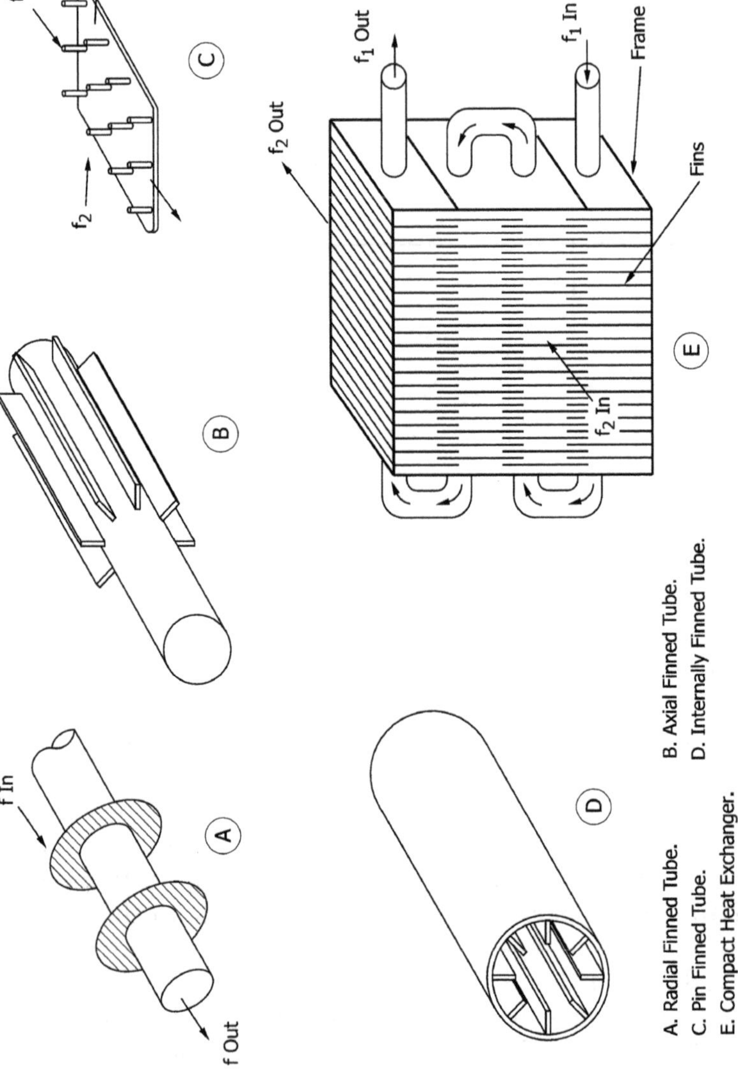

Figure 15.12 Compact heat exchangers.

Compact heat exchangers (Figure 15.12(D)) are used at low temperatures (< 400°C). The fluids are air, gas, and water.

Heat exchangers are costly and require regular maintenance as they are prone to corrosion, scaling, and clogging. They significantly increase the capital and running costs. They usually require pumps, compressors, or fan drives for fluid circulation.

Investment in heat exchangers is justified in following conditions:

1. When the investment is recovered by the saving in the cost of fuel.
2. They are mandatory by the environmental regulations.
3. Where irrespective of the costs involved, they are necessary (e.g., aviation, space, and military applications).

15.4.2 Convective Heat Transfer over Tube Banks

In the previous section we have seen how the fluid circulation and heat transfer takes place with a forced flow over a single tube. It was observed that the heat transfer coefficient depends on the circulation of the fluid around the tube, i.e., the wetted perimeter, which in turn depends on the Reynolds number Re_d, based on the tube diameter, and the arc of contact between the tube and the fluid, is small (under any Re_d). Hence, for practical purposes it can be concluded that a single tube is of very little use when large quantities of heat are required to be transferred from or to a tube.

In such situations a number of tubes are used so that the surface available for exchange is also quite large. The arrangements are called "tube banks." The tubes are arranged symmetrically in a regular pattern. Two arrangements are more common, the "inline" and the "staggered" as shown in Figure 15.13.

The tubes are arranged in rows at a certain pitch x. The pitch of tubes at right angles to the flow is called the transverse pitch x_t. The distance between the axes of tubes between

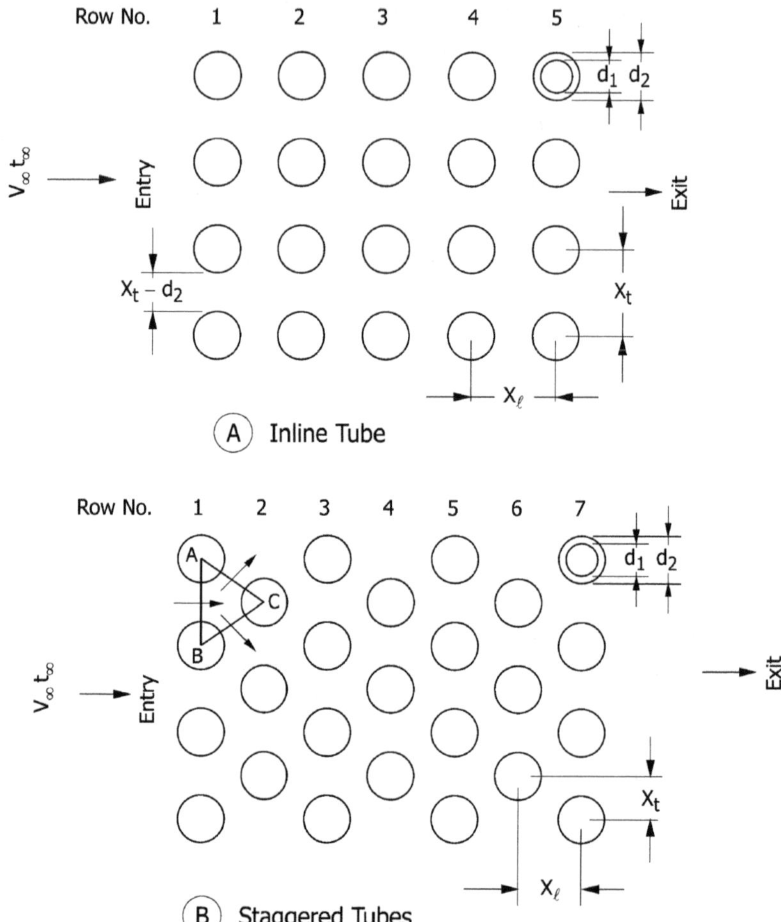

Figure 15.13 External convection over tube banks.

two adjacent rows is called the longitudinal pitch x_ℓ. Both the longitudinal and transverse pitches are related to the external diameter of the tubes d_2 so that $x_\ell = p \times d_2$ and $x_t = q \times d_2$, where p and q may be equal or different.

The Reynolds number Re_d is based on the velocity at the narrowest path in the bank. For "inline" arrangement, the narrowest path is $x_t - d_2$. For "staggered" arrangement, it depends on both pitches x_t and x_ℓ as can be seen from the triangle ABC in Figure 15.13(B). The narrowest path may be along AB or AC(BC). The velocity V is given by

$$V_\infty x_t = V(x_t - d_2)$$

and

$$\mathrm{Re}_d = \frac{Vd_s}{\upsilon}$$

where υ is the kinematic viscosity (m²/sec) taken at temperature at the entry to the bank.

The mean heat transfer coefficient is given by the correlations between Nusselt (Nu), Reynolds (Re_d), and Prandtl (Pr) numbers with a correlation factor ϵ.

For inline tube banks

$$\mathrm{Nu} = 0.26(\mathrm{Re}_d)^{0.65} \times (\mathrm{Pr})^{0.33} \times (\mathrm{Pr}/\mathrm{Pr}_w)^{0.25} \tag{15.1}$$

For staggered tube banks

$$\mathrm{Nu} = 0.41(\mathrm{Re}_d)^{0.6} \times (\mathrm{Pr})^{0.33} \times (\mathrm{Pr}/\mathrm{Pr}_w)^{0.25} \tag{15.2}$$

Both the correlations are applicable in the range $\mathrm{Re}_d = 10^3 - 2 \times 10^5$. There are correlations available for other Reynolds numbers but they are not of interest to us. Similarly, there may be small differences in the numerical constants between correlationships available from various sources. Pr_w refers to the Prandtl number at wall temperature. For air and gases the term ($\mathrm{Pr}/\mathrm{Pr}_w$) can be neglected. For rough estimates the correction factor may be taken as 1.2 to 1.5. The result obtained from the above correlations gives an accuracy of 10–15%.

The heat transfer coefficient obtained from the above correlations is the mean value applicable to the tubes after the second row. For the first two rows, the coefficient is 60 and 90%, respectively. This arises due to nonuniform flow over the first two rows.

Correlations and tables are available for various parameters and coefficients involved in the calculation of tube banks. As our purpose is only to estimate the probable heat transfer, these data are not given (see references to Chapter 3 and Chapter 4). These sources also give information on the calculation of pressure drop in the bank.

15.4.3 Heat Exchanger Calculations

There are many approaches to the design of heat exchangers. They are all based on three basic equations of heat balance of the exchanger.

Consider a simple tube in a tube exchanger, as shown earlier in Figure 15.8, involving an exchange between two fluids, f_1 (hot) and f_2 (cool).

Let t'_{f1} and t''_{f1} be the temperatures of fluid f_1 at the inlet and outlet of the exchanger.

Similarly, let t'_{f2} and t''_{f2} be the temperatures of fluid f_2. Let A be the area of contact (m²).

Heat lost by the fluid f_1 will be

$$Q = \rho_1 C_1 \dot{m}_1 (t''_1 - t'_1). \tag{15.3}$$

Assuming that the outer tube containing f_2 is well insulated, the heat taken up by f_2

$$Q = \rho_2 C_2 \dot{m}_2 (t''_2 - t'_2). \tag{15.4}$$

The heat transferred from f_1 to f_2 is Q and is given by

$$Q = hA(\Delta t). \tag{15.5}$$

Note that subscripts 1 and 2 denote the properties of fluid f_1 and f_2, respectively.

So that

$\dot{m}$ = Mass flow rate (kg/sec)

ρ = Density (kg/m³)

C = Specific heat (J/kg°C)

h = Heat transfer coefficient (w/m²°C)

λ = Thermal conductivity of wall (w/m°C)

δ = Tube wall thickness (m)

In Equation (15.5), there are three decisive terms: h, A, and Δt. Each term needs close inspection.

Heat will be transferred first from f_1 to the tube wall and the heat transfer coefficient will be h_1. Heat will then be conducted from the inner to the outer surface of the tube and will depend on the thermal conductivity λ of the wall. Finally the heat will be transfered from the wall to f_2 and this will take place with a heat transfer coefficient h_2.

Hence, the heat transfer coefficient h in the Equation (15.5) will by the *overall* heat transfer coefficient and will be given by

$$h = \frac{1}{\frac{1}{h_1} + \frac{\delta}{\lambda} + \frac{1}{h_2}} \text{ w/m}^2 {}^\circ\text{C} \tag{15.6}$$

When the fluids are passing through the exchangers, their temperatures are continuously changing. Hence, the problem arises as to which temperatures are to be used in the Equation 15.3 to Equation 15.5. We will have to use mean temperatures for the fluids. As the change in temperature between the inlet and outlet is not linear (see Figure 15.8(C), Figure 15.8(D) we can use an arithmetic mean for small temperature differences, but it is more appropriate to use "log mean temperature" (LMTD) given by

$$t^m = \frac{t_1 - t_2}{2.3\log\frac{t_1}{t_2}} \tag{15.7}$$

All the physical properties of both the fluids will be those at the mean temperatures.

The area of contact A can be easily determined from the geometry of the exchanger for simple shell tube exchangers. For finned surfaces, this area is different on each side of the wall.

When estimating the heat exchanger required for a process the required area of contact is unknown. Equation (15.5) is then used to determine the area (A). If the tube (inner) size is known, the length and number of tubes can be determined from mass flow rate m through the tube.

The heat transfer coefficients h_1 and h_2 are calculated by the methods discussed earlier in Chapter 3. Usually the transfer takes place by forced convection on both sides of the tube. If the temperatures are high (>600°C) the contribution of radiation to the heat transfer will be significant.

In detailed calculations for design there are some correction factors involved. These are not considered here as our purpose is only that of estimation.

The examples that follow will make the estimation process clear.

EXAMPLE 15.1

Design a heat exchanger for recovering heat from flue gas flowing at 2 m^3/sec at 450°C. The heat is to be recovered by using a cross-flow exchanger using a staggered steel tube bank. The external diameter of the tubes is 55 mm and the internal diameter 50 mm. The lateral and longitudinal spacing is equal and is 1.25 × external diameter.

The initial temperature of air is 25°C and the final desired temperature is 250°C. The volumetric flow rate of air is 2.5 m^3/sec.

Determine

1. The overall heat transfer coefficient
2. The length of tubes
3. Number of tubes
4. Distribution of tubes in rows

Data

Flue gas output	2m^3/sec
Flue gas density	1.3 m^3/kg
Temperature at gas intake (t_1)	450°C
Initial temperature of air (t_2)	25°C
Final temperature of air (t_2)	250°C
Volumetric flow rate of air	2.5 m^3/sec
Exchanger tubes ext/int diameter	55/50 mm
Arrangement of tubes	staggered
Lateral spacing = Longitudinal spacing	1.25d_2

Velocity of air over narrowest section of tube bank	6 m/sec
Mean velocity of the flue gas through tubes V_1	10 m/sec

Solution

Mass flow rate of flue gas $\dot{m}_1 = 1.3 \times 2 = 2.6$ kg/sec
Arithmetic mean temperature of air

$$t_2^{\,m} = \frac{250 + 25}{2} = 137.5°\text{C} \simeq 140°\text{C}$$

At this temperature, the properties of air (from the tables) are

Density $\rho_2 = 0.854$ kg/m^3, Sp. heat $C_2 = 1.013$ kJ/kg°C

Conductivity $\lambda_2 = 3.49 \times 10^{-2}$ w/m.°C, Viscosity $\upsilon_2 =$ 27.80 m^2/sec

Prandtl no. $\Pr_2 = 0.684$

Mass flow rate of air

$$\dot{m}_2 = \text{volume flow} \times \text{density}$$

$$= 2.5 \times 0.854$$

$$= 2.14 \text{ kg/sec}$$

Heat transferred from flue gas to air

$$Q = m_2 \times C_2 \times (t_2 - t_2)$$

$$= 2.14 \times 1.013 \times (250 - 25)$$

$$= 488 \text{ kW}$$

Assume the mean temperature of flue gas (t_1^{m}) is 350°C.
At 400°C, the specific heat of flue gas $C_1^{\text{m}} = 1.1356$ kJ/kg°C.

$$t_1'' = t_1' - \frac{Q}{m_1 C_1^{\dot{m}}}$$

$$= 450 - \frac{488 \times 10^3}{2.6 \times 1.1356 \times 10^3}$$

$$= 285°\text{C}$$

Mean flue gas temperature

$$= \frac{450 + 285}{2}$$

$$= 367°C$$

This is much higher than the assumed temperature (350°C).

Reiteration shows that at $t_1^m = 367°C$, the flue gas outlet temperature $t''_1 = 284°C$, and the mean temperature is 368°C. Hence, $t_1 = 450°C$, $t''_1 = 284°C$, $t_1^m = 368°C$.

At 368°C the properties of flue gas are

$\rho_1 = 0.555$ kg/m^3, $C_1 = 1.1356$ kJ/kg°C

$\lambda_1 = 5.14 \times 10^{-2}$ w/m°C, $Pr_1 = 0.64$, $\upsilon_1 = 55.30$ m^2/sec

Reynolds number of flue gas in tube

$$\text{Re}_1 = \frac{V_1 d_1}{\upsilon_1}$$

Inner cross-section of tube

$$A_1 = \frac{\pi d_1^{\,2}}{4} = \frac{\pi \times 0.05^2}{4} = 1.96 \times 10^{-2}\ \text{m}^2$$

Gas velocity in tube $V_1 = 10$ m/sec
Tube diameter $d_1 = 0.05$ m

$$\text{Re}_1 = \frac{10 \times 0.05}{55.3 \times 10^{-6}}$$

$$= 9.03 \times 10^3$$

Correlation for Nusselt number is

$$\text{Nu}_1 = 0.021(\text{Re}_1)^{0.8}(\text{Pr}_1)^{0.43}$$

$$= 0.021(9.03 \times 10^3)^{0.8} \times (0.64)^{0.43}$$

$$= 25.3$$

Heat transfer coefficient from flue gas to tube (h_1)

$$= \mathrm{Nu}_1 \frac{\lambda_1}{d_1} = \frac{25.3 \times 5.14 \times 10^{-2}}{0.05}$$

$$h_1 = 26\ w/m°C$$

Reynolds number of air flowing over tube bank is

$$\mathrm{Re}_2 = \frac{V_2 d_2}{\upsilon_2} = \frac{6 \times 0.055}{27.8 \times 10^{-6}}$$

$$= 1.19 \times 10^4$$

For flow over staggered tube bank, the correlation is

$$\mathrm{Nu}_2 = C \times (\mathrm{Re}_2)^{0.6} \times (\mathrm{Pr}_2)^{0.33}\ \epsilon$$

where ϵ is the correction factor.
For a tube bank with $x_t = x_\ell$, $\epsilon = 1$, and $C = 0.41$.
Hence,

$$\mathrm{Nu}_2 = 0.41 \times (1.19 \times 10^4)^{0.6} \times (0.684)^{0.33} = 101.$$

Heat transfer coefficient from tube surface to air is

$$h_2 = Nu_2 \frac{\lambda_2}{d_2} = \frac{101 \times 3.49 \times 10^{-2}}{0.055} = 64\ \mathrm{w/m°C}$$

Overall heat transfer coefficient for heat transfer,

Flue gas $\rightarrow$ tube $\rightarrow$ air is

$$h = \frac{1}{\frac{1}{h_1} + \frac{d_2 - d_1/2}{\lambda} + \frac{1}{h_2}}$$

λ is the conductivity of steel tube. It is about 40–50 w/m°C. Hence, the middle term is neglected.

$$h = \frac{1}{\frac{1}{h_1} + \frac{1}{h_2}} = \frac{1}{\frac{1}{26} + \frac{1}{101}}$$

$$= 20.7\ \mathrm{w/m°C}$$

Mean temperature of flue gas $t_1^m = 368°C$
Mean temperature of air $t_2^m = 140°C$
Mean temperature difference of exchange

$$\Delta t^m = 368 - 140 = 220°C$$

Heat transferred Q is

$$Q = hA\Delta t$$

(We will assume that the correction factor $\epsilon \sim 1.0$)

$A =$ Area of heat exchange surface (m^2)

$$A_1 = \frac{Q}{h\Delta t}$$

Substituting

$$A_1 = \frac{488 \times 10^3}{20.7 \times 220} = 107 \text{ m}^2$$

The mass flow rate of flue gas is $\dot{m}_1 = 2.6$ kg/sec

$$\dot{m}_1 = \rho_1 \times \frac{\pi d_1^2}{4} \times V_1 \times n$$

where n is the number of tubes.

$$n = \frac{4\,\dot{m}_1}{\rho_1 \pi d_1^2 V_1}$$

Substituting

$$n = \frac{4 \times 2.6}{0.055 \times \pi \times 0.05^2 \times 10} = 239$$

Adding 20% for uncertainties in heat transfer coefficient

$n \simeq 287$

The height (length) of one tube (single pass) can now be calculated as the total area is known.

$$A_1 = 2\pi d_1 \ell \cdot n$$

or $$\ell = \frac{A}{2\pi d_1 n} = \frac{107}{2 \times \pi \times 0.05 \times 287}$$

$$= 1.19 \simeq 1.2 \text{ m}$$

On the air side the mass flow rate m_2 and velocity V_2 is known. For the tube arrangement (staggered) and equal lateral and longitudinal spacing 1.25 d_2

$$\dot{m}_2 = \rho_2 V_2 A_2$$

where A_2 is the total narrow clear air passage available in a row

$$A_2 = \frac{\dot{m}_2}{\rho_2 V_2} = \frac{2.14}{0.854 \times 6} = 0.418$$

$$A_2 = \text{Inter tube gap} \times \text{tube length}$$

$$= (1.25 d_2 - d_2) \times \ell \times n_c, d_2 = 0.055 \text{ m}, \ \ell = 1.2 \text{ m}$$

where n_c is the number of tubes in one row (across the flow).

$$n_c = \frac{0.418}{(1.25 - 1.0) \times 0.055 \times 1.2}$$

$$= 25.3 \simeq 25$$

Total number of tubes is $n = 287$
Number of tubes placed longitudinally (along the flow)

$$n_\ell = \frac{n}{n_c} = \frac{287}{25}$$

$$\simeq 12$$

Hence the exchanger will have a matrix of 25 × 12 tubes.

15.5 DRYING OVENS

Drying, i.e., removal of moisture (complete or partial) from substances, is an important unit operation in a large number of processing industries. It is used in agriculture, raw vegetable preservation, processed food, paper, textile, dairy, painting, and many other processes.

There are a number of drying methods available such as freeze drying, solar drying, spray drying, and heat drying. Consequently, a large variety of dryers is available in the market. The method and machinery for drying will have to be carefully chosen to suit the raw material, drying conditions, and desired quality and quantity of the product.

In this book we will only consider drying by heating. For a detailed treatment of all drying treatments, refer to special literature.

Consider a solid containing moisture or water at a temperature T_2 (Figure 15.14 (A)). Hot air at temperature T_1 is flowing over the interface or boundary XY. As $T_1 > T_2$ the heat will flow from air to the interface at which heat transfer will take place by convection. From the interface the heat will flow in the solid by conduction as shown. Due to this heat, the moisture on the interface will evaporate and the vapor will be carried away by air. Moisture from the interior will then move by diffusion toward the interface. Thus, there will be a mass transfer flow set up from solid to air through the interface. This will be because of the concentration gradient between the solid (C_2) and air (C_1). As time passes, the moisture in the solid will decrease. Thus in drying, there are two opposite transfer processes taking place simultaneously. Heat will be transferred from air to solid and moisture (mass) will be transferred in the opposite direction. The typical progress of drying with time is shown in Figure 15.14(B). Water in the solid may be free, loosely bonded, or chemically bonded. Loose and free moisture will be readily removed. Chemically bonded water is difficult or sometimes impossible (uneconomical) to remove.

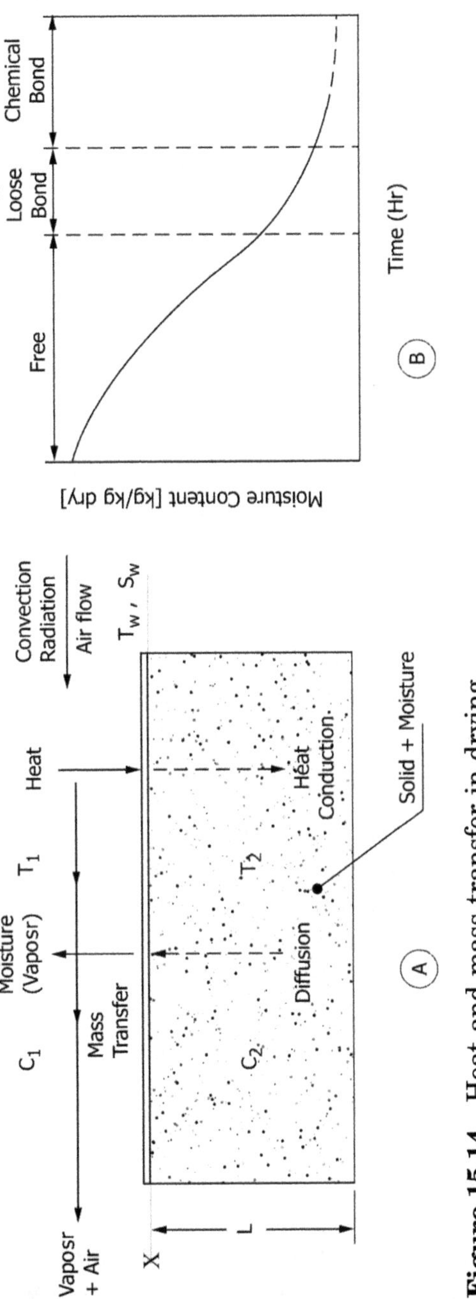

Figure 15.14 Heat and mass transfer in drying.

For heat transfer, the general equations are:

$$q = h_c(T_1 - T_w)\frac{\text{W}}{\text{m}^2\,{}^\circ\text{C}} \text{ from air to interface} \tag{15.8}$$

$$q = \lambda\frac{(T_w - T_2)}{L}\frac{\text{W}}{\text{m}^2\,{}^\circ\text{C}} \text{ from interface to interior} \tag{15.9}$$

Similarly, for mass transfer

$$j = -D\frac{(C_w - C_2)}{L} \text{ from interior to interface} \tag{15.10}$$

$$j = \beta(C_w - C_1) \text{ from interface to air} \tag{15.11}$$

Equation (15.8) and Equation (15.9) are already introduced in Chapter 3. Equation (15.10) and Equation (15.11) are mass transfer equations which are similar to Equation (15.8) and Equation (15.9). Here,

D = Diffusion coefficient m²/sec

β = Mass transfer coefficient m/sec

C_1, C_2, C_w = Moisture concentrations in air, solid, and interface (kg/m³)

j = Mass flux kg/sec

What was discussed earlier is a highly simplified version of mass transfer in a typical drying operation. Mass transfer is a separate special topic of interest to chemical engines. Our interest in this book is confined to heating design, and the discussion above should be sufficient.

The process information we need is drying curves as shown in Figure 15.14(B), established at various air temperatures. This will enable us to decide the time and the temperature required. As a rule, the highest possible temperature should be chosen without damage to the quality of the dried product. There are two reasons.

First, the moisture carrying capacity of air increases rapidly with temperature (see Appendix D). Secondly, the heat transfer to the interior and the interface of the solid will also increase with temperature. It is likely that at some air temperature, the interface will acquire a constant temperature due to the absorption of latent heat. The output desired (Kg/h at $x\%$ moisture) will lead to the choice and size of heating machinery.

If the basic process information is not available, it can be investigated by a simple laboratory procedure. Samples of a known surface area and weight are dried in an oven for a range of temperatures and time. Assuming that only simple drying takes place (i.e., without any chemical reaction) the moisture removed can be measured by loss of weight.

A commercial batch-type drying oven is shown in Figure 15.15. The wet objects are loaded in trays which may or may not be perforated. The trays are stacked on rocks. Hot air is passed over the trays. The air is heated by passing it over heaters arranged as tube bundles. Air supply is obtained by steam, electric heaters, or radiant tubes.

After passing over the trays, the air acquires the evaporated moisture but has a lot of sensible heat. In some of the exhaust, the gases may be recirculated by mixing them with fresh intake air. The rest is released into the atmosphere through a stack. The exhaust stack and intake air should be sufficiently apart so that this fresh air is relatively dry and cool. The proportion of recirculation is adjusted by using dampers.

The operation will benefit by using dry air. However, large scale drying of intake air is not economical. The same design can be modified for continuous operation by providing entry and exit doors at positions X and Y, respectively (Figure 15.15).

Another type of drier known as drum drier is shown in Figure 15.16. Here the raw material is a liquid or paste, or dough. It is dried by passing over a pair of heated rollers. These driers are used for milk, purees, mashes, etc. Note that the product is heated from the bottom while moisture is evolved at the top surface.

15.6 BAKING OVENS

Baking is defined as heating to a low temperature in an enclosure. Traditionally, "baking" refers to cooking certain food items such as bread and cakes in an oven. Technically, baking

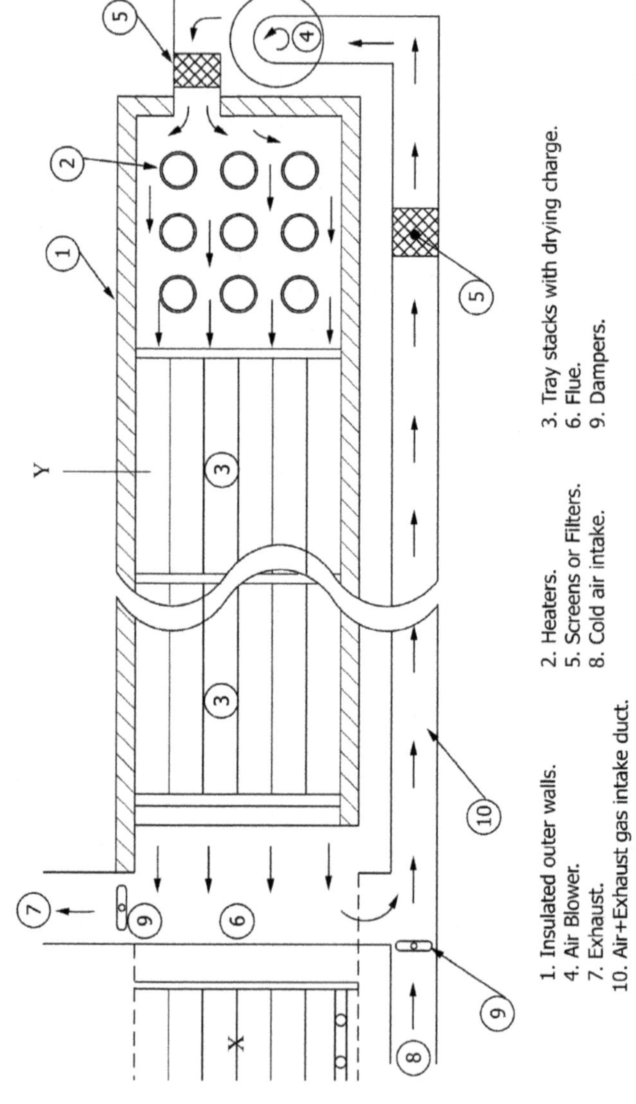

Figure 15.15 Batch Type Drying Oven.

1, 2. Drums on Rollers.
3. Feed Hopper.
4. Gate.
5. Enclosure with exhaust or vacuum.
6. Knife edge Scrapers.
7. Steam or Electric Heaters.
8. Dried Product.
9. Product Collection.
10. Waste Collection for recirculation.

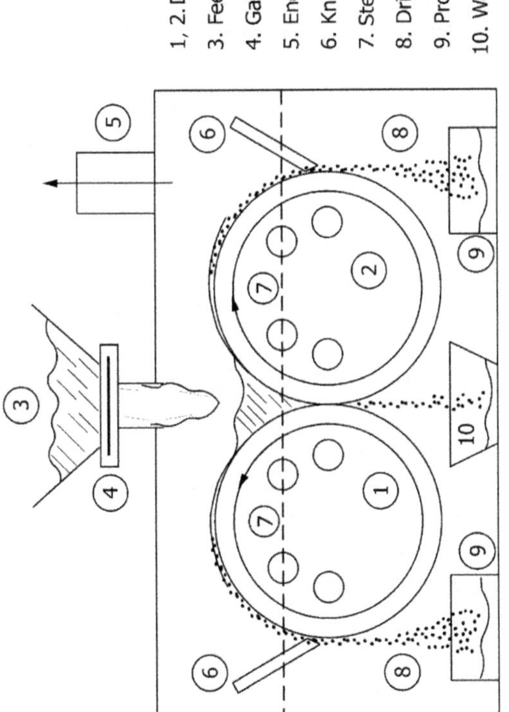

Figure 15.16 Drum drier.

is used in a large number of nonfood related processes such as foundry (core baking), adhesive joining (curing), surface coating (vitreous enameling), paint drying and so on. Bread making requires baking at about 200–250°C. Core baking (CO_2) is done at a similar temperature range. Porcelain enamel baking requires 800–850°C. Baking temperatures for painted articles are in the range 120–180°C depending upon the type of paint.

In the baking of bread and similar items, a mixture of CO_2 and steam is evolved which is harmless. In some processes such as paint drying or adhesive curing, solvents and other gases are evolved which are likely to be harmful and combustible. Consequently, in industrial ovens, arrangements to safely dispose of the evolved gases is required.

As the temperatures in baking are low (except enamelling) a large number of heat sources are available to the designer. Paint drying can be done by using high intensity or infrared bulbs. Baking can be done with wood, electricity, or indirect heating by gas or oil. Radiant tubes with gas or steam are also suitable for many applications. Microwave heating is already discussed in Chapter 10.

Due to the diversity of processes and heating techniques, a large variety of ovens is available. Some designs are shown in Figure 15.17.

A simple oven using wood or coal is shown in Figure 15.17 (A). It is built of ordinary brick masonary and consists of two chambers. The lower chamber is for burning the fuel. It may have a grate and an ash pit in the bottom. The upper chamber is used for baking. The hot combustion gases rise above and circulate around the baking chamber before exhausting at the top. There may or may not be a door to the baking chamber. Heat transfer is by conduction through the walls and convection around the chamber. Because of the brick work construction, the oven holds a lot of heat. The same design with a few modifications can be adapted for oil burning by installing a burner in the lower chamber.

A cabinet-type oven is shown in Figure 15.17(B). The articles are kept in trays which are stacked on a rollout rack. The oven may operate on an electrical heater of sealed radiant

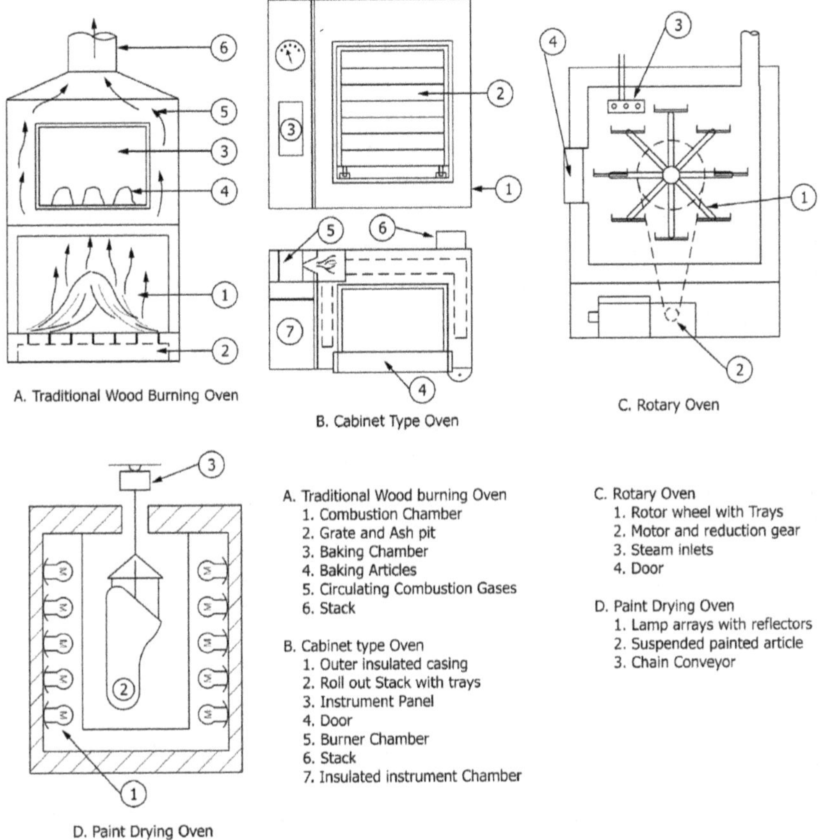

Figure 15.17 Examples of typical baking ovens.

tubes or gaseous fuel (natural gas). In both cases, the heaters or burners may be installed inside the cabinet. If oil burning is used, the burner is located outside the cabinet and combustion gases are circulated around the cabinet on the outer side. Thus heating is by convection .This avoids exposure of articles to combustion gases which may give an objectionable odor.

Food baking produces moisture, air, or carbon dioxide. Many times, additional steam is injected to improve the crust of the baked item. There is a separate exhaust for cabinet gases. Many baking ovens for food have a rotating tray arrangements as shown in Figure 15.17 (C). This arrangement gives a better quality. The temperature and rotational speed is so adjusted that baking is completed in one revolution.

Baking ovens for high volume production such as breads, cakes, and biscuits are of the continuous type. They use a moving metal band or woven mesh through the oven. The general construction is similar to a continuous furnace discussed in Section 15.2.4 (Figure 15.4). The muffle and belts are made of stainless steel or monel (Ni-Cu) alloys.

A paint drying oven is shown in Figure 15.17. It is heated by arrays of lamps with reflectors. Tungsten, carbon filament, or quartz lamps are used. The article to be dried is suspended by hooks that are moved through the oven by an overhead chain. The lamps have a limited life. The glass bulbs and reflectors are tarnished or blackened by the solvent vapors. However, the construction and maintenance is easy. Similar designs are used for enameling, with heating accomplished by radiant tubes. In both the cases, it is necessary to dispose of the solvent fumes through an exhaust fan and stack.

15.7 FANS

We use fans for agitating or moving air or gases. In furnaces or ovens, they serve as circulators, moving the atmosphere in the enclosure. In heat exchangers or drying there are used as a heat transfer medium. In some applications, ejectors or exhaust fans help remove the confined or slow moving gases. In all such applications it is necessary to choose the correct fan for satisfactory and economical operation. In this section we will take a limited review of fans suitable in heating applications. For more information consult special books or literature by manufacturers.

There are two basic types of fans. The classification is based on the flow of air through the fan. In "axial flow" fans (Figure 15.18(A)), the air enters and leaves along the axis or the shaft. The fans are made as propellers with a number of blades (usually 2–6) attached symmetrically around a hub fixed to a shaft. In passing through the blades, the air, gas, or steam acquires a velocity and pressure head.

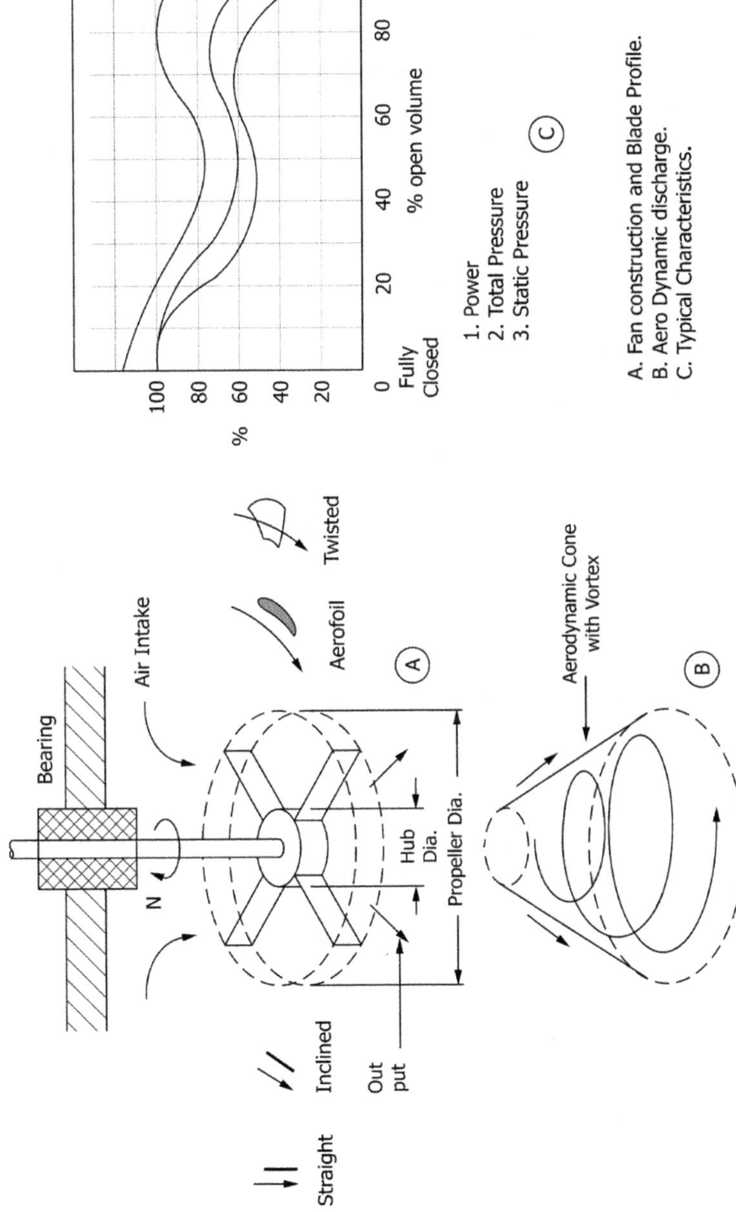

Figure 15.18 Axial or propeller fan.

Radial flow or "centrifugal" fans (Figure 15.19(A)) have a rotating cage or scroll with blades or slots at the outer periphery. The blades may be straight or inclined in, or opposite to, the direction of rotation. Air is admitted at the center. It flows toward the beds under centrifugal force and exits through the slots. The cage rotates in a volute housing which collects the existing air and discharges through a tangential outlet.

For circulation in confined places such as furnace enclosures, axial fans are used. Here circulation and velocity are more important than the volume. These fans help to homogenize the atmosphere and assist natural convection.

In applications such as waste heat exchangers, cooling beds, or drying, the volume required to be moved is large and the pressure requirements are also applicable. Here we use centrifugal fans.

The main performance requirements are volume, pressure, and velocity of discharge. One more criteria of considerable interest to our purpose is the aerodynamic shape of the discharge. This defines the extent or spread of the discharge stream. This shape is very important in the choice of an axial fan to be used as a circulator. The aerodynamic cone shape depends on the profile of the impeller blades (Figure 15.18(B)). The mechanical efficiency of fans is not very important to us as the fans usually consume only a small fraction of the total power of the furnace.

Fans of all types display typical characteristics and are governed by the "fan laws." These characteristics are shown by the manufacturer on performance curves. They show the behavior of pressure, power, and efficiency against percent rated delivery (m). If the fan is delivering to an open atmosphere without any restriction, the air delivered is 100% rated. If the fan outlet is closed the delivery is 0%. The air delivered at any delivery has a pressure (static) p and a velocity v. We have seen in Chapter 2 that the velocity can also be converted to pressure (Bernoulli's principle), hence the total pressure of a fan at output is

$$\text{Static pressure} + \text{velocity pressure} = p + \frac{\rho V^2}{2} \quad (\text{Pa})$$

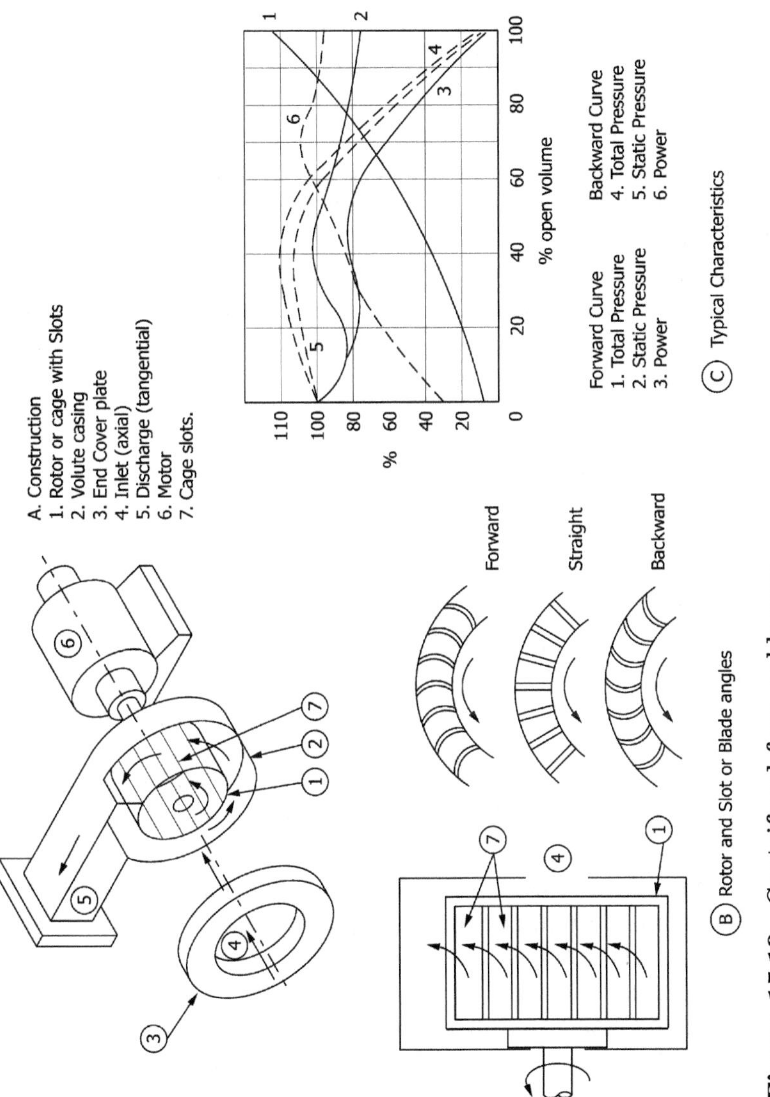
A. Construction
1. Rotor or cage with Slots
2. Volute casing
3. End Cover plate
4. Inlet (axial)
5. Discharge (tangential)
6. Motor
7. Cage slots.
Forward
Straight
Backward
B Rotor and Slot or Blade angles
%
110
100
80
60
40
20
0
20
40
60
80
100
% open volume
Forward Curve
1. Total Pressure
2. Static Pressure
3. Power
Backward Curve
4. Total Pressure
5. Static Pressure
6. Power
C Typical Characteristics

Figure 15.19 Centrifugal fans or blowers.

where ρ is the air density. The fan characteristics usually show total and static pressures. The volume handled Q (m^3) is measured at the inlet and velocity is measured at the outlet by dividing Q by the outlet area.

The typical characteristics of axial and centrifugal fans are shown in Figure 15.18(C), Figure 15.19(C). Note that these curves are not standard and will show some variation with the manufacturer's design. An important fact displayed by all fans is that they can be used over a wide range of delivery and pressure. We will have to carefully choose the desired combination.

Manufacturers usually produce a range of fans with various sizes. It will be worth noting the influence of some parameters on the characteristics. The effects are roughly the same for both types of fans.

Effect of Rotational Speed (N rpm)

Increasing the design speed will increase the swept volume and hence, the discharge Q. The total pressure and power will also increase (power $\propto N^3$, pressure $\propto N^2$).

Change in fan size (wheel or propeller diameter) D will also increase discharge Q, pressure ($P \propto N^2$), and power (power $\propto N^5$). This assumes the speed to be same, i.e., design speed.

For an axial fan to be used in an enclosure, the size of the latter will limit the maximum wheel diameter that can be accommodated. In the case of a centrifugal fan, the fan's main requirements are delivery and enough pressure to overcome ducting resistance.

Axial fans to be used at high temperatures (> 500 as in heat treatment furnaces) must be constructed from heat resistant materials such as inconel. They are usually operated at slow speeds as their main purpose is atmosphere circulation. Similar fans used in food or other ovens, can be fabricated from stainless steels. Axial fans for high temperatures are not standardized and will have to be designed by considering creep and corrosion. Their drive motor and gear are located outside, and the entry of the shaft and its bearings have to be sealed and cooled.

Centrifugal fans usually operate at ambient temperature and are available in standard size.

15.8 SOME NEW MATERIALS

15.8.1 Carbon Foams

Foams made from basic materials like pitch or tar have recently been made available in both graphite (crystalline) or carbon (amorphous) forms. Presently, carbon foams are available at a competitive price. Typical properties of these foams are:

Density	0.27 (2.2) gm/cm
Thermal conductivity	0.25–25
Electrical resistivity	—
Maximum temperature	650 air, 3000°C inert
Tensile strength	> 200 N/mm^2

The foams are available in several grades of strength and density. Presently, they are being explored for applications requiring adequate strength with low weight. Their heat resisting properties should find them suitable for high temperature applications. They can be easily machined and coated or sandwiched with metallic or ceramic materials.

Incidentally, metal foams are also available. These foams the characteristic metal properties such as electrical and thermal conductivity but have low densities. Presently, aluminium and stainless steel foams are marketed.

15.8.2 Alumina Refractory Adhesive

Ceramic fiber materials are already discussed in Chapter 7. Boards and other shapes of these materials are extensively used in the construction of furnaces for walls and roofs. The boards are anchored to each other or the furnace structure by using metallic fasteners such as hooks, staples, or wires. However, these fasteners are costly as they are made from heat resistant alloys. They also act as thermal short circuits by providing a highly conducting path in an insulating matrix.

Recently, an alumina adhesive has been made available. This is a single-part water-based adhesive. It can be used for bonding low density (fiber) insulating materials. Thus, metallic anchors can be eliminated from wall or roof constructions.

15.8.3 Cast Basalt

Basalt is an igneous (volcanic origin) rock commonly occurring all over the world. It has a greenish-gray color. Its main constituents are silicates of Ca, Mg, Fe, Al (feldspars). Some varieties of basalt can be melted and cast into a number of shapes such as tiles and pipes. This material is very hard and wear-resistant. It can withstand temperatures up to 1000–1200°C and is corrosion-resistant. It is extensively used in the lining of exhaust and flue ducts of fuel burning furnaces. They resist abrasion by ash particles in the gases. The material is also available as ready-mix concrete and can be used for masonary work of ducts.

Appendix A

Pressure

Pressure is one of the variables for describing the "state" of a gas or liquid. Another state variable is temperature. We will often come across pressure and its units. Following is a short review for refreshing the concept.

In the SI system of units adopted in this book, the unit of pressure is pascal (Pa) which has dimensions N/m^2, as pressure is defined as force per unit area. One pascal is a very small unit and we will be using its multiple as kPa (10^3 Pa) and MPa (10^6 Pa), etc.

We also use atmospheric pressure as a unit. One atmosphere (Atm) has an approximate value of 10^5 Pa and is called 1 "bar."

The simplest device for measuring the gas pressure is the U tube manometer or piezometer. The U tube is partially filled with some liquid such as mercury or water, as shown in Figure A.1(A).

If $P_{ves} > P_{atm}$ (Figure A-1(A)),

$P_{ves} - P_{atm} = \rho_{gh1}$ or,

$$P_{ves} = P_{atm} + \rho_{gh} \quad \text{(A.1)}$$

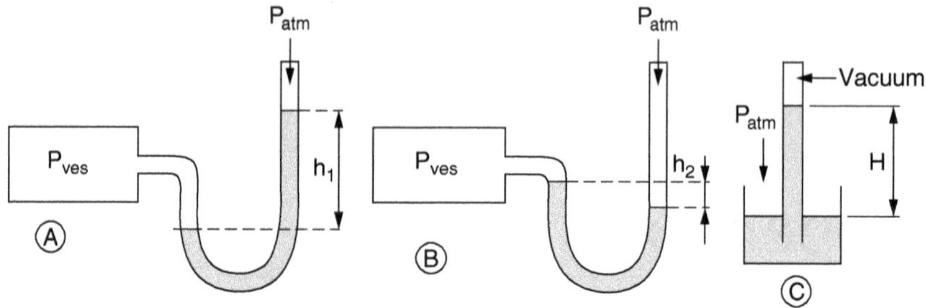

Figure A-1 U Tube manometer.

If $P_{atm} > P_{ves}$ (Figure A-1(B)),

$$P_{atm} - P_{ves} = \rho_{gh2}$$

or $$P_{ves} = P_{atm} - \rho_{gh} \qquad \text{(A-2)}$$

where ρ is the density of the liquid in the U tube.

The atmospheric pressure can also be measured in terms of liquid column as shown in (Figure A-1(C)).

where $$P_{atm} = \rho g H \qquad \text{(A-3)}$$

Thus, pressures (both positive and vacuum) can be measured and expressed in terms of the height of a liquid column.

The heights of a water column and a mercury column at 1 atmosphere pressure are 10.33 and 0.76 m, respectively.

Note that due to the density difference, the water column is much higher than the mercury column. Hence, the height of the mercury column expressed in mm is a very convenient unit for expressing low pressure.

$$1 \text{ atm} = 1 \text{ bar} = 760 \text{ mm mercury} = 10^5 \text{ Pa} \qquad \text{(A-4)}$$

or

$$1 \text{ mm Hg} = 133.32 \text{ Pa} = 0.133 \text{ kPa approx.} \qquad \text{(A-5)}$$

The height of the mercury column is a small unit and is used to express pressures which are slightly above atmospheric pressure. It is also useful for expressing small negative pressure or vacuum.

Equation (A-4) and Equation (A-5) show that the height of the liquid column is measured in comparison with the atmospheric pressure. All pressure gauges indicate pressure this way. The indicated pressure is therefore called the "gauge pressure." For absolute pressure, the atmospheric pressure must be added to the gauge pressure. Thus

$$P_{\text{abs}} = P_{\text{gauge}} + P_{\text{atm}} \tag{A-6}$$

The unit frequently used for expressing vacuum is called a torr.

$$1 \text{ torr} = 1 \text{ mm Hg} = 133.32 \text{ Pa} \tag{A-7}$$

Vacuum pressure is often expressed in millibars (m bar) so that

$$1.0 \text{ m bar} \simeq 10^2 \text{ Pa} \tag{A-8}$$

Appendix B

Viscosity

The ability of a fluid (gas or liquid) to flow under stress or pressure is represented by viscosity. High viscosity fluids flow with difficulty; low viscosity fluids (water, kerosene, and the like) flow easily.

Consider a liquid resting on a surface, as shown in Figure B-1. If a force is applied to the top surface a shear stress q will be acting on the cross-section and it will move with a velocity u. The internal layers will move with successively decreasing velocity. The layer near the bottom surface will have zero velocity, i.e., a pressure gradient du/dn will be set up in the fluid due to the friction between layers. Newton's law of viscosity states that the shear stress q will be proportional to the velocity gradient, i.e.,

$$q = \mu \frac{du}{dn} \tag{B-1}$$

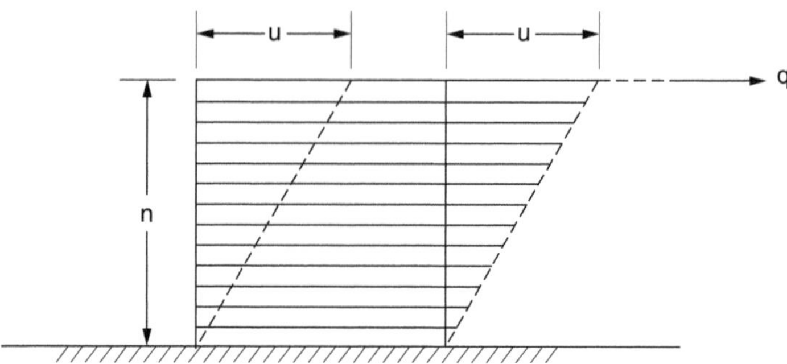

Figure B-1 Shear on a liquid, and concept of viscosity.

where μ is the constant of proportionality and is called the dynamic coefficient of viscosity, Equation (B-1) gives that

$$\mu = \frac{q}{(du/dn)} \frac{N \cdot s}{m^2} \tag{B-2}$$

In practical application we use the kinematic viscosity more frequently. The coefficient of kinematic viscosity (υ) is defined as

$$\upsilon = \frac{\mu}{\rho} \text{ m}^2/\text{sec} \tag{B-3}$$

where ρ is the density of the fluid (kg/m^3).

Common experience that liquids become less viscous (or more fluid) with increasing temperature suggests that the viscosity, and therefore the coefficients, of dynamic and kinematic viscosity are strongly temperature dependent. Hence, the coefficients must be reported along with the applicable temperature. For gases, the viscosity increases with temperature.

Numerically, the coefficients are very small in absolute units, e.g., at 15°C, the kinematic coefficients for some fluids are:

Water	0.00000114
Kerosene	0.0000025
Diesel	0.000005
Medium furnace oil	0.00005
Air (dry)	0.000018

For convenience the coefficients are generally reported as multiples of 10^6.

Measurement of viscosity in absolute units is very time-consuming. For practical purposes it is determined by the time taken for a standard sample to empty from the standardized hole in the bottom of a standard cup. There are various methods used in different countries and manufacturers. We come across Redwood number (R sec), Engler (°E), and Seybolt standard (sec), etc. Conversion tables for these to eachother, and to absolute units are available.

Practical absolute units are:

poise for dynamic viscosity = 1 g/sec.cm

stoke (or centistoke cST) for kinematic viscosity, where

$$1\ \text{m}^2/\text{sec} = 10^4\ \text{stokes} = 10^6\ \text{centistoke} = 10^6\ \text{mm}^2/\text{sec}$$

In furnace design we have to deal with flow of gases at high temperature, flow of oils through pipes, atomization of oils, and many other problems of heat transfer where the kinematic viscosity is of importance. We will follow the m^2/sec or its multiples in stokes as units for kinematic viscosity. The dynamic viscosity will not appear in our discussions.

Appendix C

Thermal Diffusivity

In many heat transfer problems (especially in transient conduction) we come across "thermal diffusivity" (α) as a coefficient in the equations. It is defined as

$$= \frac{\text{Thermal conductivity}}{\text{Density} \times \text{Specific heat}} \tag{C-1}$$

$$= \frac{\lambda}{\rho \times C} \tag{C-2}$$

Its dimension will be

$$= \frac{J}{\text{s} \cdot \text{m} \cdot {}^\circ\text{C}} \Big/ \frac{\text{kg}}{\text{m}^3} \times \frac{J}{\text{kg}^\circ\text{C}} = \frac{\text{m}^2}{s} \times (J) \tag{C-3}$$

Thermal conductivity λ gives the heat that will be conducted per second through the material (J). The product $\rho \times C$ gives the heat that will be stored in a unit volume of the

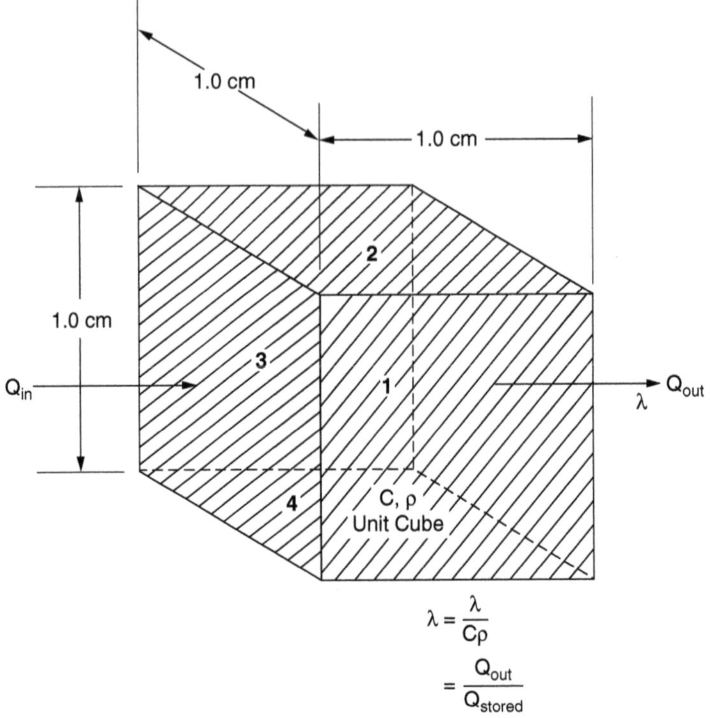

Figure C-1 Thermal diffusivity.

material. If we assume that λ, ρ, and C are constant, the thermal diffusivity will represent the ratio

$$= \frac{\text{Conducted heat}}{\text{Stored heat}}$$

It thus shows the quantity of heat (J), which flows in unit time (s) through unit area (m^2) of the substance of unit thickness and unit difference of temperature between its faces (Figure C-I).

The concept of thermal diffusivity makes it easy to compare the thermal performance of materials. Diffusivities of some selected materials are given in the Table C-1.

Note that except for good conductors such as silver, copper, aluminum, etc., the diffusivity is near or a little more than unity. For these materials heat conduction and accumulation

TABLE C-1 Thermal Properties of Selected Materials

Material	Density r kg/m3	Specific heat C kJ/kg °C	Thermal Conductivity l w/m °C	Thermal Diffusivity α cm²/sec
Silver	10,520	0.252	400	1.5
Copper	8,350	0.42	350	0.99
Aluminum	2,800	0.890	200	0.80
Brass	8,400	0.378	114	0.36
Lead	11,350	0.129	30	0.205
Mild Steel (0.2% C)	7,840	0.465	40.2	0.110
Stainless Steel	7,865	0.460	20.1	0.055
Cast iron 7274	0.42	50.5	0.165	—
Mercury	13,200	0.1351	11.7	0.065
Air 0.456	1.093	5.74×10^{-2}	115×10^{-2}	—
Glass 2200	0.67	1.1	0.007	—
High Silica Bk.1800	0.88	1.08	0.0057	—
High Alumina Bk2700	0.84	1.7	0.0075	—
Insulating Br550	0.83	0.113	0.005	—
Ceramic Fiber150-	0.05-	0.05	0.002	—
Water	965.2	4.195	68.4	—
Quenching Oil	860	1.66	0.112	7.85×10^{-2}

are almost equal. All other materials have diffusivity in the range 0.05–0.4. These will accumulate heat and do not conduct much, i.e., they are bad conductors.

Insulating materials and refractors have high density and specific heat, hence their diffusivity is of the order of 10^{-3}. The same materials in fiber form have still lower diffusivity because of the low density.

Finally, air (and other gases) has the lowest diffusivity ($\sim 10^{-6}$).

In problems involving transient conduction, the solution often contains a term involving $\sqrt{\alpha t}$. As the dimension of α is L^2/s, the term represents a linear dimension. This length, say x, is the depth to which heat has penetrated in time t. The depth is to be considered inside the body from its surface (which is heated or cooled). Alternatively, if we know the distance x we can approximately estimate the time it will take to get a certain temperature at the point.

A typical equation that we come across in time-dependent heating (or cooling) of a semi-infinite body is

$$\tau = \tau_o \operatorname{erf}\left[\frac{x}{2\sqrt{\alpha t}}\right]$$

where τ_0 is the initial temperature, and τ after time t at a point in the body at a distance x from the surface. The term $x/2\sqrt{\alpha t}$ is dimensionles. Thus, by choosing α, the temperature at any point x in different substances (bodies) can be compared. We can also choose a ratio $\rightarrow T/T_o$; the rate of heating or cooling at x can be estimated. Note that the properties such as λ, C, and ρ are assumed constant, which is not strictly true.

It is interesting to note that two more properties that we encounter have the same dimensions (m^2/s). First is the kinematics viscosity and the other is the diffusion coefficient D. Grouping these, we get a dimensionless number. For example, Lewis number, Le = α/D is used in mass transfer analysis.

Appendix D

Humidity

Air and other gases always contain moisture. This moisture plays an important role in processing. Air drawn from the atmosphere for combustion processes contains considerable moisture (H_2O or steam). This moisture does not play any part in the combustion process but incurs a lot of wastage of heat as it is heated to combustion temperature and discharged to the atmosphere. The moisture in the air cannot be economically removed. Its amount or proportion changes with the season.

In drying operations moist air, passed over a wet charge, limits its capacity of water removal.

Water may also form during the combustion of hydrogen-containing fuel, as the product of oxidation. This evolved moisture also results in wastage of heat.

Moisture enters in manufactured protective atmosphere from both air and combustion. Such moisture is harmful to the charge as it is oxidizing.

Additionally, the presence of moisture in a furnace enclosure interferes with the radiation heat exchange as it absorbs radiation in some bands.

We have come across the above phenomenon in Chapter 3, Chapter 5, and Chapter 15. Here we will review the principles of quantifying the moisture content in air or other gases.

The moisture in air (or a gas mixture) is called "humidity." The maximum amount of moisture that the air can hold, at a given temperature, is called the "saturation point." If the moisture is less than saturation then the air is "unsaturated." Any moisture in excess of saturation condenses in the form of water. The saturation point of various temperatures is available in steam tables.

According to Dalton's Law of Partial Pressures, the amount of moisture in air can be represented by its partial pressure p_w. We treat air as a single gas. If the partial pressure of air is p_a the total pressure P of the air/moisture mixture is

$$P = p_a + p_w \tag{D-1}$$

The partial pressure of moisture (p_w) can be converted to mass (kg or g). There are several ways of representing humidity.

Absolute humidity (AH) is the mass of moisture (m_w) per kilogram of mixture, i.e.,

$$\text{AH} = \frac{m_w}{1.0\,\text{kg}^3(\text{mixture})}\ \frac{\text{g}}{\text{kg}} \tag{D-2}$$

Specific humidity (SH) is the mass of moisture per kilogram of "dry" (moistureless) air

$$\text{SH} = \frac{m_w}{m_a(\text{dry})}\ \frac{\text{g}}{\text{kg}(\text{dry})} \tag{D-3}$$

Relative humidity (RH) is the mass of actual moisture in a given volume of mixture to the maximum moisture that the same volume of mixture will hold if saturated. If m_w is the actual mass and m_{ws} is the mass of saturation,

$$\text{RH} = \frac{m_w}{m_{ws}} \tag{D-4}$$

Relative humidity is usually reported in percentage (% RH) so that fully saturated air has 100% RH and absolutely dry air (which in practice never exists) has 0% RH.

Relative humidity is the most common method of quoting moisture content. It can be easily measured on a dry or wet bulb thermometer. Additionally, it is also available in daily weather reports. Remember that the dry bulb or atmospheric temperature must accompany the RH reading.

By referring to Daltons Law and the universal gas equation, it can be shown that

$$\mathrm{RH} = \frac{m_w}{m_{\mathrm{ws}}} = \frac{\rho_{\mathrm{ws}}}{\rho_w} = \frac{p_{\mathrm{ws}}}{p_w} = \frac{v_w}{v_{\mathrm{ws}}} = \frac{\mathrm{SH}}{(\mathrm{SH})_s} \tag{D-5}$$

where ρ, p, and v are, respectively, the density, partial pressure, and specific volumes of saturated and unsaturated air.

We are primarily interested in SH or AH. Specific humidity is related to relative humidity by the relation

$$\mathrm{SH} = \frac{622 \times (\mathrm{RH}) \times P_{\mathrm{ws}}}{P - (\mathrm{RH}) \times P_{\mathrm{ws}}} \frac{\mathrm{g}}{\mathrm{kg}} \tag{D-6}$$

The term 622 arises from the ratio of molecular weights (gms) of water (18) and air (28.97).

Absolute humidity can be calculated in a similar way.

The saturation partial pressure at a given temperature (p_{ws}) is available in standard steam tables. Specific humidity can be calculated at various relative humidities (100–0%) by using Equation (D-6) above. A set of such curves are shown plotted in Figure D-1 and selected tabulated values of P_{ws} at various temperatures are given in Table D-1. Detailed tables at 1°C temperature intervals and charts are available in handbooks and standard books of thermodynamics. These charts are called "pychrometric" or "humidity charts."

Consider the 100% RH curve in the figure. It can be seen that the saturation moisture content increases exponentially with temperature. At temperatures greater than about 65–70°C air can hold practically an infinite amount of moisture. The dew

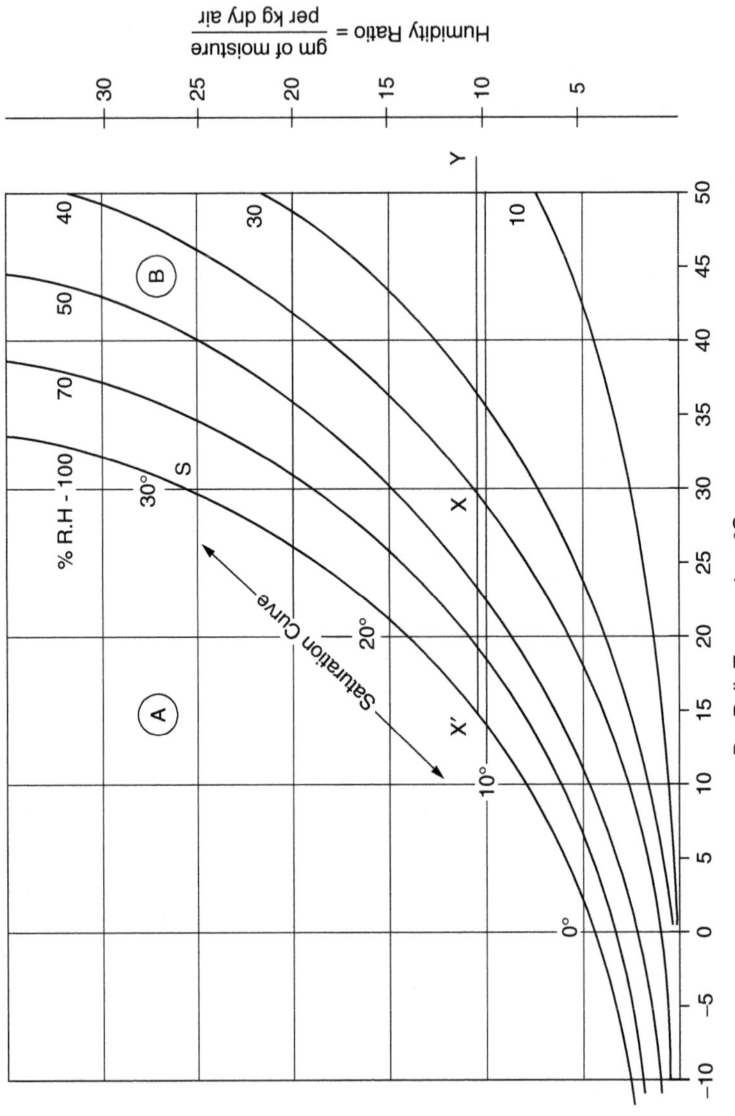

Figure D-1 Outline of a typical psychometric chart.

TABLE D-1 Partial Water Vapor Pressure and Moisture Content at Saturation of Air (Gas)/Vapor. Mixtures at 10^5 Pa Pressure

Temperature°C dry bulb	Partial pressure of saturated vapor (P_{ws}) kPa	Actual moisture content of mixture g/m³ AH	Moisture content per kg of dry air g/kg SH	Moisture content of dry gas per m³ of dry air g/m³	Volume of moisture per kg dry air m³/kg	Specification
–10	0.27	2.3	1.6		0.744	
–5	0.40	3.4	2.5		0.76	
0	0.612	4.84	3.9	4.8	0.755	
5	0.87	6.8	5.4	6.9	0.800	
10	1.22	9.4	7.6	9.80	0.805	
15	1.70	12.8	10.4	13.5	0.820	
20	2.33	17.0	14.5	18.5	0.840	
25	3.17	23.0	20.0	25.8	0.854	
30	4.23	30.4	27.3	35.2	0.873	
35	5.62	39.2	36.5	47.3	0.900	
40	7.36	51	49.0	63.5	0.912	
45	9.60	65.3	65.0	84.0	0.935	
50	12.3	82.5	86.0	110.0	0.960	
55	15.65	104	114	148	0.990	
60	20.0	130	152	196	1.020	
65	25.0	161	204	265	1.058	
70	30.7	198	240	361	—	

Table D-1 (Continued) Partial Water Vapor Pressure and Moisture Content at Saturation of Air (Gas)/ Vapor. Mixtures at 10^5 Pa Pressure

Temperature °C dry bulb	Partial pressure of saturated vapor (P_{ws}) kPa	Actual moisture content of mixture g/m³ AH	Moisture content per kg of dry air g/kg SH	Moisture content of dry gas per m³ of dry air g/m³	Volume of moisture per kg dry air m³/kg	Specification
75	38.0	242	500	500	—	
80	46.7	293	385	719	—	
85	57.0	353	470	1092	—	
90	69.2	423	580	1877	—	
95	83.4	504	690	4381	—	
100	100.0	598	820	Int.	—	

point also increases similarly. This shows that furnace gases which are at temperatures of 800–1500°C can accommodate a large amount of moisture without condensing.

The dew point is lowered with decreasing temperature. This shows the principle of drying by cooling. However, complete removal of moisture is never possible as the curve becomes parallel to the temperature axis. Thus, protective gases can be sufficiently dried (up to 1–3 g moisture per kg dry) by refrigeration to 10–15°C below zero degree celsius. However, in vacuum processes, cooling should be cryostatic, i.e., –150 to –200°C.

On the left-hand side Ⓐ of the curve the region shows super heated steam which will condense to saturation. On the right-hand side Ⓑ is the unsaturated region with a family of curves representing various percentages of saturation or the RH values. As mentioned before, the local RH values are generally available or can be read off a suitable instrument. The moisture content values can then be obtained from charts or tables in terms of AH or SH. The values of AH and RH based on weight (1 kg) of dry gas or mixture can then be converted to volume (per m) basis, from tables.

Consider for example, air at 30°C (dry bulb) temperature and 40% RH (*point X*). The specific humidity (SH) of this air is obtained at *point Y* (on vertical axis) and is approx 10.3 gm per kg. The line *XY* crosses the saturation curve at *X′* giving the dew point about 14°C at which condensation will start.

If we extend the temperature line vertically to meet the saturation curve at *point S*, the segment *XS* will show the capacity to absorb moisture up to saturation by *point S′* which is at 27.5 – 10.3 = 17.2 g/kg. This is the principle of drying operations involving evaporation.

Besides the knowledge of moisture content in a given gas mixture, other important information is the heat content. In all operations of our interest we either heat or cool the mixture, which involves addition or removal of heat. The heat content or enthalpy for a gas moisture mixture can be determined by Equation (D-7) and Equation (D-8). There are two situations possible.

First, if the heating involves evaporation such as moist air in contact with water (e.g., drying), we have to consider the latent heat of evaporation L [kj/kg].

In this case

$$H = 1.01(t_2 - t_1) + (\mathrm{SH})\,(2463 + 1.88(t_2 - t_1)\ \frac{\mathrm{kJ}}{\mathrm{kg}} \qquad \text{(D-7)}$$

where

t_2 and t_1 = Final and initial temperature (°C)
1.01 = Specific heat of air at constant pressure (kJ/kg °C)
1.88 = Specific heat of water vapor of constant pressure (kJ/kg °C)
2463 = Latent heat of evaporation of water (kJ/kg).

Second, if the gas mixture does not come in contact with water then all the moisture will be already in vapor state, hence the latent heat term from the above equation is dropped. Hence,

$$\mathrm{H} = 1.01(t_2 - t_1) + (\mathrm{SH}) + (1.88(t_2 - t_t))\ \frac{\mathrm{kJ}}{\mathrm{kg}} \qquad \text{(D-8)}$$

A third possible variation of the above two equations arises when moist air at (SH) –1 at temperature t_1 is passed over water acquiring a temperature t_2 and (SH) –2. Such conditions arise when hot moist air is passed over a wet body for drying.

The enthalpy content of moist air (at all RH) can also be directly read from psychrometric charts. What is discussed above is a very abridged account of air moisture mixtures applicable to our purpose. Detailed calculations and more precise data are available in standard thermodynamics texts or handbooks.

Appendix E

Error Function

There are many solutions of the basic conduction equation

$$\frac{\partial^2 T}{\partial x^2} = \frac{1}{\alpha} = \frac{dT}{dt}$$

Each solution is valid for a particular set of boundary conditions. In some solutions, we come across "error function" and its complementary, derivative, and integral forms. Without going into the mathematical complications of their occurrence in the solutions, we will here consider their numerical evaluation. They are basically numerical exponential functions.

Error function (erf) is defined as

$$\text{erf}(x) = \frac{2}{\sqrt{\pi}} \int_0^x e^{-\xi^2} \, d\xi \tag{E.1}$$

where ξ is a dummy variable and

$$\left.\begin{array}{l} \text{erf}(-x) = -\text{erf}(x) \\ \text{erf}(\infty) = 1 \\ \text{erf}(0) = 0 \end{array}\right\} \tag{E.2}$$

TABLE E-1 Selected Error Functions

x	erf x	erfc x	ierfc x
0	0.0	1.0	0.5642
0.1	0.112	0.888	0.4698
0.2	0.223	0.777	0.3866
0.3	0.330	0.670	0.3142
0.4	0.429	0.571	0.2521
0.5	0.521	0.479	0.1996
0.6	0.604	0.396	0.156
0.7	0.678	0.322	0.1202
0.8	0.742	0.258	0.0917
0.9	0.797	0.203	0.0682
1.0	0.843	0.157	0.0502
1.1	0.880	0.120	0.0364
1.2	0.910	0.090	0.0260
1.3	0.934	0.066	0.0183
1.4	0.952	0.048	0.0126
1.5	0.966	0.034	0.007
1.6	0.976	0.024	0.000
1.7	0.983	0.017	0.004
1.8	0.990	0.010	0.002
1.9	0.992	0.008	0.001
2.0	0.995	0.005	0.001

Complementary error function (erfc) is

$$\text{erfc}\ (x) = 1 - \text{erf}\ x \qquad \text{so that} \tag{E.3}$$

$$\text{erfc}(0) = 1 \quad \text{and erfc}\ (1) = \infty \tag{E.4}$$

Like other common functions (e.g., log x, sin x, etc.), error function can be repeatedly differentiated or the complementary error function can be integrated.

The first differential of error function, for example, will be

$$\frac{d}{dx}\text{erf}\ x = \frac{2}{\sqrt{\pi}} e^{-x^2} \tag{E.5}$$

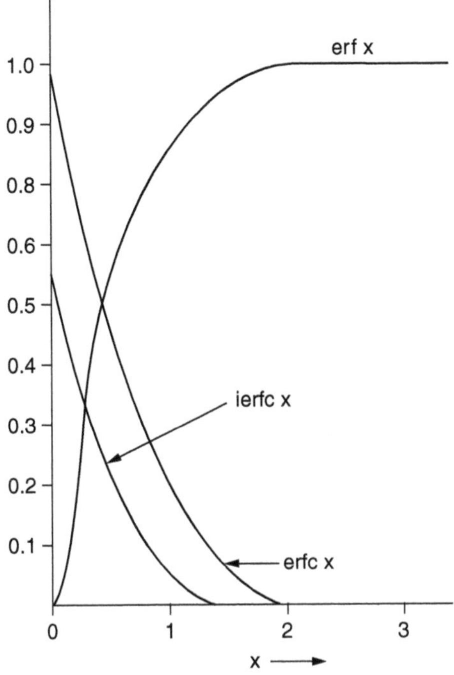

Figure E-1 Error functions, graphical.

Integration of complementary error function is denoted as "i^n erfc" where n denotes the order of integration. Thus,

$$i^1 \,\text{erfc} = i\text{erfc} = \frac{1}{\pi} e^{-x^2} - x \,\text{erfc}\, x \tag{E.6}$$

and

$$i^2 \,\text{erfc} = \frac{1}{4} (\text{erfc}\, x - 2x \,\text{ierfc}\, x) \tag{E.7}$$

and so on.

Of our interest are the functions,

erf x, erfc x, and ierfc x.

Tables of the numerical values of these functions for $x = 0$ to $x = 3$ are available in standard mathematical and statistical handbooks. An abridged extract along with a graphical plot is given in Table E-1 and (Figure E-1).

Appendix F

Properties of Air, Water, Gases

TABLE F.1 Physical Properties of Dry Air at $P_B = 760$ mm Hg

Temp. t_1 °C	Density ρ kg/m³	Sp. Heat C_p kJ/kg·K	Thermal Conductivity $\lambda \cdot 10^2$	Diffusivity $\alpha \cdot 10^6$ m²/sec	Dynamic Viscosity $\mu \cdot 10^6$ N·s/m²	Kinetic Viscosity $\nu \cdot 10^6$ m²/sec	Prandtl Number Pr
0	1.293	1.005	2.44	18.8	17.2	13.28	0.707
10	1.247	1.005	2.51	20.0	17.6	14.16	0.705
20	1.205	1.005	2.59	21.4	17.6	14.16	0.705
30	1.165	1.005	2.67	22.9	18.6	16.00	0.701
40	1.128	1.005	2.76	24.3	19.1	16.69	0.699
50	1.093	1.005	2.83	25.7	19.6	17.95	0.698
60	1.060	1.005	2.90	27.2	20.1	18.97	0.696
70	1.029	1.009	2.96	28.6	20.6	20.02	0.694
80	1.000	1.009	3.05	30.2	21.1	21.09	0.692
90	0.972	1.009	3.13	31.9	21.5	22.10	0.690
100	0.946	1.009	3.21	33.6	21.9	23.13	0.688
120	0.898	1.009	3.34	36.8	22.8	25.45	0.686
140	0.854	1.013	3.49	40.3	23.7	27.80	0.684
160	0.815	1.017	3.64	43.9	24.5	30.09	0.682
180	0.779	1.022	3.78	47.5	25.3	32.49	0.681
200	0.746	1.026	3.93	51.4	26.0	34.85	0.680
250	0.674	1.038	4.27	61.0	27.4	40.61	0.677
300	0.615	1.047	4.60	71.6	29.7	48.33	0.674
350	0.566	1.059	4.91	81.9	31.4	55.46	0.676
400	0.524	1.068	5.21	93.1	33.0	63.09	0.678
500	0.456	1.093	5.74	115.3	36.2	79.38	0.687

Table F.1 Physical Properties of Dry Air at $P_B = 760$ mm Hg (Continued)

Temp. t_1 °C	Density ρ kg/m^3	Sp. Heat C_p kJ/kg·K	Thermal Conductivity $\lambda \cdot 10^2$	Diffusivity $\alpha \cdot 10^6$ m^2/sec	Dynamic Viscosity $\mu \cdot 10^6$ N·s/m^2	Kinetic Viscosity $\nu \cdot 10^6$ m^2/sec	Prandtl Number Pr
600	0.404	1.114	6.22	138.3	39.1	96.89	0.699
700	0.362	1.135	6.71	163.4	41.8	115.4	0.706
800	0.329	1.156	7.18	188.8	44.3	134.8	0.713
900	0.301	1.172	7.63	216.2	46.7	155.1	0.717
1000	0.277	1.185	8.07	245.9	49.0	177.1	0.719
1100	0.257	1.197	8.50	276.2	51.2	199.3	0.722
1200	0.239	1.210	9.15	316.5	53.5	233.7	0.724

Table F.2 Physical Properties of Water on the Saturation Line

Temp. t_1 °C	Pressure $p \times 10^{-5}$ Pa	Density ρ kg/m^3	Sp. Enthlapy kJ/kg	Sp. Heat C_p	Thermal Conductivity $\lambda \cdot 10^2$ kJ/kg·K	Diffusivity $\alpha \cdot 10^8$ m^2/sec	Dynamic Viscosity $\mu \cdot 10^6$ N·s/m^2	Kinetic Viscosity $v \cdot 10^6$ m^2/sec	Prandtl Number Pr
0	1.013	999.9	0.0	4.212	55.1	13.1	1788	1.789	13.67
10	1.013	999.7	042.04	4.191	57.4	13.7	1306	1.306	9.52
20	1.013	998.2	083.91	4.183	59.9	14.3	1004	1.006	7.02
30	1.013	995.7	125.7	4.174	61.8	14.9	801.5	0.805	5.42
40	1.013	992.2	167.5	4.174	63.5	15.3	653.3	0.659	4.31
50	1.013	988.1	209.3	4.174	64.8	15.7	549.4	0.556	3.54
60	1.013	983.1	251.1	4.179	65.9	16.0	469.9	0.478	2.98
70	1.013	977.8	293.0	4.187	66.8	16.3	406.1	0.415	2.55
80	1.013	971.8	355.0	4.195	67.4	16.6	355.1	0.365	2.21
90	1.013	965.3	377.0	4.208	68.0	16.8	314.9	0.326	1.95
100	1.013	958.4	419.1	4.220	68.3	16.9	282.5	0.295	1.75

Latent Heat of Melting, 334 kJ/kg
Latent Heat of Evaporation, 2270 kJ/kg

Table F.3 Specific Heats of Selected Gases (kJ/m³ °C)

Temp. °C	CO_2	N_2	O_2	H_2O	Air (dry)	CO	H_2	H_2S	CH_4
0	1.620	1.333	1.308	1.491	1.300	1.252	1.277	1.515	1.767
100	1.720	1.301	1.320	1.502	1.305	1.265	1.290	1.541	2.106
200	1.808	1.303	1.337	1.517	1.310	1.302	1.297	1.574	2.330
300	1.881	1.310	1.358	1.538	1.320	1.310	1.302	1.607	2.530
400	1.944	1.317	1.380	1.560	1.330	1.323	1.304	1.645	2.721
500	2.045	1.330	1.400	1.583	1.344	1.331	1.306	1.683	2.893
600	2.060	1.342	1.415	1.608	1.360	1.344	1.310	1.721	3.048
700	2.108	1.353	1.437	1.638	1.372	1.361	1.315	1.758	3.190
800	2.152	1.370	1.453	1.660	1.382	1.373	1.318	1.796	3.341
900	2.205	1.382	1.466	1.686	1.400	1.390	1.323	1.830	3.450
1000	2.226	1.394	1.480	1.713	1.411	1.402	1.327	1.863	3.567
1100	2.263	1.405	1.493	1.740	1.424	1.415	1.336	1.892	—
1200	2.288	1.407	1.506	1.765	1.435	1.428	1.344	1.922	—
1300	2.316	1.430	1.512	1.791	1.440	1.440	1.352	1.947	—
1400	2.340	1.437	1.522	1.815	1.455	1.449	1.361	1.972	—
1500	2.366	1.447	1.531	1.840	1.464	1.461	1.369	2.000	—
1600	2.385	1.455	1.540	1.862	1.473	1.470	1.377	—	—
1700	2.404	1.462	1.548	1.884	1.481	1.478	1.386	—	—
1800	2.423	1.470	1.556	1.905	1.490	1.486	1.404	—	—
1900	2.440	1.478	1.564	1.952	1.496	1.494	1.412	—	—
2000	2.455	1.485	1.571	1.945	1.503	1.498	1.420	—	—

Appendix G

Emissivity

The total radiation (E_o) from a black body is obtained from Stefan Boltzmann's law as

$$E_o = C_o \left[\frac{T}{100}\right]^4 \text{ W/m}^2\text{K}^4$$

where T is the absolute temperature and c is the radiation constant ~ 5.67.

Most of the practical radiating bodies are considered gray and their radiation power is given by

$$E = \epsilon\, C_o \left[\frac{T}{100}\right]^4$$

where the ratio $E/E_o = \epsilon$ is the emissivity of the gray body.

Extensive collections of emissivities of practical bodies are available, and some selected values are given in the Table G.1.

Table G.1 Emissivity (ε) of Selected Materials Recommended for Design Calculations

Materials and Surface Characteristics		Temperature °C	Emissivity (ϵ)
Aluminum	polished	250	0.04
	rough	30	0.06
	oxidized	300	0.2
Copper	polished	100	0.02
	oxidized	400	0.5
Brass	rolled	30	0.06
	oxidized	200	0.55
Steel	polished	900	0.6
	oxidized	400	0.8
Cast iron	machined	500	0.7
	as cast	450	0.9
Stainless steel		800	0.6
Tungsten filament		1200	0.4
Inconel		800	0.7
Refractories		1000	0.85
Carbon		1000	0.7–0.9
Aluminum paint		60	0.2–0.7
Paints		100	0.9

There are some basic problems involved in the direct use of these values in real situations.

Emissivity depends on temperature and surface condition of the body. Values of ϵ obtained from reference books are those evaluated under carefully controlled conditions. It is generally impossible to duplicate such conditions under practical and industrial environments. Precise data about temperature dependence of ϵ are virtually nonexistent.

Flames have emissivities of 0.2–1.0. It depends on fuel, air-fuel ratio, and varies over the flame length.

Bibliography

CHAPTER 1 INTRODUCTION

Dryden, I.G.C., *The Efficient Use of Energy*. 2nd Ed., Butterworth Scientific, Oxford, 1982.

Glinkov, M.A. and Glinkov, G.M., *A General Theory of Furnaces*. Mir Publishers, Moscow, 1980.

Krivandin, V. and Markov, B., *Metallurgical Furnaces*. Mir Publishers, Moscow, 1980.

Trinks, W. and Mawhinney, M.H., *Industrial Furnaces*. 5th Ed., Vols. I and II, John Wiley & Sons, New York, 1953.

CHAPTER 2 FLUID DYNAMICS

Allen, J.E., *Aerodynamics: The Science of Air in Motion*. 3rd Ed., Allen Brothers and Father, Blythburgh, Suffolk, 1986.

Kazantsev, E.I., *Industrial Furnaces*. Mir Publishers, Moscow, 1977.

Nekrasov, B., *Hydraulics (For Aeronautical Engineers)*. Mir Publishers, Moscow, 1971.

Stampar, E. and Koral, R., (Ed.). *Handbook of Air Conditioning, Heating & Ventilation*, 3rd Ed., Industrial Press, New York, 1979.

CHAPTERS 3 AND 4 STEADY STATE HEAT TRANSFER AND TRANSIENT CONDUCTION

Bejan, A., *Heat Transfer.* John Wiley & Sons, New York, 1993.
Carlslaw, H.S. and Jaeger, C., *Conduction of Heat in Solids.* Oxford University Press, Oxford, 1959.
Heisler, M.P., *Temperature Charts for Induction and Constant Temperature Heating.* Trans ASME, Vol. 6.9, 1947.
Incropera, F. and Dewitt, D.P., *Fundamentals of Heat and Mass Transfer.* 4th Ed., John Wiley & Sons, New York, 1996.
Isachenko, V.P., Osipova, V.A., and Sukomel, A.S., *Heat Transfer.* 3rd Ed., Mir Publishers, Moscow, 1977.

CHAPTERS 5 AND 6 FUELS AND THEIR PROPERTIES AND FUEL BURNING DEVICES

Francis, W., *Fuel Technology.* Vols. 1 and 2, Pergamon Press, Oxford, 1964.
Kazantsev, E.I., *Industrial Furnaces.* Mir Publishers, Moscow, 1977.
Roddan, M., *The Use of Fuel Oil in Furnaces for Iron and Steel Fabrication Industries*, Shell B.P.-Fuel Oil Dept., U.K., 1950.
Saxon, F., Sedgewick, G., and Proffitt, R., *Industrial and Commercial Gas Installation Practice.* Vol. 3. Butterworth, Tolley Croydon, U.K., 2000.
Technical Literature (Burners), Hauck Manufacturing Company, Lebanon PA, 1990.
Technical Literature (Packaged Burners), ECOFLAM Company, Resana, Italy, 1999.

CHAPTER 7 REFRACTORIES

Brandes E.A. and Brook, G.B., *Smithells Metal Reference Book.* 7th Ed., Butterworth-Heinemann, Oxford, 1998.
Fanzott, S.M., *Modern Refractory Technique.* Springer-Verlag, Modern Industries AG & Co., Landsburgh, Germany, 1991.
Kazantsev, E.I., *Industrial Furnaces.* Mir Publishers, Moscow, 1977.
Technical Literature, Didier-Werke AG, Wiesbaden, Germany.
Technical Literature, Harbinson Walker Co.

CHAPTER 8 METALS AND ALLOYS FOR HIGH TEMPERATURE APPLICATIONS

Brandes, E.A. and Brook, G.B., *Smithells Metal Reference Book*. 7th Ed., Butterworth-Heinemann, Oxford, 1998.

Metals Handbook, Vol. I, 9th Ed., American Society for Metals, Metals Park, Ohio 1961.

Robert E. Reed Hill, *Physical Metallurgy Principles*, D. Van Nostrand Reinhold, Princeton, NJ, 1964.

Shrier, I.L., Jarman, R.A., and Burnsteirn, G.T., Ed., *Corrosion*, Vols. I & II, 3rd Ed., Butterworth-Heinemann, Oxford, 1994.

Wiggin Nickel Alloy, Wiggin Heat Resisting Alloys, Wiggin Nickel Alloys for Heat Treatment Equipment, Henry Wiggin & Co. Ltd., Hereford, U.K.

CHAPTER 9 ELECTRIC RESISTANCE HEATING

Davies, E.J., *Conduction and Induction Heating*. IEE Power Engineering Series II, Peter Peregrinus Ltd., London, 1990.

Kanthal Handbook, Resistance Heating Alloys for Appliances and Heaters, Kanthal Heating Technology, Hallstahammar, Sweden, 1986.

Kanthal Super Handbook, Kanthal Furnace Products, Hallstahammar, Sweden, 1986.

Paschkis, V. and Presson, J., *Industrial Electric Furnaces and Appliances*. 2nd Ed., Interscience, New York, 1960.

CHAPTER 10 HIGH FREQUENCY HEATING

Aguilar, J. and Rodriguey, J., Microwave as an Energy Source for Production of B-Sic *Journal of Microwave Power and Electromagnetic Energy*. Vol. 36, No. 3, 2001.

Davies, E.J., *Conduction and Induction Heating*. IEE Power Engineering Series II, Peter Peregrinus Ltd., London, 1990.

Meredith, R., *Engineering Handbook of Industrial Microwave Heating*. Institute of Electrical Engineers, London, 1998.

Paschkis, V. and Presson, J., *Industrial Electric Furnaces and Applications*. 2nd Ed., Interscience, New York, 1960.

Ryder, J.D., *Engineering Electronics*. 2nd Ed., McGraw-Hill, New York, 1967.

CHAPTER 11 CONCENTRATED HEAT SOURCES

Electron Beam Welding. Commercial Literature, Leybold Heraeus Gmbh, Hanau, Germany.

Ready, J.F., *Engineering Application of Lasers.* 2nd Ed., Academic Press, New York, 1977.

Rykalin, N., Uglov, A., Zuev, I., and Kokora, A., *Laser and Electron Beam. Material Processing Handbook,* Mir Publishers, Moscow, 1988.

Silfvast, W.T., *Laser Fundamentals.* Cambridge University Press, Cambridge, 1998.

CHAPTER 12 VACUUM ENGINEERING

Chamers, A., Fitch, R.K., and Halliday, B.S., *Basic Vacuum Technology.* 2nd Ed., Industrial and Physical Publishing, Bristol, 1998.

Roth, A., *Vacuum Technology.* 3rd Ed., Elsevier, Amsterdam, The Netherlands, 1990.

Steinherz, H.A., *Handbook of High Vacuum Engineering.* Reinhold Publishing Corporation, New York, 1963.

CHAPTER 13 PROTECTIVE ATMOSPHERES

Hotchkiss, A.G. and Weber, H.M., *Protective Atmospheres.* John Wiley & Sons, New York, 1953.

Metals Handbook. Vol. 2, 9th Ed., Metals Park, Ohio, 1980.

Trinks, W. and Mawhinny, M.H., *Industrial Furnaces.* Vol. II, 4th Ed., John Wiley & Sons, New York, 1967.

CHAPTER 14 TEMPERATURE MEASUREMENT

IR Answers and Solutions Handbook. IRCON Inc, Niles, IL, 1999.

Knight, J.R. and Rhys., D.W., *The Platinum Metals in Thermometry.* Publication No. 2244, Engelhard Industries Ltd., London, 1961.

Leigh, J.R., *Temperature Measurement and Control.* Peter Peregrinus, London, 1991.

Lide, D.R., (Editor-in-Chief), *CRC Handbook of Chemistry and Physics.* 80th Ed., CRC Press, Boca Raton, FL, 2000.

McGee, T.D., *Principles and Methods of Temperature Measurement*. Wiley Interscience, New York, 1988.
Quinn, J.J., *Temperature*. Academic Press, London, 1993.

CHAPTER 15 MISCELLANY AND FURTHER

Bird, R.B., Stewart, W.E., and Lightfoot, E.N., *Transport Phenomena*. 2nd Ed., John Wiley & Sons, Singapore, 2000.
Charms, S.E., *The Fundamentals of Food Engineering*. AVI Publishing, Westport, CT, 1998.
Perry, J.N., (Ed.), Chemical Engineers Handbook. 3rd Ed., McGraw Hill, New York, 1950.
Potter, N.N. and Hotchkiss, J.M., *Food Science*. 5th Ed., Chapman & Hall, New York, 1995.

HANDBOOKS FOR GENERAL REFERENCE AND DATA

Brands, E.A. and Brook, G.B. (Ed.), *Smithells Metals Reference Book*. 7th Ed., Butterworth-Heinemann, Oxford, 1998.
Kazanstev, E.I., *Industrial Furnaces*. Mir Publishers, Moscow, 1977.
Lide, D.R. (Ed.), *Handbook of Chemistry and Physics*. 80th Ed., CRC Press, Boca Raton, FL, 1999.
American Society for Metals, *Metals Handbook*. 9th Ed., Vols. I & II, Metals Park, Ohio, 1980.

Index

D

E

F

N

O

P

Q

R

S

T

V

Z